Advances in Natural and Technological Hazards Research

Volume 55

The book series entitled *Advances in Natural and Technological Hazards* is dedicated to serving the growing community of scholars, practitioners and policy makers concerned with the different scientific, socio-economic and political aspects of natural and technological hazards.

The series aims to provide rapid, refereed publications of topical contributions about recent advances in natural and technological hazards research. Each volume is a thorough treatment of a specific topic of importance for proper management and mitigation practices and will shed light on the fundamental and applied aspects of natural and technological hazards.

Comments or suggestions for future volumes are welcomed.

Bijeesh Kozhikkodan Veettil

The Potential of Bioshields for Coastal Protection

Springer

Bijeesh Kozhikkodan Veettil
Laboratory of Ecology and Environmental
Management, Science and Technology
Advanced Institute
Van Lang University
Ho Chi Minh City, Vietnam

ISSN 1878-9897 ISSN 2213-6959 (electronic)
Advances in Natural and Technological Hazards Research
ISBN 978-3-032-22887-1 ISBN 978-3-032-22888-8 (eBook)
https://doi.org/10.1007/978-3-032-22888-8

This Springer imprint is published by the registered company Springer Nature Switzerland AG
The registered company address is: Gewerbestrasse 11, 6330 Cham, Switzerland

Preface

The world's coastlines are under unprecedented pressure from both natural and anthropogenic forces. Rapid population growth, urbanization, industrial expansion, and climate change are reshaping coastal landscapes, increasing the vulnerability of communities and ecosystems alike. Rising sea levels, intensifying storms, and accelerated erosion pose formidable challenges to traditional coastal protection strategies, prompting an urgent search for innovative, sustainable, and adaptive solutions.

Bioshields—vegetated coastal ecosystems such as mangroves, salt marshes, seagrass meadows, and coastal dunes—offer a compelling answer to these challenges. These systems are more than ecological habitats; they are living infrastructure, performing critical protective functions while simultaneously providing ecosystem services, supporting biodiversity, and sustaining local livelihoods. Despite growing scientific evidence of their efficacy, bioshields have often been undervalued in conventional coastal management, overshadowed by engineering-based approaches.

This book is intended to bridge the gap between ecological science, coastal engineering, policy development, and community-based management. Its central aim is to provide a comprehensive, evidence-based exploration of the potential of bioshields for coastal protection, emphasizing the integration of natural processes, socio-economic considerations, and governance frameworks. While the discussion draws from global case studies, there is a particular focus on Asia, a region that combines high coastal population densities, rich ecological diversity, and increasing exposure to climatic hazards.

The structure of the book is designed to guide readers through the historical, ecological, technical, socio-economic, and policy dimensions of bioshields. Early chapters situate the discussion in a historical and ecological context, exploring how coastal vegetation has long served protective functions and how scientific understanding has evolved. Subsequent chapters examine biophysical mechanisms, climate resilience, and design principles, providing a foundation for applied management and restoration practices. Later sections address the societal, governance, and policy dimensions, highlighting community engagement, traditional knowledge, and institutional frameworks. Finally, the book presents lessons learned, successful global and regional projects, and challenges, offering guidance for future bioshield implementation.

This work is intended for a diverse readership, including researchers, policymakers, environmental planners, coastal engineers, and practitioners in the field of ecosystem-based management. By synthesizing interdisciplinary research, practical experiences, and policy frameworks, the book seeks to inform strategies that are scientifically sound, socially equitable, and ecologically resilient.

I hope that this volume will contribute to the broader understanding and application of bioshields, inspiring efforts to restore, protect, and sustainably manage coastal ecosystems. As we confront the twin imperatives of climate adaptation and coastal resilience, the integration of natural systems into human planning becomes not only desirable but essential. Bioshields embody this integration, representing a path forward that harmonizes human safety, ecological integrity, and sustainable development.

The preparation of this book has drawn upon extensive literature, field studies, and international case experiences, and it reflects the collaborative contributions of a wide community of scientists, practitioners, and policymakers who have championed nature-based coastal protection.

Ho Chi Minh City, Vietnam Bijeesh Kozhikkodan Veettil

Competing Interests The author has no competing interests to declare that are relevant to the content of this manuscript.

Contents

Part II Ecological and Biophysical Foundations of Coastal Bioshields

Part I
Coastal Bioshields: Introduction and Background

Chapter 1
Introduction: The Growing Need for Coastal Protection

Abstract Coastal zones represent one of the most dynamic and economically productive ecosystems on Earth, yet they are increasingly threatened by accelerating environmental and anthropogenic pressures. Rising sea levels, intensifying tropical storms, shoreline erosion, land subsidence, and unplanned urbanization have increased the vulnerability of coastal populations and infrastructure worldwide. Conventional hard engineering solutions, such as seawalls, groynes, and revetments, have provided localized protection but often at high ecological and financial costs, leading to habitat loss and long-term degradation of natural defenses. Consequently, global attention has shifted toward sustainable, ecosystem-based approaches that integrate ecological processes with human adaptation strategies. This chapter introduces the concept of bioshields, defined as coastal vegetation systems—mangroves, saltmarshes, seagrasses, and coastal dunes—that function as living buffers against hydrodynamic forces while providing multiple ecosystem services. It contextualizes the growing urgency of coastal protection, considering climate change projections and socio-economic dependencies on coastal resources. The discussion situates bioshields within the broader framework of nature-based solutions and resilience theory, emphasizing their capacity to deliver co-benefits including biodiversity conservation, carbon sequestration, and livelihood enhancement. The chapter also outlines the methodological and conceptual foundations of the book, linking physical, ecological, and governance dimensions of coastal management. By framing bioshields as both protective infrastructure and socio-ecological systems, the introduction establishes a scientific and policy rationale for integrating natural defenses into contemporary coastal planning and climate adaptation strategies.

Keywords Bioshields · Climate change adaptation · Coastal resilience · Ecosystem-based coastal protection · Nature-based solutions · Sustainable shoreline management

B. Kozhikkodan Veettil, *The Potential of Bioshields for Coastal Protection*, Advances in Natural and Technological Hazards Research 55,
https://doi.org/10.1007/978-3-032-22888-8_1

1.1 The Global Significance of Coastal Zones

Coastal zones are among the most productive and densely populated regions on Earth (Small and Nicholls 2003; Veettil et al. 2021). They support diverse ecosystems, including mangroves, salt marshes, coral reefs, seagrass meadows, and dunes, which are habitats that provide food, livelihoods, and critical ecosystem services to hundreds of millions of people (Barbier et al. 2011; Ayyam et al. 2019). Approximately 40% of the global population resides within 100 km of the coast, and this proportion is expected to continue rising with urbanization and economic development (Small and Nicholls 2003; Neumann et al. 2015).

Coastal areas serve as hubs for commerce, industry, fisheries, and tourism, contributing significantly to national GDPs, particularly in developing and emerging economies (Hoefle 2014; Veettil et al. 2021). Yet, these same areas face escalating threats from climate change, sea-level rise, storm surges, and anthropogenic degradation (Nichols et al. 2018; Griggs and Reguero 2021; Veettil et al. 2021). The vulnerability of coastal zones has transformed coastal protection into a critical priority for sustainable development and disaster risk reduction (Duxbury and Dickinson 2007).

In Asia, where rapid population growth and economic activity converge along low-lying deltas and estuaries, the challenge is particularly acute (Becker et al. 2024). The Mekong, Irrawaddy, and Ganges–Brahmaputra deltas are not only food baskets for millions but also some of the world's most hazard-prone landscapes (Birkmann et al. 2012; Veettil et al. 2018; Oberhagemann et al. 2020). The need for effective, adaptive, and sustainable coastal protection strategies has never been greater.

1.2 Intensifying Coastal Hazards in a Changing Climate

Several coastal hazards (Fig. 1.1), including sea level rise, storms, typhoons and extreme weather events, erosion and land subsidence, and anthropogenic pressures, are observed to have intensified due to climate change in recent decades.

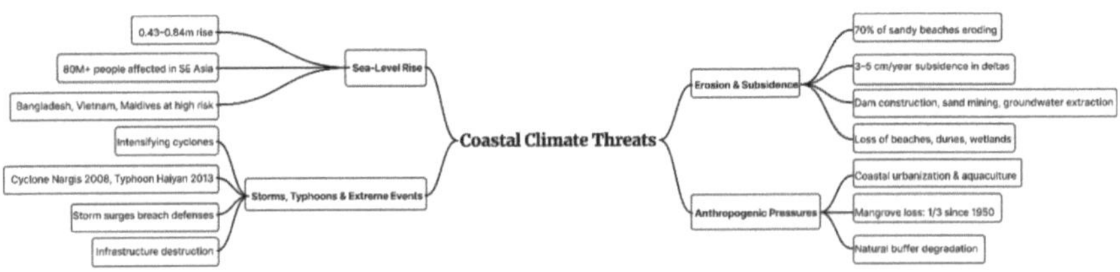

Fig. 1.1 Coastal hazards intensified due to climate change and their long-term impacts

1.2.1 Sea-Level Rise

Sea-level rise represents one of the most profound and persistent threats to coastal systems (Brown et al. 2012). The Intergovernmental Panel on Climate Change (IPCC 2021) projects a rise of 0.43–0.84 m by 2100 under high-emission scenarios, with potential for greater increase if ice-sheet collapse accelerates. Even modest rises can exacerbate coastal flooding, saltwater intrusion, and erosion, particularly in deltaic and island systems (Brown et al. 2012).

Low-lying countries, such as Bangladesh, Vietnam, and the Maldives, face existential risks, where a 1-m rise could inundate vast agricultural lands and displace millions (Veettil et al. 2019; Qadir et al. 2025). In Southeast Asia alone, sea-level rise is expected to affect over 80 million people by the mid-twenty-first century (Nicholls et al. 2021).

1.2.2 Storms, Typhoons, and Extreme Events

Climate change is also intensifying the frequency and severity of tropical cyclones and storm surges (Walsh et al. 2015). In the past two decades, catastrophic events such as Cyclone Nargis (Myanmar, 2008), Typhoon Haiyan (Philippines, 2013), and Typhoon Doksuri (Vietnam, 2023) have demonstrated the devastating impacts of coastal storms (Takagi et al. 2023). These events destroyed infrastructure, disrupted economies, and resulted in substantial human and ecological losses (Loucks 2021; Jha and Dev 2024). Storm surges, amplified by rising sea levels, can penetrate deep inland, breaching embankments and overwhelming traditional defenses (Becker et al. 2024). The increasing unpredictability of storm paths and intensities further complicates coastal management.

1.2.3 Erosion and Land Subsidence

Although coastal areas have attracted humans and human activities, shoreline erosion affects approximately 70% of the world's sandy beaches (Luijendijk et al. 2018). It is driven not only by natural processes but also by human-induced factors such as dam construction, sand mining, and groundwater extraction (Veettil et al. 2020, 2021). In deltaic systems like the Mekong and Chao Phraya, land subsidence due to over-extraction of groundwater has accelerated relative sea-level rise by up to 3–5 cm per year in some areas (Brown et al. 2012; Minderhoud et al. 2017). The resulting loss of protective beaches, dunes, and wetlands leaves coastal communities increasingly exposed to flooding and saltwater intrusion, threatening agriculture and freshwater supplies.

1.2.4 Anthropogenic Pressures

Rapid coastal urbanization, aquaculture expansion, and infrastructure development continue to degrade natural buffer systems (Yu et al. 2025). Deforestation alone has reduced global coverage of mangroves by nearly one-third since 1950, primarily due to conversion for aquaculture and agriculture (Friess et al. 2019). The destruction of these natural ecosystems has removed the first line of defense for millions of people living in coastal regions (Rahman and Rahman 2015).

1.3 Limitations of Conventional Coastal Engineering

Historically, societies have relied on hard-engineering solutions, such as seawalls, dikes, breakwaters, and groynes, to protect coasts (Morris et al. 2018). While these structures provide immediate defense against wave energy and inundation, they often generate long-term ecological and geomorphological problems (Schoonees et al. 2019).

1.3.1 Shortcomings of Engineered Structures

Hard coastal defense structures are costly to construct and maintain (Schoonees et al. 2019). They can reflect wave energy rather than absorb it, leading to scour and erosion in adjacent coastal areas (Veettil et al. 2021), which accelerates shoreline retreat in unprotected zones (French 2002). Moreover, traditional structures are typically static and inflexible, unable to adapt to the dynamic nature of coastal processes or to gradual sea-level rise (Hill 2015).

In regions with limited resources, particularly in developing Asia, hard structures are often under-designed or poorly maintained, resulting in partial failures that exacerbate flooding (Veettil et al. 2021). Additionally, the ecological footprints of these projects are significant: seawalls sever land–sea interactions, degrade habitats, and impede sediment transport (Hosseinzadeh et al. 2022).

1.3.2 The Shift Towards Integrated Coastal Management

By the late twentieth century, it became increasingly clear that purely engineered approaches could not provide sustainable solutions in coastal management (Mulder 2006). The emergence of Integrated Coastal Zone Management (ICZM) emphasized the need to balance environmental, economic, and social objectives (Maccarrone et al. 2014). Within this framework, ecosystem-based approaches, such as bioshields, gained recognition for their capacity to provide adaptive, multifunctional, and low-cost protection (Nehren et al. 2016).

1.4 The Concept of Bioshields

1.4.1 Definition and Scientific Foundation

The term "bioshield" refers to vegetated coastal ecosystems—natural or restored—that act as biological barriers against wave action, storm surges, and coastal erosion (Veettil et al. 2021; Athikalam and Vaideeswaran 2022). Bioshields include mangroves, salt marshes, seagrass meadows, and dune vegetation that collectively stabilize sediments, dissipate hydrodynamic energy, and provide ecological co-benefits (Veettil et al. 2021).

Scientifically, the protective functions of bioshields arise from a range of biophysical processes that collectively mitigate coastal hazards (Veettil et al. 2021; Athikalam and Vaideeswaran 2022). The frictional drag generated by stems and roots plays a crucial role in dissipating wave energy, thereby reducing the force of incoming waves before they reach inland areas (Denny 2021). Simultaneously, the ability of coastal bioshields to trap sediments promotes accretion and contributes to shoreline stabilization, thereby helping to counteract erosion and land loss (Veettil et al. 2021; Athikalam and Vaideeswaran 2022; Perricone et al. 2023). The extensive root networks of coastal vegetation also reinforce the soil, enhancing its cohesion and resistance to erosive forces such as waves and storm surges (Stachew et al. 2021). Additionally, the canopy structure intercepts wind and rainfall, effectively reducing wind speed and mitigating the direct impact of raindrops on the soil surface (Styczen and Morgan 2003). Together, these biophysical mechanisms enable bioshields to function as natural barriers, protecting coastal zones from both gradual and extreme environmental pressures. They contribute not only to physical protection but also to broader ecosystem resilience, allowing coasts to recover from disturbances and adapt to changing climatic conditions.

1.4.2 Multifunctionality of Bioshields

Beyond their role in hazard mitigation, coastal bioshields provide a wide array of ecosystem services that contribute to environmental sustainability and human well-being (Sutton-Grier and Sandifer 2019). They play a vital role in carbon sequestration, with mangroves and salt marshes acting as significant "blue carbon" sinks that help mitigate climate change (Alongi 2018; Hershey and Nandan 2024). Coastal bioshields also serve as crucial habitats for biodiversity conservation, supporting fish nurseries, bird populations, and a variety of pollinators that sustain coastal ecosystems (Veettil et al. 2021; Perricone et al. 2023). In addition, they underpin local livelihoods by sustaining fisheries, forestry, and eco-tourism activities that are essential for many coastal economies (Bhattacharya and Sachdev 2024). Beyond their ecological and economic value, coastal bioshields offer important cultural services, as many coastal communities maintain deep spiritual, historical, and traditional connections to their vegetated shorelines (Porri et al. 2023; Zhao et al. 2024). This multifunctionality distinguishes bioshields from conventional defenses, rendering them critical components of both ecological security and human well-being.

1.5 Global and Regional Trends in Coastal Protection

1.5.1 The Global Policy Landscape

The integration of bioshields into international environmental policy frameworks has accelerated in recent decades. The Paris Agreement (2015), the Convention on Biological Diversity (CBD), and the UN Decade on Ecosystem Restoration (2021–2030) all highlight the role of nature-based solutions (NbS) in achieving climate resilience. The Sendai Framework for Disaster Risk Reduction (2015–2030) explicitly advocates the use of ecosystem-based approaches in hazard mitigation. These frameworks have catalyzed investments in mangrove restoration, coastal afforestation, and green infrastructure worldwide.

1.5.2 Regional Focus: Asia and the Pacific

Asia is both the most vulnerable and the most active region in coastal bioshield development (Veettil et al. 2021; Athikalam and Vaideeswaran 2022). Large-scale programs in Vietnam, Bangladesh, India, Indonesia, Japan, and the Philippines illustrate the regional shift toward ecosystem-based adaptation (Tengberg and Cabanban 2013; Nguyen et al. 2017).

Vietnam's coastal mangrove restoration programs in the Mekong Delta have restored thousands of hectares of mangroves, reducing storm impacts and enhancing fisheries (Tinh et al. 2022). Bangladesh's Coastal Greenbelt Project, initiated in the 1980s, now protects over 1,000 km of coastline (Dey et al. 2021). Japan's pine forest restoration along tsunami-prone coasts demonstrates the integration of bioshields with cultural heritage and disaster risk management (Omori et al. 2022). These initiatives reflect growing recognition that bioshields are not merely ecological assets but strategic infrastructure vital for national resilience.

1.6 The Science–Policy Interface: From Evidence to Implementation

The effectiveness of bioshields is now supported by a robust body of scientific evidence. Empirical studies show that mangrove belts can reduce wave height by up to 90% over 500 m of forest (Mendez and Losada 2004; Van Wesenbeeck et al. 2025). Post-disaster analyses confirm that villages fronted by mangroves or coastal forests suffer significantly lower damage from storm surges and tsunamis (Das and Vincent 2009; Danielsen et al. 2005).

However, translating this knowledge into policy and practice requires bridging multiple disciplines—ecology, hydrodynamics, economics, and governance. The science–policy interface is increasingly supported by interdisciplinary research programs and community-based implementation models.

1.6.1 Economic and Social Dimensions

The economic valuation of bioshields underscores their cost-effectiveness. Studies estimate that the annual benefits of mangrove protection in terms of avoided flood damage and carbon sequestration can exceed restoration costs by more than an order of magnitude (Barbier 2018; Macreadie et al. 2021).

Socially, coastal bioshields enhance community resilience by providing resources, employment, and cultural continuity (Mukherjee et al. 2015). The participatory restoration of mangroves in the Philippines (Primavera and Esteban 2008) and India (Datta et al. 2012) demonstrates how community ownership ensures both ecological sustainability and long-term protection outcomes.

1.6.2 Challenges to Implementation

Despite increasing recognition of their ecological and protective value, the implementation of coastal bioshields faces several persistent challenges that hinder their widespread adoption and long-term effectiveness. A major constraint is the lack of sufficient funding and technical expertise needed for large-scale restoration and maintenance, particularly in developing regions where financial and institutional resources are limited (Hernández-Delgado 2024). Land tenure conflicts and competing economic interests—such as aquaculture, tourism, and urban expansion—often create tension between conservation goals and short-term economic gains, leading to delays or abandonment of bioshield projects (Veettil et al. 2021; Hernández-Delgado 2024). Furthermore, inadequate monitoring and weak adaptive management limit the ability to assess project performance, learn from outcomes, and make necessary adjustments to ensure resilience and sustainability (López-Portillo et al. 2017). Many coastal bioshield initiatives also rely heavily on monoculture plantations, which are highly vulnerable to diseases, pest outbreaks, and extreme weather events, thereby reducing their ecological stability and protective efficiency (Mukherjee et al. 2009; Zhu et al. 2023).

Overcoming these barriers requires a holistic and coordinated approach that combines strong policy frameworks with community engagement and intersectoral collaboration. Building scientific and technical capacity at the local level is crucial for supporting evidence-based planning and management (Chee et al. 2021). Equally important is the integration of traditional ecological knowledge with modern restoration practices, which can enhance ecological suitability, cultural acceptance, and long-term stewardship (Barman et al. 2025). By aligning policy, science, and community action, it becomes possible to establish resilient, multifunctional bioshields that effectively protect coastlines while supporting biodiversity, livelihoods, and climate adaptation.

1.7 The Role of Bioshields in Climate Adaptation and Disaster Risk Reduction

1.7.1 Nature-Based Solutions in Adaptation Policy

Bioshields are central to the broader concept of nature-based solutions (NbS)—actions that use ecosystem processes to address societal challenges. As components of ecosystem-based adaptation (EbA), bioshields reduce vulnerability to climate change by maintaining ecological integrity, buffering against hazards, and supporting livelihoods (Jaisankar et al. 2017).

Several countries, including Vietnam, India, and Bangladesh, have explicitly included bioshields in their National Adaptation Plans (NAPs) (Dey et al. 2021; Veettil et al. 2021; Athikalam and Vaideeswaran 2022). The synergy between climate adaptation and biodiversity conservation enhances both environmental and developmental outcomes.

1.7.2 Disaster Risk Reduction (DRR) Synergies

From a disaster risk reduction (DRR) perspective, coastal bioshields play a multifaceted role by contributing to prevention, mitigation, and recovery, thereby enhancing the resilience of coastal communities and ecosystems (Goncalves and Mohan 1992). In terms of prevention, bioshields stabilize coastlines by reducing erosion and maintaining sediment balance long before disasters strike, acting as a first line of defense against gradual environmental degradation (Jaisankar et al. 2017; Athikalam and Vaideeswaran 2022). During extreme events such as cyclones, tsunamis, and storm surges, they provide critical mitigation functions by attenuating wave energy and dissipating storm surges, thereby significantly reducing the force of water reaching inland areas and minimizing damage to infrastructure, agriculture, and settlements (Veettil et al. 2018, 2021).

Beyond their immediate protective role, bioshields also facilitate post-disaster recovery by supporting natural regeneration processes, such as sediment deposition, vegetation regrowth, and habitat restoration, which enable ecosystems to recover more rapidly (Naz and Chowdhury 2022; Fisher et al. 2025). Moreover, they provide essential resources for local reconstruction and livelihood recovery, including timber and non-timber forest products, as well as fisheries that sustain communities during post-disaster periods (Veettil et al. 2021; Athikalam and Vaideeswaran 2022).

The synergy between ecological function and disaster management makes bioshields a nature-based solution that complements engineered defenses while delivering co-benefits for biodiversity and climate adaptation (Thupalli and Deen 2018; Lucatello and Alcántara-Ayala 2024). Integrating coastal bioshields into DRR strategies requires cross-sectoral collaboration, community participation, and policy alignment to ensure their maintenance, protection, and long-term sustainability (Majlingova and Kádár 2025). By harnessing these synergies, coastal regions can build more adaptive and resilient landscapes capable of withstanding and recovering from both slow-onset and sudden natural hazards. Empirical data from the 2004 Indian Ocean tsunami and subsequent cyclones confirm that intact bioshields can reduce fatalities and property losses substantially (Danielsen et al. 2005; Marois and Mitsch 2015). Their self-repairing nature and regenerative capacity make them invaluable for long-term coastal resilience.

1.8 Emerging Innovations and Research Directions

1.8.1 Hybrid and Multi-Layered Protection Systems

Recent research emphasizes hybrid coastal protection, combining bioshields with engineered infrastructure (Morris et al. 2018; Veettil et al. 2021). For example, mangrove–seawall combinations in Singapore (Lai et al. 2014) and marsh–levee systems in the Netherlands (Marijnissen et al. 2020) demonstrate enhanced energy dissipation and longer structural lifespan.

Multi-layered protection strategies, such as integrating mangroves, dunes, and seagrasses, offer synergistic benefits (Dedekorkut-Howes et al. 2020; Wang et al. 2025). These systems mimic natural coastal gradients and optimize hydrodynamic and ecological functions (Veettil et al. 2020; Wang et al. 2025).

1.8.2 Remote Sensing and Monitoring

Advances in remote sensing, UAVs, and LiDAR enable accurate monitoring of bioshield extent, health, and effectiveness (Veettil et al. 2021; Masria 2024). Satellite imagery supports temporal analyses of vegetation recovery, erosion patterns, and wave attenuation (Villa et al. 2012; Ma et al. 2024). These technologies are particularly valuable in data-scarce regions, enabling evidence-based management and policy evaluation.

1.8.3 Modelling and Predictive Science

Hydrodynamic and ecological models are increasingly used to quantify bioshield performance under various climate scenarios (Obeysekera et al. 2011). Coupled models that integrate wave dynamics, vegetation drags, and sediment transport are essential tools for designing optimal bioshield width, species composition, and spatial configuration (Athikalam and Vaideeswaran 2022).

1.9 The Road Ahead: Integrating Ecology, Engineering, and Society

The growing recognition of coastal bioshields signifies a paradigm shift in coastal management—from the domination of nature through engineering to collaboration with nature through ecological design (Morris et al. 2018). The challenge currently lies in mainstreaming these approaches into policy, planning, and public perception.

The future of coastal protection lies in the seamless integration of ecology, engineering, and social sciences to create resilient and adaptive bioshield systems (Perricone et al. 2023; Bhattacharya and Sachdev 2024). Ecological science provides the fundamental understanding of biophysical processes that govern the structure and functioning of bioshields, guiding species selection, site restoration, and long-term ecosystem management (Perricone et al. 2023). Engineering complements this by offering design optimization, hydrodynamic modeling, and structural innovations that enhance the performance and stability of natural systems, ensuring that bioshields work in harmony with built infrastructure rather than in isolation (Stachew et al. 2021). Equally essential are the social sciences, which emphasize the importance of community participation, equitable governance, and cultural relevance in sustaining these initiatives over time (Jennings et al. 2024). Engaging local stakeholders not only strengthens stewardship but also ensures that interventions are socially accepted and economically viable (Turnbull et al. 2020). Only through such integration can bioshields fully fulfill their dual role as ecological infrastructure and social capital.

1.10 Conclusions

The growing need for coastal protection reflects the convergence of environmental, economic, and social imperatives in an era of climate uncertainty. Traditional hard-engineering approaches, though effective in specific contexts, are insufficient to address the complex and dynamic challenges facing modern coastlines.

Coastal bioshields, rooted in ecology yet aligned with contemporary science and policy, offer a sustainable, adaptive, and cost-effective alternative. They integrate the principles of resilience, multifunctionality, and community engagement, thereby transforming coastal protection from a purely defensive measure into a proactive framework for sustainable coastal living.

This book builds upon these foundations, exploring the scientific, technical, social, and policy dimensions of coastal bioshields. It argues that integrating bioshields into coastal protection strategies is not merely an environmental option but a strategic necessity for the twenty-first century—one that harmonizes human prosperity with the enduring strength of nature.

References

Alongi DM (2018) Blue carbon: coastal sequestration for climate change mitigation. Springer, Cham, Switzerland, pp 23–36. https://doi.org/10.1007/978-3-319-91698-9

Athikalam PT, Vaideeswaran AK (2022) Vegetation bioshield for coastal protection in South Asia: status and way forward. J Coast Conserv 26:3. https://doi.org/10.1007/s11852-022-00850-x

Ayyam V, Palanivel S, Chandrakasan S (2019) Coastal ecosystems and services. In: Coastal ecosystems of the tropics-adaptive management. Springer Singapore, Singapore, pp 21–47. https://doi.org/10.1007/978-981-13-8926-9_2

Barbier EB, Hacker SD, Kennedy C, Koch EW, Stier AC, Silliman BR (2011) The value of estuarine and coastal ecosystem services. Ecol Monogr 81(2):169–193. https://doi.org/10.1890/10-1510.1

Barbier EB (2018) The value of Coastal Wetland ecosystem services. In: Coastal Wetlands, 2nd edn, pp 947–964. https://doi.org/10.1016/B978-0-444-63893-9.00027-7

Barman A, Rajak F, Jha R (2025) Integrating traditional knowledge systems for Wetland conservation and management: a critical analysis. Nat Environ Pollut Technol 24(1):B4212. https://doi.org/10.46488/NEPT.2025.v24i01.B4212

Becker M, Seeger K, Paszkowski A, Marcos M, Papa F, Almar R, Bates P, France-Lanord C, Hossain MS, Khan MJ, Karegar MA, Syvitski J (2024) Coastal flooding in Asian megadeltas: recent advances, persistent challenges, and call for actions amidst local and global changes. Rev Geophys 62(4):e2024RG000846. https://doi.org/10.1029/2024RG000846

Bhattacharya S, Sachdev BK (2024) Impact of scientific innovation in coastal resource management: fostering blue economic development, environmental mitigation, and pollution control. In: Scientific innovations for coastal resource management. IGI Global, pp 1–32. https://doi.org/10.4018/979-8-3693-2436-3.ch001

Birkmann J, Garschagen M, Van Tuan V, Binh NT (2012) Vulnerability, coping and adaptation to water related hazards in the Vietnamese Mekong Delta. In: The Mekong Delta system: interdisciplinary analyses of a River Delta. Springer Netherlands, Dordrecht, pp 245–289. https://doi.org/10.1007/978-94-007-3962-8_10

Brown S, Nicholls RJ, Woodroffe CD, Hanson S, Hinkel J, Kebede AS, Neumann B, Vafeidis A T (2012) Sea-level rise impacts and responses: a global perspective. In: Coastal hazards. Springer Netherlands, Dordrecht, pp 117–149. https://doi.org/10.1007/978-94-007-5234-4_5
Chee SY, Firth LB, Then AYH, Yee JC, Mujahid A, Affendi YA, McQuatters-Gollop A (2021) Enhancing uptake of nature-based solutions for informing coastal sustainable development policy and planning: a Malaysia case study. Front Ecol Evol 9:708507. https://doi.org/10.3389/fevo.2021.708507
Danielsen F, Sørensen MK, Olwig MF, Selvam V, Parish F, Burgess ND, Hiraishi T, Karunagaran VM, Rasmussen MS, Hansen LB, Quarto A, Suryadiputra N (2005) The Asian tsunami: a protective role for coastal vegetation. Science 310(5748):643–643. https://doi.org/10.1126/science.1118387
Das S, Vincent JR (2009) Mangroves protected villages and reduced death toll during Indian super cyclone. Proc Natl Acad Sci USA 106(18):7357–7360. https://doi.org/10.1073/pnas.0810440106
Datta D, Chattopadhyay RN, Guha P (2012) Community based mangrove management: a review on status and sustainability. J Environ Manage 107:84–95. https://doi.org/10.1016/j.jenvman.2012.04.013
Dedekorkut-Howes A, Torabi E, Howes M (2020) When the tide gets high: a review of adaptive responses to sea level rise and coastal flooding. J Environ Planning Manage 63(12):2102–2143. https://doi.org/10.1080/09640568.2019.1708709
Denny M (2021) Wave-energy dissipation: seaweeds and marine plants are ecosystem engineers. Fluids 6(4):151. https://doi.org/10.3390/fluids6040151
Dey T, Rajib MGM, Islam MA (2021) People's perceptions about the socio-economic and environmental impact of coastal green belt in Bangladesh. Indonesian J Soc Environ Issues 2(2):155–162. https://doi.org/10.47540/ijsei.v2i2.280
Duxbury J, Dickinson S (2007) Principles for sustainable governance of the coastal zone: in the context of coastal disasters. Ecol Econ 63(2–3):319–330. https://doi.org/10.1016/j.ecolecon.2007.01.016
Fisher S, Goff J, Cundy AB, Sear D, McWhorter C (2025) Forested bioshields and tsunami impact mitigation in a Polynesian setting. Nature-Based Solutions 7:100222. https://doi.org/10.1016/j.nbsj.2025.100222
French PW (2002) Coastal defences: processes, problems and solutions. Routledge, pp 384
Friess DA, Rogers K, Lovelock CE, Krauss KW, Hamilton SE, Lee SY, Shi S (2019) The state of the world's mangrove forests: past, present, and future. Annu Rev Environ Resour 44(1):89–115. https://doi.org/10.1146/annurev-environ-101718-033302
Goncalves J, Mohan P (1992) Role of coastal bioshield in strengthening resilience in post-disaster situations stories, experience and lessons from South Asia, pp 800
Griggs G, Reguero BG (2021) Coastal adaptation to climate change and sea-level rise. Water 13(16):2151. https://doi.org/10.3390/w13162151
Hernández-Delgado EA (2024) Coastal restoration challenges and strategies for small island developing states in the face of sea level rise and climate change. Coasts 4(2):235–286. https://doi.org/10.3390/coasts4020014
Hershey NR, Nandan SB (2024) Preserving the blue carbon: the role of coastal wetlands in sustainable development. In: Ecosystem services valuation for sustainable development. Springer Nature Singapore, Singapore, pp 213–236. https://doi.org/10.1007/978-981-97-4688-0_9
Hill K (2015) Coastal infrastructure: a typology for the next century of adaptation to sea-level rise. Front Ecol Environ 13(9):468–476. https://doi.org/10.1890/150088
Hoefle SW (2014) Fishing livelihoods, seashore tourism, and industrial development in coastal Rio de Janeiro: conflict, multi-functionality, and juxtaposition. Geogr Res 52(2):198–211. https://doi.org/10.1111/1745-5871.12061

Hosseinzadeh N, Ghiasian M, Andiroglu E, Lamere J, Rhode-Barbarigos L, Sobczak J, Sealey KS, Suraneni P (2022) Concrete seawalls: a review of load considerations, ecological performance, durability, and recent innovations. Ecol Eng 178:106573. https://doi.org/10.1016/j.ecoleng.2022.106573

IPCC (2021) Summary for policymakers. In: Masson-Delmotte V, Zhai P, Pirani A, Connors SL, Péan C, Berger S, Caud N, Chen Y, Goldfarb L, Gomis MI, Huang M, Leitzell K, Lonnoy E, Matthews JBR, Maycock TK, Waterfield T, Yelekçi O, Yu R, Zhou B (eds) Climate change 2021: the physical science basis. contribution of working group I to the sixth assessment report of the intergovernmental panel on climate change. Cambridge University Press, https://doi.org/10.1017/9781009157896.001

Jaisankar I, Velmurugan A, Swarnam T (2017) Bioshield: an answer to climate change impact and natural calamities? In: Biodiversity and climate change adaptation in tropical islands, pp 667–698. https://doi.org/10.1016/B978-0-12-813064-3.00024-7

Jennings V, San Antonio KM, Brown MJ, Choice L, Simpson Q, Ford I, Robinson DM (2024) Place-based conservation in coastal and marine ecosystems: the importance of engagement with underrepresented communities. Sustainability 16(22):9965. https://doi.org/10.3390/su16229965

Jha MK, Dev M (2024) Impacts of climate change. In: Smart internet of things for environment and healthcare. Springer Nature Switzerland, Cham, pp 139–159. https://doi.org/10.1007/978-3-031-70102-3_10

Lai S, Loke LH, Hilton MJ, Bouma TJ, Todd PA (2014) The effects of urbanisation on coastal habitats and the potential for ecological engineering: a Singapore case study. Ocean Coast Manag 103:78–85. https://doi.org/10.1016/j.ocecoaman.2014.11.006

López-Portillo J, Lewis III RR, Saenger P, Rovai A, Koedam N, Dahdouh-Guebas F, Agraz-Hernández C, Rivera-Monroy VH (2017) Mangrove forest restoration and rehabilitation. In: Mangrove ecosystems: a global biogeographic perspective: structure, function, and services. Springer International Publishing, Cham, pp 301–345. https://doi.org/10.1007/978-3-319-62206-4_10

Loucks DP (2021) Impacts of climate change on economies, ecosystems, energy, environments, and human equity: a systems perspective. In: The impacts of climate change. Elsevier, pp 19–50. https://doi.org/10.1016/B978-0-12-822373-4.00016-1

Lucatello S, Alcántara-Ayala I (2024) Sustainable synergy: strengthening disaster risk reduction in Latin America and the Caribbean through nature-based solutions. Int J Disaster Risk Reduction 113:104860. https://doi.org/10.1016/j.ijdrr.2024.104860

Luijendijk A, Hagenaars G, Ranasinghe R, Baart F, Donchyts G, Aarninkhof S (2018) The state of the world's beaches. Sci Rep 8(1):6641. https://doi.org/10.1038/s41598-018-24630-6

Ma M, Huang W, Jung S, Oslon C, Yin K, Xu S (2024) Evaluating vegetation effects on wave attenuation and dune erosion during hurricane. J Mar Sci Eng 12(8):1326. https://doi.org/10.3390/jmse12081326

Maccarrone V, Filiciotto F, Buffa G, Mazzola S, Buscaino G (2014) The ICZM balanced scorecard: a tool for putting integrated coastal zone management into action. Mar Policy 44:321–334. https://doi.org/10.1016/j.marpol.2013.09.024

Macreadie PI, Costa MD, Atwood TB, Friess DA, Kelleway JJ, Kennedy H, Duarte CM (2021) Blue carbon as a natural climate solution. Nat Rev Earth Environ 2(12):826–839. https://doi.org/10.1038/s43017-021-00224-1

Majlingova A, Kádár TS (2025) From risk to resilience: integrating climate adaptation and disaster reduction in the pursuit of sustainable development. Sustainability 17(12):5447. https://doi.org/10.3390/su17125447

Marijnissen R, Kok M, Kroeze C, van Loon-Steensma J (2020) The sensitivity of a dike-marsh system to sea-level rise—a model-based exploration. J Mar Sci Eng 8(1):42. https://doi.org/10.3390/jmse8010042

Marois DE, Mitsch WJ (2015) Coastal protection from tsunamis and cyclones provided by mangrove wetlands–a review. Int J Biodivers Sci Ecosyst Serv Manage 11(1):71–83. https://doi.org/10.1080/21513732.2014.997292
Masria A (2024) Bridging coastal challenges: the role of remote sensing and future research. Reg Stud Mar Sci 73:103502. https://doi.org/10.1016/j.rsma.2024.103502
Mendez FJ, Losada IJ (2004) An empirical model to estimate the propagation of random breaking and nonbreaking waves over vegetation fields. Coast Eng 51(2):103–118. https://doi.org/10.1016/j.coastaleng.2003.11.003
Minderhoud PS, Erkens G, Pham VH, Bui VT, Erban L, Kooi H, Stouthamer E (2017) Impacts of 25 years of groundwater extraction on subsidence in the Mekong delta, Vietnam. Environ Res Lett 12(6):064006. https://doi.org/10.1088/1748-9326/aa7146
Morris RL, Konlechner TM, Ghisalberti M, Swearer SE (2018) From grey to green: efficacy of eco-engineering solutions for nature-based coastal defence. Glob Change Biol 24(5):1827–1842. https://doi.org/10.1111/gcb.14063
Mukherjee N, Dahdouh-Guebas F, Koedam N, Shanker K (2015) An interdisciplinary framework to evaluate bioshield plantations: insights from peninsular India. Acta Oecologica 63:91–100. https://doi.org/10.1016/j.actao.2014.01.005
Mukherjee N, Balakrishnan M, Shanker K (2009) Bioshields and ecological restoration in tsunami-affected areas in India. In: Integrated coastal zone management, pp 131–144. https://doi.org/10.1002/9781444316285.ch10
Mulder K (ed) (2006) Sustainable development for engineers: a handbook and resource guide, 1st edn. Routledge. https://doi.org/10.4324/9781351282925
Naz A, Chowdhury A (2022) Eco-engineering and mangrove restoration methods to stabilize earthen embankments and establishing bio-shield against natural disasters: a case study from Sundarban Ramsar Wetland, India. In: Assessing, mapping and modelling of mangrove ecosystem services in the Asia-Pacific region. Springer Nature Singapore, Singapore, pp 183–198. https://doi.org/10.1007/978-981-19-2738-6_10
Nehren U, Thai HHD, Marfai MA, Raedig C, Alfonso S, Sartohadi J, Castro C (2016) Ecosystem services of coastal dune systems for hazard mitigation: case studies from Vietnam, Indonesia, and Chile. In: Ecosystem-based disaster risk reduction and adaptation in practice. Springer International Publishing, Cham, pp 401–433. https://doi.org/10.1007/978-3-319-43633-3_18
Neumann B, Vafeidis AT, Zimmermann J, Nicholls RJ (2015) Future coastal population growth and exposure to sea-level rise and coastal flooding-a global assessment. PLoS ONE 10(3):e0118571. https://doi.org/10.1371/journal.pone.0118571
Nguyen TT, Pittock J, Nguyen BH (2017) Integration of ecosystem-based adaptation to climate change policies in Viet Nam. Clim Change 142(1):97–111. https://doi.org/10.1007/s10584-017-1936-x
Nicholls RJ, Lincke D, Hinkel J, Brown S, Vafeidis AT, Meyssignac B, Fang J (2021) A global analysis of subsidence, relative sea-level change and coastal flood exposure. Nat Clim Chang 11(4):338–342. https://doi.org/10.1038/s41558-021-00993-z
Nichols CR, Zinnert J, Young DR (2018) Degradation of coastal ecosystems: causes, impacts and mitigation efforts. In: Tomorrow's coasts: complex and impermanent. Springer International Publishing, Cham, pp 119–136. https://doi.org/10.1007/978-3-319-75453-6_8
Oberhagemann K, Haque AMA, Thompson A (2020) A century of riverbank protection and river training in Bangladesh. Water 12(11):3018. https://doi.org/10.3390/w12113018
Obeysekera J, Kuebler L, Ahmed S, Chang ML, Engel V, Langevin C, Wan Y (2011) Use of hydrologic and hydrodynamic modeling for ecosystem restoration. Crit Rev Environ Sci Technol 41(S1):447–488. https://doi.org/10.1080/10643389.2010.531226
Omori Y, Kuriyama K, Tsuge T, Onuma A, Shoji Y (2022) Coastal community preferences of gray, green, and hybrid infrastructure against tsunamis: a case study of Japan. In: Green infrastructure and climate change adaptation: function, implementation and governance. Springer Nature Singapore, Singapore, pp 415–441. https://doi.org/10.1007/978-981-16-6791-6_25

Perricone V, Mutalipassi M, Mele A, Buono M, Vicinanza D, Contestabile P (2023) Nature-based and bioinspired solutions for coastal protection: an overview among key ecosystems and a promising pathway for new functional and sustainable designs. ICES J Mar Sci 80(5):1218–1239. https://doi.org/10.1093/icesjms/fsad080

Porri F, McConnachie B, van der Walt K-A, Wynberg R, Pattrick P (2023) Eco-creative nature-based solutions to transform urban coastlines, local coastal communities and enhance biodiversity through the lens of scientific and Indigenous knowledge. Cambridge Prisms: Coastal Futures 1:e17. https://doi.org/10.1017/cft.2022.10

Primavera JH, Esteban JMA (2008) A review of mangrove rehabilitation in the Philippines: successes, failures and future prospects. Wetlands Ecol Manage 16(5):345–358. https://doi.org/10.1007/s11273-008-9101-y

Qadir H, Khursheed AW, Nazar UI, Ebru K, Cihan A (2025) Impact assessment of climate change and adaptation strategies in South Asia. Environ Claims J. https://doi.org/10.1080/10406026.2025.2522661

Rahman AM, Rahman S (2015) Natural and traditional defense mechanisms to reduce climate risks in coastal zones of Bangladesh. Weather Clim Extremes 7:84–95. https://doi.org/10.1016/j.wace.2014.12.004

Schoonees T, Gijón Mancheño A, Scheres B, Bouma TJ, Silva RCCDS, Schlurmann T, Schüttrumpf H (2019) Hard structures for coastal protection, towards greener designs. Estuaries Coasts 42(7):1709–1729. https://doi.org/10.1007/s12237-019-00551-z

Small C, Nicholls RJ (2003) A global analysis of human settlement in coastal zones. J Coastal Res 19:584–599. https://doi.org/10.2307/4299200

Stachew E, Houette T, Gruber P (2021) Root systems research for bioinspired resilient design: a concept framework for foundation and coastal engineering. Front Robot AI 8:548444. https://doi.org/10.3389/frobt.2021.548444

Styczen ME, Morgan RPC (2003) Engineering properties of vegetation. In: Slope stabilization and erosion control: a bioengineering approach. Taylor & Francis, pp 4–60

Sutton-Grier AE, Sandifer PA (2019) Conservation of wetlands and other coastal ecosystems: a commentary on their value to protect biodiversity, reduce disaster impacts, and promote human health and well-being. Wetlands 39(6):1295–1302. https://doi.org/10.1007/s13157-018-1039-0

Takagi H, Anh LT, Islam R, Hossain TT (2023) Progress of disaster mitigation against tropical cyclones and storm surges: a comparative study of Bangladesh, Vietnam, and Japan. Coast Eng J 65(1):39–53. https://doi.org/10.1080/21664250.2022.2100179

Tengberg A, Cabanban AS (2013) Lessons learned from investing in marine and coastal management initiatives in the East Asian Seas. Mar Policy 38:355–364. https://doi.org/10.1016/j.marpol.2012.06.013

Thupalli R, Deen TA (2018) an investment strategy for reducing disaster risks and coastal pollution using nature based solutions. In: Nagabhatla N, Metcalfe C (eds) Multifunctional wetlands. Environmental contamination remediation and management. Springer, Cham. https://doi.org/10.1007/978-3-319-67416-2_5

Tinh PH, MacKenzie RA, Hung TD, Vinh TV, Ha HT, Lam MH, Hanh NT, Tung NX, Hai PM, Huyen BT (2022) Mangrove restoration in Vietnamese Mekong Delta during 2015–2020: achievements and challenges. Front Mar Sci 9:1043943. https://doi.org/10.3389/fmars.2022.1043943

Turnbull JW, Johnston EL, Kajlich L, Clark GF (2020) Quantifying local coastal stewardship reveals motivations, models and engagement strategies. Biol Cons 249:108714. https://doi.org/10.1016/j.biocon.2020.108714

Van Wesenbeec BK, Van Zelst VT, Antolinez JA, De Boer WP (2025) Quantifying uncertainty in wave attenuation by mangroves to inform coastal green belt policies. Commun Earth Environ 6(1):258. https://doi.org/10.1038/s43247-025-02178-4

Veettil BK, Pereira SFR, Quang NX (2018) Rapidly diminishing mangrove forests in Myanmar (Burma): a review. Hydrobiologia 822:19–35. https://doi.org/10.1007/s10750-018-3673-1

Veettil BK, Quang NX, Thu Trang NT (2019) Changes in mangrove vegetation, aquaculture and paddy cultivation in the Mekong Delta: a study from Ben Tre Province, southern Vietnam. Estuar Coast Shelf Sci 226:106273. https://doi.org/10.1016/j.ecss.2019.106273

Veettil BK, Costi J, Marques WC, Tran X, Quang NX, Van DD, Hoai PN (2020) Coastal environmental changes in Southeast Asia: a study from Quang Nam Province, Central Vietnam. Reg Stud Mar Sci 39:101420. https://doi.org/10.1016/j.rsma.2020.101420

Veettil BK, Ward RD, Dung NTK, Van DD, Quang NX, Hoai PN, Hoang N (2021) The use of bioshields for coastal protection in Vietnam: current status and potential. Reg Stud Mar Sci 47:101945. https://doi.org/10.1016/j.rsma.2021.101945

Villa P, Boschetti M, Morse JL, Politte N (2012) A multitemporal analysis of tsunami impact on coastal vegetation using remote sensing: a case study on Koh Phra Thong Island, Thailand. Nat Hazards 64(1):667–689. https://doi.org/10.1007/s11069-012-0261-y

Walsh JE, McBride JL, Klotzbach PJ, Balachandran S, Camargo SJ, Holland G, Knutson TR, Kossin JP, Sobel A, Sugi M (2015) Tropical cyclones and climate change. Wiley Interdiscip Rev: Clim Change 7(1):65–89. https://doi.org/10.1002/wcc.371

Wang R, Zhang J, Zhang D, Wang S, Dong L, Qin G, Zhang H (2025) Problem-based techniques and practices of coastal ecological engineering construction in China. Environ Dev Sustain. https://doi.org/10.1007/s10668-025-06006-9

Yu P, Ou Y, Chen Y, Zhang H, Deng X, Xu N (2025) Nature-based solutions in coastal urbanization: addressing environmental and socio-economic challenges. Earth Crit Zone. https://doi.org/10.1016/j.ecz.2025.100032

Zhao Y, Han Z, Zhang C, Wang Y, Zhong J, Gao M (2024) Coastal cultural ecosystem services: a bridge between the natural ecosystem and social ecosystem for sustainable development. Land 13(9):1352. https://doi.org/10.3390/land13091352

Zhu H, Zhang J, Cheuk ML, Hau BC, Fischer GA, Gale SW (2023) Monoculture plantations impede forest recovery: evidence from the regeneration of lowland subtropical forest in Hong Kong. Front for Glob CHange 6:1098666. https://doi.org/10.3389/ffgc.2023.1098666

Chapter 2
Historical Perspectives on Coastal Bioshields

Abstract The concept of coastal bioshields, though increasingly recognized in modern climate adaptation and coastal management discourse, has deep historical roots embedded in traditional ecological knowledge and cultural practices. This chapter traces the evolution of vegetative coastal protection from ancient to contemporary times, illustrating how societies across Asia and beyond have long relied on mangroves, dunes, and other coastal forests as natural barriers against storm surges, erosion, and flooding. Early civilizations in India, Vietnam, China, and Japan integrated coastal vegetation into settlement planning and spiritual traditions, while indigenous communities in the Pacific Islands managed vegetated shorelines through customary law. The colonial and industrial periods marked a shift toward engineered coastal infrastructure, leading to widespread degradation of natural bioshields. However, renewed scientific understanding in the twentieth century—particularly following major disasters such as the 2004 Indian Ocean tsunami—reaffirmed the ecological and protective functions of these systems. The chapter highlights key historical turning points, from indigenous management to post-war rehabilitation and contemporary ecosystem-based adaptation policies. By synthesizing ecological, cultural, and policy histories, it emphasizes that bioshields are not a novel innovation but a revival of ancient, place-based knowledge reinforced by modern science. Understanding these historical trajectories is crucial for designing effective, culturally appropriate, and sustainable bioshield strategies in the face of escalating coastal hazards.

Keywords Colonial environmental transformation · Ecosystem-based adaptation · Indigenous coastal management · Mangrove conservation history · Nature-based coastal protection · Traditional ecological knowledge

B. Kozhikkodan Veettil, *The Potential of Bioshields for Coastal Protection*, Advances in Natural and Technological Hazards Research 55,
https://doi.org/10.1007/978-3-032-22888-8_2

2.1 Introduction

The idea of using vegetation as a natural barrier against coastal hazards is not new. Long before the term "bioshield" was coined, coastal societies across the world recognized the protective functions of mangroves, dunes, and vegetated foreshores (Porri et al. 2023; Sarthi and Michael 2025). These vegetated systems, embedded within local ecological knowledge and traditional land management practices, represented early forms of nature-based coastal defense (Salgado and Martinez 2017).

The historical development of bioshields reflects a continuous interplay among human settlement, environmental adaptation, and technological advancement (Chaturvedi 2017; NASEM 2018). From ancient Asian civilizations that revered mangrove forests as storm buffers, through colonial-era coastal reclamation, to modern integrated coastal zone management (ICZM) and ecosystem-based adaptation (EbA), bioshields have evolved both conceptually and operationally (Friess 2016; Chaturvedi 2017; Veettil et al. 2018, 2021; Zhao et al. 2024).

This chapter traces these historical trajectories across regions, highlighting key turning points that have shaped global understanding and implementation of vegetative coastal protection. The analysis draws upon ecological, cultural, and policy histories, emphasizing Asian regional experiences while situating them within a global narrative of evolving environmental governance.

2.2 Early Recognition of Coastal Vegetation as Natural Barriers

2.2.1 Ancient Civilizations and Indigenous Knowledge

Many ancient coastal civilizations (Fig. 2.1) integrated natural vegetation into their settlement planning (Pregill and Volkman 1999; Knottnerus 2005; DasGupta and Shaw 2013). In South and Southeast Asia, early records and oral traditions suggest that mangroves and coastal forests were recognized for their role in protecting against storm surges and providing resources for fisheries and livelihoods (Bandaranayake 1998; Persoon and Minter 2020; Sarthi and Michael 2025).

In India's coastal regions, references to mangrove forests appear in Sanskrit texts, such as the Charaka Samhita (ca. 600 BCE), which describes tidal forests as barriers that withstand the wrath of the sea (Krishna 2014; Chandran 2015, 2016). Similarly, Vietnamese folklore links mangroves to "green fortresses" that shielded villages from typhoons and flooding (Veettil et al. 2019). Indigenous management practices in the Mekong and Red River Deltas included maintaining belts of Rhizophora and Avicennia along embankments, reflecting centuries-old ecological wisdom (My 2017; De Meulder and Shannon 2023).

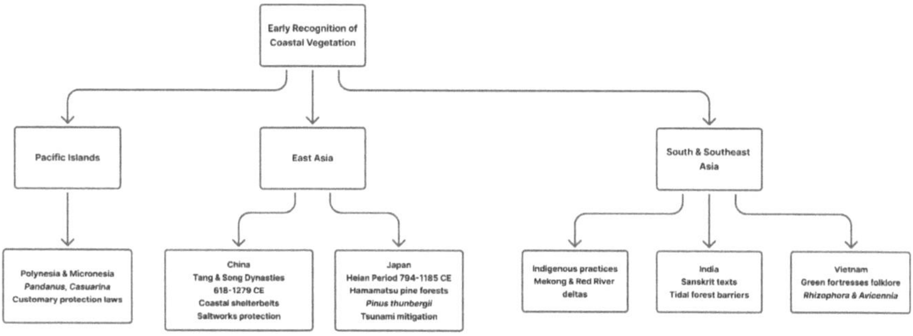

Fig. 2.1 Early recognition of coastal vegetation in different areas of the world

In Pacific Island cultures, coastal vegetation, particularly Pandanus, Casuarina, and Cocos nucifera, was cultivated intentionally to stabilize shores and provide materials for shelter and food (Thaman 1994, 2002). The Polynesians and Micronesians managed vegetated littoral zones through taboos and customary laws that restricted cutting of protective coastal trees, thereby maintaining ecosystem integrity (Gilman et al. 2006; Grantham et al. 2011).

2.2.2 *Chinese and Japanese Coastal Forest Traditions*

In China, written accounts from the Tang Dynasty (618–907 CE) and the Song Dynasty (960–1279 CE) document the establishment of coastal shelterbelts along estuaries and salt pans to protect agricultural lands and saltworks from seawater intrusion and wind erosion (Chiang 1975; Han 2002). These efforts represent some of the earliest recorded state-supported bioshield programs.

Similarly, in Japan, the concept of hamamatsu (coastal pine forests) dates to the Heian period (794–1185 CE) (Totman 1989; Coates et al. 2019). Villagers planted *Pinus thunbergii* along sandy coasts to mitigate windblown sand and to serve as natural levees during tsunamis (Matsushima and Ferreira 2022). These traditional coastal forests later became cultural landscapes symbolizing harmony between human settlements and nature (Senda 1992; Ishizawa 2018).

2.3 Medieval to Pre-industrial Period: Coastal Settlements and Environmental Transformation

Between the thirteenth and eighteenth centuries, as maritime trade and urbanization expanded, many societies began to alter coastal ecosystems for agriculture, aquaculture, and defense infrastructure (Knottnerus 2005; Williams 2010). Despite this, traditional bioshields persisted in regions where local governance and community traditions protected them (Sarthi and Michael 2025).

2.3.1 *Southeast Asian Maritime Kingdoms*

During the height of the Srivijaya and Majapahit empires (7th–fourteenth centuries), ports across Sumatra, Java, and the Malay Peninsula were situated near mangrove-fringed estuaries (Bellwood and Glover 2023). Evidence suggests these forests were deliberately preserved to reduce coastal erosion and maintain navigable waterways (Basyuni et al. 2022).

In the Philippines, pre-colonial barangay communities utilized mangroves and nipa palms (*Nypa fruticans*) for flood control and construction materials (Quintero 2022). Oral histories from the Visayas and Mindanao islands describe how these vegetated belts served as natural barriers during seasonal typhoons (Yuson 2013; Quintero 2022).

2.3.2 *The Role of Religion and Customary Law*

Religious and customary institutions played a key role in conserving coastal vegetation (Mangora and Shalli 2014; Gnansounou et al. 2024). In Buddhist and Hindu traditions, groves near temples and monasteries were protected as sacred spaces (Anthwal et al. 2010; Xie et al. 2025). In Muslim coastal settlements of the Indian Ocean, waqf (religious endowments) sometimes included mangrove forests, ensuring their communal protection (Qurrata et al. 2023). Such customary and spiritual governance mechanisms sustained bioshield functions long before formal environmental policies emerged.

2.4 Colonial Expansion and the Decline of Natural Bioshields

2.4.1 *Colonial Resource Extraction and Land Conversion*

The onset of European colonialism (16th–nineteenth centuries) led to significant ecological transformations on tropical coasts (Lightfoot et al. 2013; Seetah et al. 2022). Mangrove forests and salt marshes were widely cleared for commercial agriculture, salt production, and port infrastructure (Friess 2016). In British India and Dutch Southeast Asia, colonial authorities perceived mangroves as "wastelands" rather than protective assets (Harms and Baird 2014).

During the French colonial period in Vietnam (late 19th–early twentieth century), large tracts of mangrove forests in the Mekong Delta were converted into rice fields and shrimp ponds (Huu Nguyen et al. 2016). Similarly, Indonesia's colonial administration encouraged the clearing of tidal forests for plantations, undermining centuries-old bioshield systems (Boomgaard 2006).

2.4.2 Engineering Paradigms and Coastal Hardening

The colonial era marked a paradigm shift toward engineering-based coastal protection, favoring dikes, seawalls, and groynes over vegetative barriers (Gourlay 1996; Garschagen et al. 2018). These rigid structures, while providing short-term defense, often disrupted sediment dynamics and led to long-term erosion (Veettil et al. 2021). The loss of mangrove belts exacerbated flood risk, which is a phenomenon noted even in early hydrological surveys by Dutch engineers in Java (Marianne 2021) and British engineers in Bengal (Hossain 2020).

2.4.3 Resilience of Traditional Bioshields

Despite these transformations, local communities often resisted total degradation of bioshields. Traditional coastal forests survived in rural and remote regions, maintained through community rules or geographic isolation (Hong and Kim 2011). For instance, Casuarina shelterbelts along India's Odisha coast (Das and Sandhu 2014) and mangrove sacred groves in Thailand and Myanmar (Veettil et al. 2018; Coggins and Chen 2022) persisted as cultural and ecological refuges.

2.5 Twentieth Century: Rediscovery of the Ecological and Protective Roles

2.5.1 Post-World War II Reconstruction and Environmental Awareness

The mid-twentieth century witnessed widespread coastal reconstruction following World War II. Rapid industrialization and population growth increased vulnerability to coastal hazards, prompting a reevaluation of natural coastal systems (Burkett and Davidson 2012). In Japan, the 1959 Ise Bay Typhoon, which caused catastrophic damage, led to renewed investment in coastal forest management (Hayashi 2010). Similarly, the Bengal cyclones of the 1960s and 1970s spurred Bangladesh and India to initiate mangrove afforestation programs as part of disaster mitigation (Patra

2024). These efforts marked the transition from traditional to scientific bioshield management.

2.5.2 *Scientific Validation of Bioshield Functions*

The 1960s–1980s were transformative decades in coastal ecology. Studies by Lugo and Snedaker (1974), Chapman (1978), and Thom (1984) provided quantitative evidence that mangroves, salt marshes, and dunes attenuate wave energy, trap sediments, and reduce storm surge impacts. The recognition of the concept of ecosystem services contributed to bridging ecology with coastal engineering.

In Asia, pioneering projects in India's Bhitarkanika (Murti and Buyck 2014) and Vietnam's Can Gio (Veettil et al. 2021) regions demonstrated the dual benefits of mangroves for fisheries and storm protection. These successes laid the groundwork for integrating bioshields into national coastal defense planning.

2.5.3 *Environmental Movements and Policy Shifts*

The rise of global environmental movements in the 1970s and 1980s, following the Stockholm Conference (1972), and later the Brundtland Report (1987), catalyzed international recognition of nature-based coastal protection (Lofts 2016; Kulaeva 2025). Governments began including coastal afforestation and ecosystem management within broader sustainable development frameworks after these movements (Cicin-Sain 1992).

2.6 The 2004 Indian Ocean Tsunami: A Global Turning Point

The catastrophic Indian Ocean tsunami of December 2004 marked a defining moment in the global perception of bioshields (Tanaka et al. 2013; Veettil et al. 2018; Veettil 2022). Empirical observations revealed that areas with intact mangrove forests experienced significantly lower damage than those where vegetation had been removed (Danielsen et al. 2005; Marois and Mitsch 2015).

In India (Veettil et al. 2020), Sri Lanka (Kaplan et al. 2009), Thailand (Cochard et al. 2008), and Indonesia (Griffin et al. 2012), post-disaster assessments confirmed that mangrove belts reduced wave height and protected inland villages. These findings transformed bioshields from local ecological curiosities to central elements of disaster risk reduction (DRR) and ecosystem-based adaptation (EbA) strategies (Gonsalves and Mohan 2011).

Governments and international agencies have launched large-scale coastal rehabilitation projects that emphasize the ecological and social value of bioshields. In India, mangrove restoration was undertaken under the Integrated Coastal Zone Management Project (ICZMP) (Kumar and Suthar 2022), while Indonesia initiated nationwide replanting campaigns through the Ministry of Environment and Forestry (Supangat et al. 2023). Similarly, Sri Lanka (Dissanayaka et al. 2022) and Thailand (Falk 2014) integrated mangrove management into their post-tsunami reconstruction policies, recognizing the protective and livelihood benefits of these ecosystems. Collectively, these initiatives expanded the policy legitimacy of bioshields, establishing them as cost-effective, sustainable, and (socially) beneficial alternatives to conventional hard-engineering coastal infrastructure, such as rock walls.

2.7 The 21st Century: Institutionalization of Bioshields in Policy and Practice

2.7.1 Ecosystem-Based Adaptation and Climate Policy Integration

In the twenty-first century, bioshields gained prominence under the ecosystem-based adaptation (EbA) paradigm (Renaud et al. 2016). International conventions, such as the Convention on Biological Diversity (CBD), UNFCCC, and the Sendai Framework for Disaster Risk Reduction (2015–2030), emphasized nature-based solutions (Walz et al. 2021).

Across Asia, countries have increasingly integrated bioshield approaches into their national adaptation plans, recognizing the value of nature-based solutions for coastal resilience. Vietnam's National Target Program to Respond to Climate Change (NTP-RCC) promoted extensive mangrove rehabilitation to strengthen shoreline protection and enhance local livelihoods (Tong and Nguyen 2021). In Bangladesh, the Coastal Greenbelt Project established thousands of kilometers of mangrove plantations along embankments, providing natural buffers against cyclones and storm surges (Rahman 2023). Meanwhile, Japan's Green Infrastructure Initiative revitalized coastal pine forests as multi-functional landscapes that combine ecological restoration with disaster risk reduction (Otsuka et al. 2021). Collectively, these initiatives signify a paradigm shift from reactive coastal defense to proactive, ecosystem-based risk management that harmonizes environmental, social, and economic objectives.

2.7.2 *Knowledge Co-production and Community Participation*

Contemporary bioshield programs emphasize co-management, combining scientific expertise with traditional knowledge (Risley 2022). In the Philippines, community-managed mangrove reserves in Bohol and Palawan illustrate participatory approaches that align ecological restoration with local livelihoods (Daupan 2016). In India, women's cooperatives play central roles in mangrove planting and monitoring, enhancing both gender equity and environmental stewardship (Baker et al. 2024).

2.7.3 *Integration with Coastal Engineering*

Recent decades have also witnessed the rise of hybrid coastal protection, integrating bioshields with engineered structures, so-called "green–gray infrastructure" (Waryszak et al. 2021). Projects in Singapore (Lynette 2022), Japan (Santiago-Fandiño and Mas 2017), and the Netherlands (Schoonees et al. 2019) demonstrate how vegetated foreshores enhance the performance and lifespan of sea walls and levees by dissipating wave energy and stabilizing sediments.

2.8 Historical Lessons and Paradigm Evolution

The historical evolution of bioshields reveals several enduring lessons.

Ecological Integration over Isolation: Traditional societies viewed coastal vegetation as integral to their settlement and livelihood systems, rather than as isolated conservation zones (Nunn and Kumar 2018).

Degradation Follows Technocratic Overreach: The shift to engineered infrastructure during colonial and industrial eras led to ecological simplification and vulnerability (Hill 2015).

Resilience through Restoration and Diversity: Post-disaster rehabilitation and community stewardship demonstrate that ecological and social resilience are mutually reinforcing (Gonsalves and Mohan 2011).

Policy Maturation: Global environmental movements and scientific validation of bioshields transformed local practices into internationally endorsed strategies (Veettil et al. 2021).

Adaptive Governance: The twenty-first century's integrated approaches reflect a synthesis of historical wisdom, modern science, and participatory management (Bhattacharya and Sachdev 2024).

2.9 Synthesis: From Tradition to Modern Bioshield Concepts

The trajectory of bioshields illustrates a cyclical pattern of recognition, neglect, and rediscovery. Ancient communities intuitively understood the role of coastal vegetation as natural barriers; colonial and industrial development disrupted these systems; modern science has since reaffirmed their value. What distinguishes the modern bioshield paradigm is systemic integration, recognizing that ecological processes, socioeconomic functions, and policy frameworks must operate in harmony. The term "bioshield" today encapsulates millennia of accumulated experience reframed through the lenses of sustainability, resilience, and ecosystem services. In other words, bioshields are not a novel invention but rather a revival of ancient ecological wisdom, validated by scientific research and embedded within contemporary climate adaptation discourse.

2.10 Conclusions

The historical development of coastal bioshields reveals a profound continuity between traditional ecological knowledge and modern environmental governance. From ancient Chinese and Japanese coastal forests to South Asian mangrove belts, societies have long relied on vegetation for coastal defense. The contemporary era has expanded this understanding, placing bioshields within global frameworks for climate adaptation, disaster risk reduction, and sustainable development. Historical experiences, both of success and failure, demonstrate that the enduring effectiveness of bioshields depends on maintaining ecological integrity, community engagement, and adaptive governance. Thus, the history of bioshields is not merely a record of changing practices but evidence of the persistent dialogue between humanity and the sea, which is a dialogue that continues to shape the sustainable management of our coasts in the Anthropocene.

References

Anthwal A, Gupta N, Sharma A, Anthwal S, Kim K (2010) Conserving biodiversity through traditional beliefs in sacred groves in Uttarakhand Himalaya, India. Resour Conserv Recycl 54(11):962–971. https://doi.org/10.1016/j.resconrec.2010.02.003

Baker M, Isaac L, Alkoyak-Yildiz M et al (2024) Examining the major barriers to women's participation and employment in coastal activities of India's blue economy: a systematic literature review. Environ Dev Sustain. https://doi.org/10.1007/s10668-024-05786-w

Bandaranayake W (1998) Traditional and medicinal uses of mangroves. Mangrove Salt Marshes 2(3):133–148. https://doi.org/10.1023/A:1009988607044

Basyuni M, Sasmito SD, Analuddin K, Ulqodry TZ, Saragi-Sasmito MF, Eddy S, Milantara N (2022) Mangrove biodiversity, conservation and roles for livelihoods in Indonesia. In: Mangroves:

biodiversity, livelihoods and conservation. Springer Nature Singapore, Singapore, pp 397–445. https://doi.org/10.1007/978-981-19-0519-3_16
Bellwood P, Glover I (2023) Southeast Asia from prehistory to history. Routledge, pp. 384. https://doi.org/10.4324/9781003416609
Bhattacharya S, Sachdev BK (2024) Impact of scientific innovation in coastal resource management: fostering blue economic development, environmental mitigation, and pollution control. In: Scientific innovations for coastal resource management. IGI Global, pp 1–32. https://doi.org/10.4018/979-8-3693-2436-3.ch001
Boomgaard P (2006) The long goodbye?: trends in forest exploitation in the Indonesian Archipelago, 1600–2000. In: Muddied waters. Brill, pp 211–234. https://doi.org/10.1163/9789004454347_011
Burkett V, Davidson M (2012) Coastal impacts, adaptation, and vulnerabilities. Springer, Washington, DC, USA
Chandran MS (2016) From the shadows of legitimacy problems and prospects of folk healing in India. J Tradit Folk Pract 2(3):4
Chandran MS (2015) Glimpses into the subaltern system of natural resource management in ancient India. In: UGC sponsored national seminar on flora and fauna in ancient Indian literature, at Dr. AV Baliga College of Arts & Science, Kumta
Chapman VJ (1978) Coastal vegetation, 2nd edn. Pergamon Press, Oxford, 292 pp
Chaturvedi S (2017) Coastal Mangrove forests: micro-geopolitics of resistance and social innovation for environmental sustainability. In: Dong S, Bandyopadhyay J, Chaturvedi S (eds) Environmental sustainability from the Himalayas to the oceans. Springer, Cham. https://doi.org/10.1007/978-3-319-44037-8_8
Chiang TC (1975) The salt Industry of China, 1644–1911: a study in historical geography. Doctoral dissertation, University of Hawaii at Manoa, United States
Cicin-Sain B (1992) Sustainable development and integrated coastal management. Ocean Coast Manag 21(1–3):11–43. https://doi.org/10.1016/0964-5691(93)90019-U
Coates K, Hara K, Holroyd C, Söderberg M (eds) (2019) Japan's future and a new Meiji transformation: international reflections. Routledge, pp 226
Cochard R, Ranamukhaarachchi SL, Shivakoti GP, Shipin OV, Edwards PJ, Seeland KT (2008) The 2004 tsunami in Aceh and Southern Thailand: a review on coastal ecosystems, wave hazards and vulnerability. Perspect Plant Ecol Evol Syst 10(1):3–40. https://doi.org/10.1016/j.ppees.2007.11.001
Coggins C, Chen B (2022) Sacred forests of Asia. Routledge, pp 60
Danielsen F, Sørensen MK, Olwig MF, Selvam V, Parish F, Burgess ND, Suryadiputra N (2005) The Asian tsunami: a protective role for coastal vegetation. Science 310(5748):643–643. https://doi.org/10.1126/science.1118387
Das S, Sandhu H (2014) Role of exotic vegetation in coastal protection: an investigation into the ecosystem services of casuarina in Odisha. Econ Pol Wkly 49:42–50
DasGupta R, Shaw R (2013) Cumulative impacts of human interventions and climate change on mangrove ecosystems of South and Southeast Asia: an overview. J Ecosyst 2013(1):379429. https://doi.org/10.1155/2013/379429
Daupan SMMAV (2016) Community participation in mangrove forest management in the Philippines: management strategies, influences to participation, and socio-economic and environmental impacts. PhD Thesis, University of Michigan. United States
De Meulder B, Shannon K (2023) Topographies of resistance, resilience, bathymetrical realities, and dynamics of the Mekong and Sai Gon-Dong Nai Deltas. Landscape Architect Front 11(4):10–27
Dissanayaka KDCR, Tanaka N, Vinodh TLC (2022) Integration of Eco-DRR and hybrid defense system on mitigation of natural disasters (Tsunami and Coastal Flooding): a review. Nat Hazards 110(1):1–28. https://doi.org/10.1007/s11069-021-04965-6
Falk ML (2014) Post-tsunami recovery in Thailand: socio-cultural responses. Routledge.
Friess DA (2016) Ecosystem services and disservices of mangrove forests: insights from historical colonial observations. Forests 7(9):183. https://doi.org/10.3390/f7090183

Garschagen M, Surtiari GAK, Harb M (2018) Is Jakarta's new flood risk reduction strategy transformational? Sustainability 10(8):2934. https://doi.org/10.3390/su10082934
Gilman E, Van Lavieren H, Ellison J, Jungblut V, Wilson L, Areki F, Yuknavage K (2006) Pacific Island mangroves in a changing climate and rising sea. UNEP Reg Seas Rep Stud 179:1–58
Gnansounou SC, Salako KV, Visée C, Dahdouh-Guebas F, Glèlè Kakaï R, Kestemont P, Henry S (2024) The role of local deities and traditional beliefs in promoting the sustainable use of mangrove ecosystems. For Policy Econ 160:103145. https://doi.org/10.1016/j.forpol.2023.103145
Gonsalves J, Mohan P (eds) (2011) Strengthening resilience in post-disaster situations: stories, experience and lessons from South Asia. IDRC, pp 800
Gourlay MR (1996) History of coastal engineering in Australia. In: History and heritage of coastal engineering, pp 1–88
Grantham HS, McLeod E, Brooks A, Jupiter SD, Hardcastle J, Richardson AJ, Watson JEM (2011) Ecosystem-based adaptation in marine ecosystems of tropical Oceania in response to climate change. Pac Conserv Biol 17(3):241–258. https://doi.org/10.1071/PC110241
Griffin C, Ellis D, Beavis S, Zoleta-Nantes D (2012) Coastal resources, livelihoods and the 2004 Indian Ocean tsunami in Aceh, Indonesia. Ocean Coast Manag 71:176–186. https://doi.org/10.1016/j.ocecoaman.2012.10.017
Han M (2002) Chapter 12 Human influences on muddy coasts. Proc Mar Sci 4:293–317. https://doi.org/10.1016/S1568-2692(02)80086-9
Harms E, Baird IG (2014) Wastelands, degraded lands and forests, and the classification struggle: three critical perspectives from mainland Southeast Asia. Singap J Trop Geogr 35(3):289–294. https://doi.org/10.1111/sjtg.12074
Hayashi H (2010) Natural disasters in Japan. In: Global warming and climate change: prospects and policies in Asia and Europe. Palgrave Macmillan UK, London, pp 118–132. https://doi.org/10.1057/9780230281257_7
Hill K (2015) Coastal infrastructure: a typology for the next century of adaptation to sea-level rise. Front Ecol Environ 13(9):468–476. https://doi.org/10.1890/150088
Hong SK, Kim JE (2011) Traditional forests in villages linking humans and natural landscapes. In: Landscape ecology in Asian cultures. Springer Japan, Tokyo, pp 83–97. https://doi.org/10.1007/978-4-431-87799-8_7
Hossain M (2020) Ecology, epidemics and the colonial state: Environmental change and health in Eastern Bengal Delta, 1858–1947. Master's thesis, İİbn Haldun University, Turkey
Huu Nguyen H, Dargusch P, Moss P, Tran DB (2016) A review of the drivers of 200 years of wetland degradation in the Mekong Delta of Vietnam. Reg Environ Change 16(8):2303–2315. https://doi.org/10.1007/s10113-016-0941-3
Ishizawa M (2018) Cultural landscapes link to nature: learning from Satoyama and Satoumi. Built Heritage 2(4):7–19. https://doi.org/10.1186/BF03545680
Kaplan M, Renaud FG, Lüchters G (2009) Vulnerability assessment and protective effects of coastal vegetation during the 2004 Tsunami in Sri Lanka. Nat Hazard 9(4):1479–1494. https://doi.org/10.5194/nhess-9-1479-2009
Knottnerus OS (2005) History of human settlement, cultural change and interference with the marine environment. Helgol Mar Res 59:2–8. https://doi.org/10.1007/s10152-004-0201-7
Krishna N (2014) Sacred plants of India. Penguin UK
Kulaeva Z (2025) Narratives of change: how climate change narratives have evolved since the 1970s. Int J Clim Change Strat Manage 17(1):376–394. https://doi.org/10.1108/IJCCSM-06-2024-0089
Kumar S, Suthar C (2022) Linkages of sustainable development goals with integrated coastal zone management project in India: an approach to enhance efficacy. Ecol Environ Conserv 153–162. https://doi.org/10.53550/EEC.2022.v28i02s.026
Lightfoot KG, Panich LM, Schneider TD, Gonzalez SL (2013) European colonialism and the Anthropocene: a view from the Pacific Coast of North America. Anthropocene 4:101–115. https://doi.org/10.1016/j.ancene.2013.09.002

Lofts K (2016) Environmental governance and rights-related discourse: a study of human rights-and rights of nature-based discourses in the international climate change regime from 1992–2015. Master's Thesis, McGill University, Canada
Lugo AE, Snedaker SC (1974) The ecology of mangroves. Annu Rev Ecol Syst 5:39–64
Lynette YSM (2022) An eco-design portfolio for the urban marine environment. Master's thesis, National University of Singapore, Singapore
Mangora MM, Shalli MS (2014) Sacred mangrove forests: who bears the pride?. In: Science, policy and politics of modern agricultural system: global context to local dynamics of sustainable agriculture. Springer Netherlands, Dordrecht, pp 291–305. https://doi.org/10.1007/978-94-007-7957-0_20
Marianne P (2021) The land-water Nexus in a Sinking City: the case of Jakarta. PhD Thesis, Columbia University, United States
Marois DE, Mitsch WJ (2015) Coastal protection from tsunamis and cyclones provided by mangrove wetlands–a review. Int J Biodivers Sci Ecosyst Serv Manage 11(1):71–83. https://doi.org/10.1080/21513732.2014.997292
Matsushima H, Ferreira SM (2022) Urbanization of coastal areas: loss of coastal dune ecosystems in Japan. In: Sandy beaches as endangered ecosystems. CRC Press, pp 89–106
Murti R, Buyck C (ed) (2014) Safe havens: protected areas for disaster risk reduction and climate change adaptation. IUCN, Gland, Switzerland, xii + 168 pp
My HHT (2017) Politics of Mangrove restoration: changing policies and strategies of Mangrove restoration in the Central Coast of Vietnam. Consortium of Development Studies in Southeast Asia (CDSSEA) series, vol 16, pp 124
National Academies of Sciences, Engineering, and Medicine (NASEM) (2018) Understanding the long-term evolution of the coupled natural-human coastal system: the future of the U.S. Gulf Coast. The National Academies Press, Washington, DC. https://doi.org/10.17226/25108
Nunn P, Kumar R (2018) Understanding climate-human interactions in Small Island Developing States (SIDS) implications for future livelihood sustainability. Int J Clim Change Strat Manage 10(2):245–271. https://doi.org/10.1108/IJCCSM-01-2017-0012
Otsuka N, Abe H, Isehara Y, Miyagawa T (2021) The potential use of green infrastructure in the regeneration of brownfield sites: three case studies from Japan's Osaka Bay Area. Local Environ 26(11):1346–1363. https://doi.org/10.1080/13549839.2021.1983791
Patra A (2024) Environmental security in India and Bangladesh: a comparative study (1990s to the present). PhD Thesis, Jadavpur University, India
Persoon GA, Minter T (2020) Knowledge and practices of indigenous peoples in the context of resource management in relation to climate change in Southeast Asia. Sustainability 12(19):7983. https://doi.org/10.3390/su12197983
Porri F, McConnachie B, van der Walt K-A, Wynberg R, Pattrick P (2023) Eco-creative nature-based solutions to transform urban coastlines, local coastal communities and enhance biodiversity through the lens of scientific and Indigenous knowledge. Cambridge Prisms: Coastal Futures 1:e17. https://doi.org/10.1017/cft.2022.10
Pregill P, Volkman N (1999) Landscapes in history: design and planning in the Eastern and Western traditions. John Wiley & Sons, pp 864
Quintero RJL (2022) Longing for homes where Mangroves used to grow: narratives of internally displaced persons in Southern Philippines. PhD Thesis, Carleton University, Canada
Qurrata VA, Shafiai MHM, Alma'amun S, Wahid H, Ismail AG, Purwanti P (2023) Blue Waqf framework for blue forest sustainability: a conceptual framework. KnE Soc Sci 398–407
Rahman R (2023) Nature-based solutions in coastal disaster management in Bangladesh. In: Coastal disaster risk management in Bangladesh. Routledge, pp 297–312. https://doi.org/10.4324/978 1003253495-18
Renaud FG, Nehren U, Sudmeier-Rieux K, Estrella M (2016) Developments and opportunities for ecosystem-based disaster risk reduction and climate change adaptation. In: Ecosystem-based disaster risk reduction and adaptation in practice. Springer International Publishing, Cham, pp 1–20. https://doi.org/10.1007/978-3-319-43633-3_1

Risley SC (2022) Linking local knowledge & community science in support of Coastal Marine Stewardship. MSc Thesis, The University of Maine, United States
Salgado K, Martinez ML (2017) Is ecosystem-based coastal defense a realistic alternative? Exploring the evidence. J Coast Conserv 21:837–848. https://doi.org/10.1007/s11852-017-0545-1
Santiago-Fandiño V, Mas E (2017) The coastal environment and the reconstruction process after the Great East Japan Earthquake: a few notes. In: The 2011 Japan earthquake and tsunami: reconstruction and restoration: insights and assessment after 5 years. Springer International Publishing, Cham, pp 291–338. https://doi.org/10.1007/978-3-319-58691-5_18
Sarthi A, Michael G (2025) Protecting the traditional knowledge of coastal communities for climate resilience. In: Rivers unbound. Routledge, pp 129–137
Schoonees T, Gijón Mancheño A, Scheres B, Bouma TJ, Silva RCCDS, Schlurmann T, Schüttrumpf H (2019) Hard structures for coastal protection, towards greener designs. Estuaries Coasts 42(7):1709–1729. https://doi.org/10.1007/s12237-019-00551-z
Seetah K, Manfio S, Balbo A, Farr RH, Florens FV (2022) Colonization during colonialism: developing a framework to assess the rapid ecological transformation of Mauritius's pristine ecosystem. Front Ecol Evol 10:791539. https://doi.org/10.3389/fevo.2022.791539
Senda M (1992) Japan's traditional view of nature and interpretation of landscape. GeoJournal 26(2):129–134. https://doi.org/10.1007/bf00241206
Supangat AB, Basuki TM, Indrajaya Y, Setiawan O, Wahyuningrum N, Purwanto PPB, Savitri E, Indrawati DR, Auliyani D et al (2023) Sustainable management for healthy and productive watersheds in Indonesia. Land 12(11):1963. https://doi.org/10.3390/land12111963
Tanaka N, Yagisawa J, Yasuda S (2013) Breaking pattern and critical breaking condition of Japanese pine trees on coastal sand dunes in huge tsunami caused by Great East Japan Earthquake. Nat Hazards 65(1):423–442. https://doi.org/10.1007/s11069-012-0373-4
Thaman RR (1994) Ethnobotany of Pacific Island coastal plants. Sci Pac Island Peoples: Fauna, Flora, Food Med 3:147–184
Thaman RR (2002) Trees outside forests as a foundation for sustainable development in the small island developing states of the Pacific Ocean. Int for Rev 4(4):268–276. https://doi.org/10.1505/ifor.4.4.268.40531
Thom BG (1984) Coastal landforms and geomorphic processes. In: Snedaker SC, Snedaker JG (eds) The Mangrove ecosystem: research methods. UNESCO, Paris, pp 3–17
Tong TMT, Nguyen NH (2021) Ecosystem-based approaches and policy perspectives: towards an integrated blue–green solutions in Vietnam. In: Ecosystem-based disaster and climate resilience: integration of blue-green infrastructure in sustainable development. Springer Singapore, Singapore, pp 127–160. https://doi.org/10.1007/978-981-16-4815-1_6
Totman C (1989) The green archipelago: forestry in pre-industrial Japan. Univ of California Press, pp 320
Veettil BK (2022) Current status of mangrove vegetation in Batticaloa Lagoon, Sri Lanka, using high-resolution satellite imagery. J Coast Conserv 26(6):74. https://doi.org/10.1007/s11852-022-00923-x
Veettil BK, Pereira SFR, Quang NX (2018) Rapidly diminishing mangrove forests in Myanmar (Burma): a review. Hydrobiologia 822:19–35. https://doi.org/10.1007/s10750-018-3673-1
Veettil BK, Ward RD, Quang NX, Trang NTT, Giang TH (2019) Mangroves of Vietnam: historical development, current state of research and future threats. Estuar Coast Shelf Sci 218:212–236. https://doi.org/10.1016/j.ecss.2018.12.021
Veettil BK, Van DD, Quang NX, Hoai PN (2020) Spatiotemporal dynamics of mangrove forests in the Andaman and Nicobar Islands (India). Reg Stud Mar Sci 39:101455. https://doi.org/10.1016/j.rsma.2020.101455
Veettil BK, Ward RD, Dung NTK, Van DD, Quang NX, Hoai PN, Hoang N (2021) The use of bioshields for coastal protection in Vietnam: current status and potential. Reg Stud Mar Sci 47:101945. https://doi.org/10.1016/j.rsma.2021.101945

Walz Y, Nick F, Higuera Roa O, Nehren U, Sebesvari Z (2021) Coherence and alignment among sustainable land management, ecosystem-based adaptation, ecosystem-based disaster risk reduction and nature-based solutions. Institute for Environment and Human Security, UNCCD: Bonn, Germany
Waryszak P, Gavoille A, Whitt AA, Kelvin J, Macreadie PI (2021) Combining gray and green infrastructure to improve coastal resilience: lessons learnt from hybrid flood defenses. Coast Eng J 63(3):335–350. https://doi.org/10.1080/21664250.2021.1920278
Williams DM (2010) Humankind and the sea: the changing relationship since the mid-eighteenth century. Int J Maritime Hist 22(1):1–14. https://doi.org/10.1177/084387141002200010
Xie C, Lyu Y, Jim CY (2025) Sacred groves occluded in the urban matrix: biodiversity and conservation potential of Hangzhou's Buddhist temples, China. Reg Environ Change 25(3):1–15. https://doi.org/10.1007/s10113-025-02450-5
Yuson AA (2013) Philippines: islands of enchantment. Tuttle Publishing, pp 96
Zhao Y, Han Z, Zhang C, Wang Y, Zhong J, Gao M (2024) Coastal cultural ecosystem services: a bridge between the natural ecosystem and social ecosystem for sustainable development. Land 13(9):1352. https://doi.org/10.3390/land13091352

Part II
Ecological and Biophysical Foundations of Coastal Bioshields

Chapter 3
Ecological Characteristics of Coastal Vegetation

Abstract Coastal vegetation forms the ecological foundation of bioshields, providing both structural stability and biological resilience to dynamic coastal environments. This chapter examines the ecological characteristics, functional roles, and spatial organization of key vegetated coastal ecosystems—mangroves, salt marshes, seagrass meadows, and coastal dunes. It explores how species composition, zonation patterns, root and canopy structures, and physiological adaptations enable these systems to thrive under fluctuating salinity, tidal regimes, and sediment conditions. The discussion integrates ecological processes, including nutrient cycling, primary productivity, and sediment accretion, which collectively underpin the capacity of coastal vegetation to attenuate wave energy and stabilize shorelines. Emphasis is placed on ecological interactions within and between these habitats, including connectivity among mangrove–seagrass–coral reef complexes that enhance biodiversity and ecosystem functioning. The chapter also addresses threats to ecological integrity, including pollution, land-use changes, invasive species, and climate-induced stressors, which compromise both ecosystem health and protective functions. Drawing from case studies in tropical and subtropical Asia, it highlights the critical role of ecological understanding in the design, restoration, and management of effective bioshields. By linking ecological theory with applied coastal management, this chapter provides the scientific foundation for subsequent discussions on biophysical mechanisms, climate resilience, and bioshield design, reinforcing the central premise that ecological integrity is the cornerstone of sustainable coastal protection.

Keywords Biodiversity and resilience · Coastal vegetation ecology · Ecosystem functioning · Habitat connectivity · Mangrove ecosystems · Sediment stabilization

B. Kozhikkodan Veettil, *The Potential of Bioshields for Coastal Protection*,
Advances in Natural and Technological Hazards Research 55,
https://doi.org/10.1007/978-3-032-22888-8_3

3.1 Introduction

Coastal vegetation occupies a unique and dynamic interface between terrestrial and marine systems, forming complex ecosystems that are both ecologically productive and functionally critical to shoreline stability (Ayyam et al. 2019). These habitats, encompassing mangroves, salt marshes, seagrasses, and coastal dunes, not only support diverse biological communities but also provide essential ecosystem services including carbon sequestration, nutrient cycling, and coastal protection (Veettil et al. 2019a, 2021). Their ecological characteristics define the effectiveness of bioshields, making an understanding of their structure, productivity, and interactions with physical processes fundamental to coastal management and climate adaptation (Tanaka 2009; Veettil et al. 2021).

Gradients of salinity, inundation, sediment type, and hydrodynamic energy govern the ecology of coastal vegetation (Silvestri et al. 2004; Phillips 2018; Xin et al. 2022). Each vegetation type has evolved specialized adaptations to thrive in saline, waterlogged, and unstable substrates (Ievinsh 2006). These adaptations, in turn, determine their capacity to modify the environment, stabilizing sediments, attenuating waves, and facilitating successional processes that maintain ecosystem resilience (De Battisti 2021; Veettil et al. 2021).

This chapter examines the ecological structure and function of major coastal vegetation types, with emphasis on their distribution patterns, physiological and morphological adaptations, productivity, and ecological interactions. Comparative analyses across systems and regions highlight how ecological traits translate into protective functions, providing the biotic foundation for bioshield effectiveness.

3.2 Coastal Vegetation Types and Distribution

Climate zones, tidal amplitude, and geomorphology primarily determine the global distribution of coastal vegetation (Bird 2008). In addition, geographic location, solar radiation, ocean currents, and atmospheric chemistry also influence the distribution of coastal vegetation (Noor et al. 2020). Tropical and subtropical regions are dominated by mangrove forests (Spalding et al. 2010), temperate zones by salt marshes (Alen and Pye 1992; Worthington et al. 2024), and sandy or carbonate coasts by dune vegetation (Lane et al. 2008). Seagrass meadows occur in both tropical and temperate shallow subtidal zones (Short et al. 2007).

3.2.1 Mangroves

Mangroves are woody plants adapted to intertidal saline environments, distributed between approximately 30° N and 30° S latitude (Kathiresan and Bingham 2001).

Globally, they cover around 14–15 million ha (Global Mangrove Watch 2024), with over one-third found in Asia (Richards and Friess 2016). The Indo–West Pacific region, encompassing countries such as Indonesia, Vietnam, Malaysia, and the Philippines, represents the world's biodiversity hotspot for mangroves, harboring more than 40 of the ~70 recognized mangrove species (Alongi 2015).

In Vietnam, extensive mangrove belts occur in the Mekong and Red River deltas (Veettil et al. 2019a), while Indonesia hosts the world's largest contiguous mangrove area (>3 million ha) (Basyuni et al. 2022). Bangladesh's Sundarbans, shared with India, represent the largest single mangrove forest, renowned for its ecological productivity and storm-buffering role (Aziz and Paul 2015).

3.2.2 Salt Marshes

Salt marshes are herbaceous plant communities inhabiting intertidal mudflats of estuaries and sheltered bays (Leuschner and Ellenberg 2017a). Dominated by halophytic grasses and succulents, they occupy temperate and subtropical coastlines (Ibáñez et al. 2012; Leuschner and Ellenberg 2017a). While extensive in Europe and North America, Asian salt marshes occur mainly along the Yellow Sea and Gulf of Thailand coasts, often coexisting with mangroves at the tropical–temperate transition zone (Kirwan and Megonigal 2013).

3.2.3 Seagrass Meadows

Seagrasses are submerged marine angiosperms forming meadows in shallow coastal waters, lagoons, and estuaries (Unsworth and Cullen-Unsworth 2017). They are widely distributed across tropical Indo-Pacific and Mediterranean regions (Short et al. 2007). In Southeast Asia, meadows of *Thalassia hemprichii*, *Enhalus acoroides*, and *Halophila ovalis* contribute to sediment stabilization and water clarity, serving as nursery grounds for marine fauna (Orth et al. 2020; Veettil et al. 2022).

3.2.4 Coastal Dune Vegetation

Dune vegetation occurs on sandy shorelines and barrier islands, forming critical stabilizing communities (Stallins 2005). Species such as *Ipomoea pes-caprae*, *Spinifex littoreus*, and *Casuarina equisetifolia* dominate Asian coasts (Anbarashan et al. 2024). These plants colonize shifting sands, trap airborne particles, and initiate dune formation, offering protection against wind erosion and overwash during storms (Feagin et al. 2015).

3.3 Ecological Zonation and Environmental Gradients

Coastal ecosystems exhibit distinct zonation patterns governed (Fig. 3.1) by gradients in salinity, inundation frequency, and substrate stability (Ray 1991; Silvestri et al. 2004; Moffett et al. 2010). Ecological zonation of coastal vegetation reflects the complex balance between plant physiological adaptations and environmental gradients shaped by tidal influence, salinity, substrate type, and exposure to wave and wind energy (Fenu et al. 2013).

Along the shore-to-inland gradient, distinct vegetation zones emerge, each characterized by species uniquely adapted to specific stress regimes (Bermúdez et al. 2024; Hernández-Cordero et al. 2015). In the intertidal and supratidal zones, salt-tolerant pioneers such as mangroves, salt marsh grasses, or halophytic herbs dominate, enduring frequent inundation, high salinity, and anoxic sediments (Adam 1994; Leuschner and Ellenberg 2017a). Slightly inland, dune vegetation and coastal shrubs thrive in environments with reduced salinity and greater soil stability, facilitating succession toward more complex plant communities (Miller et al. 2010). Further inland, where tidal influence wanes and freshwater inputs increase, mixed forests or freshwater wetlands may establish, marking the transition to terrestrial ecosystems (Neubauer and Craft 2009; Williams et al. 2019). Gradients of moisture, nutrient

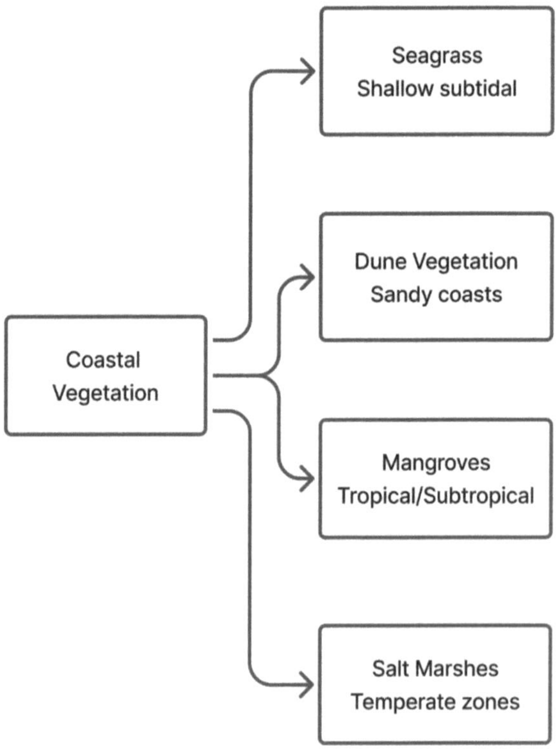

Fig. 3.1 Ecological zonation of coastal vegetation

availability, and elevation thus drive a range of ecological niches, promoting species diversity and functional complementarity across the coastal landscape (Lane et al. 2008; Phillips 2018). This zonation not only reflects adaptive responses to harsh and variable conditions but also underpins the ecological resilience and multifunctionality of coastal ecosystems.

3.3.1 Mangrove Zonation

Mangrove ecological zonation is strongly governed by gradients in tidal inundation, salinity, soil type, and oxygen availability, which together determine the spatial arrangement of species across the intertidal landscape (Snedaker 1982; Chowdhury et al. 2019). Mangrove forests typically display clear seaward-to-landward zonation (Fu et al. 2018).

In the seaward fringe, where tidal flooding and salinity are highest, pioneer species such as *Avicennia marina* and *Sonneratia alba* dominate, adapted to withstand strong wave action, saline conditions, and unstable sediments through specialized root structures and salt-excreting mechanisms (Hogarth 2015). Slightly landward, mid-intertidal zones are typically occupied by Rhizophora species with stilt roots that provide anchorage and aeration in soft, waterlogged soils as well as trapping sediments (Smith 1987; Alongi et al. 2003). In the more elevated back-mangrove areas, where inundation is infrequent and salinity may be reduced or elevated due to evaporation, species such as Bruguiera, Ceriops, and Lumnitzera predominate (Chowdhury et al. 2019; Aluri 2022). This structural stratification allows energy dissipation across tidal gradients (Tomlinson 2016).

Soil texture, nutrient availability, and freshwater inflows further refine the zonation patterns of mangroves, creating complex mosaics of vegetation adapted to microtopographic and hydrological variability (Sofawi et al. 2017; Rivera-Monroy et al. 2017). This ecological stratification not only reflects physiological adaptations to environmental stressors but also enhances overall ecosystem resilience, supporting diverse faunal communities and critical coastal functions such as sediment stabilization, nutrient cycling, and shoreline protection.

3.3.2 Saltmarsh Gradients

Saltmarsh ecological zonation arises from the interaction of tidal inundation patterns, salinity levels, sediment characteristics, and plant physiological tolerances (Silvestri et al. 2004; Moffett et al. 2010). Typically, salt marshes exhibit distinct vertical and horizontal zones that correspond to gradients in tidal frequency and soil salinity, and this vertical distribution enhances the system's capacity for wave attenuation under varying tidal regimes (Silvestri et al. 2004; Rogers and Woodroffe 2015). The low marsh, regularly flooded by tides, is dominated by highly salt-tolerant and flood-adapted species such as *Spartina alterniflora*, which can withstand anoxic and saline conditions (Malamud-Roam 2000; LaFond-Hudson and Sulman 2023). Moving landward, the middle and high marsh zones experience less frequent inundation and support less salt-tolerant but competitively stronger species such as *Spartina patens*, *Juncus gerardii*, or *Suaeda maritima* (Guo 2011; Leuschner and Ellenberg 2017a). At the upper margins, transitional communities blend with terrestrial vegetation, influenced by freshwater inputs, reduced salinity, and greater soil stability (Veldkornet et al. 2015).

Sediment texture, nutrient availability, and elevation further refine these zonation patterns, creating a mosaic of plant assemblages that respond dynamically to both physical and biological gradients (Engels 2010). This ecological stratification not only reflects adaptive responses to environmental stress but also underpins the high productivity, habitat diversity, and carbon sequestration capacity of salt marsh ecosystems.

3.3.3 Seagrass Zonation

Ecological zonation of seagrasses reflects the complex interplay of environmental gradients that shape their spatial distribution across coastal ecosystems (Van Der Heide et al. 2010). Typically, seagrass species occupy distinct zones along gradients of depth, light availability, salinity, substrate type, and hydrodynamic energy (Greve and Binzer 2004; Campbell et al. 2007; Collier et al. 2007). In shallow, high-light intertidal or nearshore areas, pioneer species such as Halodule or Halophila often dominate, adapted to frequent exposure and high sediment mobility (Strydom et al. 2017; O'Brien et al. 2018). At greater depths, where light intensity diminishes and hydrodynamic forces are weaker, more shade-tolerant and structurally robust species, such as Thalassia and Enhalus, prevail, forming dense meadows that stabilize sediments and support diverse faunal communities (Montoya 2014; O'Brien et al. 2018; Lai et al. 2020).

Salinity fluctuations, nutrient inputs, and water clarity further influence zonation patterns, particularly in estuarine or lagoonal environments where freshwater inflows and turbidity vary seasonally (Hillmann et al. 2019). In addition, anthropogenic factors, such as water pollution, were observed to impact the zonation of seagrass meadows (Veettil et al. 2022). This ecological stratification along environmental gradients ensures niche differentiation, enhances resilience to disturbance, and contributes to the overall productivity and biodiversity of coastal seascapes.

3.3.4 Dune Succession

Ecological succession on coastal dunes progresses from pioneer grasses colonizing the foredunes to shrubs and trees established further inland (Avis and Lubke 1996).

This process reflects the dynamic progression of vegetation and soil development driven by strong environmental gradients in salinity, moisture, wind exposure, and substrate stability (Olff et al. 1993; Miller et al. 2010). Pioneer species stabilize sand through rhizomes, enabling the formation of semi-fixed dunes colonized by Pandanus and Casuarina (Marod et al. 2020). These successional processes ensure a multilayered structure that enhances wind and sand control. Starting at the foredune or embryo dune zone closest to the shoreline, pioneer species such as *Ipomoea pes-caprae*, *Spinifex littoreus*, and *Uniola paniculata* colonize loose, shifting sands, stabilizing them through extensive root and rhizome networks that trap and bind sediments (Green and Miller 2019; Lee et al. 2020). As sand accumulation and stabilization increase, secondary dunes form inland, supporting grasses, herbs, and shrubs that tolerate moderate salinity and reduce wind stress (Leuschner and Ellenberg 2017a, b). Further landward, fixed or gray dunes develop with more mature soils, increased organic matter, and lower salinity, allowing the establishment of woody shrubs and dune forest species, such as Casuarina and Pandanus (Boon et al. 2016; Marod et al. 2020).

This zonation, from pioneer foredunes to stabilized backdunes, mirrors a successional gradient driven by decreasing disturbance intensity, increasing nutrient accumulation, and improving soil moisture retention (Perry 1995). Through this succession, coastal dune systems evolve from bare sand to complex vegetated landscapes that buffer coastal erosion, provide habitat diversity, and maintain ecological connectivity between marine and terrestrial environments.

3.4 Physiological and Morphological Adaptations

Coastal vegetation exhibits a range of physiological and morphological adaptations that enable it to survive in environments characterized by high salinity, waterlogging, wind exposure, and nutrient-poor substrates (Rozema et al. 1985; Hogarth 2015). Various physiological and morphological adaptations of coastal vegetation towards environmental stressors are summarized in Table 3.1. Many species have evolved specialized mechanisms for salt tolerance, such as salt excretion through glands in mangroves, like *Avicennia marina*, or salt accumulation and shedding in succulents, like Salicornia. To cope with anoxic and unstable soils, mangroves develop aerial roots, such as pneumatophores, prop roots, or knee roots, that facilitate gas exchange and provide structural support (Srikanth et al. 2016). Coastal dune plants, exposed to intense winds and sand burial, exhibit horizontal growth forms, flexible stems, and extensive rhizome systems that stabilize sediments and reduce erosion (Maun 1998; Sigren et al. 2014). Thick, waxy leaves or reduced leaf surfaces minimize water loss, while high osmotic regulation enables these plants to maintain cellular function under osmotic stress. Additionally, reproductive strategies such as buoyant seeds in mangroves (Aluri 2022) or wind-dispersed seeds in dune grasses (Valdesolo et al. 2022) ensure colonization of dynamic coastal substrates. Together, these physiological and morphological adaptations reflect evolutionary responses to the harsh coastal environment, allowing vegetation to play crucial roles in stabilizing shorelines, reducing erosion, and maintaining coastal ecosystem resilience.

3.4.1 Salt Tolerance and Osmoregulation

Halophytes possess significant physiological adaptations that enable them to survive and grow in saline environments through efficient salt tolerance and osmoregulation mechanisms (Mann et al. 2023). These plants maintain water balance under high salinity by adjusting their internal osmotic potential to counter external salt stress. In mangroves, species such as Rhizophora employ salt exclusion mechanisms, using ultrafiltration at the root membranes to prevent excessive salt uptake (Kim et al. 2016), while others like Avicennia actively secrete salts through specialized leaf glands, thereby regulating ion concentration within tissues (Suárez and Medina 2008). In contrast, salt marsh plants rely on osmotic adjustment by accumulating compatible organic solutes such as proline, glycine betaine, and sugars that stabilize cellular structures and maintain turgor pressure without interfering with metabolic processes (Rozema et al. 1985). These adaptive strategies collectively enable halophytes to sustain photosynthesis, growth, and reproduction in saline and water-limited coastal habitats (Flowers and Colmer 2015), ensuring their ecological success in challenging environments.

3.4.2 Adaptations to Anoxia

Adaptations to anoxia are vital for the survival of coastal vegetation inhabiting oxygen-deficient intertidal environments. Periodic flooding limits oxygen availability in sediments, creating hypoxic or anoxic conditions that challenge root respiration. Mangroves have evolved specialized aerial root systems such as pneumatophores, stilt roots, and knee roots, equipped with lenticels that allow direct gas exchange with the atmosphere, ensuring adequate oxygen supply to submerged tissues (Srikanth et al. 2016). These structures maintain aerobic respiration even during tidal inundation. In seagrasses, extensive aerenchyma tissues form internal air channels that transport oxygen from photosynthetic leaves to the roots and rhizomes buried in anaerobic and waterlogged sediments (Brodersen et al. 2018). This internal aeration prevents the accumulation of toxic sulfides and supports efficient nutrient uptake (Brodersen et al. 2018). Such morphological and anatomical adaptations enable both mangroves and seagrasses to thrive in waterlogged substrates, sustaining high rates of primary productivity and contributing significantly to coastal ecosystem stability and carbon sequestration.

Table 3.1 Physiological and morphological adaptations of coastal vegetation towards environmental stressors

Environmental stressor	Adaptation type	Key mechanisms/ traits	Representative vegetation types and species	Ecological function
High salinity	Physiological (salt tolerance and osmoregulation)	1. Salt exclusion at roots 2. Salt secretion via leaf glands 3. Accumulation of compatible solutes (e.g., proline, glycine betaine)	Mangroves (Rhizophora, Avicennia marina) Salt marsh plants (Salicornia, Spartina)	Maintains water balance, cellular integrity, and photosynthesis under osmotic stress
Waterlogging/ anoxia	Morphological and anatomical	1. Aerial roots (pneumatophores, prop roots, knee roots) 2. Lenticels for gas exchange 3. Aerenchyma tissues	Mangroves (Avicennia, Rhizophora) Seagrasses (Thalassia, Enhalus)	Ensures root oxygenation, prevents sulfide toxicity, sustains productivity
Hydrodynamic forces (waves, tides, currents)	Structural	1. Flexible stems and leaves 2. Strong anchoring root systems 3. Canopy structures that dissipate wind energy	Salt marshes (Spartina) Mangroves (Rhizophora)	Reduces drag and mechanical stress; stabilizes sediments
Wind exposure and sand burial	Morphological	1. Prostrate growth forms 2. Extensive rhizomes and stolons 3. Rapid vertical growth after burial	Coastal dunes (*Ipomoea pes-caprae*, *Spinifex littoreus*)	Stabilizes dunes, reduces erosion, and enables persistence on mobile substrates

(continued)

Table 3.1 (continued)

Environmental stressor	Adaptation type	Key mechanisms/traits	Representative vegetation types and species	Ecological function
Water loss and nutrient-poor substrates	Physiological and morphological	1. Thick, waxy cuticles 2. Reduced leaf surface area 3. High osmotic adjustment efficiency	Mangroves, dune vegetation, salt marsh halophytes	Minimizes transpiration; maintains metabolic function
Dynamic and unstable habitats	Reproductive	1. Buoyant propagules 2. Wind-dispersed seeds 3. Rapid colonization capacity	Mangroves (Rhizophora) Dune grasses (Spinifex)	Enhances dispersal, colonization, and ecosystem recovery

3.4.3 Structural Adaptations to Hydrodynamic Forces

Halophytes exhibit remarkable structural adaptations that enable them to withstand intense hydrodynamic forces such as waves, tides, and currents in coastal environments (Ibrahim 2021). Flexible stems and leaves allow these plants to bend and reorient with water movement, thereby minimizing mechanical stress and preventing structural damage. In salt marshes, Spartina species exemplify this flexibility by bending with tidal flow, thereby reducing drag and erosion around their root zones (Rogers and Woodroffe 2015). On coastal dunes, creeping species like *Ipomoea pes-caprae* and *Spinifex littoreus* develop elongated stolons and rhizomes that stabilize shifting sands and resist burial by extending growth points through newly deposited sediments (Lee et al. 2020). Mangroves possess strong prop and stilt roots that anchor trees firmly in soft substrates, while their complex canopy structures effectively disperse wind energy during storms (Kathiresan and Bingham 2001). Collectively, these structural traits enhance the resilience of coastal vegetation, enabling halophyte communities to function as dynamic bioshields that dissipate hydrodynamic energy and protect shorelines from erosion.

3.5 Productivity, Nutrient Cycling, and Ecosystem Function

Coastal vegetation systems, including mangroves, salt marshes, seagrass meadows, and coastal dune flora, are among the most productive and ecologically significant ecosystems on Earth (Veettil et al. 2021). They occupy the critical interface between land and sea, where energy and matter are continuously exchanged through tidal, hydrological, and biological processes. Their high productivity and efficient carbon and nutrient cycling support diverse communities of plants, animals, and microorganisms, while also performing essential ecological functions such as carbon sequestration, sediment stabilization, and coastal protection (Veettil and Puri 2025). By integrating productivity and biogeochemical cycling, these ecosystems underpin the ecological resilience and long-term sustainability of coastal zones (Turner et al. 1998). They form the foundation of bioshield systems, which are the living defenses that protect coastlines from erosion, storm surges, and climate-driven impacts while maintaining vital ecosystem services.

3.5.1 Primary Productivity

Primary productivity represents the rate at which coastal vegetation converts solar energy into organic biomass through photosynthesis. Mangrove forests are among the most productive natural ecosystems, with net primary productivity (NPP) ranging from 10 to 25 tons of dry matter per hectare per year, while salt marshes average 5-to-15 tons per hectare per year (Alongi 2015; Veettil et al. 2021). These values exceed those of tropical rainforests, reflecting the high nutrient availability and metabolic efficiency of these intertidal systems (Komiyama et al. 2008). The remarkable productivity of mangroves arises from tidal inundation, which brings nutrient-rich sediments and enhances root oxygenation. Their aerial and prop roots host diverse microbial communities that facilitate nutrient transformation and organic matter turnover (Alongi 2005).

Mangrove productivity also varies with species composition and environmental conditions (Kamruzzaman et al. 2017). For example, Rhizophora species dominate frequently in flooded areas, producing abundant aboveground biomass (He et al. 2007). On the other hand, Avicennia is adapted to higher salinity zones and contributes significantly to below-ground carbon accumulation through dense root mats (Reef et al. 2015). The continuous shedding of leaves, propagules, and woody debris generates a substantial litterfall input that decomposes into detritus, forming the base of estuarine food webs. Organic matter outwelling from mangroves sustains secondary productivity in adjacent ecosystems, such as mudflats, seagrass beds, and coral reefs, effectively linking terrestrial and marine nutrient pathways (Prasad et al. 2010).

Salt marsh productivity is similarly regulated by tidal amplitude, sediment supply, and nutrient exchange. Dominant halophytes such as *Spartina alterniflora*, *Salicornia europaea*, and *Juncus gerardii* exhibit high photosynthetic efficiency and rapid growth, driven by alternating inundation and exposure cycles (Malamud-Roam 2000; LaFond-Hudson and Sulman 2023). These plants adapt to fluctuating salinity by maintaining osmotic balance and efficiently recycling nutrients through leaf senescence and root exudation (LaFond-Hudson and Sulman 2023). Seasonal peaks in productivity coincide with warmer months, when microbial decomposition and nutrient mineralization accelerate, further enhancing plant growth (Charles and Dukes 2009). The resulting detrital material enriches estuarine sediments, supporting benthic invertebrates and fish nurseries (Charles and Dukes 2009).

Seagrass meadows, though often submerged, contribute substantially to coastal primary production (Heck et al. 2008). Their NPP can exceed 20 t dry matter ha^{-1} yr^{-1}, sustained by high light penetration and efficient nutrient uptake from sediments. Seagrasses improve water clarity by trapping suspended sediments and provide a stable substratum for epiphytes and microbial communities (Bourque et al. 2015).

Coastal dune vegetation, although less productive per unit area, plays a complementary role by fixing nitrogen through symbiotic bacteria and recycling organic matter under nutrient-poor sandy conditions (Godinho and Bhosle 2013). Collectively, these plant communities form an integrated coastal productivity network that sustains both local food webs and global biogeochemical cycles.

3.5.2 *Nutrient Retention and Cycling*

Beyond their primary production, coastal vegetation systems act as efficient nutrient processors, mediating the transformation, storage, and release of essential elements (Alongi 2020). Mangroves, salt marshes, and seagrasses function as biogeochemical filters that trap sediments, absorb pollutants, and regulate nutrient fluxes between land and sea (Ayyam et al. 2019). This capability is particularly vital in mitigating eutrophication, where excess nutrients from agriculture and urban runoff can degrade water quality and marine habitats (Devlin and Brodie 2023).

Mangrove sediments exhibit high denitrification rates and phosphorus retention, owing to their anoxic yet microbially active soils (Alongi 2005). Denitrifying bacteria in the root zone convert reactive nitrogen (nitrate and nitrite) into nitrogen gas, effectively removing it from the ecosystem (Flores-Mireles et al. 2007). Phosphorus is bound within fine sediments through precipitation with iron and calcium compounds, while the complex root structures reduce hydrodynamic energy, enhancing sediment deposition and nutrient immobilization (Sarker et al. 2021). Studies have shown that intact mangrove forests can retain up to 80% of incoming nitrogen and phosphorus, significantly improving coastal water quality (Kirwan and Megonigal 2013).

Salt marshes perform similar biogeochemical functions. Their alternating aerobic and anaerobic microsites facilitate coupled nitrification–denitrification processes, balancing nitrogen inputs and outputs (Yang et al. 2015). Plant uptake further reduces nutrient concentrations in the water column, while decaying organic matter contributes to long-term nutrient storage in peat-forming sediments. Microbial communities play a central role in these transformations, decomposing detritus and regenerating nutrients for plant uptake. The vertical accretion of salt marsh soil through sediment trapping not only stores nutrients but also counteracts sea-level rise, reinforcing its ecological and protective roles (Cahoon et al. 2021).

Seagrass ecosystems also contribute to nutrient cycling by absorbing dissolved inorganic nitrogen and phosphorus through roots and leaves. Their rhizospheres oxygenate sediment layers, stimulating microbial activity that regulates nutrient transformations (Devereux 2005). The close coupling of plant–microbe interactions ensures efficient nutrient recycling, thereby preventing nutrient leakage into adjacent marine waters.

Coastal dunes, though less nutrient-rich, support biogeochemical cycling by stabilizing soils, reducing erosion, and fostering microbial and fungal activity that recycles scarce nutrients (Nayak and Behera 2019).

The nutrient retention and cycling capacity of these ecosystems directly support biodiversity, productivity, and resilience (Ayyam et al. 2019). By maintaining nutrient equilibrium, coastal bioshields sustain diverse trophic structures and buffer the effects of anthropogenic nutrient loading (Ayyam et al. 2019; Veettil et al. 2021).

3.5.3 Carbon Sequestration

One of the most critical ecosystem functions of coastal vegetation is carbon sequestration, which is the long-term storage of atmospheric carbon dioxide in plant biomass and sediments. Collectively known as blue carbon ecosystems, mangroves, salt marshes, and seagrass meadows store disproportionately high amounts of carbon compared with terrestrial forests (Veettil and Puri 2025). Mangrove sediments contain between 500 and 1,200 megagrams (Mg) of carbon per hectare, often exceeding the carbon density of tropical rainforests (Komiyama et al. 2008). This immense storage results from continuous organic input, slow decomposition under anaerobic conditions, and the accumulation of refractory organic matter.

Salt marshes also serve as long-term carbon sinks, storing between 200 and 500 Mg C ha^{-1} in peat layers that may persist for millennia (Alongi 2018). Their ability to trap suspended sediments during tidal inundation enhances vertical accretion, burying organic carbon deeper in the soil profile. Seagrass meadows similarly sequester carbon both in their dense root mats and in underlying sediments, with some carbon pools remaining stable for thousands of years (Cahoon et al. 2021).

Beyond mitigating global climate change, blue carbon storage provides co-benefits such as sediment stabilization and shoreline protection (Veettil et al. 2021). Carbon-rich sediments increase soil cohesion, reduce erosion, and enhance coastal resilience against storm surges (Vakeri 2024). Maintaining and restoring these ecosystems, therefore, has dual ecological and climate-regulatory significance. Protecting their productivity and nutrient cycling capacity ensures not only the persistence of biodiversity and fisheries but also the long-term sustainability of coastal bioshields in an era of accelerating environmental change.

3.6 Ecological Interactions and Biodiversity Support

Bioshield ecosystems, comprising mangroves, salt marshes, seagrasses, and coastal dune vegetation, function as biodiversity hotspots and ecological connectors, supporting complex interactions among flora, fauna, and microorganisms (Perricone et al. 2023). Their structural complexity and dynamic nature allow multiple species to coexist, interact, and adapt to environmental fluctuations, thereby enhancing coastal resilience. Through a combination of habitat provision, mutualistic interactions, and ecological connectivity, bioshields underpin the functioning of entire coastal landscapes, simultaneously supporting biodiversity conservation and human well-being.

3.6.1 Habitat Provision

Mangroves, seagrasses, and salt marshes function as vital habitats that sustain diverse communities of organisms at multiple trophic levels. Mangrove forests, with their complex root networks, provide essential nursery grounds for numerous species of fish, crustaceans, and mollusks (Veettil et al. 2019a). The prop roots and pneumatophores offer refuge from predators, while their structural complexity creates niches for sessile invertebrates, epiphytic algae, and juvenile reef species. Seagrass meadows form vast underwater pastures where seahorses, pipefish, scallops, and a multitude of invertebrates thrive (Scapin 2018). Their dense rhizome mats stabilize sediments, enhancing water clarity and creating conditions favorable for benthic communities (Alsaffar et al. 2020).

The detrital material produced from leaf litter and decaying organic matter in these systems forms the foundation of rich benthic food webs, supporting detritivores and higher consumers such as commercially important fish and crustaceans (Wai et al. 2008). In Asia, fisheries associated with mangrove ecosystems alone are worth billions of dollars annually, emphasizing the intertwined nature of ecological and economic systems (Veettil et al. 2019a). Salt marshes also play a comparable role in temperate regions, hosting migratory birds and providing feeding grounds for estuarine fauna (Teixeira et al. 2014). Collectively, coastal bioshields act as living infrastructure that simultaneously supports biodiversity, food security, and livelihoods.

3.6.2 Mutualistic and Facilitative Interactions

Within bioshield ecosystems, plant–microbe and interspecific interactions are essential for maintaining ecosystem health and productivity. Mangrove rhizospheres harbor nitrogen-fixing bacteria and phosphate-solubilizing microorganisms that enhance nutrient availability in otherwise oligotrophic conditions (Flores-Mireles et al. 2007). These microbial consortia help mangrove plants tolerate saline and anaerobic environments by improving nutrient uptake and mitigating toxic sulfide accumulation (Yang et al. 2024). Similarly, salt marsh and dune plants often form symbiotic relationships with mycorrhizal fungi, which extend the root absorption area and increase plant resistance to drought and salinity stress (Bothe 2012).

Facilitative interactions also occur among plant species themselves. Pioneer species such as *Avicennia marina* or *Spartina alterniflora* stabilize sediments and reduce hydrodynamic stress (Widdows et al. 2008), thereby creating favorable conditions for the establishment of late-successional species. Such interactions contribute to habitat heterogeneity and enhance overall ecosystem resilience. The synergistic relationships among vegetation, microbes, and fauna thus maintain vital ecological processes, including primary production, decomposition, and nutrient cycling, even under fluctuating environmental stresses.

3.6.3 Successional Dynamics and Connectivity

Bioshield ecosystems form interconnected mosaics along coastal gradients, where seagrass meadows, mangroves, salt marshes, and coastal dunes interact through sediment transport, hydrological flows, and nutrient exchange. These linkages create ecological continua that sustain productivity and facilitate species movement. For instance, detritus from mangrove forests often fuels food webs in adjacent seagrass beds and coral reefs (Nagelkerken and Van Der Velde 2004), while tidal currents distribute propagules and nutrients across habitat boundaries.

Connectivity among these coastal habitats is vital for maintaining resilience against natural disturbances such as storms, erosion, and sea-level rise (Leonard et al. 2017). After damage, recolonization and regeneration often depend on the proximity of intact ecosystems that can serve as sources of propagules (Leonard et al. 2017). The successional dynamics within bioshields, where pioneer communities establish and modify environmental conditions, enable the recruitment of other species, ensuring long-term ecosystem stability and self-repair capacity.

3.7 Disturbance Ecology and Regeneration

Understanding how coastal vegetation responds to disturbances is fundamental to maintaining and restoring bioshields in the face of accelerating anthropogenic and climatic pressures. Coastal vegetation systems, including mangroves, salt marshes, seagrass meadows, and dune communities, exist in inherently dynamic environments, where periodic disturbances such as storms, sediment shifts, and flooding events shape their structure and function (Cahoon et al. 2021; Veettil et al. 2021). However, the increasing intensity and frequency of human-induced stressors, including deforestation, land conversion, and pollution, have altered natural disturbance regimes and impaired the regenerative capacity of these ecosystems (Veettil et al. 2021, 2022). Recognizing the ecological mechanisms that underpin resilience and recovery allows environmental managers to design restoration strategies that harness natural processes rather than relying solely on engineered interventions.

3.7.1 Natural Disturbances

Natural disturbances, which can cause short-term damage and create opportunities for renewal and succession, play a dual role in coastal ecosystems. Cyclones, tsunamis, storm surges, and sediment redistribution events periodically reshape coastal landscapes, removing biomass and altering topography (Veettil et al. 2020a). Despite such impacts, mangroves, salt marshes, and other bioshield components have evolved strong regenerative strategies that enable rapid recovery (Veettil et al. 2021). Mangrove species, for instance, possess specialized propagules that can float and disperse over long distances, allowing colonization of newly deposited sediments after major storm or tsunami events (Van der Stocken et al. 2019). Following the 2004 Indian Ocean tsunami, large tracts of mangroves in India and Thailand experienced extensive damage. Yet, within a decade, natural regeneration of Rhizophora and Avicennia species led to substantial forest recovery, driven by favorable sediment accretion and propagule supply (Cochard et al. 2008; Veettil et al. 2020b).

Similarly, coastal dunes regenerate through the resprouting of buried vegetation and recolonization by pioneer species such as *Ipomoea pes-caprae* and *Spinifex littoreus*, which trap sand and stabilize substrates (Lee et al. 2020). Seagrass meadows often recover from storm-induced scouring through clonal expansion of surviving rhizomes (Paulo et al. 2019).

These natural processes illustrate how disturbance and regeneration are integral components of coastal ecosystem dynamics, maintaining diversity and ecological function over time. However, when disturbance frequency exceeds natural recovery thresholds, regeneration may be incomplete, leading to long-term ecosystem degradation.

3.7.2 Anthropogenic Stressors

In contrast to natural disturbances, human-induced pressures often disrupt the ecological integrity and self-repair mechanisms of bioshields. Coastal deforestation for aquaculture, agriculture, and urban expansion has resulted in significant loss of mangroves and salt marsh cover, while pollution and eutrophication degrade water quality and suppress plant growth (Veettil et al. 2020a, 2021, 2022). Alterations to hydrological and sedimentary regimes, such as dam construction, river channelization, and groundwater extraction, further compromise the environmental conditions necessary for regeneration (Veettil et al. 2020a). Reduced sediment supply can weaken root anchorage and inhibit canopy regrowth, diminishing the protective capacity of coastal vegetation against waves and storm surges.

These anthropogenic impacts often fragment habitats and reduce species diversity, which in turn lowers ecological redundancy and resilience. Effective restoration requires adherence to ecological principles such as maintaining natural hydrological connectivity, respecting species zonation along salinity and inundation gradients, and allowing natural successional processes to unfold (Hernández-Delgado 2024). Restoration projects that overlook these ecological underpinnings frequently result in low survival rates and poor ecosystem functionality.

3.7.3 *Resilience and Recovery Mechanisms*

The resilience of coastal vegetation systems is sustained by multiple biological and geomorphological feedback mechanisms. High species diversity and structural complexity provide ecological redundancy, ensuring that if one species fails to recover, others can fulfill similar functional roles (Boström et al. 2006). Seed banks and propagule reserves act as natural buffers, enabling recolonization after disturbance. Moreover, interactions between vegetation and geomorphology reinforce stability (e.g., roots trapping sediments, building elevation, enhancing water retention) and facilitate further plant establishment (Cahoon et al. 2021).

Ecosystems with intact hydrological and sediment dynamics typically recover faster than those fragmented or modified by human activity. For example, assisted natural regeneration (ANR) approaches in Vietnam have demonstrated that protecting natural regeneration zones, managing hydrology, and allowing spontaneous colonization by native species can achieve more sustainable outcomes than rigid monoculture planting schemes (Veettil et al. 2019a). ANR influences existing ecological processes, such as seed dispersal, nutrient cycling, and microbial associations, to accelerate recovery while maintaining natural diversity and resilience (Pranchai et al. 2025).

3.8 Regional Perspectives: Asian Coastal Vegetation Ecology

Asia represents the most biologically and ecologically diverse region for coastal vegetation on Earth, encompassing vast expanses of mangroves, salt marshes, seagrass meadows, and coastal dunes (Patro et al. 2017). These ecosystems not only support extraordinary biodiversity but also serve as vital natural infrastructures that buffer coastlines from storms, erosion, and rising sea levels (Veettil et al. 2020a, 2021).

However, the same region faces unprecedented pressures from rapid coastal development, aquaculture expansion, pollution, and the accelerating impacts of climate change (Veettil et al. 2019a, 2021, 2022). The interplay between ecological richness and anthropogenic stress defines a complex regional narrative, where the future resilience of coastal systems depends on how effectively nations can harmonize conservation, restoration, and socio-economic development.

In Vietnam, coastal vegetation plays a central role in protecting densely populated deltas and sandy coasts. The Red River and Mekong Deltas, for example, rely on mangrove forests and dune vegetation as natural bioshields against tropical storms and saline intrusion (Veettil et al. 2021). The ecological resilience of these systems is rooted in maintaining hydrological connectivity and promoting mixed-species plantations (Veettil et al. 2019a). This ecological diversity enhances both structural stability and biological productivity, allowing mangrove stands to recover from disturbances and maintain sediment balance (Boström et al. 2006). However, fragmentation caused by shrimp farming, dike construction, and urban expansion continues to disrupt these hydrological linkages, reducing ecosystem functionality (Veettil et al. 2019b). Consequently, integrated management that restores tidal flow and replicates natural zonation patterns has become essential for sustaining Vietnam's coastal defenses.

In Bangladesh, the Sundarbans, which are the world's largest contiguous mangrove forest, represent a model of natural adaptation to tidal and sedimentary dynamics (Mahmood et al. 2021). Here, mangrove productivity and hydrological processes are closely intertwined, enabling the system to accrete sediments and adjust to sea-level changes. The Sundarbans not only harbor immense biodiversity but also provide crucial ecosystem services, including storm protection, carbon sequestration, and support for local fisheries (Islam 2024). However, upstream damming, salinity intrusion, and cyclonic disturbances threaten this delicate equilibrium (Das 2023). Bangladesh's management approach underscores the importance of integrating traditional ecological knowledge with scientific monitoring to ensure long-term resilience and livelihood security for local communities (Shampa et al. 2023).

In Indonesia and Malaysia, the coastal zones host some of the world's most species-rich mangrove and seagrass assemblages (Hogarth 2015). The archipelagic nature of these countries fosters ecological connectivity across thousands of islands, supporting fisheries productivity and biodiversity conservation. Yet, this richness is being rapidly undermined by aquaculture expansion, land reclamation, and oil palm cultivation in low-lying coastal areas (Song et al. 2020). Restoration initiatives have increasingly recognized that monoculture plantations offer limited ecological value and, instead, mixed-species reforestation that mimics natural successional patterns yields greater resilience against disease, pests, and climatic stressors (Boström et al. 2006).

In China, the coastal ecosystems, particularly the salt marshes and seagrass meadows along the Yellow Sea, play a pivotal role in sediment retention and nutrient cycling (Chen et al. 2019). These habitats mitigate erosion, stabilize shorelines, and provide nursery grounds for marine life. However, eutrophication, driven by agricultural runoff and urban wastewater, threatens to destabilize these systems by promoting algal blooms and deoxygenation (Omoregie et al. 2025). Recent Chinese initiatives aim to restore degraded salt marshes through ecological engineering and nutrient management, aligning with global efforts to rebuild blue carbon ecosystems (Tang et al. 2018).

Across Asia, these regional experiences converge on a shared principle: effective coastal vegetation management must be initiated in ecological understanding. Strategies that mimic natural zonation, maintain hydrological connectivity, and prioritize native species composition ensure that restoration efforts reinforce, rather than replace, natural processes. As climate change intensifies, the future of Asia's coasts will depend on integrating ecological science with adaptive governance, balancing human development with the sustaining power of nature's own coastal defenses.

3.9 Synthesis: Ecological Foundations of Bioshield Function

The effectiveness of bioshields as coastal protection systems derives directly from their ecological structure and function. Key ecological attributes, such as species diversity, zonation, productivity, and adaptive morphology, translate into physical processes of energy attenuation, sediment stabilization, and resilience.

Biodiversity enhances system redundancy, complex root architecture stabilizes sediments, and high productivity maintains vertical accretion through the accumulation of organic matter. These ecological features are not merely byproducts of vegetation but are the biophysical engines that sustain coastal defense.

Thus, understanding and conserving the ecology of coastal vegetation is not an ancillary goal but the prerequisite for sustainable bioshield design. Ecological degradation weakens these systems' capacity for protection, while restoration based on sound ecological principles ensures enduring resilience under changing climatic regimes.

3.10 Conclusions

The ecology of coastal vegetation provides the living foundation for bioshield effectiveness. These ecosystems represent evolutionary solutions to extreme environmental stress, integrating structural adaptations, high productivity, and ecological interactions that collectively protect coastlines and sustain biodiversity.

Ranging from mangrove forests that anchor tidal deltas to salt marshes and seagrass meadows that stabilize intertidal flats, and dunes that shield sandy shores, coastal vegetation constitutes a continuous ecological defense network. The interplay of species diversity, morphological adaptation, and biogeochemical processes ensures that bioshields are both self-sustaining and self-repairing.

For countries across Asia and the world, preserving and restoring the ecology of these coastal vegetation systems is not only a conservation priority but also a strategic investment in natural infrastructure. The next chapters build upon this ecological foundation to explore the biophysical mechanisms, climate resilience, and design principles that translate ecological function into measurable protection for coastal communities.

References

Adam P (1994) Saltmarsh and mangrove. In: Groves RH (ed) Australian vegetation, 2nd edn. Cambridge University Press, Cambridge, pp 395–435, 562

Allen JRL, Pye K (1992) Coastal saltmarshes: their nature and importance. In: Saltmarshes: morphodynamics, conservation and engineering significance. Cambridge University Press, pp 1–18

Alongi DM (2015) The impact of climate change on mangrove forests. Curr Clim Change Rep 1:30–39. https://doi.org/10.1007/s40641-015-0002-x

Alongi DM, Clough BF, Dixon P, Tirendi F (2003) Nutrient partitioning and storage in arid-zone forests of the mangroves *Rhizophora stylosa* and *Avicennia marina*. Trees 17(1):51–60. https://doi.org/10.1007/s00468-002-0206-2

Alongi DM (2005) Mangrove–microbe–soil relations. In: Interactions between macro-and microorganisms in marine sediments, vol 60, pp 85–103. https://doi.org/10.1029/CE060p0085

Alongi DM (2018) Salt marshes. In: Blue Carbon. SpringerBriefs in Climate Studies. Springer, Cham. https://doi.org/10.1007/978-3-319-91698-9_2

Alongi DM (2020) Coastal ecosystem processes. CRC Press, pp 448. https://doi.org/10.1201/9781003057864

Alsaffar Z, Pearman JK, Curdia J, Ellis J, Calleja ML, Ruiz-Compean P, Carvalho S (2020) The role of seagrass vegetation and local environmental conditions in shaping benthic bacterial and macroinvertebrate communities in a tropical coastal lagoon. Sci Rep 10(1):13550. https://doi.org/10.1038/s41598-020-70318-1

Aluri JSR (2022) A review of the reproductive ecology of mangrove plant species. In: Mangroves: biodiversity, livelihoods and conservation, pp 33–70. https://doi.org/10.1007/978-981-19-0519-3_3

Anbarashan M, Balachandran N, Mathevet R, Barathan N, Uma Maheswari P (2024) An evaluation of coastal sand dune flora of Cuddalore District, Tamil Nadu, India: perspectives for conservation and management. Geol Ecol Landscapes 8(2):208–221. https://doi.org/10.1080/24749508.2022.2130555

Avis A, Lubke R (1996) Dynamics and succession of coastal dune vegetation in the Eastern Cape, South Africa. Landscape Urban Planning 34(3–4):237–253. https://doi.org/10.1016/0169-2046(95)00217-0

Ayyam V, Palanivel S, Chandrakasan S (2019) Coastal ecosystems and services. In: Coastal ecosystems of the tropics-adaptive management. Springer Singapore, Singapore, pp 21–47. https://doi.org/10.1007/978-981-13-8926-9_2

Aziz A, Paul AR (2015) Bangladesh Sundarbans: present status of the environment and biota. Diversity 7(3):242–269. https://doi.org/10.3390/d7030242

Basyuni M, Sasmito SD, Analuddin K, Ulqodry TZ, Saragi-Sasmito MF, Eddy S, Milantara N (2022) Mangrove biodiversity, conservation and roles for livelihoods in Indonesia. In: Mangroves: biodiversity, livelihoods and conservation. Springer Nature Singapore, Singapore, pp 397–445. https://doi.org/10.1007/978-981-19-0519-3_16

Bermúdez R, Sánchez Vilas J, Retuerto R (2024) Species-specific functional trait responses in two species coexisting along a shore-to-inland dune gradient. Plant Biol 26(7):1162–1174. https://doi.org/10.1111/plb.13710

Bird EC (2008) Coastal geomorphology: an introduction. John Wiley & Sons, pp 448

Boon P, Keith D, Raulings E (2016) Vegetation of coastal floodplains and wetlands in south-eastern Australia. In: Vegetation of Australian riverine landscapes: biology, ecology and management, pp 145–176

Boström C, O'Brien K, Roos C, Ekebom J (2006) Environmental variables explaining structural and functional diversity of seagrass macrofauna in an archipelago landscape. J Exp Mar Biol Ecol 335(1):52–73. https://doi.org/10.1016/j.jembe.2006.02.015

Bothe H (2012) Arbuscular mycorrhiza and salt tolerance of plants. Symbiosis 58(1):7–16. https://doi.org/10.1007/s13199-012-0196-9

Bourque AS, Vega-Thurber R, Fourqurean JW (2015) Microbial community structure and dynamics in restored subtropical seagrass sediments. Aquat Microb Ecol 74:43–57. https://doi.org/10.3354/ame01725

Brodersen KE, Kühl M, Nielsen DA, Pedersen O, Larkum AW (2018) Rhizome, root/sediment interactions, aerenchyma and internal pressure changes in seagrasses. In: Seagrasses of Australia: structure, ecology and conservation. Springer International Publishing, Cham, pp 393–418. https://doi.org/10.1007/978-3-319-71354-0_13

Cahoon DR, McKee KL, Morris JT (2021) How plants influence resilience of salt marsh and mangrove wetlands to sea-level rise. Estuaries Coasts 44(4):883–898. https://doi.org/10.1007/s12237-020-00834-w

Campbell SJ, McKenzie LJ, Kerville SP, Bité JS (2007) Patterns in tropical seagrass photosynthesis in relation to light, depth and habitat. Estuar Coast Shelf Sci 73(3–4):551–562. https://doi.org/10.1016/j.ecss.2007.02.014

Charles H, Dukes JS (2009) Effects of warming and altered precipitation on plant and nutrient dynamics of a New England salt marsh. Ecol Appl 19(7):1758–1773. https://doi.org/10.1890/08-0172.1

Chen Y, Dong J, Xiao X, Ma Z, Tan K, Melville D, Liu F (2019) Effects of reclamation and natural changes on coastal wetlands bordering China's Yellow Sea from 1984 to 2015. Land Degrad Dev 30(13):1533–1544. https://doi.org/10.1002/ldr.3322

Chowdhury R, Sutradhar T, Begam MM, Mukherjee C, Chatterjee K, Basak SK, Ray K (2019) Effects of nutrient limitation, salinity increase, and associated stressors on mangrove forest cover, structure, and zonation across Indian Sundarbans. Hydrobiologia 842(1):191–217. https://doi.org/10.1007/s10750-019-04036-9

Cochard R, Ranamukhaarachchi SL, Shivakoti GP, Shipin OV, Edwards PJ, Seeland KT (2008) The 2004 tsunami in Aceh and Southern Thailand: a review on coastal ecosystems, wave hazards and vulnerability. Perspect Plant Ecol Evol Syst 10(1):3–40. https://doi.org/10.1016/j.ppees.2007.11.001

Collier CJ, Lavery PS, Masini RJ, Ralph PJ (2007) Morphological, growth and meadow characteristics of the seagrass *Posidonia sinuosa* along a depth-related gradient of light availability. Mar Ecol Prog Ser 337:103–115. https://doi.org/10.3354/meps337103

Das GK (2023) Climate change and coastal hazards in Sunderbans. In: Coastal environments of India: a coastal West Bengal perspective. Springer International Publishing, Cham, pp 175–198. https://doi.org/10.1007/978-3-031-18846-6_9

De Battisti D (2021) The resilience of coastal ecosystems: a functional trait-based perspective. J Ecol 109(9):3133–3146. https://doi.org/10.1111/1365-2745.13641

Devereux R (2005) Seagrass rhizosphere microbial communities. Interact Between Macro-and Microorgan Mar Sediments 60:199–216. https://doi.org/10.1029/CE060p0199

Devlin M, Brodie J (2023) Nutrients and eutrophication. In: Marine pollution–monitoring, management and mitigation. Springer Nature Switzerland, Cham, pp 75–100. https://doi.org/10.1007/978-3-031-10127-4_4

Engels JG (2010) Drivers of marsh plant zonation and diversity patterns along estuarine stress gradients. PhD Thesis, Staats- und Universitätsbibliothek Hamburg Carl von Ossietzky. Germany
Feagin RA, Figlus J, Zinnert JC, Sigren J, Martínez ML, Silva R, Smith WK, Cox D, Young DR, Carter G (2015) Going with the flow or against the grain? The promise of vegetation for protecting beaches, dunes, and barrier islands from erosion. Front Ecol Environ 13(4):203–210. https://doi.org/10.1890/140218
Fenu G, Carboni M, Acosta AT, Bacchetta G (2013) Environmental factors influencing coastal vegetation pattern: new insights from the Mediterranean Basin. Folia Geobot 48(4):493–508. https://doi.org/10.1007/s12224-012-9141-1
Flores-Mireles AL, Winans SC, Holguin G (2007) Molecular characterization of diazotrophic and denitrifying bacteria associated with mangrove roots. Appl Environ Microbiol 73(22):7308–7321. https://doi.org/10.1128/AEM.01892-06
Flowers TJ, Colmer TD (2015) Plant salt tolerance: adaptations in halophytes. Ann Bot 115(3):327–331. https://doi.org/10.1093/aob/mcu267
Fu H, Wang W, Ma W, Wang M (2018) Differential in surface elevation change across mangrove forests in the intertidal zone. Estuar Coast Shelf Sci 207:203–208. https://doi.org/10.1016/j.ecss.2018.03.025
Global Mangrove Watch (2024) Global Mangrove Watch v4.0. Available from: https://www.globalmangrovewatch.org/
Godinho A, Bhosle S (2013) Rhizosphere bacteria from coastal sand dunes and their applications in agriculture. In: Bacteria in agrobiology: crop productivity. Springer Berlin Heidelberg, Berlin, Heidelberg, pp 77–96. https://doi.org/10.1007/978-3-642-37241-4_4
Green MD, Miller TE (2019) Germination traits explain deterministic processes in the assembly of early successional coastal dune vegetation. Estuaries Coasts 42(4):1097–1103. https://doi.org/10.1007/s12237-019-00550-0
Greve TM, Binzer T (2004) Which factors regulate seagrass growth and distribution. In: European seagrasses: an introduction to monitoring and management. Monitoring and Managing of European Seagrasses Project (M&MS), 88pp, pp. 19–23. https://doi.org/10.25607/OBP-1997
Guo H (2011) Organization of plant communities across estuarine landscapes in low-latitude tidal marshes. PhD Thesis, University of Houston, United States
He B, Lai T, Fan H, Wang W, Zheng H (2007) Comparison of flooding-tolerance in four mangrove species in a diurnal tidal zone in the Beibu Gulf. Estuar Coast Shelf Sci 74(1–2):254–262. https://doi.org/10.1016/j.ecss.2007.04.018
Heck KL Jr, Carruthers TJ, Duarte CM, Hughes AR, Kendrick G, Orth RJ, Williams SW (2008) Trophic transfers from seagrass meadows subsidize diverse marine and terrestrial consumers. Ecosystems 11(7):1198–1210. https://doi.org/10.1007/s10021-008-9155-y
Hernández-Cordero AI, Pérez-Chacón Espino E, Hernández-Calvento L (2015) Vegetation, distance to the coast, and Aeolian geomorphic processes and landforms in a transgressive arid coastal dune system. Phys Geogr 36(1):60–83. https://doi.org/10.1080/02723646.2014.979097
Hernández-Delgado EA (2024) Coastal restoration challenges and strategies for small island developing states in the face of sea level rise and climate change. Coasts 4(2):235–286. https://doi.org/10.3390/coasts4020014
Hillmann ER, DeMarco K, La Peyre MK (2019) Salinity and water clarity dictate seasonal variability in coastal submerged aquatic vegetation in subtropical estuarine environments. Aquat Biol 28:175–186. https://doi.org/10.3354/ab00719
Hogarth PJ (2015) The biology of mangroves and seagrasses. Oxford University Press, Oxford, p 300
Ibáñez C, Morris JT, Mendelssohn IA, Day Jr JW (2012) Coastal marshes. In: Estuarine ecology. Wiley, pp 129–163. https://doi.org/10.1002/9781118412787
Ibrahim AK (2021) Exploring halophyte hydrodynamics and the role of vegetation traits on shaping salt-tolerance in coastal ecosystems. PhD Thesis, Khalifa University of Science, UAE

Ievinsh G (2006) Biological basis of biological diversity: physiological adaptations of plants to heterogeneous habitats along a sea coast. Acta Univ Latv 710:53–79

Islam MZ (2024) Implementing integrated approaches to preserve the mangrove ecosystem services: a study of Bangladesh Sundarbans. J Environ Develop 33(4):776–806. https://doi.org/10.1177/10704965241262941

Kamruzzaman M, Osawa A, Deshar R, Sharma S, Mouctar K (2017) Species composition, biomass, and net primary productivity of mangrove forest in Okukubi River, Okinawa Island, Japan. Reg Stud Mar Sci 12:19–27. https://doi.org/10.1016/j.rsma.2017.03.004

Kathiresan K, Bingham B (2001) Biology of mangroves and mangrove ecosystems. Adv Mar Biol 40:81–251. https://doi.org/10.1016/S0065-2881(01)40003-4

Kim K, Seo E, Chang SK, Park TJ, Lee SJ (2016) Novel water filtration of saline water in the outermost layer of mangrove roots. Sci Rep 6(1):20426. https://doi.org/10.1038/srep20426

Kirwan ML, Megonigal JP (2013) Tidal wetland stability in the face of human impacts and sea-level rise. Nature 504(7478):53–60. https://doi.org/10.1038/nature12856

Komiyama A, Ong JE, Poungparn S (2008) Allometry, biomass, and productivity of mangrove forests: a review. Aquat Bot 89(2):128–137. https://doi.org/10.1016/j.aquabot.2007.12.006

LaFond-Hudson S, Sulman B (2023) Modeling strategies and data needs for representing coastal wetland vegetation in land surface models. New Phytol 238(3):938–951. https://doi.org/10.1111/nph.18760

Lai S, Cheok ZY, Bouma TJ, Yaakub SM, Todd PA (2020) The relative importance of light and hydrodynamics in the decay and transport of vegetative seagrass fragments. Estuar Coast Shelf Sci 232:106430. https://doi.org/10.1016/j.ecss.2019.106430

Lane C, Wright SJ, Roncal J, Maschinski J (2008) Characterizing environmental gradients and their influence on vegetation zonation in a subtropical coastal sand dune system. J Coastal Res 24:213–224

Lee J-T, Yen L-Z, Chu M-Y, Lin Y-S, Chang C-C, Lin R-S, Chao K-H, Lee M-J (2020) Growth characteristics and anti-wind erosion ability of three tropical foredune pioneer species for sand dune stabilization. Sustainability 12(8):3353. https://doi.org/10.3390/su12083353

Leonard PB, Sutherland RW, Baldwin RF, Fedak DA, Carnes RG, Montgomery AP (2017) Landscape connectivity losses due to sea level rise and land use change. Anim Conserv 20(1):80–90. https://doi.org/10.1111/acv.12289

Leuschner C, Ellenberg H (2017a) Salt marshes and inland saline habitats. In: Ecology of Central European non-forest vegetation: coastal to alpine, natural to man-made habitats: vegetation ecology of Central Europe, vol II, pp 3–61. https://doi.org/10.1007/978-3-319-43048-5_1

Leuschner C, Ellenberg H (2017b) Sand dunes and their vegetation series. In: Ecology of Central European non-forest vegetation: coastal to alpine, natural to man-made habitats: vegetation ecology of Central Europe, vol II. Springer International Publishing, Cham, pp 63–115. https://doi.org/10.1007/978-3-319-43048-5_2

Mahmood H, Ahmed M, Islam T, Uddin MZ, Ahmed ZU, Saha C (2021) Paradigm shift in the management of the Sundarbans mangrove forest of Bangladesh: Issues and challenges. Trees, Forests and People 5:100094. https://doi.org/10.1016/j.tfp.2021.100094

Malamud-Roam KP (2000) Tidal Regimes and tide marsh hydroperiod in the San Francisco Estuary: theory and implications for ecological restoration. PhD Thesis, University of California, Berkeley, United States

Mann A, Lata C, Kumar N, Kumar A, Kumar A, Sheoran P (2023) Halophytes as new model plant species for salt tolerance strategies. Front Plant Sci 14:1137211. https://doi.org/10.3389/fpls.2023.1137211

Marod D, Sungkaew S, Mizunaga H, Thinkampheang S, Thongsawi J (2020) Woody plant community and distribution in a tropical coastal sand dune in southern Thailand. J Coast Conserv 24(4):44. https://doi.org/10.1007/s11852-020-00761-9

Maun MA (1998) Adaptations of plants to burial in coastal sand dunes. Can J Bot 76(5):713–738. https://doi.org/10.1139/b98-058

Miller TE, Gornish ES, Buckley HL (2010) Climate and coastal dune vegetation: disturbance, recovery, and succession. Plant Ecol 206(1):97–104. https://doi.org/10.1007/s11258-009-9626-z
Moffett KB, Robinson DA, Gorelick SM (2010) Relationship of salt marsh vegetation zonation to spatial patterns in soil moisture, salinity, and topography. Ecosystems 13(8):1287–1302. https://doi.org/10.1007/s10021-010-9385-7
Montoya LR (2014) The role of ocean dynamics on seagrass seed dispersal and recruitment in Western Australia. PhD Thesis, University of Western Australia, Australia
Nagelkerken I, Van Der Velde G (2004) Are Caribbean mangroves important feeding grounds for juvenile reef fish from adjacent seagrass beds? Mar Ecol Prog Ser 274:143–151. https://doi.org/10.3354/meps274143
Nayak S, Behera S, Dash PK (2019) Potential of microbial diversity of coastal sand dunes: need for exploration in Odisha Coast of India. Sci World J 2019(1):2758501. https://doi.org/10.1155/2019/2758501
Neubauer SC, Craft CB (2009) Global change and tidal freshwater wetlands: scenarios and impacts. In: Tidal freshwater wetlands. Margraf Publishers GmbH, pp 253–266
Noor M, Rehman NU, Jalil A, Fahad S, Adnan M, Wahid F, Hassan S (2020) Climate change and costal plant lives. In: Environment, climate, plant and vegetation growth. Springer International Publishing, Cham, pp 93–108. https://doi.org/10.1007/978-3-030-49732-3_4
O'Brien KR, Adams MP, Ferguson AJ, Samper-Villarreal J, Maxwell PS, Baird ME, Collier C (2018) Seagrass resistance to light deprivation: implications for resilience. In: Seagrasses of Australia: structure, ecology and conservation. Springer International Publishing, Cham, pp 287–311. https://doi.org/10.1007/978-3-319-71354-0_10
Olff H, Huisman J, Van Tooren BF (1993) Species dynamics and nutrient accumulation during early primary succession in coastal sand dunes. J Ecol 81:693–706
Omoregie AI, Silini MO, Wong LS, Rajasekar A (2025) Nitrogen eutrophication in Chinese aquatic ecosystems: drivers, impacts, and mitigation strategies. Nitrogen 6(4):92. https://doi.org/10.3390/nitrogen6040092
Orth RJ, Lefcheck JS, McGlathery KS, Aoki L, Luckenbach MW, Moore KA, Lusk B (2020) Restoration of seagrass habitat leads to rapid recovery of coastal ecosystem services. Sci Adv 6(41):eabc6434. https://doi.org/10.1126/sciadv.abc6434
Patro S, Krishnan P, Deepak Samuel V, Purvaja R, Ramesh R (2017) Seagrass and salt marsh ecosystems in South Asia: an overview of diversity, distribution, threats and conservation status. In: Wetland science: perspectives from South Asia, pp 87–104. https://doi.org/10.1007/978-81-322-3715-0_5
Paulo D, Diekmann O, Ramos AA, Alberto F, Serrão EA (2019) Sexual reproduction vs. clonal propagation in the recovery of a seagrass meadow after an extreme weather event. Scientia Marina 83(4):357–363
Perricone V, Mutalipassi M, Mele A, Buono M, Vicinanza D, Contestabile P (2023) Nature-based and bioinspired solutions for coastal protection: an overview among key ecosystems and a promising pathway for new functional and sustainable designs. ICES J Mar Sci 80(5):1218–1239. https://doi.org/10.1093/icesjms/fsad080
Perry DA (1995) Forests, competition and succession. Encyclopedia of environmental biology, vol 2. Academic Press Inc., New York, pp 135–153
Phillips JD (2018) Environmental gradients and complexity in coastal landscape response to sea level rise. CATENA 169:107–118. https://doi.org/10.1016/j.catena.2018.05.036
Pranchai A, Jenke M, Thongmee J, Sutthinon P, Karuna S, Wongkamhaeng K, Berger U (2025) Artificial macropores enhance mangrove establishment in hypersaline environments. Restor Ecol 33(4):e70015. https://doi.org/10.1111/rec.70015
Prasad MBK, Dittmar T, Ramanathan AL (2010) Organic matter and mangrove productivity. In: Management and sustainable development of coastal zone environments. Springer Netherlands, Dordrecht, pp 175–193. https://doi.org/10.1007/978-90-481-3068-9_12

Ray GC (1991) Coastal-zone biodiversity patterns. Bioscience 41(7):490–498. https://doi.org/10.2307/1311807

Reef R, Markham HL, Santini NS, Lovelock CE (2015) The response of the mangrove *Avicennia marina* to heterogeneous salinity measured using a split-root approach. Plant Soil 393(1):297–305. https://doi.org/10.1007/s11104-015-2489-2

Richards DR, Friess DA (2016) Rates and drivers of mangrove deforestation in Southeast Asia, 2000–2012. Proc Natl Acad Sci 113(2):344–349. https://doi.org/10.1073/pnas.1510272113

Rivera-Monroy VH, Osland MJ, Day JW, Ray S, Rovai A, Day RH, Mukherjee J (2017) Advancing mangrove macroecology. In: Mangrove ecosystems: a global biogeographic perspective: structure, function, and services. Springer International Publishing, Cham, pp 347–381. https://doi.org/10.1007/978-3-319-62206-4_11

Rogers K, Woodroffe CD (2015) Tidal flats and salt marshes. In: Coastal environments and global change. Wiley, pp 227–250. https://doi.org/10.1002/9781119117261.ch10

Rozema J, Bijwaard P, Prast G, Broekman R (1985) Ecophysiological adaptations of coastal halophytes from foredunes and salt marshes. Vegetatio 62(1):499–521. https://doi.org/10.1007/BF00044777

Sarker S, Masud-Ul-Alam M, Hossain MS, Rahman Chowdhury S, Sharifuzzaman SM (2021) A review of bioturbation and sediment organic geochemistry in mangroves. Geol J 56(5):2439–2450. https://doi.org/10.1002/gj.3808

Scapin L (2018) Habitat use of fish in seagrass meadows of the Venice lagoon: implications for seagrass restoration and for conservation of the lagoon seascape. PhD Thesis, Ca' Foscari University of Venice, Italy

Shampa MTA, Shimu NJ, Chowdhury KA, Islam MM, Ahmed MK (2023) A comprehensive review on sustainable coastal zone management in Bangladesh: present status and the way forward. Heliyon 9(8). https://doi.org/10.1016/j.heliyon.2023.e18190

Short F, Carruthers T, Dennison W, Waycott M (2007) Global seagrass distribution and diversity: a bioregional model. J Exp Mar Biol Ecol 350(1–2):3–20. https://doi.org/10.1016/j.jembe.2007.06.012

Sigren J, Figlus J, Armitage A (2014) Coastal sand dunes and dune vegetation: restoration, erosion, and storm protection. Shore Beach 82(4):5–12

Silvestri S, Defina A, Marani M (2004) Tidal regime, salinity and salt marsh plant zonation. Estuar Coast Shelf Sci 62(1–2):119–130. https://doi.org/10.1016/j.ecss.2004.08.010

Smith TJ (1987) Effects of light and intertidal position on seedling survival and growth in tropical tidal forests. J Exp Mar Biol Ecol 110(2):133–146. https://doi.org/10.1016/0022-0981(87)90024-4

Snedaker SC (1982) Mangrove species zonation: why? Contributions to the ecology of halophytes. Springer, Netherlands, Dordrecht, pp 111–125

Sofawi AB, Nazri MN, Rozainah MZ (2017) Nutrient variability in mangrove soil: anthropogenic, seasonal and depth variation factors. Appl Ecol Environ Res 15(4). https://doi.org/10.15666/aeer/1504_19831998

Song Y, Li D, Hou X (2020) Characteristics of mainland coastline changes in Southeast Asia during the 21st century. J Coastal Res 36(2):261–275. https://doi.org/10.2112/JCOASTRES-D-19-00018.1

Spalding M, Kainuma M, Collins L (2010) World atlas of mangroves. Routledge, pp 336

Srikanth S, Lum SKY, Chen Z (2016) Mangrove root: adaptations and ecological importance. Trees 30(2):451–465. https://doi.org/10.1007/s00468-015-1233-0

Stallins JA (2005) Stability domains in barrier island dune systems. Ecol Complex 2(4):410–430. https://doi.org/10.1016/j.ecocom.2005.04.011

Strydom S, McMahon K, Kendrick GA, Statton J, Lavery PS (2017) Seagrass *Halophila ovalis* is affected by light quality across different life history stages. Mar Ecol Prog Ser 572:103–116. https://doi.org/10.3354/meps12105

Suárez N, Medina E (2008) Salinity effects on leaf ion composition and salt secretion rate in *Avicennia germinans* (L.) L. Braz J Plant Physiol 20:131–140. https://doi.org/10.1590/S1677-04202008000200005

Tanaka N (2009) Vegetation bioshields for tsunami mitigation: review of effectiveness, limitations, construction, and sustainable management. Landscape Ecol Eng 5(1):71–79. https://doi.org/10.1007/s11355-008-0058-z

Tang J, Ye S, Chen X, Yang H, Sun X, Wang F, Chen S (2018) Coastal blue carbon: concept, study method, and the application to ecological restoration. Sci China Earth Sci 61(6):637–646. https://doi.org/10.1007/s11430-017-9181-x

Teixeira A, Duarte B, Caçador I (2014) Salt marshes and biodiversity. In: Sabkha ecosystems: volume IV: cash crop halophyte and biodiversity conservation. Springer Netherlands, Dordrecht, pp 283–298. https://doi.org/10.1007/978-94-007-7411-7_20

Tomlinson PB (2016) The botany of mangroves. Cambridge University Press, pp 432

Turner RK, Adger WN, Lorenzoni I, Bateman IJ (1998) Towards integrated modelling and analysis in coastal zones: principles and practice (vol 11). LOICZ Core Project Office, Netherlands Institute for Sea Research (NIOZ)

Unsworth RK, Cullen-Unsworth LC (2017) Seagrass meadows. Curr Biol 27(11):R443–R445. https://doi.org/10.1016/j.cub.2017.01.021

Vakeri SM (2024) Biological carbon sequestration strategies: a case study on microbial exopolysaccharides in arid soils. PhD Thesis, Khalifa University of Science. UAE

Valdesolo T, Del Vecchio S, Buffa G (2022) Patterns of seed dispersal in coastal dune plant communities. Sustainability 14(17):10983. https://doi.org/10.3390/su141710983

Van Der Heide T, Bouma TJ, Van Nes EH, Van De Koppel J, Scheffer M, Roelofs JG, Smolders AJ (2010) Spatial self-organized patterning in seagrasses along a depth gradient of an intertidal ecosystem. Ecology 91(2):362–369. https://doi.org/10.1890/08-1567.1

Van der Stocken T, Wee AK, De Ryck DJ, Vanschoenwinkel B, Friess DA, Dahdouh-Guebas F, Webb EL (2019) A general framework for propagule dispersal in mangroves. Biol Rev 94(4):1547–1575. https://doi.org/10.1111/brv.12514

Veettil BK, Puri V (2025) Blue carbon ecosystems in Vietnam: a review. Ocean Coast Manag 267:107711. https://doi.org/10.1016/j.ocecoaman.2025.107711

Veettil BK, Quang NX, Thu Trang NT (2019a) Changes in mangrove vegetation, aquaculture and paddy cultivation in the Mekong Delta: a study from Ben Tre Province, southern Vietnam. Estuar Coast Shelf Sci 226:106273. https://doi.org/10.1016/j.ecss.2019.106273

Veettil BK, Ward RD, Quang NX, Trang NTT, Giang TH (2019b) Mangroves of Vietnam: historical development, current state of research and future threats. Estuar Coast Shelf Sci 218:212–236. https://doi.org/10.1016/j.ecss.2018.12.021

Veettil BK, Costi J, Marques WC, Tran X, Quang NX, Van DD, Hoai PN (2020a) Coastal environmental changes in Southeast Asia: a study from Quang Nam Province, Central Vietnam. Reg Stud Mar Sci 39:101420. https://doi.org/10.1016/j.rsma.2020.101420

Veettil BK, Van DD, Quang NX, Hoai PN (2020b) Spatiotemporal dynamics of mangrove forests in the Andaman and Nicobar Islands (India). Reg Stud Mar Sci 39:101455. https://doi.org/10.1016/j.rsma.2020.101455

Veettil BK, Ward RD, Dung NTK, Van DD, Quang NX, Hoai PN, Hoang N (2021) The use of bioshields for coastal protection in Vietnam: current status and potential. Reg Stud Mar Sci 47:101945. https://doi.org/10.1016/j.rsma.2021.101945

Veettil BK, Ward RD, Van DD, Quang NX, Hoai PN (2022) Seagrass ecosystems along the Vietnamese coastline: current state of research and future perspectives. Estuar Coast Shelf Sci 277:108085. https://doi.org/10.1016/j.ecss.2022.108085

Veldkornet D, Adams J, Potts A (2015) Where do you draw the line? Determining the transition thresholds between estuarine salt marshes and terrestrial vegetation. S Afr J Bot 101:153–159. https://doi.org/10.1016/j.sajb.2015.05.003

Wai TC, Ng JS, Leung KM, Dudgeon D, Williams GA (2008) The source and fate of organic matter and the significance of detrital pathways in a tropical coastal ecosystem. Limnol Oceanogr 53(4):1479–1492. https://doi.org/10.4319/lo.2008.53.4.1479

Widdows J, Pope ND, Brinsley MD (2008) Effect of *Spartina anglica* stems on near-bed hydrodynamics, sediment erodability and morphological changes on an intertidal mudflat. Mar Ecol Prog Ser 362:45–57. https://doi.org/10.3354/meps07448

Williams T, Amatya D, Conner W, Panda S, Xu G, Dong J, Wang H (2019) Tidal forested wetlands: mechanisms, threats, and management tools. In: Wetlands: ecosystem services, restoration and wise use. Springer International Publishing, Cham, pp 129–158. https://doi.org/10.1007/978-3-030-14861-4_6

Worthington TA, Spalding M, Landis E, Maxwell TL, Navarro A, Smart LS, Murray NJ (2024) The distribution of global tidal marshes from earth observation data. Glob Ecol Biogeogr 33(8):e13852. https://doi.org/10.1111/geb.13852

Xin P, Wilson A, Shen C, Ge Z, Moffett KB, Santos IR, Barry DA (2022) Surface water and groundwater interactions in salt marshes and their impact on plant ecology and coastal biogeochemistry. Rev Geophys 60(1):e2021RG000740. https://doi.org/10.1029/2021RG000740

Yang WH, Traut BH, Silver WL (2015) Microbially mediated nitrogen retention and loss in a salt marsh soil. Ecosphere 6(1):1–15. https://doi.org/10.1890/ES14-00179.1

Yang X, Yuan R, Yang S, Dai Z, Di N, Yang H, Wei M (2024) A salt-tolerant growth-promoting phyllosphere microbial combination from mangrove plants and its mechanism for promoting salt tolerance in rice. Microbiome 12(1):270. https://doi.org/10.1186/s40168-024-01969-9

Chapter 4
Biophysical Mechanisms of Coastal Protection

Abstract Coastal ecosystems such as mangroves, salt marshes, seagrass meadows, and dune vegetation play a pivotal role in mitigating the impacts of waves, currents, and storm surges through a range of interconnected biophysical mechanisms. This chapter explores the scientific principles underpinning the ability of vegetated coastal systems to dissipate hydrodynamic energy, reduce erosion, and promote sediment deposition. It examines how vegetation structure—root density, stem flexibility, canopy complexity, and spatial arrangement—influences water flow resistance and sediment trapping efficiency. Emphasis is placed on quantitative models and empirical studies that examine the interactions between vegetation and physical processes, including wave attenuation, turbulence reduction, and shoreline stabilization. The chapter further discusses the influence of geomorphological and hydrological conditions, including substrate type, tidal regime, and coastal slope, on the efficiency of bioshields. Comparative analyses between natural and restored bioshields illustrate the dynamic feedback between ecological succession and physical stability. Ultimately, understanding these mechanisms is fundamental to optimizing the design, placement, and long-term sustainability of bioshields as nature-based solutions for coastal defense. By integrating ecological and engineering perspectives, this chapter establishes a scientific framework for the functional assessment of bioshields in diverse coastal settings.

Keywords Hydrodynamic energy dissipation · Nature-based coastal defense · Sediment dynamics · Shoreline stabilization · Vegetation–hydrodynamics interaction · Wave attenuation

4.1 Introduction

Coastal zones represent the dynamic interface between terrestrial and marine environments, where the interaction of waves, tides, currents, and sediment transport processes shapes shoreline morphology (Chowdhury et al. 2023). These areas support dense human populations, critical ecosystems, and economic infrastructure, but they

B. Kozhikkodan Veettil, *The Potential of Bioshields for Coastal Protection*, Advances in Natural and Technological Hazards Research 55,
https://doi.org/10.1007/978-3-032-22888-8_4

are also highly vulnerable to natural hazards such as storm surges, erosion, and flooding (Veettil et al. 2020). Traditional approaches to coastal defense, such as seawalls, groynes, and revetments, have proven effective in localized contexts but often lead to unintended ecological degradation and sediment imbalance (Veettil et al. 2021). In recent decades, attention has increasingly shifted toward nature-based solutions (NbS) that influence the biophysical properties of ecosystems for sustainable protection (Schoonees et al. 2019; Temmerman et al. 2013, 2023).

Among these, bioshields—natural or restored vegetative barriers composed of mangroves, dune vegetation, salt marshes, or coastal forests—play a pivotal role in mitigating the impacts of hydrodynamic forces. They function by reducing wave energy, stabilizing sediments, trapping particulate matter, and facilitating accretion processes that maintain coastal elevation (Cahoon et al. 2021). Their protective capacity arises not from static physical structures but from complex biophysical interactions between vegetation, hydrodynamics, and sedimentary processes (Veettil et al. 2021).

Understanding the mechanisms underlying bioshield function is fundamental to their effective design and management (Perricone et al. 2023). This chapter dissects these mechanisms from a scientific perspective, detailing how vegetation characteristics, such as root structure, canopy density, and species composition, mediate physical forces to reduce coastal vulnerability. The analysis integrates empirical data from global and Asian contexts, drawing insights from coastal vegetation bioshields in Vietnam, Bangladesh, India, and the Philippines. The theoretical basis of coastal protection by bioshield vegetation, such as mangroves, seagrass meadows, salt marshes, and coastal dune vegetation, is discussed in detail.

4.2 Theoretical Basis of Coastal Protection by Vegetation

The coastal protection function of bioshields is grounded in fundamental principles of fluid mechanics, sediment transport, and eco-geomorphology. Vegetation acts as a roughness element that modifies the momentum and energy of flowing water. The presence of stems, trunks, and roots within the intertidal or supratidal zones increases drag, dissipating kinetic energy from waves and currents through turbulence and friction (Mazda et al. 2006; Perricone et al. 2023).

From a hydrodynamic perspective, when waves propagate into a vegetated area, their wave height (H) and velocity are attenuated according to the relationship:

$$\frac{dH}{dx} = -kH$$

where k is the attenuation coefficient determined by vegetation density, stem diameter, and flow depth (Kobayashi et al. 1993). This coefficient quantifies the rate of energy dissipation per unit distance, illustrating how denser or stiffer vegetation enhances energy loss. H is the wave height at a given position x (in m), which represents the local wave amplitude as waves move through vegetation; x is the horizontal distance to the vegetated zone (in m), which is measured from the seaward edge (where x = 0). dH/dx is the rate of change of wave height with distance (dimensionless), which quantifies how rapidly the wave height decreases as it propagates inland.

Vegetation also influences sediment transport by altering near-bed shear stress (τ_b), which determines whether particles are eroded, transported, or deposited (Neary et al. 2012). By reducing flow velocity, vegetation facilitates the deposition of fine sediments and organic matter, promoting vertical accretion and shoreline stabilization (Nepf 2012). Over time, these processes enable coastal landscapes to adapt naturally to sea-level changes through feedback mechanisms between biological productivity and sediment dynamics.

Moreover, bioshields contribute to hydrological regulation by storing and slowly releasing water during tidal cycles or storm events (Cahoon et al. 2021; Ferreira et al. 2023). Root systems and soil porosity facilitate infiltration and water retention, buffering against flash floods and maintaining groundwater-surface water equilibrium (Chu et al. 2024). Thus, the theoretical framework of bioshield protection is inherently multidisciplinary, integrating physics, ecology, and geomorphology into a coherent system of natural defense.

4.3 Wave Attenuation Mechanisms

One of the most well-documented functions of coastal bioshields is wave attenuation, which is the reduction of wave height and energy as water passes through vegetated zones (Veettil et al. 2021; Athikalam and Vaideeswaran 2022). Empirical and modeling studies consistently demonstrate that bioshields, particularly mangrove forests and coastal vegetation belts, can substantially diminish wave forces, thereby protecting hinterland areas from erosion and inundation (Veettil et al. 2021).

4.3.1 Physical Process of Wave Energy Dissipation

As waves enter a vegetated zone, part of their energy is converted into turbulence and drag within the canopy layer. The drag force (F_d) in N exerted by vegetation elements (Kobayashi et al. 1993; Nepf 2012) is expressed as:

$$F_d = \frac{1}{2} C_d \rho A u^2$$

where ρ is the water density in kg m^{-3}; C_d is the drag coefficient (dimensionless, typically 0.8–1.2 for cylindrical stems); A is the projected frontal area of the vegetation per unit volume normal to flow in m^2; u is the instantaneous flow velocity relative to the vegetation (in m s^{-1}). This drag induces localized turbulence and energy loss, resulting in a measurable reduction in wave height. The magnitude of attenuation depends on various factors, including vegetation density, system flexibility, and the relative water depth to canopy height (He et al. 2019).

4.3.2 Influence of Vegetation Type and Structure

Different vegetation types exhibit varying attenuation efficiencies. Mangroves, with their complex root structures and high stem density, are particularly effective at dissipating short-period waves (up to 50% reduction over 100 m of forest) (Mazda et al. 2006). Salt marshes, dominated by flexible grasses (e.g., *Spartina alterniflora*), perform best during moderate water levels, where the plant stems remain partially submerged (Anderson and Smith 2013). Dune vegetation, though less effective in direct wave attenuation, stabilizes sand ridges that act as secondary barriers against overtopping.

4.3.3 Empirical Evidence from Asia

In Vietnam's Mekong Delta, field measurements have shown that a 150 m-wide *Rhizophora apiculata* mangrove belt can reduce wave height by 60–80% under moderate tidal conditions (Phan et al. 2015; Spalding and Leal 2021). Similar findings in Bangladesh's Sundarbans indicate a 20–30% reduction in wave energy per 100 m of forest during cyclonic events (Das and Vincent 2009). In Indonesia, restoration projects in the Demak District, employing hybrid mangrove–bamboo structures, have achieved up to 70% attenuation during monsoon storms (Van Zelst et al. 2021; van Bijsterveldt et al. 2022).

These examples confirm that bioshields act as first-line defenses, reducing the energy of incident waves before they reach engineered structures or populated areas. The cumulative effect across large coastal segments can substantially reduce maintenance costs for conventional infrastructure.

4.4 Sediment Trapping and Shoreline Stabilization

Beyond their role in attenuating hydrodynamic forces, coastal bioshields, such as mangroves, salt marshes, seagrasses, and dune vegetation, play a fundamental role in shoreline stabilization through sediment trapping and erosion control (Gracia et al.

2018). These vegetated systems act as natural sediment filters, capturing, retaining, and stabilizing particulate matter transported by tides, currents, and waves. The interplay between vegetation structure and sediment transport establishes positive feedback mechanisms (Fig. 4.1) that maintain or even build land elevation relative to sea level (Cahoon et al. 2021). By promoting vertical accretion and lateral stability, bioshields enhance the resilience of coastal landscapes against both chronic (sea-level rise) and episodic (storms, surges) disturbances.

In essence, the long-term stability of vegetated shorelines depends on maintaining the integrity of vegetation–sediment feedbacks. Bioshields not only trap and stabilize sediment but also regulate geomorphic processes that sustain coastal landforms under changing environmental conditions (Veettil et al. 2021). By promoting natural accretion, reinforcing soil cohesion, and enabling adaptive morphodynamics, these ecosystems offer a living alternative to hard engineering structures, one that combines ecological resilience, carbon storage, and sustainable shoreline protection.

4.4.1 Sediment Deposition Processes

As tidal or fluvial flows move through vegetated zones, the dense structure of stems, roots, and litter slows down water velocity, effectively reducing the flow's capacity to transport suspended sediments. This decline in flow energy lowers the critical shear stress needed to keep fine sediment particles in suspension, allowing them to settle out of the water column. The result is localized sediment accumulation, especially of fine-grained and organic-rich material, within and around root zones. Over time, this process leads to vertical accretion, enhancing surface elevation and supporting vegetation persistence even under rising sea levels.

Mangroves are among the most efficient natural sediment traps. Their complex aerial root systems, composed of prop roots, pneumatophores, and buttresses, generate micro-turbulent flow fields that favor the deposition of silt and clay particles (Alongi 2025). Field measurements have shown that mangrove forests can induce sediment deposition rates of more than 6–10 mm per year, as observed in Malaysia and Indonesia, often surpassing local sea-level rise rates (Hoque et al. 2015; Ginting 2018). Such rates enable mangroves to maintain or increase surface elevation, preventing inundation and converting unstable mudflats into resilient coastal forests. In Vietnam's Red River Delta, restoration projects combining artificial sediment fences with mangrove replanting have yielded particularly impressive results (Quang and Van An 2025). By trapping suspended sediment before it escapes to deeper waters, these interventions have achieved shoreline stabilization with a gradual seaward shift rate of +17 m per year within a single decade (2009–2022) (Quang and Van An 2025).

Seagrass meadows and salt marshes similarly enhance sediment deposition in their respective tidal environments. Seagrasses, with their flexible leaves and dense rhizome mats, attenuate near-bed currents and promote the settling of suspended materials, thereby stabilizing the seabed and enhancing water clarity (Widdows et al.

Fig. 4.1 Coastal vegetation structure and feedback mechanisms

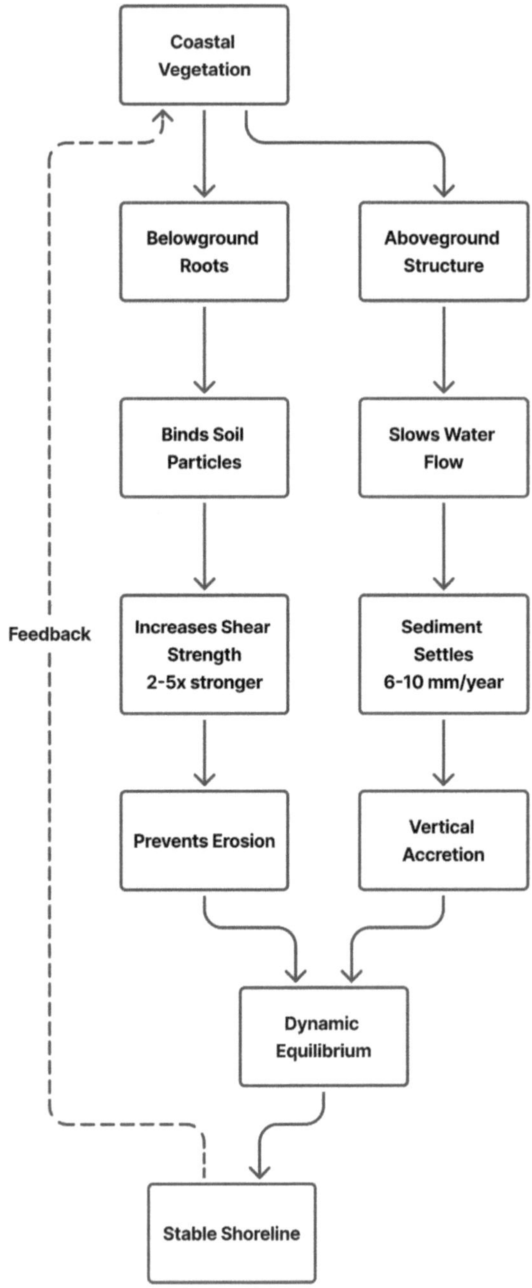

2008). Saltmarsh vegetation, such as *Spartina anglica*, acts as a frictional barrier in intertidal zones, reducing tidal current velocities and fostering the accumulation of fine sediments and organic detritus (Swales et al. 2004). These processes not only build the physical substrate for plant expansion but also store large quantities of blue carbon within the sediment profile, linking shoreline stabilization with climate change mitigation.

4.4.2 Root Reinforcement and Soil Cohesion

While aboveground structures slow water flow and trap sediment, belowground biomass provides mechanical reinforcement to the substrate, enhancing soil cohesion and resistance to erosion. The dense networks of roots and rhizomes act as a natural reinforcing matrix, binding soil particles and increasing the shear strength of sediments (Stachew et al. 2021). This biological stabilization significantly reduces vulnerability to scouring and undercutting caused by waves, currents, and rainfall runoff.

In salt marshes, the dense root mats of Spartina species bind fine-grained sediments, preventing resuspension and minimizing lateral erosion (Neumeier and Ciavola 2004). Similarly, in mangrove forests, fine feeder roots and pneumatophores anchor the upper sediment layer, while deeper structural roots stabilize the substrate against shear forces (Alongi 2025). Studies have shown that soils reinforced by tree roots exhibit two to five times greater shear strength compared to unvegetated mudflats, substantially reducing sediment loss during storm events (Li et al. 2022).

Coastal dune systems depend on vegetation for stabilization in a different hydrodynamic context. Here, the primary erosive agent is wind rather than water. Dune plants, such as *Ipomoea pes-caprae*, *Spinifex littoreus*, and *Pandanus odorifer*, develop extensive fibrous root systems that bind loose sand grains and prevent aeolian transport (Lee et al. 2020). As the plants trap moving sand, they promote vertical dune growth, creating a self-reinforcing feedback loop between vegetation cover and dune morphology. The combined effect of root reinforcement and sand accretion establishes the structural integrity of dune systems, enabling them to function as effective bioshields against coastal erosion and storm surges (Feagin et al. 2015).

4.4.3 Shoreline Morphodynamics and Equilibrium

Over time, the interactions between sediment trapping, root reinforcement, and hydrodynamic forces give rise to complex shoreline morphodynamics. Vegetated coastal zones exhibit a capacity for dynamic equilibrium, wherein erosion and accretion processes balance each other naturally (Prasetya 2007). Rather than resisting all change, as rigid engineered structures attempt to do, bioshields facilitate a self-adjusting shoreline that evolves in harmony with environmental forcing (Veettil et al. 2021).

This adaptive capacity is evident in large deltaic and estuarine systems such as the Sundarbans and the Mekong Delta. In these regions, extensive mangrove belts have been directly correlated with the persistence of intertidal flats and the buffering of inland erosion, even amid increased storm frequency and intensity (Aziz and Paul 2015; Veettil et al. 2019a). By continuously trapping sediment and dissipating energy, these ecosystems maintain coastal profiles and prevent the landward migration of erosion zones (Quang and Van An 2025). Conversely, where vegetation has been cleared for aquaculture, agriculture, or urban expansion, erosion rates have accelerated dramatically (Veettil et al. 2019b, 2020). The absence of vegetative binding and sediment trapping leads to the loss of elevation capital, triggering a feedback loop of subsidence, erosion, and habitat degradation.

4.5 Reduction of Storm Surge and Flood Impacts

One of the most critical protective functions of coastal bioshields is their ability to reduce storm surge and flood severity during extreme weather events (Veettil et al. 2021). Unlike regular wave attenuation, storm surge mitigation involves large-scale hydrodynamic processes influenced by topography, vegetation density, and storm intensity (Hu et al. 2014).

4.5.1 Mechanistic Understanding

Storm surges result from a combination of wind stress and low atmospheric pressure, which push seawater toward the coast. As the surge propagates over a vegetated coastal plain, frictional drag exerted by stems and canopies slows water movement, leading to reduced surge height and delayed inland penetration (Sheng et al. 2012; Montgomery et al. 2019).

The extent of attenuation depends on the vegetation roughness coefficient (n) in Manning's equation:

$$V = \frac{1}{n} R^{2/3} S^{1/2}$$

where V is the velocity, R is the hydraulic radius, and S is the energy slope. Higher roughness values associated with dense vegetation translate into lower flow velocities (Krauss et al. 2009).

4.5.2 Quantitative Evidence from Case Studies

Modeling and field data indicate that coastal bioshields can substantially diminish surge levels (Chen et al. 2021). In Bangladesh, simulations suggested surge height reductions of 4–16.5 cm for mangrove belts of widths from about 50 m to 2 km and a significant reduction in water flow velocity (29–92%) (Dasgupta et al. 2019). Similarly, in Vietnam's northern coast, mangrove zones along Hai Phong and Nam Dinh provinces have been observed to reduce floods during typhoons (Pham and Bui 2020; Duc et al. 2025). In India's Odisha coast, regions with intact mangroves experienced significantly lower casualty rates and property damage during the 1999 Super Cyclone compared to deforested areas (Das and Vincent 2009). Economic analyses attribute millions of dollars in avoided damages to the protective role of these natural buffers (Barbier 2016).

4.5.3 Synergy with Coastal Topography

The effectiveness of storm surge attenuation by coastal bioshields also depends on topographic configuration. Gradual, low-lying plains with extensive intertidal vegetation allow greater dissipation of energy over long distances (James et al. 2021). In contrast, steep or urbanized coastlines reduce the spatial extent for energy dispersion (James et al. 2021). Integrating vegetation restoration with land-use planning and maintaining buffer zones of at least 100–500 m enhances the synergy between bioshields and coastal geomorphology (Prasetya 2007). It is worth noting that, within mangrove forests that are below 500 m wide, wave dissipation is strongly dependent on local water levels, wave characteristics, and forest density (Van Wesenbeeck et al. 2025).

By moderating flood depth and duration, coastal bioshields not only protect infrastructure but also reduce saltwater intrusion and soil salinization, which is critical for sustaining agricultural productivity in deltaic regions.

4.6 Energy Dissipation and Coastal Morphodynamics

The interaction between hydrodynamic energy and vegetation forms the foundation of bioshields' protective function (Tanaka 2009). Coastal vegetation mediates energy dissipation through a combination of frictional resistance, turbulence generation, and flow redirection, leading to reductions in both kinetic and potential energy of incoming waves and currents (Denny 2021). These processes influence coastal morphodynamics, which is the spatial and temporal evolution of shoreline morphology under hydrodynamic forcing.

4.6.1 Mechanisms of Energy Transformation

As wave energy encounters vegetation, a portion of it is converted into turbulent kinetic energy (TKE) through wake formation behind stems and roots (Xiang et al. 2023). The turbulent eddies generated increase local mixing but significantly reduce the coherence of wave motion (Norris 2019). This transformation leads to an exponential decay in wave height and energy flux with increasing distance into the vegetation belt (Xiang et al. 2023). Laboratory flume experiments show that energy decay rates in vegetated zones are typically an order of magnitude higher than those in unvegetated areas (Shi et al. 2013; Yiping et al. 2015).

In mangrove forests, aerial root systems such as pneumatophores and prop roots act as vertical drag elements, disrupting flow continuity (Alongi 2005). This complex root morphology enhances form drag, producing additional dissipation even under submerged conditions. In contrast, salt marsh and seagrass ecosystems rely more on flexural motion of stems that oscillate with flow, absorbing and elastically redistributing energy (Mendez and Losada 2004; Denny 2021).

4.6.2 Morphodynamic Feedbacks

Energy dissipation within vegetated zones has long-term implications for shoreline evolution. Reduced wave energy promotes sediment deposition and limits erosional scour, fostering the formation of tidal flats, mudbanks, and beach ridges (Nardin and Fagherazzi 2018). These geomorphic features, in turn, enhance conditions for vegetation establishment, reinforcing a positive feedback loop between biological and physical processes (Fan 2012).

For example, along the northern Java coast, energy reduction by newly established mangrove stands has allowed suspended sediments to accumulate, creating mudflats that further stabilize the coast (van Wesenbeeck et al. 2015, 2025). Over decadal timescales, such bio-geomorphic feedbacks contribute to coastal resilience, enabling natural adaptation to moderate sea-level rise without external engineering.

4.6.3 Interaction with Extreme Events

During high-energy storm events, bioshields absorb and redistribute energy that would otherwise cause catastrophic erosion. While complete attenuation is impossible under extreme surge conditions, even a partial reduction in energy flux significantly decreases the force exerted on levees and seawalls. Observational data from Typhoon Haiyan (2013) in the Philippines indicate that villages fronted by mangrove forests suffered notably lower structural damage and erosion than those directly exposed to open coasts (Delfino et al. 2015). Hence, energy dissipation by bioshields serves not only as a day-to-day stabilizer but as an essential buffer against catastrophic energy pulses.

4.7 Role of Root Systems and Vegetation Structure

The morphological and anatomical traits of coastal vegetation directly determine their effectiveness in providing physical protection. Bioshield species exhibit diverse adaptations that allow them to anchor, filter, and reinforce sediment, as well as to withstand hydrodynamic stress.

4.7.1 Root System Diversity and Function

Mangrove species display a remarkable range of root architectures that serve specialized protective roles. Prop roots (e.g., *Rhizophora apiculata*) extend laterally and vertically, providing strong anchorage and forming complex networks that trap sediment. Pneumatophores (e.g., *Avicennia marina*) act as vertical projections enhancing soil aeration but also disrupt flow, creating micro-barriers that slow water movement. Buttress and stilt roots in species such as *Sonneratia alba* further stabilize substrates in highly dynamic tidal environments (Krauss et al. 2009).

In salt marshes, dense rhizome mats and fibrous roots of Spartina and Phragmites species bind sediment grains, substantially increasing shear strength. Experimental shear vane tests show that root-reinforced soils possess several times higher erosion thresholds than unvegetated muds (Wani and Mir 2024).

4.7.2 Aboveground Structure and Canopy Dynamics

Canopy structure influences both aerodynamic and hydrodynamic drag. The vertical density profile, which is the distribution of stems and leaves along plant height, controls the extent of momentum extraction from the flow (Yoshikai et al. 2022). Species with dense lower canopies (e.g., mangroves) generate greater drag at lower water depths, whereas tall trees in coastal forests intercept wind-driven stress (Kaimal and Finnigan 2020). Flexibility of stems also matters; flexible plants dissipate energy by bending rather than breaking, while rigid species transfer more force to the sediment surface.

4.7.3 Species Zonation and Functional Gradients

In natural bioshield systems, species zonation along tidal gradients enhances functional diversity. Seaward zones dominated by Avicennia or Sonneratia absorb wave energy; mid-zones with Rhizophora roots trap sediment; and landward zones with Bruguiera and Lumnitzera reinforce soils and prevent inland erosion (Ma et al. 2020; Irawan and Chikmawati 2021). This spatial functional layering optimizes protection across hydrodynamic conditions.

Vietnam's Can Gio Mangrove Biosphere Reserve exemplifies this multi-zonal structure, where varied root morphologies and canopy layers jointly stabilize the deltaic front (Veettil et al. 2019a; Le et al. 2020). Restoration strategies that replicate this zonation pattern demonstrate superior survival and protection outcomes compared to monoculture plantations (Nguyen et al. 2018).

4.8 Comparative Bioshield Effectiveness Across Ecosystem Types

Not all bioshield systems function identically. The type, structure, and spatial arrangement of coastal vegetation determine how effectively it mitigates hydrodynamic energy under various environmental conditions. Each ecosystem (mangrove forests, salt marshes, seagrass meadows, and coastal dune vegetation) plays a distinct yet complementary role in coastal defense. Their collective contributions depend on their morphology, sediment interaction, and capacity to adapt to site-specific climatic and geomorphological settings. Understanding these comparative dynamics is crucial for designing effective, nature-based coastal protection strategies that optimize both ecological and engineering outcomes.

4.8.1 Mangroves

Mangroves are widely recognized as the most effective natural bioshields against wave and storm surge energy, particularly in tropical and subtropical deltas (Veettil et al. 2021). Their aerial root systems (prop roots in Rhizophora, pneumatophores in Avicennia, and buttress roots in Sonneratia) create high drag forces that dissipate hydrodynamic energy. Empirical studies demonstrate that mangroves can reduce wave heights by more than 50% over a few hundred meters of forest width, depending on species composition, density, and tidal elevation (Best et al. 2022; Van Wesenbeeck et al. 2025). Their dense canopy also reduces wind energy at the surface, while the root–sediment interface traps fine particles, enhancing vertical accretion and shoreline stability. In regions such as the Mekong and Ayeyarwady deltas, mangroves have been shown to offset moderate sea-level rise through sediment trapping and organic matter accumulation (Seeger et al. 2023). However, their protective capacity diminishes along steep or eroding coasts, or where hydrodynamic stress exceeds species' tolerance thresholds (Besset et al. 2019; Cannon et al. 2020). Consequently, successful bioshield performance requires site-specific matching of mangrove species to local sediment, salinity, and hydrodynamic conditions (Juanico 2022).

4.8.2 Saltmarshes

Salt marshes are dominant along temperate and subtropical coastlines, where they provide robust stabilization of fine-grained sediments and attenuation of tidal currents (Rogers and Woodroffe 2015). Their root and rhizome systems, particularly those of *Spartina alterniflora* and *Phragmites australis*, enhance soil shear strength and prevent erosion (Feher and Hester 2018; Li et al. 2024). Vertical accretion is maintained through both sediment deposition and organic matter accumulation, allowing marshes to keep pace with moderate rates of sea-level rise. Salt marshes along the Yellow Sea coast (China) offer valuable protection against tidal flooding, functioning as buffer zones that absorb wave energy before it reaches coastal infrastructure (Yang et al. 2012). Their efficacy, however, depends strongly on landscape slope and hydrological regime; marshes on gentle gradients exhibit greater attenuation potential than those constrained by dikes or steep shorelines. Moreover, excessive inundation or nutrient loading can weaken vegetation structure, reducing long-term stability (Besset et al. 2019).

4.8.3 Seagrass Meadows

Seagrass meadows function primarily within subtidal environments, where they attenuate near-bed hydrodynamics rather than direct wave crests (Reidenbach and Thomas 2018). Their flexible blades slow current velocity and dampen wave orbital motion close to the seabed, while rhizome networks trap and bind sediments, promoting sediment stability and improving water clarity by reducing resuspension (Denny 2021). Although seagrasses such as *Thalassia hemprichii* and *Enhalus acoroides* have less capacity for surge attenuation compared to mangroves or salt marshes, they play a crucial complementary role. By stabilizing subtidal sediments, they enhance the effectiveness of adjacent intertidal vegetation and contribute to the overall resistance of the coastal profile (Potouroglou et al. 2017). The combined function of seagrass meadows with mangroves and coral reefs forms a multi-layered green belt that sustains biodiversity while protecting coastlines from chronic erosion (Guannel et al. 2016).

4.8.4 Coastal Dune Vegetation

Coastal dune vegetation forms the first terrestrial line of defense in sandy coastal systems. Species such as *Ipomoea pes-caprae*, *Spinifex littoreus*, and *Pandanus tectorius* play key geomorphic roles by trapping wind-blown sand and stabilizing dune ridges. These dunes act as natural levees, providing elevation above surge and wave impact zones. While dune vegetation offers limited direct hydrodynamic attenuation, its significance lies in maintaining coastal elevation and reducing overwash during storm events (Long et al. 2014). In many barrier island systems across Southeast Asia and the Pacific, dune vegetation has been integral in post-cyclone recovery and in maintaining the morphological resilience of beaches (Tomasicchio et al. 2022). The protective efficiency of dunes increases when they are vegetated continuously and connected with inland mangrove or forest systems (Feher and Hester 2018), ensuring continuity between aeolian and hydrodynamic protection processes.

4.8.5 Integrative Approaches

Modern coastal management increasingly emphasizes hybrid bioshield systems that integrate multiple vegetation types across tidal gradients. For example, Feher and Hester (2018) observed that soil shear strength was significantly higher in experimental sites with *Spartina alterniflora* and *Avicennia germinans* compared to unvegetated soils. In Bangladesh and Vietnam, restoration efforts combining mangrove–saltmarsh–dune mosaics have demonstrated synergistic benefits, with wave energy reductions exceeding 50% compared to single-species stands (Veettil et al. 2021; Gijón Mancheño et al. 2022). Such multi-zonal configurations replicate natural coastal transitions, from subtidal seagrasses to intertidal mangroves and supratidal dunes, maximizing multifunctionality. Beyond physical protection, they support fisheries, carbon sequestration, and biodiversity conservation, aligning ecological resilience with socio-economic benefits. Integrative bioshield strategies thus represent a key paradigm in climate adaptation and coastal planning, promoting sustainable coexistence between human development and natural coastal processes (Perricone et al. 2023).

4.9 Case Studies from Asia and Other Regions

Empirical case studies from across Asia and other coastal regions provide compelling evidence of the biophysical and socioeconomic efficacy of bioshields in mitigating erosion, dissipating wave energy, and enhancing coastal resilience (Veettil et al. 2021). While the structural and functional characteristics of bioshields vary from mangrove forests to dune vegetation and hybrid green-gray systems, their performance consistently demonstrates the power of ecosystem-based coastal protection (Huynh et al. 2024). The following regional case studies highlight successful implementations, the mechanisms driving their effectiveness, and lessons for scaling such nature-based solutions globally.

4.9.1 Vietnam: Mangroves and Mudflat Dynamics

Vietnam's Mekong Delta represents one of the most dynamic and vulnerable coastal systems in Southeast Asia, shaped by high sediment loads, tidal fluxes, and anthropogenic interventions (Tu et al. 2019; Veettil et al. 2019a, 2021). In this deltaic environment, bioshields composed of mangroves have become critical components of coastal defense strategies (Veettil et al. 2021; Thieu et al. 2023). Decades of deforestation for aquaculture and agriculture initially led to rapid shoreline retreat and land subsidence (Veettil et al. 2019a, b). However, restoration initiatives since the early 2000s have shown how mangrove rehabilitation can reverse erosion trends and enhance geomorphic stability (Tinh et al. 2022). However, it should be acknowledged that in the Mekong Delta, mangroves cannot fully perform a coastal protection function under conditions of diminished sediment supply (Besset et al. 2019).

Field observations in Soc Trang and Ca Mau provinces suggest that mangrove belts with widths of 100–200 m can attenuate 60–80% of incident wave energy during typical tidal conditions (Phan et al. 2015; Spalding and Leal 2021). This reduction in hydrodynamic energy decreases sediment resuspension and promotes deposition on the sheltered mudflats landward of the forest fringe (Gijsman et al. 2024). Monitoring data suggest sediment accumulation rates of approximately 5–10 mm yr^{-1}, contributing to the partial offsetting of local subsidence and the maintenance of relative surface elevation above tidal inundation thresholds (Thi Ha et al. 2018; Minderhoud et al. 2019).

Beyond the purely physical benefits, mangrove bioshields in the Mekong Delta provide ecological and livelihood co-benefits. Their root systems create critical habitat for juvenile fish and crustaceans, whereas the aboveground canopy limits inland tidal water penetration, thereby reducing salinity intrusion (Veettil et al. 2019a). Integrated restoration approaches combining sediment fencing and natural regeneration in the Red River Delta have enhanced mudflat accretion, thereby facilitating mangrove colonization of formerly degraded shorelines (Quang and Van An 2025). Collectively, these findings highlight the critical role of bioshields as living infrastructure capable of both adapting to and buffering against environmental change.

4.9.2 Bangladesh: Cyclone Mitigation and Socioeconomic Protection

Bangladesh, located in the world's largest deltaic system (Ganges–Brahmaputra-Meghna Delta), faces recurrent cyclones, storm surges, and tidal flooding. Following devastating events such as the 1970 Bhola cyclone and 1991 super cyclone, national initiatives prioritized mangrove reforestation along embankments and coastal polders. These large-scale interventions have since provided vital evidence for the disaster risk reduction potential of bioshields (World Bank 2012).

A comprehensive analysis of Cyclone Amphan (2020) demonstrated that villages shielded by mangrove belts experienced markedly lower damage compared to adjacent unprotected settlements, and protected areas recorded 50% fewer structural losses and significantly lower inundation depths (Dasgupta et al. 2019; Kumar et al. 2021). The wave-damping and drag forces produced by mangrove stems and root networks substantially reduced current velocity and surge height before waves reached inland villages (Dasgupta et al. 2019).

Economically, the benefits are equally striking. Akber et al. (2018) report that villages in mangrove-protected zones experienced losses of US $1,025 per household during Cyclone Sidr (2007) in protected villages, which was about half the loss compared to unprotected villages. Another study (Sarker et al. 2024) estimates the total storm-protection value of the Sundarbans at approximately US$543.3 million for Cyclone Sidr and further indicates that a 1 km-wide belt of intact mangrove forest can avert damages of about US$396 per household (2015 prices) during cyclones and storm surges. These figures represent the tangible monetary value of nature-based coastal protection, highlighting the cost-effectiveness of ecosystem restoration compared with traditional engineering structures. Moreover, mangrove belts also sustain local livelihoods, providing wood, honey, fish, and fuel, creating a multi-functional protective landscape that integrates ecology, economy, and human safety (Rahman et al. 2018).

These experiences position Bangladesh as a global leader in integrating ecosystem-based adaptation (EbA) into coastal management policy. The success of these bioshields illustrates that effective disaster mitigation need not rely solely on concrete infrastructure; rather, ecological restoration can form the backbone of resilient community protection systems.

4.9.3 India: Dune and Casuarina Forest Systems

India's extensive coastline faces diverse geomorphological challenges, from cyclone impacts in the Bay of Bengal to erosion and salinity intrusion along sandy coasts. In Tamil Nadu, one of the most compelling demonstrations of bioshield effectiveness occurred during the 2004 Indian Ocean tsunami. Coastal stretches with vegetated dunes and *Casuarina equisetifolia* plantations experienced significantly less inland inundation and structural damage compared to deforested zones (Danielsen et al. 2005).

Field assessments revealed that vegetated dune complexes reduced wind speed and wave run-up, dissipating hydrodynamic energy before waves reached residential areas (Nigam et al. 2024). The Casuarina trees acted as aerodynamic barriers, converting wave momentum into turbulence and friction, while dune vegetation trapped sand and prevented its remobilization during the tsunami drawdown. The cumulative effect is a substantial reduction in momentum flux, which is the energy transmitted by waves inland (Mascarenhas and Jayakumar 2008).

Following this tsunami event, India has expanded bioshield plantations across Tamil Nadu, Andhra Pradesh, and Odisha states (Roy 2025). In addition to protecting lives and infrastructure, these systems stabilize sandy shorelines, mitigate salt spray, and support biodiversity. The integration of native dune grasses and Casuarina windbreaks represents an effective hybrid bioshield, balancing ecological integrity with practical protective functions (Mascarenhas and Jayakumar 2008).

4.9.4 Indonesia and the Philippines: Hybrid Protection Systems

The rapidly eroding coasts of Indonesia and the Philippines have become testbeds for hybrid coastal protection systems that combine natural and engineered components (Solihuddin et al. 2021; Juanico 2022). In Demak District, Java, severe subsidence and erosion, driven by aquaculture expansion and groundwater extraction, had destroyed extensive mangrove areas (Triyanti et al. 2017). To restore coastal stability, engineers and ecologists implemented a system of bamboo wave barriers (permeable structures) placed parallel to the shoreline (Armono et al. 2021). These barriers reduce wave energy while allowing suspended sediment to pass through and deposit behind them.

Once sediment accumulation reached suitable levels, natural mangrove colonization followed, forming a self-sustaining vegetated belt. This approach, often referred to as building with nature, accelerated tidal flat formation and reduced wave energy reaching the hinterland by up to 70% (Widyaningtias et al. 2022). Similar hybrid systems have been adopted in the Visayas region of the Philippines, where mangroves combined with brushwood fences and community-managed buffer zones have produced rapid accretion and shoreline stabilization (Gevaña et al. 2022).

These case studies demonstrate that bioshields can be successfully integrated with engineered elements to address severe sediment deficits. The key innovation lies in using nature's processes, such as sediment trapping, root stabilization, and vegetation regeneration, to achieve engineering goals sustainably. Beyond the biophysical results, these projects have also fostered community participation and environmental stewardship, ensuring the long-term maintenance of the bioshield systems.

4.9.5 Global Comparisons

While Asia provides the most extensive real-world testing of bioshields, parallel evidence from other continents reinforces the universal biophysical principles underlying them. In Europe, comprehensive research on salt marshes in the Netherlands and the United Kingdom has demonstrated that vegetated foreshores can reduce nearshore wave energy by up to 50%, enabling smaller and more cost-effective dike reinforcements (Vuik et al. 2016, 2019). By dissipating energy before it reaches hard infrastructure, these nature-based defenses enhance the overall resilience of flood protection systems (Temmerman et al. 2013).

In the Caribbean, mangrove forests provide dual ecosystem services, including shielding coastal communities from hurricanes while protecting adjacent coral reefs from sedimentation and nutrient overloading (Miranda et al. 2021). A coastal flooding study in Cuba (Ciego de Ávila Province) indicates that areas protected by mangroves had smaller inland penetration of flood waters compared to other areas, suggesting a protective effect (Matos-Pupo et al. 2023). Similarly, in Florida and Belize, mangroves have been estimated to reduce storm surge heights by up to 50 cm per kilometer of forest width, illustrating their critical buffering role (Zhang et al. 2012).

Across the Pacific Islands, traditional coastal management practices often incorporate bioshields of *Pandanus, Scaevola*, and *Tournefortia* species to stabilize dunes and protect settlements from typhoons (Thaman 1992). These systems demonstrate that the underlying mechanisms, such as wave energy dissipation, sediment retention, and root reinforcement, are functionally consistent across ecosystems, regardless of regional climate or geomorphology.

4.10 Conclusions

Bioshields represent one of the most effective and ecologically sustainable mechanisms for coastal protection. Their efficiency derives from a combination of biophysical processes, including energy dissipation, sediment trapping, root reinforcement, and morphodynamic feedback, which collectively enhance shoreline stability and resilience. The interplay between vegetation structure and hydrodynamic forces underpins their capacity to attenuate waves, mitigate storm surges, and reduce erosion.

Empirical evidence from across Asia, particularly Vietnam, Bangladesh, India, Indonesia, and the Philippines, demonstrates that well-designed bioshield systems can reduce wave energy by 50–90%, facilitate accretion rates exceeding local sea-level rise, and prevent substantial economic losses during extreme weather events. These natural systems outperform or complement engineered coastal defense structures by offering adaptive, self-repairing, and multifunctional protection.

However, the protective potential of bioshields depends on appropriate species selection, spatial configuration, and ecological maintenance. Monocultures, inadequate buffer width, or degradation can compromise the performance of bioshields. Integrating bioshields into broader ecosystem-based coastal management frameworks, supported by scientific monitoring and policy coordination, ensures long-term sustainability.

Ultimately, the biophysical mechanisms outlined in this chapter form the scientific foundation for bioshield design and implementation, connecting ecological function to tangible protective outcomes. As global climate change intensifies, the restoration and conservation of coastal vegetation stand as a scientific imperative as well as a pragmatic strategy for safeguarding vulnerable coastlines and communities worldwide.

References

Akber MA, Patwary MM, Islam MA, Rahman MR (2018) Storm protection service of the Sundarbans mangrove forest, Bangladesh. Nat Hazards 94(1):405–418. https://doi.org/10.1007/s11069-018-3395-8

Alongi DM (2005) Mangrove–microbe–soil relations. In: Interactions between macro-and microorganisms in marine sediments, vol 60, pp 85–103. https://doi.org/10.1002/gj.3808

Alongi DM (2025) Global Meta-analysis of mangrove primary production: Implications for carbon cycling in mangrove and other coastal ecosystems. Forests 16(5):747. https://doi.org/10.3390/f16050747

Anderson M, Smith J (2013) Wave attenuation by flexible, idealized salt marsh vegetation. Coast Eng 83:82–92. https://doi.org/10.1016/j.coastaleng.2013.10.004

Armono HD, Bromo BH, Sujantoko S (2021) Numerical study of bamboo breakwater for wave reduction. Fluids 7(1):14. https://doi.org/10.3390/fluids7010014

Athikalam PT, Vaideeswaran AK (2022) Vegetation bioshield for coastal protection in South Asia: status and way forward. J Coast Conserv 26(1):3. https://doi.org/10.1007/s11852-022-00850-x

Aziz A, Paul AR (2015) Bangladesh Sundarbans: present status of the environment and biota. Diversity 7(3):242–269. https://doi.org/10.3390/d7030242

Barbier EB (2016) The protective service of mangrove ecosystems: a review of valuation methods. Mar Pollut Bull 109(2):676–681. https://doi.org/10.1016/j.marpolbul.2016.01.033

Besset M, Gratiot N, Anthony EJ, Bouchette F, Goichot M, Marchesiello P (2019) Mangroves and shoreline erosion in the Mekong River delta, Viet Nam. Estuar Coast Shelf Sci 226:106263. https://doi.org/10.1016/j.ecss.2019.106263

Best ÜSN, van der Wegen M, Dijkstra J, Reyns J, van Prooijen BC, Roelvink D (2022) Wave attenuation potential, sediment properties and mangrove growth dynamics data over Guyana's intertidal mudflats: assessing the potential of mangrove restoration works. Earth Syst Sci Data 14:2445–2462. https://doi.org/10.5194/essd-14-2445-2022

Cahoon DR, McKee KL, Morris JT (2021) How plants influence resilience of salt marsh and mangrove wetlands to sea-level rise. Estuar Coasts 44(4):883–898. https://doi.org/10.1007/s12237-020-00834-w

Cannon D, Kibler K, Donnelly M, McClenachan G, Walters L, Roddenberry A, Phagan J (2020) Hydrodynamic habitat thresholds for mangrove vegetation on the shorelines of a microtidal estuarine lagoon. Ecol Eng 158:106070. https://doi.org/10.1016/j.ecoleng.2020.106070

Chen Q, Li Y, Kelly DM, Zhang K, Zachry B, Rhome J (2021) Improved modeling of the role of mangroves in storm surge attenuation. Estuar Coast Shelf Sci 260:107515. https://doi.org/10.1016/j.ecss.2021.107515

Chowdhury P, Lakku NKG, Lincoln S, Seelam JK, Behera MR (2023) Climate change and coastal morphodynamics: Interactions on regional scales. Sci Total Environ 899:166432. https://doi.org/10.1016/j.scitotenv.2023.166432

Chu L, Yuan S, Chen D, Kang Y, Shaghaleh H, El-Tayeb MA, Hamoud YA et al (2024) Influences of vegetation rehabilitation on soil infiltrability and root morphological characteristics in coastal saline soil. Land 13(6):849. https://doi.org/10.3390/land13060849

Danielsen F, Sørensen MK, Olwig MF, Selvam V, Parish F, Burgess ND, Hiraishi T, Karunagaran VM, Rasmussen MS, Hansen LB, Quarto A, Suryadiputra N (2005) The Asian Tsunami: a protective role for coastal vegetation. Science 310(5748):643–643

Das S, Vincent JR (2009) Mangroves protected villages and reduced death toll during Indian super cyclone. Proc Natl Acad Sci 106(18):7357–7360. https://doi.org/10.1073/pnas.0810440106

Dasgupta S, Islam MS, Huq M, Huque Khan Z, Hasib MR (2019) Quantifying the protective capacity of mangroves from storm surges in coastal Bangladesh. PLoS ONE 14(3):e0214079. https://doi.org/10.1371/journal.pone.0214079

Delfino RJ, Carlos C, David L, Lasco R, Juanico DE (2015) Perceptions of Typhoon Haiyan-affected communities about the resilience and storm protection function of mangrove ecosystems in Leyte and Eastern Samar, Philippines. Clim Disaster Dev J 1(1):15–24. https://doi.org/10.18783/cddj.v001.i01.a03

Denny M (2021) Wave-energy dissipation: seaweeds and marine plants are ecosystem engineers. Fluids 6(4):151. https://doi.org/10.3390/fluids6040151

Duc TN, Trong NM, Viet DL, Danh KN, Dang TN, Tai TN, Anh TN et al (2025) Assessing the resilience of the coastal ecosystem under the impact of climate change: research area from Quang Ninh to Hai Phong. In: Green transformation in the context of global change. CRC Press, pp 219–228. https://doi.org/10.1201/9781003602941-23

Fan D (2012) Open-coast tidal flats. In: Principles of tidal sedimentology. Springer, Dordrecht, pp 187–229. https://doi.org/10.1007/978-94-007-0123-6_9

Feagin RA, Figlus J, Zinnert JC, Sigren J, Martínez ML, Silva R, Smith WK, Cox D, Young DR, Carter G (2015) Going with the flow or against the grain? The promise of vegetation for protecting beaches, dunes, and barrier islands from erosion. Front Ecol Environ 13(4):203–210. https://doi.org/10.1890/140218

Feher LC, Hester MW (2018) The interactive effects of created salt marsh substrate type, hydrology, and nutrient regime on *Spartina alterniflora* and *Avicennia germinans* productivity and soil development. Wetl Ecol Manag 26(4):715–728. https://doi.org/10.1007/s11273-018-9603-1

Ferreira CS, Kašanin-Grubin M, Solomun MK, Sushkova S, Minkina T, Zhao W, Kalantari Z (2023) Wetlands as nature-based solutions for water management in different environments. Curr Opin Environ Sci Health 33:100476. https://doi.org/10.1016/j.coesh.2023.100476

Gevaña DT, Villanueva CMM, Garcia JE, Camacho LD (2022) Mangroves sustaining biodiversity, local livelihoods, blue carbon, and local resilience in Verde Island Passage in Luzon, Philippines. In: Mangroves: Biodiversity, livelihoods and conservation. Springer Nature Singapore, Singapore, pp 447–461. https://doi.org/10.1016/B978-0-12-810473-6.00014-5

Gijón Mancheño A, Vuik V, van Wesenbeeck BK, Jonkman SN (Bas), Moll R, Kazi S, Urrutia Duarte IM, Van Ledden M (2022) Wave reduction by mangroves during cyclones in Bangladesh: implementing nature-based solutions for coastal resilience. Policy Research Working Paper No. 10240. World Bank

Gijsman R, Horstman EM, Swales A, MacDonald IT, Bouma TJ, van der Wal D, Wijnberg KM (2024) Mangrove forest drag and bed stabilisation effects on intertidal flat morphology. Earth Surf Proc Land 49(3):1117–1134. https://doi.org/10.1002/esp.5758

Ginting YRS (2018) Mangrove distribution, sedimentation, and soil carbon accumulation in North Sumatra, Indonesia. PhD thesis, Bogor Agricultural University (IPB), Indonesia

Gracia A, Rangel-Buitrago N, Oakley JA, Williams A (2018) Use of ecosystems in coastal erosion management. Ocean Coast Manag 156:277–289. https://doi.org/10.1016/j.ocecoaman.2017.07.009

Guannel G, Arkema K, Ruggiero P, Verutes G (2016) The power of three: coral reefs, seagrasses and mangroves protect coastal regions and increase their resilience. PLoS ONE 11(7):e0158094. https://doi.org/10.1371/journal.pone.0158094

He F, Chen J, Jiang C (2019) Surface wave attenuation by vegetation with the stem, root and canopy. Coast Eng 152:103509. https://doi.org/10.1016/j.coastaleng.2019.103509

Hoque MM, Abu Hena MK, Ahmed OH, Idris MH, Hoque ATMR, Billah MM (2015) Can mangroves help combat sea level rise through sediment accretion and accumulation? Malays J Sci 34:78–86

Hu K, Chen Q, Wang H (2014) A numerical study of vegetation impact on reducing storm surge by wetlands in a semi-enclosed estuary. Coast Eng 95:66–76. https://doi.org/10.1016/j.coastaleng.2014.09.008

Huynh LT, Su J, Wang Q, Stringer LC, Switzer AD, Gasparatos A (2024) Meta-analysis indicates better climate adaptation and mitigation performance of hybrid engineering-natural coastal defence measures. Nat Commun 15(1):2870. https://doi.org/10.1038/s41467-024-46970-w

Irawan A, Chikmawati T (2021) Diversity and zonation of mangrove flora in Belitung Island, Indonesia. Biodivers J Biol Divers 22(5). https://doi.org/10.13057/biodiv/d220563

James RK, Lynch A, Herman PM, Van Katwijk MM, Van Tussenbroek BI, Dijkstra HA, Bouma TJ et al (2021) Tropical biogeomorphic seagrass landscapes for coastal protection: persistence and wave attenuation during major storms events. Ecosystems 24(2):301–318. https://doi.org/10.1007/s10021-020-00519-2

Juanico DEO (2022) Does mangrove restoration imply coastal protection? A prospective simulation study. Front Mar Sci 9:968420. https://doi.org/10.3389/fmars.2022.968420

Kaimal JC, Finnigan JJ (2020) Flow over plant canopies. In: Atmospheric boundary layer flows: their structure and measurement. Oxford Academic, pp 66–108. https://doi.org/10.1093/oso/9780195062397.003.0006

Kobayashi N, Raichle AW, Asano T (1993) Wave attenuation by vegetation. J Waterw Port Coast Ocean Eng 119:30–48. https://doi.org/10.1061/(ASCE)0733-950X(1993)119:1(30)

Krauss KW, Doyle TJ, Doyle TW, Swarzenski CM, From AS, Day RH, Conner WH (2009) Water-level observations in mangrove swamps during two hurricanes in Florida. Wetlands 29(1):142–149. https://doi.org/10.1672/07-232.1

Kumar S, Lal P, Kumar A (2021) Influence of super cyclone "Amphan" in the Indian subcontinent amid COVID-19 pandemic. Remote Sens Earth Syst Sci 4:96–103. https://doi.org/10.1007/s41976-021-00048-z

Le HT, Tran TV, Gyeltshen S, Nguyen CPT, Tran DX, Luu TH, Duong MB (2020) Characterizing spatiotemporal patterns of mangrove forests in can gio biosphere reserve using Sentinel-2 imagery. Appl Sci 10(12):4058. https://doi.org/10.3390/app10124058

Lee JT, Yen LZ, Chu MY, Lin YS, Chang CC, Lin RS, Lee MJ et al (2020) Growth characteristics and anti-wind erosion ability of three tropical foredune pioneer species for sand dune stabilization. Sustainability 12(8):3353. https://doi.org/10.3390/su12083353

Li P, Xiao X, Wu L, Li X, Zhang H, Zhou J (2022) Study on the shear strength of root-soil composite and root reinforcement mechanism. Forests 13(6):898. https://doi.org/10.3390/f13060898

Li X, Jin K, Qin P, Liu C, Zhu X, Zhang Y, Zong Q (2024) Enhancement effect of Phragmites australis roots on soil shear strength in the Yellow River Delta. Sustainability 16(23):10657. https://doi.org/10.3390/su162310657

Long J, de Bakker AT, Plant NG (2014) Scaling coastal dune elevation changes across storm-impact regimes. Geophys Res Lett 41:5390–5396. https://doi.org/10.1002/2014GL059616

Ma W, Wang W, Tang C, Chen G, Wang M (2020) Zonation of mangrove flora and fauna in a subtropical estuarine wetland based on surface elevation. Ecol Evol 10(14):7404–7418. https://doi.org/10.1002/ece3.6467

Mascarenhas A, Jayakumar S (2008) An environmental perspective of the post-tsunami scenario along the coast of Tamil Nadu, India: role of sand dunes and forests. J Environ Manag 89(1):24–34. https://doi.org/10.1016/j.jenvman.2007.01.053

Matos-Pupo F, Peros MC, González-De Zayas R, Valero-Jorge A, Pérez-López OE, Álvarez-Taboada F, Sorí R (2023) Coastal flooding associated with Hurricane Irma in Central Cuba (Ciego de Ávila Province). Atmosphere 14(9):1445. https://doi.org/10.3390/atmos14091445

Mazda Y, Magi M, Ikeda Y, Kurokawa T, Asano T (2006) Wave reduction in a mangrove forest dominated by Sonneratia sp. Wetl Ecol Manag 14(4):365–378. https://doi.org/10.1007/s11273-005-5388-0

Mendez FJ, Losada IJ (2004) An empirical model to estimate the propagation of random breaking and nonbreaking waves over vegetation fields. Coast Eng 51(2):103–118. https://doi.org/10.1016/j.coastaleng.2003.11.003

Minderhoud PSJ, Coumou L, Erkens G, Middelkoop H, Stouthamer E (2019) Mekong delta much lower than previously assumed in sea-level rise impact assessments. Nat Commun 10(1):3847. https://doi.org/10.1038/s41467-019-11602-1

Miranda JJ, Butron L, Pantoja C, Gunasekera R (2021) Mangroves as coastal protection for local economic activities from hurricanes in the Caribbean. Policy Research Working Paper Series 9863, The World Bank

Montgomery JM, Bryan KR, Mullarney JC, Horstman EM (2019) Attenuation of storm surges by coastal mangroves. Geophys Res Lett 46(5):2680–2689. https://doi.org/10.1029/2018GL081636

Nardin W, Fagherazzi S (2018) The role of waves, shelf slope, and sediment characteristics on the development of erosional chenier plains. Geophys Res Lett 45(16):8435–8444. https://doi.org/10.1029/2018GL078694

Neary VS, Constantinescu SG, Bennett SJ, Diplas P (2012) Effects of vegetation on turbulence, sediment transport, and stream morphology. J Hydraul Eng 138(9):765–776. https://doi.org/10.1061/(ASCE)HY.1943-7900.0000168

Nepf HM (2012) Hydrodynamics of vegetated channels. J Hydraul Res 50(3):262–279. https://doi.org/10.1080/00221686.2012.696559

Neumeier U, Ciavola P (2004) Flow resistance and associated sedimentary processes in a *Spartina maritima* salt-marsh. J Coast Res 20(2):435–447

Nguyen LV, To TT, Trinh HX, Hoang TTL (2018) The analysis of mangrove forest changes period of 20 years in Can Gio Biosphere Reserve, Viet Nam using remote sensing and GIS technology. SF J Environ Earth Sci 1:1–7. https://doi.org/10.25316/IR-14917

Nigam R, Luis AJ, Gagnon AS, Vaz E, Damasio B, Kotha M (2024) Assessing coastal vulnerability at the village level using a robust framework, the example of Canacona in South Goa, India. Iscience 27(4). https://doi.org/10.1016/j.isci.2024.109129

Norris B (2019) Small-scale turbulence and its influence on forest-scale morphodynamics within a coastal mangrove forest. PhD thesis, The University of Waikato, New Zealand

Perricone V, Mutalipassi M, Mele A, Buono M, Vicinanza D, Contestabile P (2023) Nature-based and bioinspired solutions for coastal protection: an overview among key ecosystems and a promising pathway for new functional and sustainable designs. ICES J Mar Sci 80(5):1218–1239. https://doi.org/10.1093/icesjms/fsad080

Pham HT, Bui TTT (2020) A study on the effects of sea level rise on mangrove ecosystem in Giao Thuy district, Nam Dinh province, Vietnam. J Sci Technol Eng 62(3):90–96. https://doi.org/10.31276/VJSTE.62(3).90-96

Phan LK, van Thiel de Vries JS, Stive MJ (2015) Coastal mangrove squeeze in the Mekong Delta. J Coast Res 31(2):233–243. https://doi.org/10.2112/JCOASTRES-D-14-00049.1

Potouroglou M, Bull JC, Krauss KW, Kennedy HA, Fusi M, Daffonchio D, Mangora MM, Githaiga MN, Diele K, Huxham M (2017) Measuring the role of seagrasses in regulating sediment surface elevation. Sci Rep 7(1):1–11. https://doi.org/10.1038/s41598-017-12354-y

Prasetya G (2007) Protection from coastal erosion. Thematic paper: the role of coastal forests and trees in protecting against coastal erosion. FAO-UN—Food and Agriculture Organization of the United Nations

Quang NH, Van An N (2025) Four decades of mangrove shoreline changes driven by human activities in the Red River Delta. Estuar Coast Shelf Sci 314:109145. https://doi.org/10.1016/j.ecss.2025.109145

Rahman MM, Jiang Y, Irvine K (2018) Assessing wetland services for improved development decision-making: a case study of mangroves in coastal Bangladesh. Wetl Ecol Manag 26:563–580. https://doi.org/10.1007/s11273-018-9592-0

Reidenbach MA, Thomas EL (2018) Influence of the Seagrass, Zostera marina, on wave attenuation and bed shear stress within a shallow coastal bay. Front Mar Sci 5:380088. https://doi.org/10.3389/fmars.2018.00397

Rogers K, Woodroffe CD (2015) Tidal flats and salt marshes. In: Coastal environments and global change, pp 227–250. https://doi.org/10.1002/9781119117261.ch10

Roy SD (2025) Coastal ecosystem of India: recent development and perspective planning. In: Marine and coastal resources of India: selected case studies. Springer Nature Switzerland, Cham, pp 393–405. https://doi.org/10.1007/978-3-031-81276-7_22

Sarker AHMR, Nobi MN, Roskaft E, Chivers DJ, Suza M (2024) Value of the storm-protection function of Sundarban mangroves in Bangladesh. J Sustain Dev Can Center Sci Educ 3(3):128–128

Schoonees T, Gijón Mancheño A, Scheres B, Bouma TJ, Silva RCCDS, Schlurmann T, Schüttrumpf H (2019) Hard structures for coastal protection, towards greener designs. Estuar Coasts 42(7):1709–1729. https://doi.org/10.1007/s12237-019-00551-z

Seeger K, Minderhoud PS, Peffeköver A, Vogel A, Brückner H, Kraas F, Brill D et al (2023) Assessing land elevation in the Ayeyarwady Delta (Myanmar) and its relevance for studying sea level rise and delta flooding. Hydrol Earth Syst Sci 27(11):2257–2281. https://doi.org/10.5194/hess-27-2257-2023

Sheng YP, Lapetina A, Ma G (2012) The reduction of storm surge by vegetation canopies: three-dimensional simulations. Geophys Res Lett 39(20). https://doi.org/10.1029/2012GL053577

Shi JZ, Li YH, Hughes JM, Zhao M (2013) Hydrological characteristics of vegetated river flows: a laboratory flume study. Hydrol Sci J 58(5):1047–1058. https://doi.org/10.1080/02626667.2013.797580

Solihuddin T, Husrin S, Salim HL, Kepel TL, Mustikasari E, Heriati A, Berliana B et al (2021) Coastal erosion on the north coast of Java: adaptation strategies and coastal management. IOP Conf Ser Earth Environ Sci 777:012035. https://doi.org/10.1088/1755-1315/777/1/012035

Spalding MD, Leal M (2021) The state of the world's mangroves 2021. Technical report, Global Mangrove Alliance, p 79

Stachew E, Houette T, Gruber P (2021) Root systems research for bioinspired resilient design: a concept framework for foundation and coastal engineering. Front Robot AI 8:548444. https://doi.org/10.3389/frobt.2021.548444

Swales A, MacDonald IT, Green MO (2004) Influence of wave and sediment dynamics on cordgrass (*Spartina anglica*) growth and sediment accumulation on an exposed intertidal flat. Estuaries 27(2):225–243. https://doi.org/10.1007/BF02803380

Tanaka N (2009) Vegetation bioshields for tsunami mitigation: review of effectiveness, limitations, construction, and sustainable management. Landscape Ecol Eng 5(1):71–79. https://doi.org/10.1007/s11355-008-0058-z

Temmerman S, Horstman EM, Krauss KW, Mullarney JC, Pelckmans I, Schoutens K (2023) Marshes and mangroves as nature-based coastal storm buffers. Ann Rev Mar Sci 15(1):95–118. https://doi.org/10.1146/annurev-marine-040422-092951

Temmerman S, Meire P, Bouma TJ, Herman PM, Ysebaert T, De Vriend HJ (2013) Ecosystem-based coastal defence in the face of global change. Nature 504(7478):79–83. https://doi.org/10.1038/nature12859

Thaman RR (1992) Vegetation of Nauru and the Gilbert Islands: case studies of poverty, degradation, disturbance, and displacement. Pac Sci 46:128–158

Thi Ha D, Ouillon S, Van Vinh G (2018) Water and suspended sediment budgets in the lower Mekong from high-frequency measurements (2009–2016). Water 10(7):846. https://doi.org/10.3390/w10070846

Thieu TQ, Phan LK, Truong SH (2023) Erosion of mangrove-mud coasts, resilience of coastal protection, and outlook on nature-based solutions in the Vietnamese Mekong Delta. In: The Mekong River Basin, pp 479–525. https://doi.org/10.1016/B978-0-323-90814-6.00015-2

Tinh PH, MacKenzie RA, Hung TD, Vinh TV, Ha HT, Lam MH, Huyen BT et al (2022) Mangrove restoration in Vietnamese Mekong Delta during 2015–2020: achievements and challenges. Front Mar Sci 9:1043943. https://doi.org/10.3389/fmars.2022.1043943

Tomasicchio GR, Frega F, Leone E, Francone A, Pantusa D, Barbaro G, Foti G (2022) Beach-dune system morphodynamics. J Mar Sci Eng 10(5):627. https://doi.org/10.3390/jmse10050627

Triyanti A, Walz Y, Marfai MA, Renaud F, Djalante R (2017) Ecosystem-based disaster risk reduction in Indonesia: unfolding challenges and opportunities. In: Disaster risk reduction in Indonesia: progress, challenges, and issues, pp 445–467. https://doi.org/10.1007/978-3-319-54466-3_18

Tu LX, Thanh VQ, Reyns J, Van SP, Anh DT, Dang TD, Roelvink D (2019) Sediment transport and morphodynamical modeling on the estuaries and coastal zone of the Vietnamese Mekong Delta. Cont Shelf Res 186:64–76. https://doi.org/10.1016/j.csr.2019.07.015

van Bijsterveldt CE, Debrot AO, Bouma TJ, Maulana MB, Pribadi R, Schop J, van Wesenbeeck BK et al (2022) To plant or not to plant: when can planting facilitate mangrove restoration? Front Environ Sci 9:690011. https://doi.org/10.3389/fenvs.2021.690011

Van Wesenbeeck B, Balke T, Van Eijk P, Tonneijck F, Siry H, Rudianto M, Winterwerp J (2015) Aquaculture induced erosion of tropical coastlines throws coastal communities back into poverty. Ocean Coast Manag 116:466–469. https://doi.org/10.1016/j.ocecoaman.2015.09.004

Van Wesenbeeck BK, Van Zelst VT, Antolinez JA, De Boer WP (2025) Quantifying uncertainty in wave attenuation by mangroves to inform coastal green belt policies. Commun Earth Environ 6(1):1–8. https://doi.org/10.1038/s43247-025-02178-4

Van Zelst VT, Dijkstra JT, Van Wesenbeeck BK, Eilander D, Morris EP, Winsemius HC, Ward PJ, De Vries MB (2021) Cutting the costs of coastal protection by integrating vegetation in flood defences. Nat Commun 12(1):1–11. https://doi.org/10.1038/s41467-021-26887-4

Veettil BK, Quang NX, Thu Trang NT (2019a) Changes in mangrove vegetation, aquaculture and paddy cultivation in the Mekong Delta: a study from Ben Tre Province, southern Vietnam. Estuar Coast Shelf Sci 226:106273. https://doi.org/10.1016/j.ecss.2019.106273

Veettil BK, Ward RD, Quang NX, Trang NTT, Giang TH (2019b) Mangroves of Vietnam: historical development, current state of research and future threats. Estuar Coast Shelf Sci 218:212–236. https://doi.org/10.1016/j.ecss.2018.12.021

Veettil BK, Costi J, Marques WC, Tran X, Quang NX, Van DD, Hoai PN (2020) Coastal environmental changes in Southeast Asia: a study from Quang Nam Province, Central Vietnam. Reg Stud Mar Sci 39:101420. https://doi.org/10.1016/j.rsma.2020.101420

Veettil BK, Ward RD, Dung NTK, Van DD, Quang NX, Hoai PN, Hoang N (2021) The use of bioshields for coastal protection in Vietnam: current status and potential. Reg Stud Mar Sci 47:101945. https://doi.org/10.1016/j.rsma.2021.101945

Vuik V, Borsje BW, Willemsen PW, Jonkman SN (2019) Salt marshes for flood risk reduction: quantifying long-term effectiveness and life-cycle costs. Ocean Coast Manag 171:96–110. https://doi.org/10.1016/j.ocecoaman.2019.01.010

Vuik V, Jonkman SN, Borsje BW, Suzuki T (2016) Nature-based flood protection: the efficiency of vegetated foreshores for reducing wave loads on coastal dikes. Coast Eng 116:42–56. https://doi.org/10.1016/j.coastaleng.2016.06.001

Wani KS, Mir BA (2024) Experimental study on root-reinforced soils (R2S) and soil root (SR) interaction in nature. J Chin Inst Eng 47(6):655–673. https://doi.org/10.1080/02533839.2024.2368469

Widdows J, Pope ND, Brinsley MD, Asmus H, Asmus RM (2008) Effects of seagrass beds (*Zostera noltii* and Z. marina) on near-bed hydrodynamics and sediment resuspension. Mar Ecol Prog Ser 358:125–136. https://doi.org/10.3354/meps07338X

Widyaningtias ISB, Adityawan MB, Nugroho J, Kuntoro AA (2022) Implementation of building with nature (BwN) as adaptive concept to prevent coastal erosion in Demak Regency, Central Java, Indonesia. Int J Adv Sci Eng Inf Technol 12:2023–2029

World Bank (2012) Project information document (PID)—coastal afforestation & embankment protection, Bangladesh. World Bank, Washington DC

Xiang K, Gao W, Yang Z (2023) Estimation of the turbulent kinetic energy in vegetated channels from the perspective of vortex dynamics. Phys Fluids 35(9). https://doi.org/10.1063/5.0160481

Yang SL, Shi BW, Bouma TJ, Ysebaert T, Luo XX (2012) Wave attenuation at a salt marsh margin: a case study of an exposed coast on the Yangtze Estuary, Estuar. Coasts 35:169–182. https://doi.org/10.1007/s12237-011-9424-4

Yiping L, Anim DO, Wang Y, Tang C, Du W, Lixiao N, Chen L et al (2015) Laboratory simulations of wave attenuation by an emergent vegetation of artificial Phragmites australis: an experimental study of an open-channel wave flume. J Environ Eng Landsc Manag 23(4):251–266. https://doi.org/10.3846/16486897.2015.1055750

Yoshikai M, Nakamura T, Bautista DM, Herrera EC, Baloloy A, Suwa R, Basina R, Primavera-Tirol YH, Blanco AC, Nadaoka K (2022) Field measurement and prediction of drag in a planted Rhizophora mangrove forest. J Geophys Res Oceans 127(11):e2021JC018320. https://doi.org/10.1029/2021JC018320

Zhang K, Liu H, Li Y, Xu H, Shen J, Rhome J, Smith TJ (2012) The role of mangroves in attenuating storm surges. Estuar Coast Shelf Sci 102–103:11–23. https://doi.org/10.1016/j.ecss.2012.02.021

Chapter 5
Climate Change and the Resilience of Bioshields

Abstract Climate change poses unprecedented challenges to coastal ecosystems and the effectiveness of bioshields as nature-based coastal protection strategies. This chapter examines the multifaceted impacts of climate-induced stressors—rising sea levels, increased storm frequency and intensity, ocean warming, and salinity intrusion—on the structure, function, and resilience of bioshields such as mangroves, saltmarshes, and coastal dunes. It explores ecological and geomorphological feedbacks that determine the adaptive capacity of these systems to withstand and recover from disturbances. Particular attention is given to mechanisms of ecological resilience, including species diversity, genetic variability, recruitment dynamics, and landscape connectivity, which collectively underpin the persistence of bioshields under changing climatic conditions. The chapter synthesizes empirical findings and modeling approaches that evaluate bioshield responses to projected climate scenarios across tropical and subtropical regions. It highlights both vulnerabilities and adaptive potentials, emphasizing the role of climate-smart management practices—such as assisted regeneration, sediment nourishment, and landward migration corridors—in sustaining bioshield functionality. Furthermore, the discussion links resilience theory with practical coastal planning, offering insights into integrating bioshields into national adaptation frameworks and disaster risk reduction policies. By bridging ecological resilience and climate adaptation science, this chapter underscores that maintaining the health and adaptability of bioshields is central to safeguarding coastal communities and ecosystems in an era of accelerating climate change.

Keywords Adaptive management · Climate change adaptation · Coastal vulnerability · Ecosystem resilience · Nature-based solutions · Sea-level rise

5.1 Introduction

Coastal ecosystems are at the forefront of the climate crisis, acting as both buffers against and victims of climate-induced hazards (Veettil et al. 2020). Rising sea levels, intensifying storms, saltwater intrusion, and increasing coastal erosion threaten

B. Kozhikkodan Veettil, *The Potential of Bioshields for Coastal Protection*, Advances in Natural and Technological Hazards Research 55, https://doi.org/10.1007/978-3-032-22888-8_5

the stability of shorelines and the communities that depend on them (Burkett and Davidson 2012). Such climate-induced hazards are highly visible in Southeast Asia (Dong et al. 2024). Against this backdrop, bioshields, which comprise vegetative structures such as mangroves, coastal forests, dune vegetation, seagrass beds, and salt marshes, have emerged as vital nature-based solutions (NbS) that integrate ecological restoration with climate adaptation (Veettil et al. 2019a, 2021). Beyond serving as physical barriers, coastal bioshields provide multiple ecosystem services such as carbon sequestration, shoreline stabilization, nutrient cycling, biodiversity conservation, and livelihood enhancement, aligning with global policy frameworks including the Paris Agreement, the Sendai Framework for Disaster Risk Reduction, and the UN Sustainable Development Goals (SDGs 13, 14, and 15) (Lee et al. 2023).

From a biophysical perspective, coastal bioshields reduce wave energy and storm surge through surface roughness and drag forces created by vegetation density and structure (Athikalam and Vaideeswaran 2022). Mangroves, for instance, attenuate wave heights by 50–60% over just 100 m of forest width, while saltmarshes and dune vegetation trap sediments that enhance coastal elevation and stability (Temmerman et al. 2013). These processes are particularly important in deltaic and low-lying coasts where vertical accretion rates must keep pace with relative sea-level rise. The synergistic interactions between plant root systems and sediment dynamics play a critical role in maintaining geomorphic resilience, allowing ecosystems to adapt naturally to changing hydrodynamic conditions (Elsey-Quirk and Unger 2018; Wimmler et al. 2021; Infantes et al. 2022). Furthermore, the high productivity and substantial organic carbon storage capacity of mangroves and marshes contribute significantly to global blue carbon sinks, thereby mitigating greenhouse gas emissions and strengthening national climate commitments under Nationally Determined Contributions (NDCs) (Taillardat et al. 2018; Morrissette et al. 2023).

However, the resilience of bioshields under changing climatic conditions depends on complex feedbacks among ecological processes, environmental gradients, and human interventions. In tropical Asia, which is the home to some of the world's most densely populated and rapidly developing coastlines, bioshields face mounting pressures from anthropogenic drivers such as land reclamation, aquaculture expansion, upstream damming, and pollution, as well as climate-induced stressors like cyclones, droughts, and salinity intrusion (Veettil et al. 2020, 2021). For example, mangrove ecosystems in Vietnam's Mekong Delta (Besset et al. 2019; Veettil et al. 2019b) and Bangladesh's Sundarbans (Das and Vincent 2009) are simultaneously influenced by sediment depletion from river regulation (e.g., dams), subsidence, and increased storm frequency. These compounded pressures not only reduce the regenerative capacity of coastal bioshields but also compromise their ecological functions and socio-economic benefits (Besset et al. 2019).

To ensure the long-term functionality of bioshields, adaptive management and restoration strategies must be informed by robust ecological monitoring, hydrodynamic modeling, and participatory governance (Craig and Ruhl 2010). Integrating remote sensing technologies, such as LiDAR and multispectral satellite imagery, enables large-scale mapping of vegetation cover, sediment accretion, and coastal vulnerability (Veettil et al. 2020). Meanwhile, community-based co-management approaches, developing upon traditional ecological knowledge (TEK), can enhance local stewardship and resilience (Stori et al. 2019). Ultimately, maintaining and restoring bioshields represents a cost-effective, multi-benefit approach to climate-resilient coastal planning, bridging the gap between disaster risk reduction, biodiversity conservation, and sustainable development (Narayan et al. 2016; Veettil et al. 2021; Huynh et al. 2024).

This chapter explores the interlinkages between climate change and the resilience of coastal bioshields. It assesses how climatic variables affect bioshield ecosystems, examines the mechanisms underpinning their resilience, and highlights global and Asian case studies that illustrate adaptive responses. Ultimately, it provides insights into integrating climate adaptation into bioshield management, ensuring these nature-based defenses remain effective amid accelerating climate uncertainty.

5.2 Climate Change and Coastal Vulnerability

Coastal regions are among the most climate-sensitive environments on Earth, functioning as dynamic interfaces between terrestrial, marine, and atmospheric systems. The Intergovernmental Panel on Climate Change in its Sixth Assessment Report underscores that global mean sea level has risen by approximately 20 cm since 1900, primarily due to thermal expansion of seawater, melting of glaciers, and ice sheet loss (Lee et al. 2023). Depending on greenhouse gas emission trajectories, projections indicate an additional 0.3–1.1 m rise by 2100, which poses a profound threat to coastal ecosystems and human livelihoods (IPCC 2021). Rising sea levels exacerbate saltwater intrusion, coastal flooding, and erosion—processes that destabilize both natural habitats and built infrastructure. For low-lying deltaic regions such as the Mekong (Vietnam), Irrawaddy (Myanmar), and Ganges–Brahmaputra (Bangladesh and India), even modest sea-level increases can inundate extensive areas of agricultural land and aquaculture ponds, undermining food security and forcing population displacement (Becker et al. 2020; Nguyen et al. 2023a, b).

In addition to rising sea levels, climate change amplifies the frequency and intensity of extreme weather events, including tropical cyclones, storm surges, and monsoon variability (Kropf et al. 2024). These phenomena inflict direct and indirect damage on coastal ecosystems, infrastructure, and communities (Fig. 5.1). For instance, the 2020 Super Cyclone Amphan, one of the most powerful storms ever recorded in the Bay of Bengal, devastated the Sundarbans mangrove forest, uprooting millions of trees and eroding hundreds of kilometers of embankments in India and Bangladesh (Bhargava and Friess 2022). This catastrophic event illustrated how climate-driven disturbances can reverse decades of ecological restoration in a few hours. Similar patterns are evident in Southeast Asia, where Typhoons Haiyan (2013) and Rai (2021) caused widespread destruction across the Philippines, damaging not only coastal infrastructure but also the bioshield ecosystems, including mangroves, dunes, and coastal forests, that serve as natural defenses (Long et al. 2016; Esteban et al. 2023).

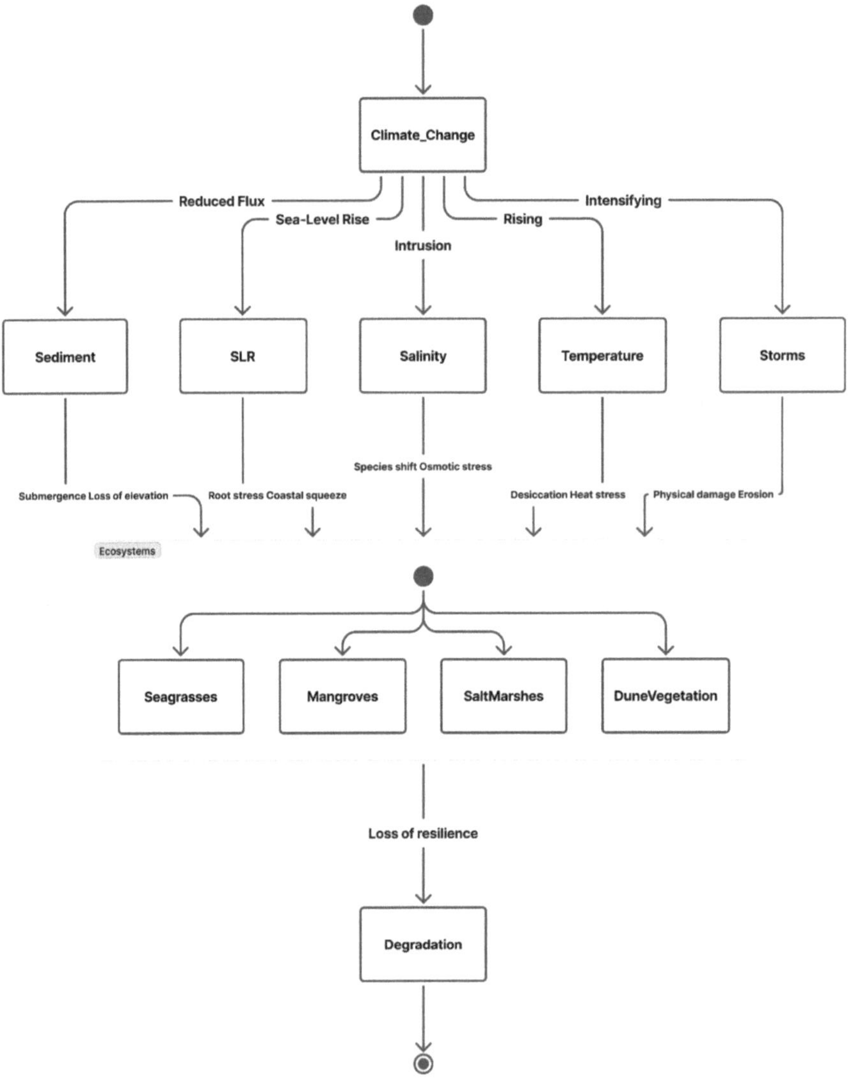

Fig. 5.1 Impacts of climate change on various coastal vegetation

The consequences of these climate hazards are magnified by anthropogenic stressors, including coastal reclamation, deforestation, aquaculture expansion, and river damming (Veettil et al. 2021). These human activities disrupt sediment transport, reducing the natural replenishment of coastal zones. In the Mekong Delta, for example, sediment loads have declined by over 50% due to upstream dam construction, leading to accelerated coastal erosion and subsidence (Tamura et al. 2020; Bussi et al. 2021). This sediment deficit weakens the capacity of mangroves and salt marshes to maintain their elevation relative to sea level, making them increasingly vulnerable to inundation and dieback. The interplay between climatic and anthropogenic pressures thus creates a compound risk, whereby the degradation of natural buffers amplifies the exposure and sensitivity of coastal communities to hazards.

Bioshields, which include vegetated coastal buffers composed of mangroves, salt marshes, dune vegetation, and coastal forests, have emerged as nature-based solutions (NbS) that provide multifunctional protection. They attenuate wave energy, reduce storm surge heights, and stabilize sediments through their root structures (Perricone et al. 2023). The synergistic interactions between plant root systems and sediment dynamics are critical for maintaining geomorphic resilience, allowing ecosystems to adjust naturally to hydrodynamic changes and sea-level fluctuations (Kazemi et al. 2021).

However, the protective capacity of coastal bioshields against climate hazards is not uniform. It depends on species composition, vegetation density, age structure, and spatial configuration (Tanaka 2009). For example, Rhizophora species, with their prop roots, are highly effective in dissipating wave energy, whereas Avicennia species are more important for sediment stabilization through pneumatophore-mediated trapping (Mori et al. 2022). Similarly, dune vegetation, such as *Ipomoea pes-caprae* and *Spinifex littoreus*, binds loose sand and reduces aeolian erosion, enhancing dune stability during high-wind events (Lee et al. 2020). The functional diversity within these vegetative assemblages thus determines the overall resilience and recovery potential of bioshields following disturbance.

Under conditions of accelerated sea-level rise and altered sediment budgets, the equilibrium between sediment deposition and erosion can shift rapidly. When sediment supply is insufficient or hydrodynamic energy exceeds vegetative resistance, coastal ecosystems can transition from accretional to erosional states, leading to vegetation loss and shoreline retreat (Wright and Thom 2023). Such threshold dynamics often result in regime shifts from vegetated to unvegetated mudflats, reducing both ecological function and protective capacity. Furthermore, salinity intrusion due to rising sea levels and reduced freshwater inflows can alter species composition, replacing salt-tolerant pioneer vegetation with less protective assemblages (Perri et al. 2023).

Climate-induced changes in hydrodynamics and biogeochemistry also affect the carbon balance of coastal ecosystems (IPCC 2022). While mangroves and salt marshes are known for their high primary productivity and capacity for storing organic carbon, stress-induced degradation can transform these ecosystems from carbon sinks to carbon sources, releasing stored carbon into the atmosphere (Taillardat et al. 2018). Maintaining the structural integrity of bioshields, therefore, contributes not only to disaster risk reduction but also to climate change mitigation, thereby reinforcing national commitments under Nationally Determined Contributions (NDCs).

Addressing coastal vulnerability requires a multidisciplinary and adaptive management framework. Traditional engineering approaches, such as sea walls and groynes, provide immediate protection but often disrupt natural sediment processes and reduce long-term resilience (Mohamed Rashidi et al. 2021). In contrast, hybrid systems combining green (ecosystem-based) and grey (engineered) infrastructures, such as mangrove restoration integrated with breakwaters, have shown enhanced performance and cost-effectiveness in tropical Asia (Narayan et al. 2016). For these measures to succeed, management strategies must be grounded in robust ecological monitoring, hydrodynamic modeling, and participatory governance, enabling iterative learning and adaptive responses to emerging climate risks.

Furthermore, community-based and co-management approaches that integrate local ecological knowledge have proven crucial in sustaining restoration efforts. In Vietnam (Hai et al. 2020) and the Philippines (Valenzuela et al. 2020), for example, participatory mangrove rehabilitation projects have demonstrated higher survival rates and broader socioeconomic benefits compared to top-down interventions. Local engagement enhances compliance, stewardship, and resilience, ensuring that restoration is both ecologically sound and socially equitable.

5.3 Biophysical Impacts of Climate Change on Coastal Ecosystems

Climate change exerts profound and multifaceted effects on coastal ecosystems that function as bioshields, which are the natural vegetative barriers composed of mangroves, salt marshes, seagrasses, and coastal dune vegetation. These ecosystems are critical for mitigating storm surges, preventing coastal erosion, and sustaining biodiversity. However, their stability and functionality are increasingly compromised by a set of climate-driven biophysical stressors, including hydrological alterations, salinity intrusion, temperature rise, and changing sediment dynamics (White and Kaplan 2017). Each of these stressors modifies the growth, productivity, and regeneration potential of bioshield vegetation, influencing both their structural integrity and ecological services (Sharma and Naik 2024).

5.3.1 Sea-Level Rise and Inundation

One of the most direct biophysical consequences of climate change is sea-level rise, which subjects coastal ecosystems to prolonged inundation and waterlogging. Excessive submergence reduces oxygen availability in the root zone, inhibiting root respiration and impairing nutrient uptake. For mangroves and dune vegetation, this physiological stress often results in stunted growth, canopy dieback, and eventual mortality (Kahnouj et al. 2023). Tolerance levels, however, vary considerably among species. *Avicennia marina* exhibits moderate tolerance to periodic inundation, enabling it to colonize intertidal zones with fluctuating water levels (Chen and Ye 2013). In contrast, *Rhizophora apiculata*, typically found in more stable hydrological regimes, is highly sensitive to prolonged submergence, leading to higher mortality under rising sea-level conditions (Nor 2017).

When sediment accretion cannot keep pace with sea-level rise, mangrove forests and other coastal vegetative systems undergo vertical submergence and horizontal landward migration (Alongi 2015). In many regions, this natural retreat is constrained by artificial embankments, seawalls, and human settlements, a phenomenon termed coastal squeeze (Doody 2013). As a result, bioshields lose critical intertidal space, diminishing their buffering capacity and habitat value. Coastal squeeze is particularly severe in densely populated deltas such as the Mekong and Ganges–Brahmaputra, where rapid urbanization and infrastructure development impede natural adaptive responses (Truong et al. 2017; Rahman et al. 2020). The long-term outcome is a net loss of protective green belts as well as increased vulnerability of coastal communities to flooding and storm surges.

5.3.2 *Salinity Intrusion and Altered Water Chemistry*

Changes in salinity regimes represent another major driver of biophysical transformation in coastal ecosystems. Sea-level rise, coupled with reduced freshwater inflow from upstream due to dam construction and altered rainfall patterns, intensifies saline intrusion into estuarine and deltaic regions (Chen 2005). Elevated salinity levels can disrupt osmotic balance in plant tissues, reduce photosynthetic efficiency, and modify species composition.

In the Mekong Delta, increasing salinity during prolonged dry seasons has already triggered shifts in mangrove species dominance (Fagherazzi et al. 2017; Trân et al. 2024). Brackish-water species such as *Bruguiera gymnorhiza* and *Ceriops decandra* are progressively being replaced by more salt-tolerant taxa like *Avicennia alba* and *Sonneratia caseolaris* (Su et al. 2022). While such adaptive compositional changes may sustain a degree of coastal protection, they often come at the cost of biodiversity loss and altered ecological function. Salinity-induced stress can also impair seed germination and propagule establishment, thereby reducing regeneration potential and compromising long-term resilience (Ball and Sobrado 2002). In salt marshes, similar processes lead to dieback of less tolerant halophytes and the dominance of a few resistant species, which diminishes the ecosystem's capacity to buffer storm surges and store carbon (Craft et al. 2009).

5.3.3 *Rising Temperatures and Thermal Stress*

The temperature rise associated with global warming has pervasive effects on the physiology and productivity of coastal vegetation (Zhou et al. 2018). Elevated temperatures alter phenological cycles, increase respiration rates, and modify water balance. In tropical regions, sustained heat stress intensifies evapotranspiration, leading to desiccation and water deficits in coastal dune systems (Dos Santos et al.

2022). These conditions weaken plant root networks that stabilize dune sediments, thereby accelerating erosion and landscape instability.

For mangroves, increased temperature can enhance enzymatic activity and metabolic respiration, reducing net primary productivity and carbon sequestration potential (Clough 1992). Photosynthetic efficiency declines beyond species-specific thermal thresholds, leading to reduced growth and increased leaf senescence. Moreover, higher air and soil temperatures can influence reproductive timing, seedling viability, and propagule dispersal patterns, thereby affecting forest recruitment dynamics (Hernandez et al. 2023). The combined effects of thermal stress and water imbalance may also amplify vulnerability to pests and diseases, further destabilizing bioshield ecosystems.

5.3.4 Storm Intensity, Damage, and Coastal Erosion

Climate change is expected to intensify the frequency and severity of tropical cyclones and storm surges, directly impacting the structure and stability of bioshields (Blankespoor et al. 2017; Kropf et al. 2024). High-velocity winds and wave action uproot trees, defoliate canopies, and erode sediment substrates, exposing root systems to desiccation and nutrient loss. The immediate aftermath of such events often includes massive deforestation in intertidal zones and loss of critical habitat for coastal fauna.

The recovery capacity of these ecosystems is largely determined by the availability of viable propagules (Vanderklift et al. 2020), soil stability (Jiba et al. 2024), and post-disturbance hydrological conditions (Purandare et al. 2024). In cyclone-prone regions such as the eastern Philippines, mangrove restoration programmes often prioritise robust, wind- and wave-resistant species such as *Avicennia marina* and *Sonneratia alba*, rather than rhizophorids that are indiscriminately planted on exposed coastlines (Primavera et al. 2014). However, repeated storm disturbances can deplete the seed bank and degrade soil structure, creating feedback loops that hinder natural regeneration. Coastal dunes experience similar erosion-driven degradation, where vegetation loss accelerates sand displacement, reducing the effectiveness of these systems as physical barriers (Broome 2015).

5.3.5 Sediment Dynamics and Geomorphological Feedbacks

Sediment supply plays a foundational role in maintaining the elevation and integrity of coastal bioshields. Climate-induced shifts in river discharge, compounded by anthropogenic factors such as damming, land-use change, and upstream sediment trapping, have drastically reduced sediment flux to many deltaic systems (Truong et al. 2017). This reduction limits natural accretion processes necessary for maintaining surface elevation relative to sea-level rise.

In deltas such as those of Vietnam, reduced sediment input has been identified as a critical vulnerability factor for the persistence of mangrove belts and salt marshes (Allison et al. 2017). Without sufficient sediment replenishment, these ecosystems cannot build vertical elevation, resulting in chronic submergence and eventual conversion into tidal flats or open water. Changes in sediment dynamics also affect nutrient transport and substrate composition, influencing species distribution and root anchorage. For example, finer sediments promote the establishment of Rhizophora species, while coarse sandy substrates favor Avicennia dominance (Balke et al. 2013). Alterations in these conditions can therefore cascade into changes in community structure and geomorphological stability.

5.3.6 Interconnected Biophysical Feedbacks and System Resilience

The biophysical impacts of climate change on coastal ecosystems are not isolated; they interact through complex feedback mechanisms that collectively determine bioshield resilience. Sea-level rise, for instance, amplifies saline intrusion and alters sediment deposition patterns, while temperature increases can exacerbate desiccation stress and hinder post-storm recovery. These interconnected processes create nonlinear ecological responses that challenge prediction and management.

Resilience in bioshield systems arises from their capacity for ecological and geomorphological adjustment through species turnover, sediment trapping, and adaptive hydrological feedbacks (Seddon et al. 2011; Willemsen et al. 2016; Alongi 2022). However, this adaptive capacity is finite and increasingly constrained by anthropogenic pressures such as land reclamation, pollution, and infrastructure encroachment (Veettil et al. 2019a). Maintaining or restoring these adaptive processes is therefore essential for sustaining the protective and ecological functions of coastal bioshields under a changing climate.

5.4 Mechanisms of Coastal Bioshield Resilience to Climate Stressors

Coastal bioshields play a vital role in mitigating the impacts of climate change by stabilizing shorelines, attenuating waves, and sequestering carbon. Their resilience, defined as the capacity to absorb, recover, and adapt to climatic disturbances without losing key structural and functional attributes, underpins their long-term protective value (Turki et al. 2025). This resilience is inherently multi-dimensional, operating through interconnected biological, ecological, geomorphological, and human-mediated processes that enable bioshields to persist and evolve under changing environmental conditions (Turki et al. 2025).

5.4.1 Biological Adaptation

At the most fundamental level, bioshield resilience originates from the physiological and morphological adaptations of constituent species that allow survival in highly dynamic coastal environments (Naskar and Palit 2015). Many mangrove taxa, for instance, possess specialized root systems such as pneumatophores and aerial prop roots that facilitate oxygen uptake in waterlogged and anoxic substrates. Species like *Avicennia marina* and *Rhizophora apiculata* are equipped with salt-excreting glands and ultrafiltration mechanisms that regulate internal ionic balance, conferring tolerance to hypersaline conditions and fluctuating inundation regimes (Guo et al. 2023). In addition to morphological and physiological adaptations, several complex genetic and molecular mechanisms control mangrove adaptations to changing intertidal environments (Nizam et al. 2021). Similarly, salt marsh halophytes, including *Spartina alterniflora*, display osmotic regulation and succulence that allow persistence in saline sediments (Xin et al. 2022).

In coastal dune systems, plant species such as *Ipomoea pes-caprae* and *Spinifex littoreus* exemplify mechanical and reproductive adaptations to extreme wind and sand dynamics (Lee et al. 2020). Their creeping stems and extensive rhizome networks anchor the substrate and promote dune stability, while clonal propagation enables rapid recolonization after storm-induced erosion (Rivera et al. 2025). These traits collectively enhance the biophysical resilience of the vegetation cover, ensuring that even after severe disturbances, vegetative structure and ecological function can recover swiftly.

5.4.2 Ecological Redundancy and Functional Diversity

At the ecosystem scale, biodiversity and functional redundancy are key determinants of bioshield resilience. Systems with high species and structural diversity possess multiple organisms capable of performing similar ecological roles, thereby maintaining critical functions even when individual species decline (Konefal et al. 2024). In mixed mangrove assemblages, for example, if salt-sensitive species such as *Bruguiera gymnorhiza* are lost under increasing salinity, more tolerant taxa like *Avicennia alba* or *Sonneratia caseolaris* can sustain sediment trapping and hydrological buffering functions (Cheng et al. 2025).

This concept of redundancy supports the insurance hypothesis (Yachi and Loreau 1999), wherein species diversity provides stability against environmental perturbations. Accordingly, restoration and management programs across Southeast Asia increasingly emphasize multi-species plantations rather than monocultures (Veettil et al. 2021; Gerona-Daga and Salmo 2022). Such approaches enhance resilience not only through ecological complementarity but also by fostering habitat heterogeneity that supports broader faunal communities. Diverse root structures and canopy morphologies also increase the system's ability to dissipate wave energy, filter sediments, and support nutrient cycling across varying environmental gradients (Twomey and Lovelock 2025).

5.4.3 Sediment Accretion and Biogeomorphic Feedbacks

Resilience in bioshields is also governed by biogeomorphic feedbacks that link vegetation dynamics with sedimentary processes (Gourgue et al. 2022). Vegetation acts as an effective sediment trap, where stems, roots, and litter slow water flow, encouraging the deposition of fine sediments and organic matter. This process contributes to vertical accretion, elevating the substrate and enabling ecosystems such as mangroves and salt marshes to keep pace with sea-level rise. When sediment accumulation rates exceed or match the rate of relative sea-level rise, the bioshield maintains its functional elevation and continues to buffer coastal hazards (Mudd et al. 2009). However, if sediment supply or productivity is insufficient, submergence and degradation may occur (Besset et al. 2019).

The magnitude of this resilience mechanism depends on external sediment inputs, hydrodynamic conditions, and biological productivity. For instance, in deltas with high sediment loads, such as the Mekong or Ganges–Brahmaputra, vegetation–sediment feedbacks often promote rapid accretion and land-building (Anthony et al. 2015; Auerbach et al. 2015). Conversely, sediment starvation due to upstream damming or embankments can weaken these feedbacks, rendering bioshields more vulnerable to erosion and sea-level rise (Besset et al. 2019). Thus, maintaining natural sediment connectivity is essential for sustaining long-term bioshield resilience.

5.4.4 Regeneration and Recovery Dynamics

Another critical mechanism underpinning resilience is the capacity for natural regeneration following disturbances such as cyclones, tsunamis, or extreme flooding. Many coastal vegetation species have evolved reproductive strategies that enable them to colonize disturbed substrates rapidly. Mangroves, for example, produce buoyant propagules that disperse via tidal currents, thereby allowing recolonization of adjacent mudflats or eroded areas (Van der Stocken et al. 2019). In the Sundarbans, natural regrowth of *Avicennia marina* and *Excoecaria agallocha* has been observed within two years of major cyclone impacts, provided that hydrological conditions remain favorable (Chowdhury et al. 2023).

The success of this regenerative process depends on several factors, including seed availability, substrate stability, and the absence of anthropogenic barriers such as embankments or aquaculture ponds (Balke et al. 2011). Disturbance-driven regeneration not only restores vegetative cover but also reinitiates biogeomorphic feedbacks that stabilize sediments and promote habitat recovery (Balke et al. 2013). Therefore, maintaining ecological connectivity and minimizing post-disturbance land-use conversion are vital to ensuring natural resilience pathways remain functional (Aguilera et al. 2020).

5.4.5 *Human-Assisted Adaptation and Hybrid Solutions*

While natural adaptive mechanisms are powerful, human interventions can substantially enhance bioshield resilience in the face of accelerating climate change. Human-assisted adaptation strategies, such as managed realignment, artificial sedimentation, and targeted replanting, seek to restore ecological function and geomorphic stability through a combination of engineering and ecosystem-based approaches (French 2006; Temmerman et al. 2013). In Vietnam's Mekong Delta, for instance, sediment embankments and wooden fencing structures (Fig. 5.2) have been installed to attenuate waves, promote sediment deposition, and facilitate mangrove colonization on eroding mudflats (Duy et al. 2025). These hybrid solutions effectively accelerate natural recovery processes while maintaining ecological integrity.

In addition, adaptive management practices, including periodic thinning, species diversification, and hydrological restoration, help maintain optimal conditions for growth and regeneration (Bosire et al. 2008; Alongi 2015; Van Loon et al. 2016). Participatory approaches involving local communities further strengthen resilience by integrating traditional knowledge and ensuring long-term stewardship (Stori et al. 2019). When combined with robust monitoring and policy support, such interventions transform bioshields from static conservation targets into dynamic adaptive systems capable of evolving alongside climatic and geomorphological changes.

Fig. 5.2 Sediment embankments (Bac Lieu province) and wooden fences (Ben Tre province) to facilitate mangrove colonization in Vietnam

5.5 Integrating Climate Adaptation into Bioshield Design

The integration of climate adaptation principles into coastal bioshield design marks a transformative shift from static, one-time restoration efforts toward dynamic, learning-oriented management (Huynh et al. 2024). In the past, bioshield development largely emphasized initial plantation density, linear layouts, and structural uniformity aimed at providing immediate coastal protection (Bosire et al. 2008). However, as climate change intensifies sea-level rise, storm surges, salinity intrusion, and erosion processes, such static approaches have proven insufficient. Achieving long-term resilience requires an anticipatory framework that integrates design flexibility, continuous feedback learning, and ecological redundancy, thereby enabling bioshields to adjust dynamically to evolving climatic and geomorphic conditions (Folke et al. 2010). This approach moves beyond simple replanting efforts to establish adaptive bioshield systems capable of evolving alongside environmental change (Temmerman et al. 2013).

5.5.1 Site Selection and Elevation Gradients

Adaptive bioshield design begins with meticulous site assessment grounded in future climate projections. Elevation gradients and sedimentation potential are key determinants of long-term ecosystem persistence (Zhang et al. 2025). Elevation influences inundation frequency, soil salinity, and vegetation zonation, all of which control the success of mangrove or dune vegetation establishment (Lara and Cohen 2006). In low-lying coastal areas projected to face permanent submergence, highly salt-tolerant species such as *Avicennia marina* or *Sonneratia alba* may dominate (Tomlinson 2016). Conversely, slightly elevated sites can support *Rhizophora apiculata, Bruguiera gymnorhiza*, or mixed mangrove stands that provide greater canopy complexity and structural diversity (Alongi 2009). Integrating high-resolution elevation models derived from LiDAR or remote sensing platforms enhances predictive capacity by identifying stable sediment surfaces and areas prone to erosion or accretion (Mohamad et al. 2021). These data-driven insights enable planners to match species and structural configurations with expected hydrodynamic conditions, ensuring the long-term viability of the coastal bioshield.

5.5.2 Species Diversity and Functional Traits

Ecological resilience in bioshields is closely tied to biodiversity and functional complementarity among species (Elmqvist et al. 2003). Monospecific plantations, while easier to manage, often lack redundancy and are more vulnerable to pest outbreaks, salinity shifts, or extreme climatic events (Friess and Webb 2014). By contrast, mixed-species systems integrate a range of traits that collectively strengthen the system's stability and functionality. For example, Rhizophora species contribute robust prop roots that enhance sediment trapping and wave attenuation, while Avicennia and Sonneratia species stabilize upper soil layers through pneumatophores that facilitate oxygen exchange under anaerobic conditions.

Functional diversity also broadens the ecosystem services provided, ranging from carbon sequestration and nutrient cycling to habitat provision for fish and crustaceans (Mori et al. 2013). The inclusion of both pioneer and late-successional species facilitates rapid colonization and long-term structural development, thereby creating bioshields that can self-organize and adapt after disturbance (Alongi 2008).

5.5.3 *Hydrological Connectivity and Sediment Dynamics*

A critical, yet often overlooked, element of climate-adaptive bioshield design is maintaining natural hydrological connectivity. Mangrove and salt marsh systems depend on tidal flushing for nutrient exchange, sediment replenishment, and propagule dispersal (Veettil et al. 2021). Overly engineered interventions, such as impermeable embankments, culverts, or aquaculture barriers, can disrupt these flows, leading to stagnation, hypersalinity, and vegetation dieback (Alongi 2008). Adaptive approaches favor permeable or semi-natural boundaries that enable sediment transport and water circulation (Friess and Webb 2014). Incorporating features such as tidal creeks, sediment bypasses, and controlled inlets promotes the self-sustaining dynamics of accretion and regeneration (Van Loon et al. 2016). In deltaic regions such as the Mekong and Ganges–Brahmaputra, where sediment supply is high, maintaining open hydrological pathways enhances natural elevation gain and helps bioshields keep pace with rising sea levels (Schuerch et al. 2018).

5.5.4 *Integration with Gray Infrastructure*

Hybrid or green–gray strategies are increasingly being adopted to enhance the durability of coastal defenses under climate stress (Huynh et al. 2024). In this model, natural and engineered components function synergistically: vegetative buffers reduce hydrodynamic energy before it reaches hard infrastructure, thereby lowering maintenance costs and extending the lifespan of built structures (Narayan et al. 2016). A practical example comes from Vietnam's Ca Mau Province, where sediment fencing and mangrove belts are established seaward of earthen dykes to promote sediment deposition and facilitate mangrove colonization on eroding mudflats (Schmitt et al. 2013). These bioshields act as the first line of defense, dissipating wave energy and stabilizing the foreshore, while the engineered dyke provides a secondary safeguard (Veettil et al. 2021). Such integrative designs embody the building with nature philosophy, wherein natural processes are used to strengthen engineering outcomes rather than being constrained by them.

5.5.5 Adaptive Management and Monitoring Frameworks

A defining feature of climate-adaptive bioshield design is its reliance on continuous learning through monitoring and iterative adjustment. Environmental conditions and vegetation responses must be regularly assessed using a combination of field surveys and remote sensing tools (Veettil et al. 2019a, b). Image series, such as Sentinel-2 MSI, UAV-based multispectral mapping, and LiDAR, allow for precise tracking of canopy cover, vegetation vigor, and sediment morphology over time. Data derived from these monitoring systems inform adaptive management interventions, such as selective thinning to improve light penetration, replanting to fill canopy gaps, sediment augmentation in areas of subsidence, or restoration of blocked tidal creeks (Pham et al. 2019). This cyclical process of feedback and adjustment ensures that bioshields maintain ecological functionality and protective performance even as climatic drivers evolve.

5.6 Modeling and Monitoring Climate Resilience

Quantitative modeling and long-term monitoring are fundamental tools for understanding, predicting, and strengthening the climate resilience of coastal bioshields. These approaches allow scientists and policymakers to visualize how mangroves, salt marshes, and dune systems respond to sea-level rise, storm surges, and human pressures. Modeling provides foresight by simulating potential ecosystem trajectories under different climate and management scenarios (Marzouk and Azab 2024), while monitoring offers the empirical validation needed to refine these projections and support evidence-based adaptation strategies (Willcock et al. 2019). In essence, the fusion of quantitative modeling and participatory monitoring transforms bioshield management from a static conservation activity into a dynamic, evidence-driven process. Such integration ensures that coastal ecosystems remain functional, adaptive, and resilient in the face of accelerating climate change.

5.6.1 Ecosystem Process Models

Ecosystem models play a critical role in predicting how vegetated coastal systems evolve under changing environmental conditions. Tools such as the Sea Level Affects Marshes Model (SLAMM) simulate vegetation dynamics in response to sea-level rise, salinity fluctuations, and hydrological changes (Dang et al. 2022). By integrating field data with topography, sediment accretion rates, and tidal range, these models can project spatial shifts in mangrove and marsh zones over the coming decades. For example, Schuerch et al. (2018) demonstrated that process-based modeling could identify areas where coastal vegetation is likely to persist, migrate inland, or face submergence. Such insights inform adaptive land-use and zoning policies by pinpointing zones where restoration or managed retreat would yield the most sustainable outcomes.

5.6.2 Wave Attenuation and Storm-Surge Modeling

Hydrodynamic modeling provides a quantitative understanding of how bioshields mitigate coastal hazards. Models such as SWAN (Simulating WAves Nearshore) and Delft3D simulate wave transformation and energy dissipation across vegetated coastlines (Vinh 2015; Baron-Hyppolite et al. 2018). Parameters such as vegetation density, stem diameter, root structure, and canopy height are incorporated to assess their combined effects on wave attenuation. For example, Mazda et al. (2006) and Quartel et al. (2007) mentioned that mangrove belts with widths of 100 m or more can reduce incoming wave heights by 30–50%, depending on vegetation density, species composition, and stand structure. Integrating these findings into coastal defense planning allows engineers and ecologists to optimize the width, density, and species structure of bioshields, combining natural and engineered defenses for maximum protection efficiency (Gijsman et al. 2021).

5.6.3 Carbon and Sediment Flux Modeling

The capacity of bioshields to store carbon and trap sediment directly contributes to their resilience and to broader climate mitigation goals (Smoak et al. 2013). Carbon and sediment models estimate net ecosystem carbon balance (NECB), soil organic accumulation, and vertical accretion rates, revealing whether mangrove and marsh systems are keeping pace with sea-level rise (Breithaupt et al. 2012; Ishtiaq et al. 2022). Saintilan et al. (2020) showed that mangrove persistence is highly probable when vertical accretion exceeds approximately 5–6 mm yr^{-1} but declines sharply beyond this threshold as relative sea-level rise accelerates. Similarly, Zhang et al. (2025) reported mean accretion rates of 6.7 ± 1.9 mm yr^{-1} in China's Beilun Estuary, suggesting that mangroves in regions with comparable sediment dynamics are likely to remain resilient under moderate sea-level-rise scenarios. Integrating carbon and sediment flux modeling with field-based Surface Elevation Table—Marker Horizon (SET-MH) allows researchers to link biogeomorphic processes (such as root production, sediment accretion, and subsidence) to climate-resilience indicators (surface elevation gain and carbon burial) in vegetated coastal systems

(McIvor et al. 2013). These analyses not only support adaptive restoration but also inform national greenhouse gas inventories and blue carbon accounting.

5.6.4 Socioeconomic and Participatory Modeling

While biophysical processes are crucial, social resilience ultimately determines the long-term success of bioshield management. Socioeconomic and participatory modeling frameworks simulate how communities respond to ecological and policy changes. Agent-based models, for example, can replicate decision-making dynamics related to land-use change, aquaculture expansion, and resource extraction (Truong et al. 2022). Barbier (2020) highlights how such models help identify potential conflicts or synergies between conservation and livelihood objectives. Participatory modeling, involving local stakeholders in scenario building, strengthens co-management and promotes shared ownership of adaptation measures (Johnson et al. 2021). This approach captures local knowledge and behavioral responses often overlooked in purely biophysical models, thereby bridging the gap between ecological forecasting and community-based implementation.

5.6.5 Long-Term Monitoring Systems

Model projections are only as reliable as the data used to calibrate and validate them. Continuous, long-term monitoring provides this critical feedback loop. Many Southeast Asian countries, including Indonesia, the Philippines, and Vietnam, have established community-based monitoring networks that combine traditional ecological knowledge with modern technologies (Veettil et al. 2019a; Arifanti et al. 2022; Kusumadewi et al. 2024). Through mobile applications, drones, and low-cost sensors, local volunteers can record vegetation changes, sediment accumulation, salinity fluctuations, and shoreline movements (Koh and Wich 2012). These citizen-science initiatives generate real-time datasets that feed into national adaptation databases, supporting near-term management decisions while building local capacity for stewardship (Menon et al. 2020). The integration of remote sensing platforms, such as satellite-derived vegetation indices and LiDAR-based elevation models, further enhances the spatial and temporal resolution of monitoring efforts, allowing for the detection of subtle ecological changes that precede major degradation events (Guo et al. 2017).

5.6.6 Integrating Modeling and Monitoring for Adaptive Management

Together, modeling and monitoring form the scientific backbone of climate-resilient bioshield management. Modeling offers predictive power, identifying thresholds, vulnerabilities, and optimal intervention points, while monitoring ensures continuous learning through observation and feedback (World Bank 2016). When integrated, these systems support adaptive management cycles, where predictions are tested, outcomes are evaluated, and strategies are adjusted based on empirical evidence (Powell et al. 2019). This iterative process enables managers to refine restoration designs, adjust protection zoning, and anticipate long-term ecosystem responses to climate variability (Valente and Pinho 2025).

5.7 Policy and Management Responses

Enhancing the climate resilience of coastal bioshields requires coherent governance, policy integration, community participation, and sustained financial investment, in addition to a sound ecological design. Across Asia, countries are increasingly incorporating bioshields into their broader climate adaptation and coastal management strategies (Morita and Matsumoto 2021; Veettil et al. 2021; Tun et al. 2025). These efforts reflect a recognition that nature-based solutions are not just ecological interventions but critical components of national resilience strategies. Nature-based solutions and coastal bioshields are being incorporated into Nationally Determined Contributions (NDCs), National Adaptation Plans (NAPs), and Integrated Coastal Zone Management (ICZM) frameworks, thereby providing both policy legitimacy and financial pathways for long-term implementation in several coastal countries across Asia. By aligning ecological principles with governance innovation, countries across Asia are laying the groundwork for climate-resilient coastlines that protect both people and nature. In this way, bioshields evolve from temporary restoration projects into enduring, adaptive systems at the heart of national climate resilience planning (Wetlands International 2020).

5.7.1 Mainstreaming Bioshields into Climate Policy

Mainstreaming bioshields into climate policy ensures that mangrove restoration, salt marsh rehabilitation, and dune stabilization receive consistent support and visibility (Bhatnagar and Kumar 2024; Hernández-Delgado 2024). Vietnam's National Strategy on Climate Change to 2050 explicitly identifies mangrove rehabilitation as a key pillar of coastal adaptation, linking it with blue carbon commitments and disaster risk reduction goals (MONRE 2023; Trang and Trinh 2025). The strategy prioritizes restoration in vulnerable deltas such as the Mekong and Red River systems, aligning ecological restoration with socio-economic development. Similarly, Bangladesh's Delta Plan 2100 treats bioshields as a cost-effective measure to mitigate cyclone and storm-surge impacts, reducing the need for expensive hard-engineering structures (Rahman et al. 2019). Embedding coastal bioshields within these high-level policy frameworks ensures not only national commitment but also access to international climate finance and technical assistance. When bioshields are integrated into multiple policy domains, such as climate change, biodiversity conservation, and disaster management, they gain institutional durability that extends beyond the project funding period (Islam et al. 2021).

5.7.2 Cross-sectoral Coordination

Coastal ecosystems are commonly managed by a group of institutions managing forestry, fisheries, water resources, environment, and disaster management. Without coordination, overlapping mandates often lead to fragmented or conflicting actions (Veettil et al. 2019a). Effective bioshield governance requires cross-sectoral collaboration to harmonize objectives, reduce duplication, and align budgets. Multi-sectoral task forces, such as those established under Indonesia's Mangrove and Coastal Rehabilitation Plan, serve as models of institutional coherence (CMMAI, MoEF and BRGM 2022). These platforms enable integrated planning that links ecological restoration with aquaculture zoning, flood management, and livelihood programs. By fostering communication among ministries, local governments, and non-state actors, such frameworks ensure that bioshield restoration aligns with national development priorities while maintaining ecological integrity.

5.7.3 Innovative Financing Mechanisms

Securing long-term, reliable funding remains one of the greatest challenges in bioshield management. Restoration and maintenance demand ongoing investment in monitoring, replanting, and capacity building. Traditional public budgets alone are insufficient, and, therefore, innovative financing models are gaining traction. Payment for Ecosystem Services (PES) schemes reward communities for conserving mangrove forests that provide carbon sequestration, coastal protection, and fisheries enhancement (Thompson et al. 2017). Blue carbon credit programs offer a promising avenue for linking bioshields with global carbon markets (Macreadie et al. 2021). Projects like Mikoko Pamoja in Kenya have demonstrated how carbon finance can support mangrove protection while delivering community benefits (Kairo et al. 2018). In Vietnam's Mekong Delta, pilot programs are testing similar models, channeling carbon revenues into community-based restoration and monitoring (Vu et al. 2023; Veettil and Puri 2025). Public–private partnerships (PPP) also play a growing role, with corporate actors investing in bioshield rehabilitation to offset emissions and enhance coastal sustainability branding (Nguyen et al. 2023a, b). Establishing transparent financial mechanisms, combined with benefit-sharing arrangements, ensures that restoration efforts remain economically sustainable and socially equitable.

5.7.4 Legal and Regulatory Frameworks

A strong legal foundation is essential for protecting bioshields against degradation and conversion. The formal recognition of bioshields as critical natural infrastructure anchors them within national development planning and justifies investment in their maintenance. India's Coastal Regulation Zone (CRZ) notification restricts development in ecologically sensitive coastal areas, providing a legal shield against habitat loss (MoEFCC 2025). Similarly, the Philippines' National Greening Program includes mangrove rehabilitation under its reforestation mandate, linking legal protection with large-scale restoration (Gregorio et al. 2017). However, enforcement remains a persistent challenge, especially in rapidly urbanizing or industrializing coastal zones where economic incentives for land conversion are high (Veettil et al. 2019a; Long et al. 2025). Strengthening compliance mechanisms through environmental courts, community monitoring, and transparent permitting systems can help bridge the gap between policy intent and on-ground effectiveness (Long et al. 2025).

5.7.5 Community-Based Governance and Participation

Empowering local communities is central to ensuring that bioshield management aligns with socioecological realities. Top-down interventions often falter when they ignore traditional knowledge, livelihood needs, or local governance structures (CIFOR 2017). In contrast, participatory approaches foster stewardship and long-term commitment (CIFOR 2017). The Philippines' Community-Based Forest Management (CBFM) model illustrates how devolved authority can enhance both ecological outcomes and social resilience (Pulhin et al. 2024). Similarly, Vietnam's co-management systems, where communities share responsibility with local authorities, have improved mangrove survival rates and reduced illegal exploitation (Veettil et al. 2019a). Participatory monitoring using mobile applications and local training programs further strengthens accountability and adaptive learning (Danielsen et al. 2022). When communities are involved from planning through evaluation, they not only protect ecosystems but also benefit from enhanced fisheries, ecotourism, and climate-adaptive livelihoods (Kasten et al. 2021).

5.7.6 *Institutionalizing Coastal Bioshields for Long-Term Resilience*

Integrating ecological science with policy, law, finance, and governance transforms coastal bioshields from isolated conservation projects into institutionalized pillars of national resilience (Mendes et al. 2019). This integration ensures that restoration efforts are underpinned by stable funding, clear mandates, and community ownership. It also facilitates synergies with broader sustainability goals, such as achieving Sustainable Development Goals (SDGs) 13 (Climate Action), 14 (Life Below Water), and 15 (Life on Land) (Gómez Martín et al. 2020). The goal is to embed bioshields within adaptive management systems, where monitoring data, policy feedback, and local participation continuously refine strategies in response to environmental change.

5.8 Conclusions

Climate change poses one of the greatest challenges to the stability and sustainability of coastal systems, placing both human and ecological communities under increasing pressure. In this context, bioshields—living coastal defenses composed of mangroves, salt marshes, dunes, and other vegetated systems—represent a sustainable and adaptive alternative to traditional hard-engineered structures. Their ability to attenuate waves, stabilize sediments, store carbon, and support biodiversity makes them indispensable components of nature-based coastal protection.

The resilience of bioshields lies in their ecological complexity and adaptive capacity. However, sustaining these systems requires more than ecological understanding; it calls for an integrated strategy that unites scientific modeling, evidence-based restoration, participatory governance, and coherent policy frameworks. In Asia, countries such as Vietnam, Bangladesh, and Indonesia have demonstrated how bioshields can be effectively embedded within national adaptation plans and coastal management strategies. These examples illustrate the power of aligning ecological design with local knowledge, institutional coordination, and long-term financing to achieve durable climate resilience.

Looking ahead, the management of bioshields must embrace flexibility and adaptive learning. As sea levels rise and storm intensities shift, static restoration approaches will be insufficient. Instead, continuous monitoring, modeling, and stakeholder engagement must guide iterative adjustments in management practices. Investing in education, research, and technology will further enhance the capacity of communities and institutions to respond to emerging climatic realities.

Ultimately, bioshields exemplify how ecological restoration can serve as both a shield and a bridge—protecting vulnerable coastlines while fostering harmony between people and nature. Through sustained commitment, innovation, and collaboration, bioshields can evolve into enduring instruments of resilience, ensuring that coastal societies remain safe, productive, and adaptive in an era of accelerating climate change.

References

Aguilera MA, Tapia J, Gallardo C, Núñez P, Varas-Belemmi K (2020) Loss of coastal ecosystem spatial connectivity and services by urbanization: natural-to-urban integration for bay management. J Environ Manag 276:111297. https://doi.org/10.1016/j.jenvman.2020.111297

Allison MA, Nittrouer CA, Ogston AS, Mullarney JC, Nguyen TT (2017) Sedimentation and survival of the Mekong Delta: a case study of decreased sediment supply and accelerating rates of relative sea level rise. Oceanography 30(3):98–109. https://doi.org/10.5670/oceanog.2017.318

Alongi DM (2008) Mangrove forests: resilience, protection from tsunamis, and responses to global climate change. Estuar Coast Shelf Sci 76(1):1–13. https://doi.org/10.1016/j.ecss.2007.08.024

Alongi DM (2009) The energetics of mangrove forests. Springer, Netherlands, Dordrecht, p 216

Alongi DM (2015) The impact of climate change on mangrove forests. Curr Clim Change Rep 1:30–39. https://doi.org/10.1007/s40641-015-0002-x

Alongi DM (2022) Climate change and mangroves. In: Mangroves: biodiversity, livelihoods and conservation. Springer Nature Singapore, Singapore, pp 175–198. https://doi.org/10.1007/978-981-19-0519-3_8

Anthony EJ, Brunier G, Besset M, Goichot M, Dussouillez P, Nguyen VL (2015) Linking rapid erosion of the Mekong River delta to human activities. Sci Rep 5(1):1–12. https://doi.org/10.1038/srep14745

Arifanti VB, Sidik F, Mulyanto B, Susilowati A, Wahyuni T, Yuniarti N, Aminah A, Novita N et al (2022) Challenges and strategies for sustainable mangrove management in Indonesia: a review. Forests 13(5):695. https://doi.org/10.3390/f13050695

Athikalam PT, Vaideeswaran AK (2022) Vegetation bioshield for coastal protection in South Asia: status and way forward. J Coast Conserv 26(1):3. https://doi.org/10.1007/s11852-022-00850-x

Auerbach LW, Goodbred SL Jr, Mondal DR, Wilson CA, Ahmed KR, Roy K, Steckler MS, Small C, Gilligan JM, Ackerly BA (2015) Flood risk of natural and embanked landscapes on the Ganges-Brahmaputra tidal delta plain. Nat Clim Change 5(2):153–157. https://doi.org/10.1038/nclimate2472

Balke T, Bouma TJ, Horstman EM, Webb EL, Erftemeijer PL, Herman PM (2011) Windows of opportunity: thresholds to mangrove seedling establishment on tidal flats. Mar Ecol Prog Ser 440:1–9. https://doi.org/10.3354/meps09364

Balke T, Bouma TJ, Herman PMJ, Horstman EM, Sudtongkong C, Webb EL (2013) Cross-shore gradients of physical disturbance in mangroves: implications for seedling establishment. Biogeosciences 10(8):5411–5419. https://doi.org/10.5194/bg-10-5411-2013

Ball MC, Sobrado MA (2002) Ecophysiology of mangroves: challenges in linking physiological processes with patterns. In: Physiological plant ecology: 39th symposium of the British Ecological Society. Cambridge University Press, p 331

Barbier EB (2020) Estuarine and coastal ecosystems as defense against flood damages: an economic perspective. Front Clim 2:594254. https://doi.org/10.3389/fclim.2020.594254

Baron-Hyppolite C, Lashley CH, Garzon J, Miesse T, Ferreira C, Bricker JD (2018) Comparison of implicit and explicit vegetation representations in SWAN hindcasting wave dissipation by coastal wetlands in Chesapeake Bay. Geosciences 9(1):8. https://doi.org/10.3390/geosciences9010008

Becker M, Papa F, Karpytchev M, Delebecque C, Durand F (2020) Water level and subsidence changes in the Ganges–Brahmaputra–Meghna Delta: a robust regional estimate. Earth Syst Dyn 11(2):515–533. https://doi.org/10.5194/esd-11-515-2020

Besset M, Gratiot N, Anthony EJ, Bouchette F, Goichot M, Marchesiello P (2019) Mangroves and shoreline erosion in the Mekong River delta, Viet Nam. Estuar Coast Shelf Sci 226:106263. https://doi.org/10.1016/j.ecss.2019.106263

Bhargava R, Friess DA (2022) Previous shoreline dynamics determine future susceptibility to cyclone impact in the Sundarban mangrove forest. Front Mar Sci 9:814577. https://doi.org/10.3389/fmars.2022.814577

Bhatnagar DG, Kumar S (2024) Ecological conservation and sustainable management of marine environments. In: Nature-based solutions in achieving sustainable development goals: harmonizing nature and progress. Springer Nature Switzerland, Cham, pp 233–261. https://doi.org/10.1007/978-3-031-76128-7_8

Blankespoor B, Dasgupta S, Lange GM (2017) Mangroves as a protection from storm surges in a changing climate. Ambio 46:478–491. https://doi.org/10.1007/s13280-016-0838-x

Bosire J, Dahdouh-Guebas F, Walton M, Crona B, Lewis R, Field C, Kairo J, Koedam N (2008) Functionality of restored mangroves: a review. Aquat Bot 89(2):251–259. https://doi.org/10.1016/j.aquabot.2008.03.010

Breithaupt JL, Smoak JM, Smith TJ, Sanders CJ, Hoare A (2012) Organic carbon burial rates in mangrove sediments: strengthening the global budget. Glob Biogeochem Cycles 26(3). https://doi.org/10.1029/2012GB004375

Broome S (2015) Restoration and management of coastal dune vegetation. NC Cooperative Extension Service. https://content.ces.ncsu.edu/restoration-and-management-ofcoastal-dune-vegetation

Burkett V, Davidson M (2012) Coastal impacts, adaptation, and vulnerabilities. Springer, Washington, DC, p 217

Bussi G, Darby SE, Whitehead PG, Jin L, Dadson SJ, Voepel HE, Vasilopoulos G, Hackney CR, Hutton C, Berchoux T, Parsons DR, Nicholas A (2021) Impact of dams and climate change on suspended sediment flux to the Mekong Delta. Sci Total Environ 755:142468. https://doi.org/10.1016/j.scitotenv.2020.142468

Chen J (2005) Dams, effect on coasts. In: Schwartz ML (ed) Encyclopedia of coastal science. Encyclopedia of earth science series. Springer, Dordrecht. https://doi.org/10.1007/1-4020-3880-1_108

Chen Y, Ye Y (2013) Early responses of Avicennia marina (Forsk.) Vierh. To intertidal elevation and light level. Aquat Bot 112:33–40. https://doi.org/10.1016/j.aquabot.2013.07.006

Cheng H, Chen Y, Peng Y, Wei M, Niu J (2025) Hydraulic traits constrain salinity-dependent niche segregation in mangroves. Plants 14(12):1850. https://doi.org/10.3390/plants14121850

Chowdhury A, Naz A, Sharma SB, Dasgupta R (2023) Changes in salinity, mangrove community ecology, and organic blue carbon stock in response to cyclones at Indian Sundarbans. Life 13(7):1539. https://doi.org/10.3390/life13071539

CIFOR (2017) Study shows that involving communities in mangrove management makes them more sustainable. https://www.cifor.org/press-release/study-shows-that-involving-communities-in-mangrove-management-makes-them-more-sustainable/. Accessed 02 Feb 2025

Clough BF (1992) Primary productivity and growth of mangrove forests. Trop Mangrove Ecosyst 41:225–249. https://doi.org/10.1029/CE041p0225

Coordinating Ministry for Maritime Affairs and Investment (CMMAI), Ministry of Environment and Forestry (MoEF), Peat and Mangrove Restoration Agency (BRGM), Government of Indonesia (2022) Mangroves for Coastal Resilience (M4CR) project: Institutional framework and implementation arrangement. Ministry of Foreign Affairs of Denmark, Danida. https://um.dk/en/-/media/websites/umen/danida/about-danida/danida-transparency/documents/bevilling-til-mangrove-i-indonesien.ashx. Accessed 02 Feb 2025

Craft C, Clough J, Ehman J, Joye S, Park R, Pennings S, Guo H, Machmuller M (2009) Forecasting the effects of accelerated sea-level rise on tidal marsh ecosystem services. Front Ecol Environ 7(2):73–78. https://doi.org/10.1890/070219

Craig RK, Ruhl JB (2010) Governing for sustainable coasts: complexity, climate change, and coastal ecosystem protection. Sustainability 2(5):1361–1388. https://doi.org/10.3390/su2051361

Dang ATN, Reid M, Kumar L (2022) Assessing potential impacts of sea level rise on mangrove ecosystems in the Mekong Delta, Vietnam. Reg Environ Change 22:70. https://doi.org/10.1007/s10113-022-01925-z

Danielsen F, Eicken H, Funder M, Johnson N, Lee O, Theilade I, Burgess ND et al (2022) Community monitoring of natural resource systems and the environment. Ann Rev Environ Resour 47(1):637–670. https://doi.org/10.1146/annurev-environ-012220-022325

Das S, Vincent JR (2009) Mangroves protected villages and reduced death toll during Indian super cyclone. Proc Natl Acad Sci USA 106(18):7357–7360. https://doi.org/10.1073/pnas.0810440106

Dong WS, Ismailluddin A, Yun LS, Ariffin EH, Saengsupavanich C, Maulud KN A, Yunus K et al (2024) The impact of climate change on coastal erosion in Southeast Asia and the compelling need to establish robust adaptation strategies. Heliyon 10(4). https://doi.org/10.1016/j.heliyon.2024.e25609

Doody JP (2013) Coastal squeeze and managed realignment in southeast England, does it tell us anything about the future? Ocean Coast Manag 79:34–41. https://doi.org/10.1016/j.ocecoaman.2012.05.008

Dos Santos TB, Ribas AF, de Souza SGH, Budzinski IGF, Domingues DS (2022) Physiological responses to drought, salinity, and heat stress in plants: a review. Stresses 2(1):113–135. https://doi.org/10.3390/stresses2010009

Duy DV, Ty TV, Phat LT, Minh HV, Nam ND, Downes NK, Avtar R, Tanaka H (2025) Simulating the coastal protection performance of breakwaters in the Mekong Delta: insights from the Western Coast of Ca Mau Province, Vietnam. J Mar Sci Eng 13(8):1559. https://doi.org/10.3390/jmse13081559

Elmqvist T, Folke C, Nyström M, Peterson G, Bengtsson J, Walker B, Norberg J (2003) Response diversity, ecosystem change, and resilience. Front Ecol Environ 1(9):488–494. https://doi.org/10.1890/1540-9295(2003)001[0488:RDECAR]2.0.CO;2

Elsey-Quirk T, Unger V (2018) Geomorphic influences on the contribution of vegetation to soil C accumulation and accretion in Spartina alterniflora marshes. Biogeosciences 15(1):379–397. https://doi.org/10.5194/bg-15-379-2018

Esteban M, Valdez J, Tan N, Rica A, Vasquez G, Jamero L, Shibayama T et al (2023) Field survey of 2021 Typhoon Rai—Odette—in the Philippines. J Coast Riverine Flood Risk 1:1–19. https://doi.org/10.48438/jcrfr.2023.0001

Fagherazzi S, Bryan KR, Nardin W (2017) Buried alive or washed away: the challenging life of mangroves in the Mekong Delta. Oceanography 30(3):48–59. https://doi.org/10.5670/oceanog.2017.313

Folke C, Carpenter SR, Walker B, Scheffer M, Chapin T, Rockström J (2010) Resilience thinking: integrating resilience, adaptability and transformability. Ecol Soc 15(4):20

French PW (2006) Managed realignment—the developing story of a comparatively new approach to soft engineering. Estuar Coast Shelf Sci 67(3):409–423. https://doi.org/10.1016/j.ecss.2005.11.035

Friess DA, Webb EL (2014) Variability in mangrove change estimates and implications for the assessment of ecosystem service provision. Glob Ecol Biogeogr 23(7):715–725. https://doi.org/10.1111/geb.12140

Gerona-Daga MEB, Salmo SG III (2022) A systematic review of mangrove restoration studies in Southeast Asia: challenges and opportunities for the United Nation's decade on ecosystem restoration. Front Mar Sci 9:987737. https://doi.org/10.3389/fmars.2022.987737

Gijsman R, Horstman EM, Friess DA, Swales A, Wijnberg KM (2021) Nature-based engineering: a review on reducing coastal flood risk with mangroves. Front Mar Sci 8:702412. https://doi.org/10.3389/fmars.2021.702412

Gómez Martín E, Giordano R, Pagano A, Van der Keur P, Máñez Costa M (2020) Using a system thinking approach to assess the contribution of nature based solutions to sustainable development goals. Sci Total Environ 738:139693. https://doi.org/10.1016/j.scitotenv.2020.139693

Gourgue O, van Belzen J, Schwarz C, Vandenbruwaene W, Vanlede J, Belliard J-P, Fagherazzi S, Bouma TJ, van de Koppel J, Temmerman S (2022) Biogeomorphic modeling to assess the resilience of tidal-marsh restoration to sea level rise and sediment supply. Earth Surf Dyn 10:531–553. https://doi.org/10.5194/esurf-10-531-2022

Gregorio N, Herbohn J, Harrison S, Pasa A, Ferraren A (2017) Regulating the quality of seedlings for forest restoration: lessons from the National Greening Program in the Philippines. Small-Scale For 16(1):83–102. https://doi.org/10.1007/s11842-016-9344-z

Guo M, Li J, Sheng C, Xu J, Wu L (2017) A review of wetland remote sensing. Sensors 17(4):777. https://doi.org/10.3390/s17040777

Guo Z, Wei MY, Zhong YH, Wu X, Chi BJ, Li J, Zheng HL et al (2023) Leaf sodium homeostasis controlled by salt gland is associated with salt tolerance in mangrove plant Avicennia marina. Tree Physiol 43(5):817–831. https://doi.org/10.1093/treephys/tpad002

Hai NT, Dell B, Phuong VT, Harper RJ (2020) Towards a more robust approach for the restoration of mangroves in Vietnam. Ann For Sci 77(1):18. https://doi.org/10.1007/s13595-020-0921-0

Hernandez JO, Naeem M, Zaman W (2023) How does changing environment influence plant seed movements as populations of dispersal vectors decline? Plants 12(7):1462. https://doi.org/10.3390/plants12071462

Hernández-Delgado EA (2024) Coastal restoration challenges and strategies for small island developing states in the face of sea level rise and climate change. Coasts 4(2):235–286. https://doi.org/10.3390/coasts4020014

Huynh LT, Su J, Wang Q, Stringer LC, Switzer AD, Gasparatos A (2024) Meta-analysis indicates better climate adaptation and mitigation performance of hybrid engineering-natural coastal defence measures. Nat Commun 15(1):2870. https://doi.org/10.1038/s41467-024-46970-w

Infantes E, Hoeks S, Adams MP, van der Heide T, van Katwijk MM, Bouma TJ (2022) Seagrass roots strongly reduce cliff erosion rates in sandy sediments. Mar Ecol Prog Ser 700:1–12. https://doi.org/10.3354/meps14196

Intergovernmental Panel on Climate Change (IPCC) (2021) Climate change 2021: the physical science basis. contribution of working group I to the sixth assessment report of the intergovernmental panel on climate change. Cambridge University Press. https://doi.org/10.1017/9781009157896

Intergovernmental Panel on Climate Change (IPCC) (2022) Chapter 3: Oceans and coastal ecosystems and their services. In: Climate change 2022: impacts, adaptation and vulnerability (working group II). Cambridge University Press

Ishtiaq KS, Troxler TG, Lamb-Wotton L, Wilson BJ, Charles SP, Davis SE, Kominoski JS, Rudnick DT, Sklar FH (2022) Modeling net ecosystem carbon balance and loss in coastal wetlands exposed to sea-level rise and saltwater intrusion. Ecol Appl 32(8):e2702. https://doi.org/10.1002/eap.2702

Islam S, Anzum F, Kamal AB, Akter M, Rezaie AM (2021) Policy analysis: the integration of nature-based solutions into climate adaptation policy and planning in Bangladesh. J Sci Policy Gov 18(02). https://doi.org/10.38126/JSPG18020

Jiba W, Manyevere A, Mashamaite CV (2024) The impact of ecological restoration on soil quality in humid region forest habitats: a systematic review. Forests 15(11):1941. https://doi.org/10.3390/f15111941

Johnson BA, Magcale-Macandog DM, Kumar P, Dasgupta R, Kawai M (2021) Participatory coastal land-use management (PCLM) guidebook: participatory approaches and geospatial modeling for increased resilience to climate change at the local level. Institute for Global Environmental Strategies

Kahnouj VF, Rezai M, Mahdavi R, Moslehi M, Eskandari S (2023) Effects of dieback on the vegetative, chemical, and physiological status of mangrove forests, Iran. J Arid Land 15(11):1391–1404. https://doi.org/10.1007/s40333-023-0031-6

Kairo JG, Hamza AJ, Wanjiru C (2018) Mikoko Pamoja: a demonstrably effective community-based blue carbon project in Kenya. In: A blue carbon primer. CRC Press, pp 341–350

Kasten P, Jenkins SR, Christofoletti RA (2021) Participatory monitoring—a citizen science approach for coastal environments. Front Mar Sci 8:681969. https://doi.org/10.3389/fmars.2021.681969

Kazemi A, Castillo L, Curet OM (2021) Mangrove roots model suggest an optimal porosity to prevent erosion. Sci Rep 11(1):1–14. https://doi.org/10.1038/s41598-021-88119-5

Koh LP, Wich SA (2012) Dawn of drone ecology: low-cost autonomous aerial vehicles for conservation. Trop Conserv Sci 5(2):121–132. https://doi.org/10.1177/194008291200500202

Konefal A, Kirkland A, Gilpin R, Wyssmann K, Anthony NM, Cebrian J, Cox TE (2024) The relationship between genetic diversity, function, and stability in marine foundation species. Bioscience 74(3):187–206. https://doi.org/10.1093/biosci/biad123

Kropf CM, Vaterlaus L, Bresch DN, Pellissier L (2024) Tropical cyclone risk for global ecosystems in a changing climate. Nat Clim Change 15(1):92–100. https://doi.org/10.1038/s41558-024-02194-w

Kusumadewi SD, Purnomo H, Nadhira S, Putrizulfan R (2024) Systematic review on the implementation of mangrove community-based restoration in Indonesia and beyond. IOP Conf Ser Earth Environ Sci 1315(1):012052. https://doi.org/10.1088/1755-1315/1315/1/012052

Lara RJ, Cohen MCL (2006) Sediment porewater salinity, inundation frequency and mangrove vegetation height in Bragança, North Brazil: an ecohydrology-based empirical model. Wetl Ecol Manag 14:349–358. https://doi.org/10.1007/s11273-005-4991-4

Lee J, Yen L, Chu M, Lin Y, Chang C, Lin R, Chao K, Lee M (2020) Growth characteristics and anti-wind erosion ability of three tropical foredune pioneer species for sand dune stabilization. Sustainability 12(8):3353. https://doi.org/10.3390/su12083353

Lee H, Calvin K, Dasgupta D, Krinner G, Mukherji A, Thorne P, Park Y et al (2023) IPCC, 2023: climate change 2023: synthesis report, summary for policymakers. Contribution of working groups I, II and III to the sixth assessment report of the Intergovernmental Panel on Climate Change. IPCC, Geneva

Long J, Giri C, Primavera J, Trivedi M (2016) Damage and recovery assessment of the Philippines' mangroves following Super Typhoon Haiyan. Mar Pollut Bull 109(2):734–743. https://doi.org/10.1016/j.marpolbul.2016.06.080

Long Z, Zhou J, Feng S (2025) The climate mitigation effect of legal institutions: environmental courts and urban carbon balance. Front Environ Sci 13:1622148. https://doi.org/10.3389/fenvs.2025.1622148

Macreadie PI, Costa MD, Atwood TB, Friess DA, Kelleway JJ, Kennedy H, Lovelock CE, Serrano O, Duarte CM (2021) Blue carbon as a natural climate solution. Nat Rev Earth Environ 2(12):826–839. https://doi.org/10.1038/s43017-021-00224-1

Marzouk M, Azab S (2024) Modeling climate change adaptation for sustainable coastal zones using GIS and AHP. Environ Monit Assess 196(2):147. https://doi.org/10.1007/s10661-023-12287-2

Mazda Y, Magi M, Ikeda Y, Kurokawa T, Asano T (2006) Wave reduction in a mangrove forest dominated by Sonneratia sp. Wetl Ecol Manag 14(4):365–378. https://doi.org/10.1007/s11273-005-5388-0

McIvor AL, Spencer T, Möller I, Spalding M (2013) The response of mangrove soil surface elevation to sea level rise. Natural coastal protection series: report 3. Cambridge coastal research unit working paper 42. Published by The Nature Conservancy and Wetlands International. 59 pp. ISSN 2050-7941http://coastalresilience.org/science/mangroves/surface-elevation-and-sea-level-rise

Mendes R, Fidélis T, Roebeling P, Teles F (2019) The institutionalization of nature-based solutions—a discourse analysis of emergent literature. Resources 9(1):6. https://doi.org/10.3390/resources9010006

Menon N, George G, Ranith R, Sajin V, Murali S, Abdulaziz A, Brewin RJ, Sathyendranath S (2020) Citizen science tools reveal changes in estuarine water quality following demolition of buildings. Remote Sens 13(9):1683. https://doi.org/10.3390/rs13091683

Ministry of Environment, Forest and Climate Change (MoEFCC) (2025) Coastal regulation zone notification 2019 (with amendments). Government of India. https://envt.kerala.gov.in/wp-content/uploads/2025/03/CRZ-Notification-2019_with-Amendments_19.pdf. Accessed 02 Feb 2025

Ministry of Natural Resources and Environment (MONRE) (2023) Dual solutions for climate change response. MONRE Vietnam. https://en.monre.gov.vn/climate-change/dual-solutions-for-climate-change-response-8340.htm. Accessed 02 Feb 2025

Mohamad N, Ahmad A, Khanan MF, Din AH (2021) Surface elevation changes estimation underneath mangrove canopy using SNERL filtering algorithm and DoD technique on UAV-derived DSM data. ISPRS Int J Geo Inf 11(1):32. https://doi.org/10.3390/ijgi11010032

Mohamed Rashidi AH, Jamal MH, Hassan MZ, Mohd Sendek SS, Mohd Sopie SL, Abd Hamid MR (2021) Coastal structures as beach erosion control and sea level rise adaptation in Malaysia: a review. Water 13(13):1741. https://doi.org/10.3390/w13131741

Mori AS, Furukawa T, Sasaki T (2013) Response diversity determines the resilience of ecosystems to environmental change. Biol Rev 88(2):349–364. https://doi.org/10.1111/brv.12004

Mori N, Chang C, Inoue T, Akaji Y, Hinokidani K, Baba S, Takagi M, Mori S, Koike H, Miyauchi M, Suganuma R, Sabunas A, Miyashita T, Shimura T (2022) Parameterization of mangrove root structure of Rhizophora stylosa in coastal hydrodynamic model. Front Built Environ 7:782219. https://doi.org/10.3389/fbuil.2021.782219

Morita K, Matsumoto KI (2021) Governance challenges for implementing nature-based solutions in the Asian region. Polit Gov 9(4):102–113. https://doi.org/10.17645/pag.v9i4.4420

Morrissette HK, Baez SK, Beers L, Bood N, Martinez ND, Novelo K, Andrews G, Balan L, Beers CS, Betancourt SA, Blanco R, Bowden E, Burns-Perez V, Carcamo M, Chevez L, Crooks S, Feller IC, Galvez G, Garbutt K, Canty SW (2023) Belize blue carbon: establishing a national carbon stock estimate for mangrove ecosystems. Sci Total Environ 870:161829. https://doi.org/10.1016/j.scitotenv.2023.161829

Mudd SM, Howell SM, Morris JT (2009) Impact of dynamic feedbacks between sedimentation, sea-level rise, and biomass production on near-surface marsh stratigraphy and carbon accumulation. Estuar Coast Shelf Sci 82(3):377–389. https://doi.org/10.1016/j.ecss.2009.01.028

Narayan S, Beck MW, Reguero BG, Losada IJ, van Wesenbeeck B, Pontee N, Sanchirico JN, Ingram JC, Lange GM, Burks-Copes KA (2016) The effectiveness, costs and coastal protection benefits of natural and nature-based defences. PLoS ONE 11(5):e0154735. https://doi.org/10.1371/journal.pone.0154735

Naskar S, Palit PK (2015) Anatomical and physiological adaptations of mangroves. Wetl Ecol Manag 23(3):357–370. https://doi.org/10.1007/s11273-014-9385-z

Nguyen H, Harper R, Dell B (2023a) Examining local community understanding of mangrove carbon mitigation: a case study from Ca Mau province, Mekong River Delta, Vietnam. Mar Policy 148:105398. https://doi.org/10.1016/j.marpol.2022.105398

Nguyen TC, Le TPQ, Vo TT, Pham HV, Nguyen QD, Pham TTH (2023b) Water-level changes and subsidence rates along the East Sea coastline of the Saigon-Dong Nai River Estuary and the Mekong Delta. Vietnam J Earth Sci 45(1):1–17. https://doi.org/10.15625/0866-7187/17939

Nizam A, Meera SP, Kumar A (2021) Genetic and molecular mechanisms underlying mangrove adaptations to intertidal environments. Iscience 25(1):103547. https://doi.org/10.1016/j.isci.2021.103547

Nor SMM (2017) Physiological ecology of Malaysian mangroves in response to sea level rise. PhD thesis, University of Edinburgh, UK

Perri S, Detto M, Porporato A, Molini A (2023) Salinity-induced limits to mangrove canopy height. Glob Ecol Biogeogr 32(9):1561–1574. https://doi.org/10.1111/geb.13720

Perricone V, Mutalipassi M, Mele A, Buono M, Vicinanza D, Contestabile P (2023) Nature-based and bioinspired solutions for coastal protection: an overview among key ecosystems and a promising pathway for new functional and sustainable designs. ICES J Mar Sci 80(5):1218–1239. https://doi.org/10.1093/icesjms/fsad080

Pham TD, Yokoya N, Bui DT, Yoshino K, Friess DA (2019) Remote sensing approaches for monitoring mangrove species, structure, and biomass: opportunities and challenges. Remote Sens 11(3):230. https://doi.org/10.3390/rs11030230

Powell EJ, Tyrrell MC, Milliken A, Tirpak JM, Staudinger MD (2019) A review of coastal management approaches to support the integration of ecological and human community planning for climate change. J Coast Conserv 23(1):1–18. https://doi.org/10.1007/s11852-018-0632-y

Primavera JH, Yap WG, Savaris JP, Loma RJA, Moscoso ADE, Coching JD, Montilijao CL, Poingan RP, Tayo ID (2014) Manual on mangrove reversion of abandoned and illegal brackishwater fishponds—mangrove manual series No. 2. ZSL, London, pp xii + 108 pp

Pulhin JM, Ramirez MAM, Garcia JE, Pangilinan MJQ, Evaristo MBS, Catudio MRO, Magpantay AT, Tasico SL, Pulhin FB, Abes JL, Mariano BJS (2024) Contextualizing sustainable forest management and social justice in community-based forest management (CBFM) program in the Philippines. Trees For People 16:100589. https://doi.org/10.1016/j.tfp.2024.100589

Purandare J, Bayraktarov E, Boström-Einarsson L, Carnell PE, Eger AM, Port AL, Macreadie PI, Reeves SE, Waltham NJ, Wartman M, McLeod IM (2024) Database for marine and coastal restoration projects in Australia and New Zealand. Ecological Management & Restoration 25(1):14–20. https://doi.org/10.1111/emr.12596

Quartel S, Kroon A, Augustinus PG, Van Santen P, Tri NH (2007) Wave attenuation in coastal mangroves in the Red River Delta, Vietnam. J Asian Earth Sci 29(4):576–584. https://doi.org/10.1016/j.jseaes.2006.05.008

Rahman M, Nicholls RJ, Hanson SE, Salehin M, Alam S (2019) Integrated assessment for the Bangladesh Delta Plan 2100: analysis of selected interventions. University of Southampton, General Economic Division, Government of Bangladesh, BUET

Rahman MM, Ghosh T, Salehin M, Ghosh A, Haque A, Hossain MA, Hutton CW et al (2020) Ganges-Brahmaputra-Meghna delta, Bangladesh and India: a transnational mega-delta. In: Deltas in the anthropocene, pp 23–51. https://doi.org/10.1007/978-3-030-23517-8_2

Rivera BJ, Belone J, Mathew AR, Vega K, Mosqueda SN, Sommer DS, Maciel-Martinez PD (2025) How you dune-ing? A systematic review of coastal dune plant community assembly. J Coast Conserv 29(3):23. https://doi.org/10.1007/s11852-025-01107-z

Saintilan N, Khan NS, Ashe E, Kelleway JJ, Rogers K, Woodroffe CD, Horton BP (2020) Thresholds of mangrove survival under rapid sea level rise. Science 368(6495):1118–1121. https://doi.org/10.1126/science.aba2656

Schmitt K, Albers T, Pham TT et al (2013) Site-specific and integrated adaptation to climate change in the coastal mangrove zone of Soc Trang Province, Viet Nam. J Coast Conserv 17:545–558. https://doi.org/10.1007/s11852-013-0253-4

Schuerch M, Spencer T, Temmerman S, Kirwan ML, Wolff C, Lincke D, McOwen CJ, Pickering MD, Reef R, Vafeidis AT, Hinkel J, Nicholls RJ, Brown S (2018) Future response of global coastal wetlands to sea-level rise. Nature 561(7722):231–234. https://doi.org/10.1038/s41586-018-0476-5

Seddon AWR, Froyd CA, Leng MJ, Milne GA, Willis KJ (2011) Ecosystem resilience and threshold response in the Galápagos Coastal Zone. PLoS ONE 6(7):e22376. https://doi.org/10.1371/journal.pone.0022376

Sharma LK, Naik R (2024) Saline wetland ecosystems. In: Conservation of saline wetland ecosystems: An initiative towards UN decade of ecological restoration. Springer Nature Singapore, Singapore, pp 33–75. https://doi.org/10.1007/978-981-97-5069-6_2

Smoak JM, Breithaupt JL, Smith TJ, Sanders CJ (2013) Sediment accretion and organic carbon burial relative to sea-level rise and storm events in two mangrove forests in Everglades National Park. CATENA 104:58–66. https://doi.org/10.1016/j.catena.2012.10.009

Stori FT, Peres CM, Turra A, Pressey RL (2019) Traditional ecological knowledge supports ecosystem-based management in disturbed coastal marine social-ecological systems. Front Mar Sci 6:452035. https://doi.org/10.3389/fmars.2019.00571

Su BY, Wang YS, Sun CC (2022) Response adaptive mechanisms of three mangrove (Avicennia marina, Aegiceras corniculatum, and Bruguiera gymnorrhiza) plants to waterlogging stress

revealed by transcriptome analysis. Front Mar Sci 9:929649. https://doi.org/10.3389/fmars.2022.929649
Taillardat P, Friess DA, Lupascu M (2018) Mangrove blue carbon strategies for climate change mitigation are most effective at the national scale. Biol Lett 14(10):20180251. https://doi.org/10.1098/rsbl.2018.0251
Tamura T, Nguyen VL, Ta TKO, Bateman MD, Gugliotta M, Anthony EJ, Saito Y et al (2020) Long-term sediment decline causes ongoing shrinkage of the Mekong megadelta, Vietnam. Sci Rep 10(1):8085. https://doi.org/10.1038/s41598-020-64630-z
Tanaka N (2009) Vegetation bioshields for tsunami mitigation: review of effectiveness, limitations, construction, and sustainable management. Landsc Ecol Eng 5:71–79. https://doi.org/10.1007/s11355-008-0058-z
Temmerman S, Meire P, Bouma TJ, Herman PMJ, Ysebaert T, De Vriend HJ (2013) Ecosystem-based coastal defence in the face of global change. Nature 504(7478):79–83. https://doi.org/10.1038/nature12859
Thompson BS, Primavera JH, Friess DA (2017) Governance and implementation challenges for mangrove forest Payments for Ecosystem Services (PES): empirical evidence from the Philippines. Ecosyst Serv 23:146–155. https://doi.org/10.1016/j.ecoser.2016.12.007
Tomlinson PB (2016) The botany of mangroves. Cambridge University Press, p 418
Trân N, Diễm P, Tín N, Kim T (2024) The trend of salinity changes at coastal stations in The Mekong Delta during the period 1996–2018. VNUHCM J Earth Sci Environ 8(2):988–1002. https://doi.org/10.32508/stdjsee.v8i2.771
Trang BTQ, Trinh NTH (2025) Exploring the potential for climate change litigation in Vietnam: a forward-looking assessment. Chin J Trans Law 2(2). https://doi.org/10.1177/2753412X251357715
Truong SH, Ye Q, Stive MJ (2017) Estuarine mangrove squeeze in the Mekong Delta, Vietnam. J Coast Res 33(4):747–763. https://doi.org/10.2112/JCOASTRES-D-16-00087.1
Truong QC, Drogoul A, Gaudou B, Taillandier P, Huynh NQ, Nguyen TH, Minderhoud P, Espagne E (2022) An agent-based model for land-use change adaptation strategies in the context of climate change and land subsidence in the Mekong Delta. Sustainability 15(6):5355. https://doi.org/10.3390/su15065355
Tun KZ, Pramanik M, Chakrabortty R, Chowdhury K, Halder B, Pande CB, Zhran M et al (2025) Mainstreaming nature-based solutions for climate adaptation in Southeast Asia: a systematic review. Earth Syst Environ 9(2):1333–1351. https://doi.org/10.1007/s41748-024-00473-2
Turki EI, Ojeda E, Mendoza ET, Medellín G, Price TD, Salameh E, Wang XH, Li L, Franklin GL (2025) Monitoring resilience in coastal systems: a comprehensive assessment. J Mar Sci Eng 13(11):2113. https://doi.org/10.3390/jmse13112113
Twomey AJ, Lovelock CE (2025) Variation in mangrove geometric traits among genera and climate zones. Estuar Coasts 48:55. https://doi.org/10.1007/s12237-025-01487-3
Valente S, Pinho P (2025) Adaptive planning approaches for coastal climate adaptation: process and key-elements. Environ Manag 75:1013–1038. https://doi.org/10.1007/s00267-025-02117-1
Valenzuela BR, Yeo-Chang Y, Park MS, Chun JN (2020) Local people's participation in mangrove restoration projects and impacts on social capital and livelihood: a case study in the Philippines. Forests 11(5):580. https://doi.org/10.3390/f11050580
Van der Stocken T, Wee AK, De Ryck DJ, Vanschoenwinkel B, Friess DA, Dahdouh-Guebas F, Webb EL et al (2019) A general framework for propagule dispersal in mangroves. Biol Rev 94(4):1547–1575. https://doi.org/10.1111/brv.12514
Van Loon AF, Te Brake B, Van Huijgevoort MHJ, Dijksma R (2016) Hydrological classification, a practical tool for mangrove restoration. PLoS ONE 11(3):e0150302. https://doi.org/10.1371/journal.pone.0150302
Vanderklift MA, Doropoulos C, Gorman D, Leal I, Minne AJ, Statton J, Steven AD, Wernberg T (2020) Using propagules to restore coastal marine ecosystems. Front Mar Sci 7:565403. https://doi.org/10.3389/fmars.2020.00724

Veettil BK, Puri V (2025) Blue carbon ecosystems in Vietnam: a review. Ocean Coast Manag 267:107711. https://doi.org/10.1016/j.ocecoaman.2025.107711

Veettil BK, Quang NX, Thu Trang NT (2019a) Changes in mangrove vegetation, aquaculture and paddy cultivation in the Mekong Delta: a study from Ben Tre Province, southern Vietnam. Estuar Coast Shelf Sci 226:106273. https://doi.org/10.1016/j.ecss.2019.106273

Veettil BK, Ward RD, Quang NX, Trang NTT, Giang TH (2019b) Mangroves of Vietnam: historical development, current state of research and future threats. Estuar Coast Shelf Sci 218:212–236. https://doi.org/10.1016/j.ecss.2018.12.021

Veettil BK, Costi J, Marques WC, Tran X, Quang NX, Van DD, Hoai PN (2020) Coastal environmental changes in Southeast Asia: a study from Quang Nam Province, Central Vietnam. Reg Stud Mar Sci 39:101420. https://doi.org/10.1016/j.rsma.2020.101420

Veettil BK, Ward RD, Dung NTK, Van DD, Quang NX, Hoai PN, Hoang N (2021) The use of bioshields for coastal protection in Vietnam: current status and potential. Reg Stud Mar Sci 47:101945. https://doi.org/10.1016/j.rsma.2021.101945

Vinh VD (2015) Applying numerical models to assess role of mangrove forests in wave attenuation in Hai Phong coastal area. Vietnam J Mar Sci Technol 15(1):67–76. https://doi.org/10.15625/1859-3097/15/1/6082

Vu TP, Pham TT, Tran NMH, Nguyen TTA, Nguyen TVA, Tang TKH (2023) The blue carbon market in Vietnam: future development, potential and challenges. Occasional Paper 8. Center for International Forestry Research (CIFOR), Bogor, Indonesia; and World Agroforestry (ICRAF), Nairobi, Kenya

Wetlands International (2020) Accelerating adaptation through building with nature in Asia. Wetlands International. https://www.wetlands.org/case-study/building-with-nature-in-asia/. Accessed 02 Feb 2025

White E, Kaplan D (2017) Restore or retreat? Saltwater intrusion and water management in coastal wetlands. Ecosyst Health Sustain 3(1):e01258. https://doi.org/10.1002/ehs2.1258

Willcock S, Hooftman DA, Balbi S, Blanchard R, Dawson TP, O'Farrell PJ, Bullock JM et al (2019) A continental-scale validation of ecosystem service models. Ecosystems 22(8):1902–1917. https://doi.org/10.1007/s10021-019-00380-y

Willemsen P, Horstman E, Borsje B, Friess D, Dohmen-Janssen C (2016) Sensitivity of the sediment trapping capacity of an estuarine mangrove forest. Geomorphology 273:189–201. https://doi.org/10.1016/j.geomorph.2016.07.038

Wimmler MC, Bathmann J, Peters R, Jiang J, Walther M, Lovelock CE, Berger U (2021) Plant–soil feedbacks in mangrove ecosystems: establishing links between empirical and modelling studies. Trees 35(5):1423–1438. https://doi.org/10.1007/s00468-021-02182-z

World Bank (2016) Managing coasts with natural solutions: Guidelines for measuring and valuing the coastal protection services of mangroves and coral reefs. In: Beck MW, Lange G-M (eds) Wealth accounting and the valuation of ecosystem services partnership (WAVES). World Bank, Washington, DC

Wright LD, Thom BG (2023) Coastal morphodynamics and climate change: a review of recent advances. J Mar Sci Eng 11(10):1997. https://doi.org/10.3390/jmse11101997

Xin P, Wilson A, Shen C, Ge Z, Moffett KB, Santos IR, Barry DA et al (2022) Surface water and groundwater interactions in salt marshes and their impact on plant ecology and coastal biogeochemistry. Rev Geophys 60(1):e2021RG000740. https://doi.org/10.1029/2021RG000740

Yachi S, Loreau M (1999) Biodiversity and ecosystem productivity in a fluctuating environment: the insurance hypothesis. Proc Natl Acad Sci USA 96(4):1463. https://doi.org/10.1073/pnas.96.4.1463

Zhang J, Jiang Z, Li D, Zhou P (2025) Mangrove vertical soil accretion and potential risk—resilience assessment of sea-level rise in the Beilun Estuary and Guangxi Coastal Zone, China. Sustainability 17(18):8099. https://doi.org/10.3390/su17188099

Zhou C, Wong K, Zhao J (2018) Coastal wetland vegetation in response to global warming and climate change. In: Sea level rise and coastal infrastructure, pp 61–81. https://doi.org/10.5772/intechopen.73509

Part III
Design, Implementation, and Management of Coastal Bioshields

Chapter 6
Principles of Coastal Bioshield Design

Abstract The effective design of bioshields requires a comprehensive understanding of ecological, hydrodynamic, and geomorphological principles that govern coastal processes. This chapter outlines the scientific and engineering foundations of bioshield design, emphasizing the integration of ecosystem functionality with structural protection objectives. It explores key design parameters such as vegetation type, species composition, planting density, zonation, and spatial configuration relative to tidal regimes and shoreline geomorphology. By analyzing how different plant morphologies influence wave energy dissipation, sediment stabilization, and erosion control, the chapter establishes the importance of site-specific design approaches tailored to local environmental conditions. The discussion highlights the role of multidisciplinary assessment—combining ecological surveys, hydrodynamic modeling, and risk-based planning—to ensure bioshields achieve optimal performance under variable climatic and anthropogenic pressures. Case studies from Asia and other tropical regions illustrate successful design frameworks that integrate traditional ecological knowledge with modern restoration techniques. Attention is also given to scalability, maintenance, and monitoring as critical components of long-term bioshield efficiency. This chapter provides a practical blueprint for designing resilient and sustainable bioshields by synthesizing principles from coastal engineering, landscape ecology, and environmental management. It underscores that successful bioshield design must balance protection, biodiversity enhancement, and socioeconomic co-benefits to serve as a holistic nature-based solution for coastal resilience.

Keywords Bioshield design principles · Ecosystem-based engineering · Hydrodynamic modeling · Site-specific restoration · Vegetation zonation

B. Kozhikkodan Veettil, *The Potential of Bioshields for Coastal Protection*, Advances in Natural and Technological Hazards Research 55,
https://doi.org/10.1007/978-3-032-22888-8_6

6.1 Introduction

The design of bioshields, which are the natural vegetative barriers established along coastal zones, represents a cornerstone in contemporary coastal ecosystem management and climate adaptation planning. Bioshields integrate ecological integrity with engineering functionality, serving as dynamic and self-sustaining defenses against coastal hazards such as storm surges, cyclones, and erosion (Veettil et al. 2021; Perricone et al. 2023). Unlike hard-engineering structures such as seawalls, groynes, or breakwaters, which impose static resistance to coastal forces, bioshields operate as living infrastructures, evolving, regenerating, and adapting in response to changing environmental and socio-economic conditions (Narayan et al. 2016; Powell et al. 2019). The concept emphasizes harmony between natural processes and human interventions, creating a resilient buffer that protects both livelihoods and ecosystems.

Designing effective bioshields demands a multidisciplinary understanding that bridges ecology, geomorphology, hydrodynamics, and social planning (Perricone et al. 2023). Ecologists focus on species selection and community structure (Fernández-Pascual et al. 2017); geomorphologists assess sediment dynamics and coastal morphology (Feng et al. 2023); hydrologists analyze tidal regimes, wave energy, and salinity gradients (Parkinson et al. 2024); and social scientists evaluate community needs, land-use practices, and governance frameworks (Powell et al. 2019; Wu and Barret 2022). Integrating these domains ensures that bioshields are not only biologically viable but also physically stable and socially acceptable. Failure in any one of these dimensions can compromise the entire system, leading to reduced ecological performance or even total collapse.

Bioshields are fundamentally ecosystem-based solutions that employ natural processes to achieve both human and ecological objectives simultaneously (Perricone et al. 2023). When properly designed, they attenuate wave energy, trap sediments, reduce shoreline retreat, and provide habitat for diverse flora and fauna (Veettil et al. 2021). As mentioned in previous chapters, bioshields contribute to carbon sequestration, nutrient cycling, and the maintenance of fisheries productivity. Their multifunctionality distinguishes them from conventional coastal protection measures, offering a spectrum of co-benefits that enhance overall coastal resilience (Veettil et al. 2021; Huynh et al. 2024). However, realizing these benefits requires an understanding of the underlying ecological principles and physical processes that govern coastal systems.

Across tropical and subtropical Asia, particularly in Vietnam, Bangladesh, India, Indonesia, and the Philippines, bioshields (e.g., mangroves, casuarinas, and other salt-tolerant species) were planted extensively along shorelines, river mouths, and estuarine fringes to promote them as cost-effective and sustainable alternatives to hard infrastructure (Veettil et al. 2021; Juanico 2022; Damastuti et al. 2022). These initiatives aimed to create vegetative barriers capable of reducing storm surge impact, stabilizing sediments, and protecting inland settlements. Nevertheless, the outcomes of these projects have been mixed. While some succeeded in restoring ecological function and reducing disaster vulnerability, others failed to establish sustainable vegetation cover or deliver tangible protective benefits. A critical reason for these mixed outcomes lies in the deficiencies of design. Many early bioshield projects were conceived as reforestation or plantation programs rather than as integrated ecological-engineering systems (Vannucci 2004). Monocultures of fast-growing but ecologically unsuitable species were often planted without adequate consideration of local hydrological conditions or sediment dynamics (Veettil et al. 2018, 2019a). Inappropriate site selection, such as planting mangroves on sandy beaches or high-energy shorelines, where natural establishment is impossible, led to widespread mortality (Hai et al. 2020; Mori et al. 2025). Similarly, inadequate spacing, poor soil preparation, or lack of community participation resulted in limited long-term survival and functionality (Nguyen et al. 2016; Hai et al. 2020; Phong and Luom 2021; Damastuti et al. 2022). These experiences underscore that bioshields cannot be designed using a one-size-fits-all approach; rather, they must be tailored to site-specific environmental, biophysical, and socio-economic contexts.

Establishing robust design principles is therefore essential to ensure that bioshields fulfill both their ecological and protective functions (Lewis 2005; Charoenlerkthawin et al. 2024). A sound coastal bioshield design process begins with a comprehensive site analysis that evaluates geomorphological setting, tidal amplitude, wave climate, sediment characteristics, and salinity regimes (Rudianto et al. 2022; Perri et al. 2023). Understanding these parameters is crucial for identifying suitable vegetation zones varying from pioneer salt-tolerant grasses and mangrove fringes in intertidal flats to backshore tree species in higher elevations. Equally important is species selection, which should prioritize native species adapted to local conditions and capable of withstanding periodic inundation, salinity fluctuations, and storm exposure. Mixed-species configurations generally offer greater ecological resilience and structural diversity than monocultures, reducing the risk of large-scale dieback from pests, diseases, or climatic extremes (He et al. 2021; Edwards and Cerullo 2024).

Another key component of bioshield design is functional zoning to achieve multi-layer coastal protection (Syahid et al. 2023). This involves the spatial arrangement of different vegetation types (i.e., site-specific) according to their ecological tolerances and protective roles (Schmitt et al. 2013). For example, seaward zones may consist of dense mangrove stands that dissipate wave energy, followed by mid-zones dominated by salt-tolerant shrubs that stabilize sediments, and landward zones comprising larger trees that serve as windbreaks and provide shade or timber. This stratification creates a multi-layered coastal defense system, distributing stress across various vegetative components and enhancing overall system robustness.

Equally vital is the spatial configuration of the bioshield—its width, density, and orientation relative to dominant wind and wave directions. Research indicates that wider and denser bioshields generally offer greater wave attenuation (Schmitt et al. 2013), but excessive planting density can hinder growth and natural regeneration (Van Wesenbeeck et al. 2025). Optimal configuration thus depends on balancing protective efficiency with ecological sustainability. In some contexts, alternating vegetated and open corridors may improve hydrological connectivity and facilitate fish migration or sediment exchange (Sheaves 2005; McLachlan et al. 2020). Design must also consider the potential impacts of sea-level rise and shoreline migration, allowing room for landward movement of vegetation zones–a concept known as "coastal squeeze mitigation" (Chávez et al. 2021).

Beyond the biophysical parameters, the social and institutional dimensions of bioshield design are equally crucial. Coastal communities are not passive beneficiaries but active stewards whose knowledge, participation, and long-term commitment are critical to project success (Damastuti et al. 2022; Risley 2022). Traditional ecological knowledge, such as local understanding of tidal rhythms, soil salinity, and species performance, can provide invaluable insights for site selection and management (Damastuti et al. 2022; Ravaoarinorotsihoarana et al. 2023). Moreover, participatory planning fosters local ownership, ensuring that bioshields are maintained beyond initial project funding cycles. Integrating coastal bioshield design with livelihood opportunities (e.g., sustainable aqua-silviculture, eco-tourism, or honey production) can further enhance community support and economic sustainability.

In the context of climate change adaptation, bioshields serve as both mitigation and adaptation tools. By sequestering significant amounts of blue carbon, particularly in mangrove soils, they contribute to national greenhouse gas reduction targets (Zeng et al. 2021; Choudhary et al. 2024). Simultaneously, they reduce the vulnerability of coastal populations to sea-level rise, storm surges, and saline intrusion—key threats identified in the national adaptation plans of many Asian countries (Blankespoor et al. 2017; Hochard et al. 2021; Hernández-Delgado 2024). For instance, Vietnam's Nationally Determined Contributions (NDCs) explicitly emphasize mangrove restoration and bioshield establishment as priority adaptation strategies in the Mekong and Red River deltas (UNFCCC 2022). Designing these systems effectively, therefore, carries implications not only for local resilience but also for achieving broader global climate objectives.

Finally, the design of coastal bioshields must embrace adaptive planning. Coastal environments are inherently dynamic, characterized by continuous feedback among hydrology, sediment transport, and vegetation growth (Veettil et al. 2021). Rigid or static designs risk becoming outdated as environmental conditions shift. Instead, bioshields should be managed as evolving systems, monitored and adjusted through iterative learning (Ellison et al. 2020). Incorporating flexibility in species composition, zoning, and maintenance schedules allows for adjustment to emerging challenges such as increased salinity, invasive species, or altered tidal regimes (Xing et al. 2025). Continuous scientific monitoring coupled with community-based observation can inform timely interventions and ensure long-term functionality.

In summary, the design of bioshields constitutes an integrated synthesis of scientific understanding, engineering practice, and participatory social processes. Advancing beyond narrow, tree-planting–oriented approaches requires the appreciation of ecological dynamics, geomorphic contexts, and human dimensions. This chapter examines the core scientific, structural, and socio-ecological principles that inform effective bioshield design, with particular attention to site assessment, species selection, functional zoning, spatial configuration, and adaptive management. By clarifying these foundational elements, it seeks to support the development of resilient, multifunctional coastal bioshields capable of sustaining both human well-being and ecological integrity amid accelerating climatic change.

6.2 Functional Objectives of Bioshields

The functional objectives of coastal bioshields can be classified broadly into three categories: protection and disaster reduction (Powell et al. 2019), ecosystem services (Khan et al. 2025), and social and aesthetic functions (Garmaeepour et al. 2025).

6.2.1 *Protection and Risk Reduction*

A central objective of bioshield design is the provision of reliable, nature-based coastal protection. Vegetative barriers, composed of species such as mangroves, casuarina, nipa palm, or mixed coastal woodlands, function as living buffers that dissipate hydrodynamic forces (Powell et al. 2019; Veettil et al. 2021; Daro Justine and Seenath 2025). By increasing surface roughness, enhancing drag, and promoting sediment accretion, bioshields can attenuate wave energy, reduce wind velocities, and stabilize vulnerable shorelines (Quang Bao 2011; Kamil et al. 2021; Zhang et al. 2022a; Jayson-Quashigah et al. 2025a). Numerous empirical and modelling studies have demonstrated that well-structured bioshields can reduce the height and force of storm surges and tsunami waves by substantial margins, with reductions commonly ranging between 30 and 90% depending on stand width, vegetation density, structural complexity, and local geomorphology (e.g., Doan 2012; Zhang et al. 2022a; Jayson-Quashigah et al. 2025a; Tien et al. 2025; Van Wesenbeeck et al. 2025). These effects are not uniform; rather, they emerge from context-specific interactions between species traits, tidal regimes, sediment characteristics, and landscape configuration.

6.2.2 *Ecosystem Services*

While physical protection remains the most recognized function of bioshields, their ecological contributions are equally fundamental. Coastal vegetative systems deliver a wide array of ecosystem services that extend beyond hazard mitigation (Khan et al. 2025). As highly productive environments, they play a significant role in carbon sequestration and storage, thereby contributing to climate change mitigation goals (Choudhary et al. 2024). Their complex structures provide habitats for diverse assemblages of fish, crustaceans, birds, and invertebrates, supporting both biodiversity conservation and local food security (Macintosh and Ashton 2002; Veettil et al. 2018, 2019a; Zu Ermgassen et al. 2025). Nutrient cycling, shoreline stabilization, and improvements in water quality are additional benefits that arise from healthy bioshield systems (Veettil et al. 2019a; Li et al. 2025a). Designing bioshields with ecosystem service optimization in mind requires attention to ecological heterogeneity, species complementarity, and long-term successional trajectories. A narrowly engineered bioshield focused solely on wave attenuation may fall short in delivering broader ecological functions, whereas a multi-layered system with varied species composition and structural diversity can generate co-benefits across ecological, economic, and social domains (Li et al. 2025b; Xie et al. 2025).

In numerous coastal areas, including parts of Vietnam's deltaic provinces, bioshields have been co-designed with aquaculture ponds, agroforestry systems, and eco-tourism initiatives (Veettil et al. 2019a, b). Such arrangements reflect the importance of aligning bioshield design with local development aspirations and land-use practices. Multifunctional planning can enhance stakeholder buy-in, reduce land-use conflicts, and promote equitable benefit-sharing. Incorporating accessible pathways, viewpoints, or designated community spaces can further enhance recreational and educational values without compromising protective functions (Xiong et al. 2024; Dulyakasem et al. 2026; Ebeler et al. 2025).

6.2.3 *Social and Aesthetic Functions*

The social dimensions of bioshields are critical to their long-term sustainability. Coastal vegetation often holds cultural, spiritual, or symbolic significance for local communities, influencing public acceptance and stewardship (UNEP 2014; Dookie 2025; Garmaeepour et al. 2025; Sobhani et al. 2025). Moreover, bioshields contribute to the aesthetic value of coastal landscapes, improving scenic quality and recreational opportunities. These attributes can strengthen community support, facilitate participatory management, and generate economic benefits when integrated with compatible land uses (UNEP 2014). The success of a coastal bioshield depends on its ability to function as a socially legitimate and ecologically robust system. Recognizing social acceptability, cultural relevance, and aesthetic contributions as core design criteria, and not peripheral considerations, ensures that bioshields are embraced, maintained, and adapted by the communities they are intended to protect.

6.3 Characterization and Design Constraints of Bioshield Sites

Designing an effective coastal bioshield begins with a comprehensive understanding of the site's biophysical conditions and socio-cultural context. Each coastal setting presents a unique interplay of geomorphological processes, hydrodynamic forces, ecological conditions, and human activities (Alcántara and Goudie 2010; Binh et al. 2019; Veettil et al. 2021). A bioshield design that fails to recognize these complexities risks ecological failure, poor survival rates, or even social conflict over land use. Therefore, site characterization, which is the systematic assessment of environmental and socioeconomic parameters, forms the scientific foundation for all subsequent design decisions. This section elaborates on three key dimensions of site characterization and associated design constraints: (i) geomorphological and hydrodynamic setting, (ii) hydrology and salinity gradients, and (iii) socioeconomic and land-use factors. Site characterization and recognition of design constraints form the scientific and social foundation of bioshield planning. By systematically evaluating geomorphological settings, hydrological conditions, and socioeconomic realities (Le et al. 2022; Veettil et al. 2022; Yan et al. 2025), designers can align ecological potential with human priorities, ensuring that bioshields evolve as resilient, multifunctional components of the coastal landscape.

6.3.1 Geomorphological and Hydrodynamic Setting

The geomorphological context of a coastal zone determines its stability, sediment transport regime, and capacity to support vegetation (Jackson et al. 2013; Ciarletta et al. 2019). Coastal morphology varies widely from extensive tidal mudflats and estuarine deltas to sandy beaches and rocky shorelines, each posing distinct opportunities and limitations for bioshield establishment. Geomorphology and hydrodynamics define the fundamental design envelope for bioshields. A mismatch between vegetation type and coastal setting is among the most common reasons for bioshield failure (Veettil et al. 2021). Understanding the physical template (using elevation data, sediment sampling, and hydrodynamic modeling) of the coast is therefore a prerequisite for selecting appropriate vegetation types and structural arrangements.

Mudflats and Deltaic Plains

Mudflats and deltaic coasts, such as those found in the Mekong and Red River Deltas of Vietnam, the Ganges–Brahmaputra Delta of Bangladesh and India, and the Ayeyarwady Delta of Myanmar, represent ideal environments for bioshield establishment. These low-gradient zones are characterized by high sediment availability, gentle tidal slopes, and dynamic accretion–erosion processes (Bird 2008; Shi et al. 2017; Xie et al. 2018; Xing et al. 2022). In such areas, mangrove ecosystems, particularly species like *Avicennia alba*, *Rhizophora apiculata*, and *Sonneratia caseolaris*, play a dual role: they stabilize sediments through complex root systems and simultaneously promote further accretion by trapping suspended particles (Kathiresan 2003; Hongwise et al. 2022). The resulting positive feedback between vegetation and sedimentation contributes to vertical resilience, enabling these systems to keep pace with moderate sea-level rise (Zhang et al. 2025a).

However, not all deltaic environments are uniformly suitable for coastal bioshield installation. The degree of tidal energy, sediment supply, and subsidence rate can vary significantly within short distances (Veettil et al. 2024). For example, in the Mekong Delta, the eastern coast facing the South China Sea experiences higher erosion rates due to wave exposure (Phan et al. 2017), while the western coast bordering the Gulf of Thailand remains relatively stable and accreting (Anthony et al. 2015; Phan et al. 2017). Such variability necessitates site-specific assessments before bioshield placement. Furthermore, large-scale upstream dam construction or river regulation may reduce sediment delivery, thereby diminishing the natural accretion potential of deltaic plains, which is an important constraint that must be integrated into design planning (Binh et al. 2020; Schmitt et al. 2021).

Sandy Coasts

Sandy shorelines, such as those along the Kerala coast of India (Fig. 6.1) or some areas of the central Vietnamese coast, pose distinct challenges. High wave energy, shifting dunes, and poor water retention make sandy shoreline environments unsuitable for mangrove establishment (Veettil et al. 2019a, b, 2021; Van Bijsterveldt et al. 2023). Instead, bioshields on sandy coasts rely on salt-tolerant trees and shrubs such as *Casuarina equisetifolia*, *Pandanus odoratissimus*, *Ipomoea pes-caprae*, and *Spinifex littoreus* (Veettil et al. 2021; Athikalam and Vaideeswaran 2021). These species form shelterbelts that protect the locality from wind erosion, sand drift, and salt spray.

Bioshield design for sandy coasts requires an emphasis on root anchorage and dune stabilization. Species must be selected for their ability to establish on loose substrates and withstand desiccation (Martínez and Psuty 2004; Sigren et al. 2014). The spacing and orientation of vegetation rows are critical: dense windbreaks perpendicular to prevailing wind directions help trap moving sand and facilitate dune formation, while staggered planting patterns reduce wind tunnel effects (Yang et al. 2019; Guo et al. 2021). Maintenance is equally important, as sand burial and grazing can severely affect plant survival in the initial years.

Lagoonal and Estuarine Environments

In lagoonal systems and estuaries, such as those common in Sri Lanka (Veettil 2022; Veettil et al. 2023) and central Vietnam (Vinh and Quan 2015), hydrodynamics are

Fig. 6.1 Attempt to plant *Pandanus* and *Casuarina* along the central coast of Kerala State, India. Poor monitoring has affected the succession of seedlings. *Photographs* Paul Thomas Athikalam

influenced by both marine and freshwater inflows (Guo 2022). These areas support diverse vegetation assemblages that combine mangroves, salt marshes, and riparian forests. Bioshield design here must therefore account for fluctuating salinity regimes, periodic freshwater flushing, and potential eutrophication from upstream agriculture or aquaculture areas. Excessive nutrient loading may promote algal growth and hypoxia, hindering bioshield vegetation seedling establishment (Wang 2024; Hu et al. 2024). Therefore, integrating buffer zones and maintaining hydrological connectivity between bioshields and adjacent wetlands can sustain long-term functionality.

Rocky and Cliffed Coasts

On rocky or cliffed shorelines, the potential for bioshield development is limited due to minimal sediment deposition and shallow soils (Airoldi 2003; Davidson-Arnott et al. 2009). Nonetheless, vegetation can still play a supplementary role by reducing splash erosion and stabilizing shallow debris. Pioneer species such as *Sesuvium portulacastrum* or *Scaevola taccada* can colonize rock crevices and help dissipate wave spray energy (Maun 1998). In such contexts, bioshields primarily serve as micro-protective systems—enhancing slope stability rather than acting as large-scale storm barriers.

6.3.2 Hydrology and Salinity

Hydrological conditions form the ecological backbone of coastal bioshield design. Coastal vegetation must cope with tidal inundation, salinity gradients, and variable freshwater inputs—all of which dictate species distribution, growth performance, and overall ecosystem function (Redelstein et al. 2018; Kumbier et al. 2021; Medina-Calderón et al. 2021; Perri et al. 2023). Hydrological characterization involves understanding both the vertical (depth and duration of flooding) and horizontal (salinity and moisture gradients from sea to land) dimensions of site hydrology.

Salinity Regimes and Species Tolerance

Salinity is one of the most decisive factors governing species viability. Each mangrove or coastal plant species has a specific salt tolerance range, often shaped by physiological adaptations such as salt excretion, succulence, or selective ion uptake (Wang et al. 2025). For instance, *Avicennia marina* is an extremophile capable of surviving salinities up to 40 parts per thousand (ppt), making it ideal for seaward fringe zones or hypersaline lagoons (Parida and Jha 2010; Dhawi 2025). In contrast, *Bruguiera gymnorrhiza* and *Heritiera littoralis* require lower salinity (<25 ppt) and are best suited for intermediate or landward zones with greater freshwater influence (Erasto 2021; Huang et al. 2022).

Designing hydrological zoning that mirrors these natural gradients is essential for ensuring species survival and ecological stability. The seaward edge, subjected to daily inundation, should host salt-tolerant pioneers such as *Sonneratia alba* and *Avicennia alba* (Tomlinson 2016). The mid-intertidal zone may support *Rhizophora apiculata* (Giesen et al. 2007), while the upper intertidal and supratidal zones can transition to *Bruguiera, Ceriops*, or terrestrial coastal forest species (Saenger 2002). This zonation not only enhances ecological realism but also ensures that bioshields can withstand hydrological variability, including extreme events such as storm surges and king tides (Robertson and Alongi 2013).

Tidal Amplitude and Inundation Frequency

Tidal dynamics govern seed dispersal (Zainol et al. 2022), oxygen exchange (De Santana et al. 2021), and nutrient cycling (Wang et al. 2021a) within bioshields. Areas with high tidal amplitudes offer broader intertidal zones suitable for mangrove establishment, whereas microtidal coasts may restrict mangrove colonization to narrow strips (Veettil et al. 2019a). Over-inundation can lead to anaerobic soil conditions and seedling mortality, while insufficient inundation limits propagule dispersal, seed germination, and nutrient supply. Therefore, hydrological modeling—integrating tidal range, inundation duration, and seasonal water level fluctuations—is vital for optimizing planting elevations and predicting survivability (Van Loon et al. 2016; Balke et al. 2021; Gijsman et al. 2024).

Freshwater Inflows and Upstream Influences

Freshwater inflow from rivers or rainfall dilutes salinity and delivers essential sediments and nutrients. However, anthropogenic alterations such as dam construction, irrigation diversion, or drainage canalization can drastically alter coastal hydrology (Zhang et al. 2022b). In Vietnam's Mekong Delta, for example, reduced freshwater discharge during the dry season and saline intrusion during high tides have altered mangrove growth patterns (Veettil et al. 2019b, 2022). Adaptive design must account for these shifts by selecting tolerant species, modifying planting elevation, or incorporating engineered features such as shallow drainage channels to regulate water flow.

6.3.3 *Socioeconomic and Land-Use Constraints*

While biophysical parameters (e.g., tidal regime, salinity, elevation) determine where a bioshield can grow, socioeconomic factors (such as governance, land tenure, local livelihoods, community engagement, incentives, and policy support) often determine whether it can persist over time (Walters et al. 2008; Dahdouh-Guebas et al. 2021; Powlen et al. 2023). Coastal zones in tropical Asia are among the most densely populated and economically contested areas, supporting aquaculture, agriculture, industry, tourism, and settlement (Veettil et al. 2021). The competition for land and resources introduces a complex layer of social and governance constraints that directly influence bioshield feasibility (Merven et al. 2023). In essence, bioshield design is as much a social negotiation as it is an ecological or engineering exercise. Physical feasibility alone cannot guarantee success; long-term functionality depends on governance frameworks, community engagement, and the equitable distribution of benefits.

Land-Use Conflicts and Tenure Security

Conversion of mangroves into shrimp ponds, paddy fields, or urban infrastructure has been widespread across Asia (e.g., Veettil et al. 2019a). In such contexts, designating land for bioshields can provoke disputes unless land tenure and usage rights are clearly defined. Participatory planning processes, where local communities, government agencies, and private stakeholders jointly determine bioshield locations, help minimize future conflicts (Holland et al. 2022; Jansen and Kalas 2022). Establishing co-management or community stewardship agreements can further enhance long-term maintenance and enforcement.

Livelihood Dependencies and Incentives

Bioshields must complement rather than compete with local livelihoods. Integrating economic incentives such as sustainable aqua-silviculture, crab fattening, or mangrove honey production can transform bioshields into productive landscapes (Laynes-Magaña et al. 2025; Zhang et al. 2025b). When communities perceive direct benefits—both protective and economic—they are more likely to invest labor and care in maintaining the system. Conversely, neglecting livelihood considerations can lead to encroachment, illegal harvesting, or abandonment (Veettil et al. 2019a).

Infrastructure and Development Pressures

In rapidly urbanizing coasts, infrastructures such as roads, ports, and seawalls often fragment or replace potential bioshield areas (Veettil et al. 2021). Strategic zoning and environmental impact assessments should therefore identify buffer corridors where vegetation can thrive without impeding development (Lan and Hsu 2021; Hegazy 2024; Jin et al. 2024). Integrating bioshields into broader coastal zone management plans ensures alignment with other land uses and prevents short-lived projects (Carter et al. 2015; Veettil et al. 2021).

6.4 Zonation and Spatial Configuration

The structural arrangement of vegetation within a bioshield is central to its effectiveness in mitigating coastal hazards. Bioshields are not simply plantations of trees or shrubs; they are carefully organized ecological systems designed to interact with hydrodynamic and atmospheric forces (Schmitt et al. 2013; Perricone et al. 2023). Their performance depends on how plant species are distributed spatially (across the shore profile) and vertically (within the canopy layers). Optimal design must therefore consider functional zonation, width and density parameters, and vertical structure, ensuring that each component contributes synergistically to the overall protective function (Qiu et al. 2019; Veettil et al. 2021; Ciccarelli and Bona 2022).

6.4.1 Concept of Functional Zonation

The principle of functional zonation is rooted in the natural structure of coastal ecosystems such as mangroves, beach forests, and dune vegetation. These systems typically display distinct bands of vegetation running parallel to the shoreline, each adapted to specific hydrological and sedimentary conditions (Ma et al. 2020; Veettil et al. 2021). In bioshield design, replicating this natural pattern enhances both ecological realism and engineering efficiency. Each zone performs complementary roles in dissipating energy, stabilizing sediments, and buffering inland areas from storm and wind damage (Veettil et al. 2021). Together, the three zones—seaward, mid, and landward—create a multi-layered ecological defense system. Their collective functioning depends on careful spatial planning that aligns species characteristics with tidal range, sediment type, and exposure level.

Seaward Zone—Wave Dissipation and Sediment Accretion

The seaward or fringe zone forms the first line of defense against waves, tides, and storm surges. Vegetation in this zone experiences the most frequent inundation, the highest salinity, and the strongest hydrodynamic forces (Liu et al. 2018). Pioneer mangrove species such as *Avicennia marina*, *Avicennia alba*, *Sonneratia alba*, and *Sonneratia caseolaris* are well adapted to these extreme conditions (Tomlinson 2016). Their pneumatophores and stilt roots create a dense, irregular structure that enhances bottom friction and reduces wave energy (Wei et al. 2024).

Empirical studies have demonstrated that wave height can decrease by 13–66% over 100 m of mangrove forest, depending on root density and species composition (Quartel et al. 2007; McIvor et al. 2012). The seaward zone also traps suspended sediments, promoting vertical accretion that helps coastal landforms keep pace with moderate sea-level rise (Adame et al. 2010; Xie et al. 2022). In deltaic environments such as the Mekong Delta (Vietnam) and the Ayeyarwady Delta (Myanmar), maintaining a healthy seaward mangrove fringe is crucial to counteract subsidence and erosion (Veettil et al. 2018, 2019a).

Mid Zone—Sediment Stabilization and Energy Dissipation

Behind the pioneer belt lies the mid-zone, typically occupied by mangrove species with robust, interlocking prop roots such as *Rhizophora apiculata*, *Rhizophora mucronata*, and *Bruguiera gymnorrhiza* (Giesen et al. 2007). This zone plays a vital role in sediment stabilization, nutrient retention, and the consolidation of newly accreted substrates (Yoshikai et al. 2023; Partani et al. 2024). The intricate aerial root systems create high drag coefficients, further attenuating waves and currents that penetrate the seaward fringe (Zhang et al. 2015; Reef and Sayers 2025).

In addition to their physical function, these species enhance ecological stability by providing habitat for fish, crabs, and mollusks, thereby supporting coastal fisheries. Mixed planting of *Rhizophora* and *Bruguiera* species increases structural diversity, which in turn improves the bioshield's resilience to disturbances such as pest outbreaks, storm damage, or localized dieback (Naz and Chowdhury 2022; Su et al.

2022). Where natural hydrology allows, the mid-zone may also support understory shrubs like *Acanthus ilicifolius* and *Clerodendrum inerme*, which contribute to soil binding and biodiversity enrichment (Bhanja et al. 2025).

Landward Zone—Windbreak and Transition to Human Land Use

The landward or supratidal zone acts as a windbreak and transition area between natural vegetation and human-dominated landscapes (Trujillo and Thurman 2017; Lan and Hsu 2021). This zone is less frequently inundated but exposed to strong onshore winds and salt spray (Keith et al. 2020). Species such as *Bruguiera*, *Ceriops*, or terrestrial coastal forest species such as *Casuarina equisetifolia*, *Thespesia populnea*, *Pandanus odorifer*, *Terminalia catappa*, and *Calophyllum inophyllum* are well suited for this environment (Saenger 2002). Their tall, dense canopies reduce wind velocity and protect inland settlements, agricultural fields, and aquaculture ponds from salt intrusion and desiccation (Lin et al. 2023; Benazir et al. 2024).

In some regions, the landward zone also supports economically valuable multipurpose species. For example, *Casuarina equisetifolia* provides timber and fuelwood, Pandanus leaves are used for weaving, and *Thespesia populnea* offers medicinal and ornamental benefits. Integrating such species strengthens local support for bioshield conservation. Importantly, the landward zone must be ecologically connected to the inner mangrove and dune systems to maintain hydrological continuity and prevent waterlogging or salinization of adjacent land uses (Pérez-Ceballos et al. 2020; Yang et al. 2022; Dyring et al. 2023).

6.4.2 Width and Density Parameters

The effectiveness of a bioshield is not solely determined by species composition but also by its spatial dimensions, particularly the width and vegetation density (Laso Bayas et al. 2011; Kamil et al. 2021; Fisher et al. 2025). The appropriate width depends on the dominant hazard (e.g., tsunami, cyclone, erosion), coastal topography, and sediment supply, while vegetation density determines the degree of hydrodynamic resistance and energy dissipation (Benazir et al. 2024; Daro Justine and Seenath 2025; Fisher et al. 2025).

Determining Bioshield Width

Field observations and hydrodynamic models provide general guidance for bioshield width under different hazard scenarios:

a. ***Tsunami-prone coasts***: A belt of 100–500 m of dense mangrove forest is typically recommended (Quang Bao 2011; Van Wesenbeeck et al. 2025). Wider belts enhance frictional resistance and wave energy attenuation. For example, simulations after the 2004 Indian Ocean tsunami showed that a 500 m wide mangrove belt could reduce wave height and flow velocity by up to 50%, substantially lowering inland damage (Teh et al. 2009).

b. ***Cyclone-prone coasts***: In regions such as the Bay of Bengal or the South China Sea, 50–200 m wide bioshields composed of mixed mangrove and coastal forest species provide effective storm surge reduction (Veettil et al. 2018; Dasgupta et al. 2019; Van Wesenbeeck et al. 2025). During Cyclone Amphan (2020), villages shielded by mangrove belts in Bangladesh experienced significantly less damage than those without vegetative buffers (Chowdhury et al. 2021). Zhang et al. (2022a, b) assessed the resilience of mangrove trees to the impact of Typhoon Mangkhut (2018) and observed a 77% reduction in wave height across a 100 m-wide mangrove forest.
c. ***Erosion control zones***: Where the main objective is shoreline stabilization rather than storm protection, narrower belts (20–50 m) may be adequate. Such belts (Fig. 6.2) are often used along riverbanks, aquaculture and fishpond embankments, or coastal roads to prevent undercutting and scouring (Smith et al. 2020).

While these values offer broad benchmarks, site-specific modeling using tools like SWAN (Wang and Hu 2023) or Delft3D (Vinh 2015) can refine design width according to local wave energy, slope, and tidal characteristics.

Vegetation Density and Energy Dissipation

Planting density determines the degree of drag and turbulence within the bioshield (Zhang et al. 2021a). Vegetation that is too sparse allows waves to penetrate more easily, whereas excessively dense planting can intensify competition for light and nutrients, leading to stunted growth. Observational and modeling studies, such as Li et al. (2023), suggest that wave attenuation increases with mangrove density

Fig. 6.2 Narrow mangrove belts used for riverbanks and fishpond embankments in Ben Tre, Vietnam. *Photographs* Bijeesh Kozhikkodan Veettil

and forest width—denser forests attenuate waves more effectively than sparse ones. However, optimal density may vary by species and environmental condition. *Rhizophora* species with large prop roots may provide more drag per individual tree than slender *Avicennia* saplings, allowing for slightly lower planting density in mixed stands.

Designers should also account for age-dependent structure: young plantations may initially offer limited wave attenuation, with performance improving as trees mature and root systems expand. Adaptive thinning and enrichment planting can help maintain appropriate density as the bioshield evolves (Phong and Nuong 2023).

6.4.3 Vertical Structure and Layering

Beyond horizontal arrangement, the vertical structure of a bioshield plays a crucial role in energy absorption and aerodynamic resistance (Suzuki et al. 2011; Daro Justine and Seenath 2025). Multi-layered vegetation enhances surface roughness, creating turbulence that dissipates wind and wave energy more effectively than uniform canopies (Derakhshan et al. 2022).

Canopy Stratification

An ideal bioshield exhibits three primary canopy layers:

(a) Ground and shrub layer: Composed of low-lying species such as *Acanthus ilicifolius*, *Ipomoea pes-caprae*, or *Clerodendrum inerme*. These species cover bare soil, reduce splash erosion, and intercept salt spray.
(b) Mid-canopy layer: Dominated by medium-height mangrove trees like *Bruguiera gymnorrhiza*, *Ceriops tagal*, or *Sonneratia caseolaris*. Their dense foliage and root systems offer the main physical barrier to water flow (Nguyen et al. 2023).
(c) Emergent layer: Formed by tall species such as *Rhizophora apiculata* in mangrove zones or *Casuarina equisetifolia* and *Terminalia catappa* in coastal forests. This upper tier reduces wind velocity, provides shade, and contributes to microclimatic regulation.

This vertical layering ensures continuous aerodynamic resistance from ground level to canopy top, reducing both wind and wave impacts. In cyclone-prone areas, multi-tiered casuarina and coastal forest belts have been shown to deflect high winds and reduce damage to inland infrastructure (Danesh Miah et al. 2013; Lin et al. 2023).

Structural Complexity and Resilience

Vertical heterogeneity not only enhances energy absorption but also contributes to ecological resilience (Quang Bao 2011; Zhang et al. 2023; Wei et al. 2024). Layered canopies provide niches for diverse fauna, promote nutrient recycling through litter fall, and buffer temperature and humidity extremes. In mixed bioshields, varying root morphologies—such as prop roots, pneumatophores, and fibrous root systems—stabilize sediments at multiple depths, minimizing the risk of uprooting during extreme events (Nguyen et al. 2023).

From an engineering perspective, multi-layered vegetation acts similarly to a graded filter in hydraulic structures: it gradually dissipates force rather than reflecting it abruptly. This characteristic is vital for maintaining long-term shoreline stability and preventing erosion in front of the vegetated belt.

Application to Casuarina-Based Systems

In non-mangrove settings, such as sandy coasts dominated by *Casuarina equisetifolia*, vertical structure can be achieved by combining species of different heights (e.g., underplanting *Pandanus odorifer* or *Spinifex littoreus*) beneath *Casuarina* rows (Fig. 6.1). This combination increases ground roughness and reduces wind tunnel effects. Regular pruning and staggered spacing help maintain canopy porosity, allowing air to diffuse gradually rather than forming destructive eddies behind the bioshield (An et al. 2022).

6.5 Species Selection and Ecological Compatibility

The success and longevity of bioshields depend fundamentally on selecting appropriate plant species that can withstand local environmental conditions while performing essential ecological and protective functions (Fig. 6.3). Species selection is not a simple matter of choosing fast-growing vegetation—it involves aligning ecological compatibility, functional traits, and community dynamics with the physical setting and long-term objectives of the bioshield (Giannini et al. 2017; Fu et al. 2023). The process must therefore consider the ecological history of the site, species-specific tolerance thresholds, and the synergistic interactions among different plants (Wang et al. 2021b). Poor species selection has been a leading cause of bioshield failure across tropical and subtropical Asia, often resulting in low survival rates, invasive outbreaks, or structural instability (Veettil et al. 2021). This section outlines three major principles guiding effective species selection: (i) preference for native and regionally adapted species, (ii) emphasis on functional traits linked to protection and survival, and (iii) design based on mixed and multi-species assemblages to enhance resilience and ecosystem services. The species selection for bioshield design is an exercise in ecological precision and adaptive planning. Prioritizing native and regionally adapted species ensures genetic and environmental compatibility (Thomas et al. 2014; Houde et al. 2015). Selecting species based on functional traits ensures that vegetation performs its intended protective roles under dynamic coastal conditions. In addition, adopting mixed and multi-species configurations enhances ecological resilience, biodiversity, and community benefits (Xie et al. 2025). When combined, these principles create bioshields that are not only effective barriers against coastal hazards but also thriving ecosystems that support livelihoods, biodiversity, and long-term climate resilience.

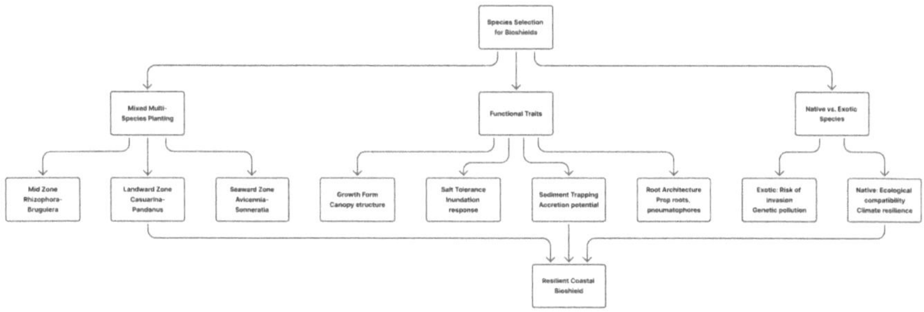

Fig. 6.3 Different methods for selecting suitable species for the establishment of coastal bioshields

6.5.1 Native Versus Exotic Species

A central principle of bioshield design is the preference for native species that are plants naturally occurring within the coastal ecosystems of a given region. Native species have evolved under local climatic, hydrological, and sedimentary conditions, providing them with inherent resilience to site-specific stressors such as salinity, tidal inundation, and seasonal drought (Gairola et al. 2023). Moreover, native vegetation maintains ecological compatibility with local fauna, supporting pollinators, fisheries, and other wildlife integral to ecosystem functioning.

For example, in Vietnam's Mekong Delta, native mangrove species such as *Rhizophora apiculata*, *Avicennia alba*, and *Bruguiera gymnorrhiza* are well adapted to the delta's complex sediment and salinity gradients (Veettil et al. 2019a). These species exhibit stable growth patterns, high sediment trapping efficiency, and long-term structural integrity (Chaudhuri et al. 2019). Similarly, in India and Bangladesh, *Avicennia marina* and *Sonneratia caseolaris* have proven resilient in highly saline or disturbed estuarine conditions (Chowdhury et al. 2023).

By contrast, exotic or non-native species, while sometimes offering rapid early growth or tolerance to specific stressors, pose ecological and management challenges. A notable example is *Sonneratia apetala*, a fast-growing mangrove species introduced from South Asia to Vietnam and parts of China during the 1990s. Although its early performance was impressive, the species demonstrated invasive tendencies, outcompeting the native *Avicennia alba* and altering local soil chemistry and faunal composition (Fazlioglu and Chen 2020; Feng et al. 2025). Over time, such dominance reduces biodiversity, destabilizes community structure, and undermines the ecological balance that coastal bioshields are meant to sustain.

The introduction of exotic species also raises concerns about genetic pollution and the loss of locally adapted genotypes. Native populations often contain genetic variations fine-tuned to micro-environmental conditions such as salinity fluctuations or sediment type. Replacing them with non-native species can reduce the adaptive capacity of the broader landscape (Busoms et al. 2015; Crosby et al. 2024). Consequently, bioshield design should prioritize native or regionally adapted species that have demonstrated success within similar climatic and geomorphological contexts.

Exotics should only be considered under exceptional circumstances, such as highly degraded sites where no native species can survive, and always under controlled experimental conditions with strict monitoring.

6.5.2 Functional Traits

Beyond origin, species selection must focus on functional traits—the morphological, physiological, and ecological characteristics that determine how a plant interacts with its environment and contributes to bioshield performance. Functional traits translate directly into the protective functions that bioshields are designed to provide, such as wave attenuation, sediment stabilization, and wind resistance. Table 6.1 illustrates various functional traits of coastal bioshields. By evaluating and combining these functional traits, designers can create bioshields tailored to site-specific hazards and environmental gradients.

Root Architecture

Root structure is among the most critical determinants of bioshield stability and sediment trapping. Mangrove species display a remarkable diversity of root forms, each adapted to specific hydrodynamic conditions. *Rhizophora* species develop large stilt or prop roots that provide mechanical anchorage and create a physical barrier to wave flow (Méndez-Alonzo et al. 2015). Avicennia and Sonneratia produce dense pneumatophores that trap sediments and oxygenate anaerobic soils (Nguyen et al. 2023). Bruguiera and Ceriops develop knee roots that stabilize mid-intertidal zones (Quadros et al. 2021). For sandy or dune systems, fibrous root networks of *Casuarina equisetifolia*, *Pandanus odorifer*, and *Spinifex littoreus* are ideal for binding loose substrates and reducing wind erosion (Pattanaik et al. 2008).

Salt Tolerance and Inundation Response

Coastal environments expose vegetation to varying degrees of salinity and inundation. Different Species exhibit specific physiological adaptations to cope with these stresses. *Avicennia marina* possesses specialized salt glands on its leaves for excreting excess salt, allowing it to thrive in hypersaline conditions exceeding 40 ppt

Table 6.1 Functional traits of coastal vegetation bioshields with example species

Function	Example species	Trait
Wave attenuation	*Rhizophora apiculata*	Prop roots, dense stem density
Sediment accretion	*Avicennia marina*	Pneumatophores, high colonization rate
Windbreak	*Casuarina equisetifolia*	Tall, narrow canopy
Dune stabilization	*Ipomoea pes-caprae, Spinifex littoreus*	Creeping roots, sand binding

(Dhawi 2025). *Bruguiera gymnorrhiza*, in contrast, prefers brackish environments and performs poorly under prolonged saline exposure (Tada and Kashimura 2009). *Sonneratia caseolaris* and *Heritiera littoralis* tolerate more freshwater influence, making them suitable for upper estuarine and riverine fringes. Understanding these thresholds is essential to assigning species to appropriate hydrological zones within the bioshield (Wang et al. 2011; Tatongjai et al. 2021) (as discussed in Sect. 6.4).

Growth Form and Canopy Architecture

Species growth patterns influence aerodynamic resistance and shading dynamics. Tall, columnar trees such as *Casuarina equisetifolia* are effective windbreaks, while spreading species like *Thespesia populnea* and *Terminalia catappa* create low, dense canopies that intercept salt spray and reduce ground-level desiccation (De Sedas et al. 2019; Maria et al. 2021; Lin et al. 2023). Layered growth forms, combining shrubs, understory trees, and emergents, maximize energy absorption and habitat diversity.

Sediment Trapping and Accretion Potential

Species vary in their ability to trap sediments depending on root density, stem spacing, and canopy drag. Studies from the Mekong and Irrawaddy deltas have shown that dense stands of *Rhizophora apiculata* can enhance sedimentation rates, accelerate mudflat formation, and support natural colonization of secondary species (Veettil et al. 2018; Zoccarato et al. 2018). This sediment accretion function is crucial for bioshield sustainability under sea-level rise scenarios.

Regenerative Capacity and Longevity

Resilience to disturbance, whether from storms, pests, or anthropogenic impacts, depends on regenerative capacity. Species such as *Avicennia marina* regenerate rapidly from dispersed propagules, while *Rhizophora* species rely more heavily on mature propagules that require specific conditions for establishment (Clarke and Allaway 1993; Zhang et al. 2021b). Designing with a mix of regenerative strategies ensures the bioshield recovers after disturbances without external replanting efforts.

6.5.3 Mixed and Multi-species Planting

Diversity is a cornerstone of resilience in natural systems, and bioshields are no exception. Mixed and multi-species designs mimic the structure of natural coastal forests, promoting ecological balance, functional redundancy, and long-term stability (Huang et al. 2024). Monocultures, though easier to plant and manage, are inherently vulnerable to pests, diseases, and environmental fluctuations (Huang et al. 2024).

Ecological and Structural Benefits

Mixed species stands exhibit complementary functional traits. For example, *Avicennia alba* establishes quickly and traps initial sediments, creating favorable conditions for slower-growing but structurally robust *Rhizophora apiculata* (Nguyen et al. 2020). This facilitative interaction mirrors natural successional dynamics in mangrove ecosystems, where pioneer species pave the way for climax communities. Similarly, combining *Rhizophora* with *Bruguiera* enhances below-ground complexity, improving sediment retention and resistance to uprooting during storms (Nguyen et al. 2023) as well as carbon sequestration (Li et al. 2025a, b, c).

Empirical evidence supports the superior performance of mixed coastal bioshield plantations (Dai et al. 2018; Su et al. 2021; Kremer et al. 2025). Their combined root structures dissipate hydrodynamic energy more effectively, while their mixed canopy layers buffer wind and salt spray (Yoshikai et al. 2023). Moreover, the ecological diversity of mixed bioshields supports a wider range of fauna, including crustaceans, mollusks, and birds, thereby enhancing ecosystem services such as nutrient cycling and fisheries productivity (Nagelkerken et al. 2008).

Enhancing Resilience to Climate and Disturbance

Biodiversity also provides insurance against disturbance. In the event of extreme salinity spikes, disease outbreaks, or cyclone damage, differential tolerance among species ensures that some components of the bioshield remain functional (Veettil et al. 2021; Chowdhury et al. 2023; Halpin-McCormick et al. 2025). For instance, if storm surges uproot *Rhizophora* trees, *Avicennia* saplings may quickly recolonize the gaps, maintaining ground cover and sediment stabilization. This redundancy is vital for sustaining protective function during recovery periods.

Designing Mixed Assemblages

The configuration of mixed species should follow ecological zonation principles. Seaward zones can feature *Avicennia–Sonneratia* combinations for early sediment trapping (Tomlinson 2016), mid-zones can integrate *Rhizophora–Bruguiera* for mechanical strength (Giesen et al. 2007), and landward zones can incorporate *Casuarina–Pandanus* or *Thespesia* species for wind protection (Saenger 2002). Planting should alternate species in clusters or strips rather than in uniform rows, promoting natural intermixing and variable canopy density (Pretzsch and Zenner 2017).

Socioeconomic Integration

Mixed plantations also enable multiple livelihood opportunities. While some species provide timber, poles, or fuelwood (e.g., *Rhizophora*, *Casuarina*), others yield non-timber products such as honey (*Avicennia*, *Sonneratia*), edible fruits (*Pandanus*), or medicinal resources (*Thespesia*). This multifunctionality increases community support and provides incentives for long-term maintenance.

Incorporating local ecological knowledge enhances species selection and combination strategies. Coastal communities often possess a generational understanding of which species thrive under specific tidal or soil conditions (Stori et al. 2019). Participatory design workshops involving local stakeholders and scientists can help match this knowledge with modern ecological assessment to ensure both scientific rigor and social acceptance (Risley et al. 2025).

6.6 Structural and Hydraulic Design Considerations

The structural and hydraulic dimensions of bioshield design form the scientific foundation for quantifying their effectiveness in dissipating wave energy, stabilizing sediments, and integrating with engineered systems (Veettil et al. 2021; Huynh et al. 2024). While ecological factors determine the viability of plant growth, it is the structural and hydrodynamic parameters that translate biological assemblages into measurable coastal protection outcomes. Coastal bioshields operate at the interface of vegetation mechanics and fluid dynamics, where plant form, density, and spatial arrangement interact with water motion, wave propagation, and sediment transport (Yang et al. 2017; Torabbeigi et al. 2024). A well-designed bioshield should therefore be guided not only by ecological appropriateness but also by hydraulic functionality, ensuring that vegetation structure aligns with the intensity and frequency of coastal hazards.

This section explores three core aspects of structural and hydraulic design: (i) modeling wave energy dissipation through vegetation, (ii) understanding sediment trapping and accretion processes, and (iii) integrating bioshields with engineered infrastructure for hybrid coastal defense. The structural and hydraulic considerations translate the ecological potential of bioshields into quantifiable protective performance. Wave energy dissipation models guide the determination of optimal width, density, and species composition. Sediment trapping dynamics ensure that bioshields evolve and maintain elevation under changing sea levels. Integration with engineered structures enhances resilience and provides immediate protection during establishment phases.

6.6.1 *Wave Energy Dissipation Models*

Quantifying wave attenuation is vital for coastal bioshield design validation. The fundamental principle is that as waves propagate through vegetated areas, energy is lost through drag and turbulence induced by stems, roots, and canopy elements. A widely cited relationship by Mazda et al. (2006) suggests that wave height attenuation follows an exponential decay model. Empirical and semi-empirical models estimate wave height reduction as a function of vegetation density (N), stem diameter (d), and forest width (W):

$$H_x = H_0 e^{-kNdW}$$

where H_0 and H_x are incident and transmitted wave heights, respectively, and *k* is the species-specific coefficient. Horstman et al. (2012) measured wave attenuation through mangrove forests in Trang Province of Thailand and found 63–72% reduction in incident wave energy along transects of ~246 m and ~98 m, illustrating effective wave damping by coastal mangroves. Field measurements indicate that for dense mangrove strands, wave height can be reduced by over 60% over 80 m of forest width (Wei et al. 2024).

The drag coefficient (C_d), representing the resistance imposed by vegetation, is influenced by both plant morphology and hydrodynamic conditions. Flexible stems, such as those of *Avicennia marina*, bend under flow, thereby reducing drag during high-energy events while recovering upright positions as energy dissipates (Wei et al. 2024). Rigid roots of *Rhizophora apiculata*, on the other hand, maintain structural resistance across varying wave intensities (Mugilan et al. 2024). This interplay between rigidity and flexibility is critical for maintaining effective dissipation without vegetation failure.

Wave attenuation is also frequency-dependent. High-frequency, short-period waves (e.g., wind waves) experience stronger attenuation than long-period waves (e.g., tsunamis), which require broader and denser vegetation belts for substantial reduction (Best et al. 2022). Numerical models that integrate vegetation drag into the shallow-water equations allow designers to simulate specific coastal profiles and optimize bioshield dimensions accordingly (e.g., Mendez and Losada 2004; Suzuki et al. 2011).

In practical design, wave dissipation modeling should be site-specific, calibrated with field data on stem geometry, roughness, and forest porosity. This helps determine the minimum bioshield width for target attenuation levels. For instance, in cyclone-prone regions like Odisha (India) or southern Vietnam, achieving substantial reductions in storm surge height typically requires broad mangrove belts, often on the order of 100–200 m or more of mature forest with mixed species composition for meaningful attenuation of surge and wave energy (Zhang et al. 2012).

Moreover, hydrodynamic feedback loops must be considered. As vegetation reduces flow velocity, sediment deposition increases, thereby enhancing root stability and forest expansion, which in turn further improves wave attenuation capacity over time (Baaij et al. 2021; Gomes et al. 2025). Thus, bioshields represent self-reinforcing systems, where biological growth enhances hydraulic function in a positive feedback mechanism.

6.6.2 Sediment Trapping and Accretion

Sediment dynamics form the foundation for the long-term sustainability of coastal bioshields. Root systems and soil surface roughness act as natural sediment traps, slowing down water flow, promoting deposition, and facilitating vertical accretion (Nguyen et al. 2023; Ahmed and Mohanadhas 2025). The capacity to capture and retain sediments determines whether a bioshield can keep pace with sea-level rise and maintain its elevation relative to tidal datums (Krauss et al. 2017).

Field studies across tropical deltas have demonstrated that mangrove ecosystems can exhibit sedimentation rates of ~5–10 mm yr^{-1} —sufficient to offset or even exceed current regional sea-level rise rates of approximately 3–4 mm yr^{-1} (Sanders et al. 2010; McIvor et al. 2013; Saintilan et al. 2020). In the Sundarbans mangroves of India, this natural accretion process has been observed to stabilize extensive low-lying areas, reducing the frequency of tidal inundation and maintaining soil fertility for adjacent agricultural lands (Sarkar et al. 2024).

Sediment trapping efficiency increases exponentially with vegetation density and structural complexity. The relationship between flow velocity (u) and sediment deposition rate (D) is often expressed as:

$$D = D_0 e^{-\alpha u}$$

where α represents the roughness coefficient determined by vegetation morphology, and D_0 is a reference deposition rate at very low flow velocity. Denser root systems (e.g., *Rhizophora*, *Bruguiera*) increase α, leading to greater sediment retention (Yoshikai et al. 2023). Moreover, pneumatophores and buttress roots of *Avicennia* create micro-eddies that facilitate fine sediment deposition and organic matter accumulation, thereby improving substrate stability and nutrient cycling (Hongwiset et al. 2022; Partani et al. 2024).

The spatial pattern of sedimentation within bioshields typically shows a gradient—highest near the seaward edge and tapering landward (Adame et al. 2010). Over time, this gradient contributes to natural elevation zonation, allowing the colonization of additional vegetation strata and increasing forest width. This process transforms the bioshield into a self-building landscape, enhancing both ecological succession and protection capacity (Adame et al. 2010).

However, the sediment balance can be disrupted by upstream alterations such as dam construction, riverbank erosion, or channel dredging, which reduce sediment supply to the coast and may compromise the success of coastal wetland restoration (Liu et al. 2021). Effective bioshield design must therefore integrate sediment budget assessments and coordinate with catchment-level management to ensure long-term stability.

In high-energy coasts, where natural sediment accretion is insufficient, bioengineering interventions, such as brushwood fences, coir logs, or permeable bamboo barriers, can be deployed temporarily to trap sediments and facilitate natural regeneration (Veettil et al. 2021). This strategy, which uses bamboo and *Melaleuca* fences, was successfully implemented in Vietnam's Mekong Delta, accelerating the formation of stable mudflats suitable for mangrove colonization (Albers and Schmitt 2015; Van Cuong et al. 2015; Mai Van et al. 2021).

6.6.3 Integration with Engineered Structures

While coastal bioshields offer multiple ecosystem services and long-term sustainability, they are not always sufficient as stand-alone defenses, particularly in regions with extreme hydrodynamic forces, severe erosion, or heavily modified coastlines (Veettil et al. 2021). In such contexts, hybrid coastal protection systems, which combine ecological and engineered components, offer the most effective and adaptable solutions (Huynh et al. 2024).

The integration of bioshields with engineered structures merges the strengths of both approaches. Engineered elements, such as seawalls, breakwaters, floating barges, or bamboo fences, provide immediate physical stability and reduce initial erosion, while bioshields contribute long-term adaptability, sediment trapping, and ecological restoration (Van Cuong et al. 2015; Mai Van et al. 2021; Raju and Arockiasamy 2022). The synergy between these components allows a gradual transition from rigid protection to flexible, self-maintaining natural barriers.

A widely cited example of this hybrid approach is the coastal restoration project in Demak District, Central Java, Indonesia (Winterwerp et al. 2014; Molnar-Tanaka and Surminski 2024). Here, bamboo fences and brushwood dams were strategically installed to reduce wave energy and trap sediments, thereby creating sheltered environments for mangrove establishment (Winterwerp et al. 2014). Over several years, this integration facilitated the natural recovery of the degraded coastline.

Similarly, in Vietnam's Ca Mau province, hybrid designs have combined earthen embankments with mangrove belts to provide dual protection—engineered structures guard against extreme storm surges, while mangroves reduce daily wave energy, prevent erosion, and maintain ecological productivity (Nguyen et al. 2023; Duy et al. 2025; MARD 2025). This approach aligns with nature-based solutions (NbS) frameworks advocated by the International Union for Conservation of Nature (IUCN) and the Global Mangrove Alliance, emphasizing that the integration of green and grey infrastructure enhances both safety and sustainability.

From a design perspective, integration requires careful spatial and functional planning. The placement of engineered structures should not obstruct tidal exchange or sediment flow, which are necessary for vegetation health (Morris et al. 2018). Permeable barriers are preferred over solid ones, as they allow hydrodynamic connectivity while reducing energy intensity. The distance between hard structures and vegetated zones should be sufficient to enable root growth and canopy expansion (Schoonees et al. 2019).

Maintenance regimes must also be coordinated—while engineering structures may need periodic repair, bioshields evolve naturally and require minimal intervention once established (Sutton-Grier et al. 2018; Duvat et al. 2025). Long-term monitoring of sediment dynamics, vegetation growth, and wave attenuation ensures that both components continue to function synergistically (Veettil et al. 2021).

Hybrid systems offer significant co-benefits. They improve biodiversity, enhance carbon sequestration, support local livelihoods through fisheries and forestry, and contribute to national commitments under frameworks such as the UN Sustainable Development Goals (SDG 13: Climate Action, SDG 14: Life Below Water, and SDG 15: Life on Land) (Lecerf et al. 2023; Huynh et al. 2024).

6.7 Design for Climate Adaptation and Resilience

Designing bioshields in the context of accelerating climate change requires a forward-looking and adaptive approach. Traditional coastal protection designs often rely on static parameters and fixed infrastructure, but climate change introduces dynamic challenges, such as rising sea levels, intensifying storms, changing sediment regimes, and shifts in salinity and temperature, that demand flexibility and resilience (Griggs and Reguero 2020; Dodman et al. 2023). Coastal bioshields, as living systems, inherently possess the capacity to adjust and evolve in response to environmental change, but only when they are designed with these processes in mind (Duvat et al. 2025).

The concept of climate-resilient bioshield design integrates ecological succession, hydrological feedback, and community-based management into a unified framework (Schueler 2017). It aims to enhance not just immediate coastal protection but also long-term adaptability of both ecosystems and human communities. This section discusses three major design dimensions essential for climate resilience: (i) anticipating and accommodating sea-level rise through vertical accretion and hydrological management, (ii) enhancing structural resilience against cyclones and storms, and (iii) ensuring ecological connectivity to sustain biodiversity and adaptive capacity across coastal landscapes. Designing bioshields for climate adaptation and resilience requires integrating geomorphic, biological, and social dimensions into a cohesive framework. Anticipating sea-level rise through sediment-based elevation strategies ensures longevity. Enhancing structural and ecological resilience against cyclones and storms safeguards coastal communities. Maintaining ecological connectivity supports biodiversity and adaptive capacity. These principles together position bioshields not just as defensive barriers, but as a dynamic, living infrastructure, which is self-sustaining, regenerative, and inherently adaptive to the changing climate realities of the twenty-first century.

6.7.1 Anticipating Sea-Level Rise

Sea-level rise (SLR) represents one of the most pervasive threats to coastal ecosystems and infrastructure worldwide (Veettil et al. 2021). For bioshields to remain effective under projected increases of 3–5 mm per year (and higher under high-emission scenarios), they must be designed to sustain vertical growth—that is, to maintain or increase surface elevation relative to rising water levels (Liu et al. 2021; Zhang et al. 2025a). This capacity depends largely on the balance between sediment accretion and local subsidence rates, collectively referred to as the sediment budget. In general, anticipating sea-level rise requires integrating geomorphic and ecological processes into bioshield design, ensuring that these living systems can build with nature and maintain protective capacity as coastlines evolve.

Sediment Accretion and Root Elevation

Healthy mangroves and coastal vegetation systems naturally promote vertical growth through sediment trapping, root accumulation, and organic matter buildup. The root systems of mangroves, such as *Rhizophora apiculata* and *Avicennia marina*, contribute to soil elevation by producing below-ground biomass and trapping suspended sediments (Nguyen et al. 2023). Empirical studies from the Mekong Delta (MacKenzie et al. 2016), the Sundarbans (McIvor et al. 2013), and the Niger Delta (Hopley and Jones 2022) demonstrate that sediment accretion rates between 5 and 10 mm per year can offset or exceed local sea-level rise, enabling bioshields to maintain their protective elevation over time.

Design Implications

Designers must quantify and model these processes to ensure that the sediment budget of bioshields exceeds projected sea-level rise. This can be achieved through:

a. Sediment Budget Analysis: Estimating annual sediment deposition versus erosion losses to identify potential deficits.
b. Hydrological Zonation: Ensuring that tidal and riverine flows facilitate sediment delivery to the vegetated zone.
c. Adaptive Width Planning: Allowing bioshields to migrate landward as sea levels rise by maintaining setback zones free from permanent infrastructure.

Where natural sediment supply is insufficient due to upstream dams or land-use changes, hydrological restoration, such as reconnecting tidal creeks or removing barriers, can restore sediment dynamics (Li et al. 2025c). In severely sediment-deprived environments, artificial sediment nourishment (placement of dredged materials or sediment slurry) can accelerate accretion processes (Gonçalves et al. 2024).

Monitoring and Feedback

Continuous monitoring using sediment elevation tables (SETs) and marker horizons (MH) helps assess accretion rates and detect early signs of elevation loss (Cahoon 2024). Incorporating adaptive management ensures that interventions can be adjusted based on observed trends, thereby sustaining bioshield functionality under changing conditions.

6.7.2 Cyclone and Storm Resilience

Coastal storms and tropical cyclones pose recurrent threats to life, infrastructure, and ecosystems in tropical and subtropical regions (Feehan et al. 2024; Kropf et al. 2025). The frequency and intensity of such events are projected to increase under climate change, necessitating bioshields designed for mechanical resilience and rapid recovery (Feehan et al. 2024; Hülsen et al. 2025). Structural integrity, aerodynamic design, and species-specific traits all play critical roles in ensuring that bioshields withstand these extreme forces.

Wind- and Wave-Resistant Species

Certain species exhibit superior tolerance to high winds, salt spray, and inundation. *Casuarina equisetifolia*, for instance, possesses flexible trunks and deep roots, allowing it to withstand strong cyclone winds without uprooting (GISD 2025). *Thespesia populnea* and *Pandanus odorifer* have low, spreading canopies that minimize wind drag (Saenger 2002), while mangrove species such as *Rhizophora mucronata* and *Bruguiera gymnorrhiza* resist storm surges through robust prop and knee root systems (Sunkur et al. 2023; Mugilan et al. 2024). Selecting species with such morphological and physiological resilience ensures that the bioshield remains structurally intact during extreme events.

Spatial Configuration and Planting Patterns

Planting arrangement strongly influences bioshield performance under storm stress. Multi-row configurations of coastal shelterbelts, such as *Casuarina equisetifolia* plantations established as windbreaks perpendicular to prevailing winds, can markedly reduce local wind velocities by creating a cascading drag effect across rows, with the sheltered zone extending multiple times the vegetation height downwind (Chen et al. 2025). Such windbreak plantations can be seen in Hambantota City in Sri Lanka, where Casuarina and other tree coastal shelterbelts are deployed along cyclone-prone coasts to reduce wind forces on inland infrastructure and communities (De Zoysa 2008). The cumulative drag of multiple rows increases the effective sheltered distance and wind speed reduction compared with single-row plantings, thereby enhancing coastal bioshield performance under storm conditions (Wang et al. 2024). For mangrove bioshields, as discussed previously, gradual elevation zoning (seaward to landward) ensures that each species occupies its optimal hydrodynamic niche (seaward fringe: dominated by *Avicennia* and *Sonneratia* absorbs initial wave impact; mid-zone: composed of *Rhizophora* and *Bruguiera* provides mechanical stability; landward zone: *Casuarina*, *Pandanus*, or *Terminalia catappa* buffers residual wind energy).

Root Interlocking and Soil Stability

Resilient bioshields rely on extensive below-ground networks that interlock and anchor soil. Mixed plantations enhance root interconnection—*Avicennia*'s shallow pneumatophores complement *Rhizophora*'s deep prop roots, improving both surface binding and subsurface anchorage (Giesen et al. 2007). This synergy prevents large-scale uprooting during storm surges and accelerates post-disturbance recovery.

Post-disturbance Regeneration

Designing for resilience also means planning for recovery. Species with high regenerative potential, such as *Avicennia marina* and *Sonneratia caseolaris*, should be included to facilitate natural recolonization after cyclonic damage (Carlos et al. 2015; Gao et al. 2025). Maintaining seed banks and propagule sources nearby enhances the capacity for spontaneous regeneration, thereby reducing restoration costs and downtime between disturbance events (Wambugu et al. 2023; Mulloy et al. 2025). Ultimately, cyclone- and storm-resilient design transforms bioshields from passive barriers into adaptive defense systems that absorb energy, recover rapidly, and sustain ecosystem functions even under recurrent climatic shocks (Huynh et al. 2024).

6.7.3 Ecological Connectivity

Bioshields should not be conceived as isolated green belts but rather as integral components of larger coastal landscape mosaics (Preston et al. 2025). Designing for ecological connectivity, which is the continuity of habitats and ecological processes across spatial scales, ensures that bioshields support biodiversity, facilitate species migration, and enhance adaptive capacity under climatic stressors (Preston et al. 2025).

Linking Coastal Ecosystems

Natural coastal systems function as interconnected networks of habitats, including mangroves, salt marshes, seagrass meadows, coral reefs, and dune vegetation, each contributing complementary ecological and protective functions (Barbier 2017). For example, mangroves attenuate wave energy and trap sediments, salt marshes buffer against tidal inundation and filter nutrients, seagrass beds stabilize subtidal sediments and support fisheries, and coral reefs dissipate offshore wave energy, reducing the burden on near-shore bioshields (Veettil et al. 2021). Designing bioshields to connect or overlap with these ecosystems enhances overall system performance (Guannel et al. 2016). Integrated coastal green belts, which link mangrove forests with adjacent seagrass beds and coral reefs, have demonstrated superior protection and biodiversity outcomes compared to fragmented habitats (Guannel et al. 2016).

Corridors for Biodiversity and Migration

As climate change drives shifts in temperature, salinity, and hydrological regimes, species must migrate to suitable habitats (Antão et al. 2022). Continuous or stepping-stone bioshields provide migration pathways for flora and fauna, preventing population isolation and genetic bottlenecks (Mengzhi et al. 2025). For example, fish and crustacean species utilize mangrove–seagrass corridors for breeding and feeding, supporting both ecosystem stability and local fisheries (Veettil et al. 2019a).

Landscape-Level Planning

Ensuring ecological connectivity requires planning at a landscape scale rather than at isolated project sites (Lee et al. 2014). This involves identifying existing natural habitats, potential corridors, and anthropogenic barriers (Bennet 2003). Bioshields should be positioned to bridge fragmented habitats and restore hydrological continuity across coastal plains, lagoons, and estuaries. Incorporating connectivity metrics, such as least-cost path analysis and habitat suitability modeling, into design frameworks can help optimize placement and composition (Lee et al. 2014).

Socio-Ecological Integration

Connectivity also extends to human systems. Bioshields can link ecological functions with livelihood activities such as aquaculture, eco-tourism, and sustainable forestry (Da Silva et al. 2025). Participatory design ensures that community needs align with conservation objectives, strengthening stewardship and adaptive management (Nishi et al. 2025). By promoting ecological connectivity, bioshields evolve into resilient socio-ecological networks capable of adapting to changing climatic conditions while sustaining biodiversity and ecosystem services.

6.8 Socioeconomic and Cultural Design Considerations

The long-term success and sustainability of bioshield projects depend not only on their ecological and engineering soundness but also on their alignment with the socioeconomic and cultural fabric of coastal communities. Designing coastal bioshields as socially integrated systems rather than as externally imposed interventions is essential for ensuring both community participation and adaptive management (Veettil et al. 2021; Siddiqui and Ishtiaque 2025). This section examines the human dimensions of bioshield design—participatory approaches, livelihood integration, and the incorporation of cultural and aesthetic values, each of which plays a vital role in sustaining these natural defense systems. Participatory design enhances inclusiveness and legitimacy (Vasseur 2020), livelihood integration underpins economic viability (Dulyakasem et al. 2026), and cultural alignment fosters emotional and ethical attachment (Nienaber 2024). When these principles are systematically embedded in planning and implementation processes, bioshield initiatives extend beyond their immediate protective function to serve as catalysts for sustainable development and cultural resilience.

6.8.1 Participatory Design

Participatory design is the cornerstone of effective and equitable bioshield planning. It recognizes that coastal residents, including fishers, farmers, and indigenous communities, are not passive beneficiaries but active stewards of coastal ecosystems (Anbleyth-Evans 2022). Their local ecological knowledge, accumulated through generations of interaction with coastal landscapes, can substantially enhance the precision of site selection, species choice, and maintenance strategies (Loch and Riechers 2021; Rivers et al. 2023). In Vietnam, for instance, participatory coastal management programs under the Mangrove for the Future (MFF) initiative have demonstrated how engaging local communities from project inception leads to higher survival rates of planted mangroves and better conflict resolution regarding land use (Reed et al. 2015).

Participatory mapping tools enable communities to delineate areas prone to flooding, erosion, or saline intrusion, and to propose suitable sites for bioshield establishment (Klain and Chan 2012). This bottom-up process fosters a sense of ownership and ensures that design decisions reflect both ecological realities and local livelihood needs. The participatory approach also helps reconcile multiple uses of coastal space, such as aquaculture, tourism, and conservation, through negotiated zoning agreements (Calado et al. 2025).

Institutionalizing participation further strengthens governance. Establishing village-level coastal management committees allows for continuous monitoring, adaptive decision-making, and equitable benefit-sharing (Sinare et al. 2025). In contrast, top-down projects often suffer from neglect or failure when community interests are overlooked, leading to low survival rates and subsequent land-use conflicts (Lopes et al. 2025).

Ultimately, participatory design transforms bioshield projects from externally funded interventions into community-driven enterprises (Milupi et al. 2017; Syamsiyah et al. 2025). This social inclusion ensures long-term continuity, as community members become the custodians of both ecological function and socioeconomic benefit.

6.8.2 Livelihood Integration

The integration of bioshields with livelihood opportunities enhances economic resilience and encourages local support for restoration and protection initiatives (Dulyakasem et al. 2026; Ebeler et al. 2025). Purely conservation-focused bioshield projects may face resistance if they restrict access to traditional livelihood zones (Ebeler et al. 2025; Lopes et al. 2025). Conversely, multi-functional designs that incorporate sustainable economic activities can harmonize ecological protection with human well-being (Dulyakasem et al. 2026).

Silvo-fisheries, which is a practice combining mangrove planting with aquaculture, offer a prime example of livelihood integration. In the Mekong Delta, mangrove-based shrimp farming systems have achieved both ecological stability and improved household incomes (IUCN 2018; Nguyen et al. 2022). The mangrove component filters effluents, stabilizes pond embankments, and enhances biodiversity, while shrimp culture provides direct economic returns (Veettil et al. 2019a, b, 2021). Such systems in Bangladesh and Vietnam have shown that a nearly 50:50 ratio of mangrove cover to aquaculture area can optimize both production and ecological function (Ahmed et al. 2023; Anh et al. 2024).

Bioshields can also support alternative livelihoods such as apiculture, crab collection, and eco-tourism (Veettil et al. 2019a). In Vietnam's Ca Mau province, community-managed mangrove areas have been linked to honey production and carbon credit programs, generating diversified income streams (Mani and Nguyen

2025). Integrating such enterprises into bioshield design requires planning for access pathways, buffer zones, and sustainable harvest regimes to avoid overexploitation.

Moreover, livelihood integration serves as a form of social insurance against environmental shocks. Different communities engaged in coastal bioshield management often exhibit greater adaptive capacity during storms or floods because their livelihood base is diversified and more closely aligned with ecosystem health (Pal et al. 2023; Bui et al. 2025). This alignment fosters a self-reinforcing feedback loop—healthy ecosystems sustain livelihoods, and sustainable livelihoods, in turn, motivate ecological stewardship.

6.8.3 Cultural and Aesthetic Values

Coastal bioshields are not merely ecological or economic entities; they are also embedded within cultural landscapes and social identities (Scott et al. 2024; Dulyakasem et al. 2026). Recognizing and incorporating cultural values into bioshield design can significantly enhance their social acceptance and long-term maintenance (Müller et al. 2019; Breyne et al. 2021). In many parts of Southeast Asia, coastal vegetation carries spiritual significance. For example, in Indonesia and the Philippines, mangrove groves and beach forests are often regarded as sacred sites associated with ancestral spirits or community deities (Buenavista and Purnobasuki 2023). These traditional beliefs foster a sense of respect and moral obligation to protect coastal vegetation, which can be strategically harnessed for conservation.

Cultural perceptions also shape aesthetic preferences and land-use behavior. Communities that view coastal vegetation as an integral part of their heritage are more likely to support restoration projects. In Sri Lanka, eco-aesthetic landscaping has been incorporated into bioshield designs to enhance visual appeal and promote ecotourism (De Zoysa 2022). Planting arrangements that combine functional zoning with aesthetic layout, such as alternating mangrove belts with pandanus and hibiscus lines, may create visually pleasing landscapes that also deliver protective benefits.

Integrating cultural narratives into education and outreach further strengthens bioshield stewardship. Community festivals, storytelling, and school-based mangrove planting campaigns can be effective in embedding conservation ethics in local consciousness. In Vietnam's coastal communes, cultural events celebrating mangrove restoration have become annual traditions, reinforcing collective responsibility and pride (Veettil et al. 2021).

Additionally, bioshields can serve as cultural buffers that maintain traditional ways of life. For indigenous fishing communities, maintaining access to natural coastal buffers ensures the continuity of customary practices tied to the rhythm of tides, fishing seasons, and coastal rituals (Delaney and Frangoudes 2024). Thus, bioshields contribute not only to ecological resilience but also to the preservation of intangible cultural heritage (Thi 2021).

6.9 Design Process and Implementation Framework

Designing and implementing effective bioshields requires a structured, science-based, and participatory approach that bridges ecological theory, engineering practice, and social engagement. A well-defined framework ensures that bioshields are not only ecologically functional but also economically feasible and socially acceptable. This section presents a detailed overview of the stepwise design framework and the tools and technologies that support evidence-based planning, emphasizing the need for adaptive and iterative processes that evolve with environmental feedback and community experience. A systematic, stepwise framework ensures that interventions are context-specific, technically sound, and socially inclusive (Veerkamp et al. 2021; Pérez et al. 2024). Meanwhile, modern tools, ranging from GIS-based mapping to hydrodynamic modeling, provide the analytical precision needed to design bioshields that are both protective and adaptive (Son et al. 2024; Marino et al. 2025). By embedding this framework within an adaptive management cycle, bioshield projects can evolve with changing environmental conditions and social priorities (McQueen et al. 2025).

6.9.1 Stepwise Design Framework

The bioshield design process typically follows a sequence of interrelated steps, each building on the knowledge and outcomes of the previous phase. This structured approach enhances coherence, transparency, and flexibility, allowing designers to integrate ecological science with community-based management and long-term sustainability goals.

a. ***Baseline Assessment—Environmental, Hydrological, and Socioeconomic Profiling***

The foundation of any coastal bioshield project lies in a comprehensive baseline assessment (IUCN 2020; Jayson-Quashigah et al. 2025b). This step involves gathering data on coastal geomorphology, soil characteristics, hydrology, biodiversity, and socioeconomic context (Beeston et al. 2023; Cutts et al. 2024). Key parameters include tidal amplitude, sediment supply, salinity gradients, erosion rates, and land-use patterns (Mukherjee et al. 2015; Monty et al. 2017).

Environmental profiling helps determine which vegetative species and configurations are compatible with the local ecosystem (Veettil et al. 2021). For example, the sediment-rich deltas of southern Vietnam support dense mangrove systems dominated by *Rhizophora apiculata* (Binh and Houser 2025), whereas the sandy coasts of Tamil Nadu, India, require wind-resistant species like *Casuarina equisetifolia* (Thamizoli and Rengalakshmi 2022).

Socioeconomic assessment is equally crucial; it identifies local livelihood patterns, land tenure systems, and potential conflicts between conservation and resource use

(Mangora 2022; Arifanti et al. 2025). Participatory rural appraisal (PRA) methods are often employed to integrate local knowledge, perceptions, and expectations into the design process (Mondal and Paul 2022). This ensures that bioshields align with community priorities and long-term economic interests.

b. ***Zonation Mapping—Delineating Functional Zones***

Once baseline data are compiled, zonation mapping defines the spatial organization of the bioshield. Functional zones are demarcated parallel to the coastline to perform complementary ecological roles. The seaward zone acts as the first line of defense, composed of pioneer and salt-tolerant species that absorb wave energy and promote sediment deposition. The mid zone stabilizes sediments through complex root systems and supports successional vegetation that enhances structural diversity. The landward zone functions as a windbreak and transition to agricultural or settlement areas, integrating native coastal forest species for additional resilience. Zonation mapping also incorporates topographic and hydrological features such as tidal creeks, mudflats, and drainage channels to preserve natural hydrodynamics (Chen et al. 2022). GIS tools facilitate accurate delineation of these zones and their spatial connectivity (Zhao et al. 2019; Chen et al. 2023; Kang et al. 2023).

c. ***Species Selection—Matching Ecological and Functional Objectives***

Species selection must be ecologically appropriate and functionally relevant to the protection goal. This involves identifying native or regionally adapted species that correspond to salinity levels, inundation frequency, and sediment type. For example, *Avicennia marina* thrives in high-salinity seaward areas, while *Bruguiera gymnorrhiza* performs better in lower-salinity mid zones (Cheng et al. 2025). Incorporating multiple species with complementary traits, such as root structures, canopy heights, and salt tolerance, enhances ecosystem resilience (Williams et al. 2017; McAfee et al. 2022). The species selection stage also integrates local preferences and cultural values, ensuring community acceptance and participation in maintenance activities (Hemmerling et al. 2023).

d. ***Spatial Design—Defining Width, Density, and Pattern***

The spatial configuration of bioshields determines their effectiveness against hydrodynamic forces. Design parameters such as width, planting density, and spatial arrangement are tailored to local exposure and hazard intensity (Daro Justine and Seenath 2025). For example, tsunami-prone areas may require belts 300–500 m wide with high-density planting, whereas cyclone-prone zones perform effectively with 50–200 m widths of mixed vegetation (Van Wesenbeeck et al. 2025). Erosion-prone coasts may suffice with narrower belts (20–50 m), strategically positioned along vulnerable stretches (Quang Bao 2011). Row spacing, stem density, and alignment are designed to balance light penetration, growth competition, and hydraulic resistance (Van Hespen et al. 2023). Computer models simulating wave attenuation can predict wave height reduction relative to vegetation density, helping optimize these parameters (e.g., Mazda et al. 2006).

Fig. 6.4 The use of bamboo fences for complementing mangrove colonization in Bau Ca Cai mangrove forest, Quang Ngai Province (Central Vietnam) (top) and recently planted mangrove seedlings with wooden support in Nha Mat, Bac Lieu Province (Mekong Delta, Vietnam) (bottom). *Photographs* Bijeesh Kozhikkodan Veettil

e. ***Hydrological Integration—Maintaining Natural Flow Regimes***

Bioshields are living systems that depend on the continuous interaction between vegetation and hydrology. Ensuring natural tidal exchange and sediment delivery is vital for long-term survival (Bryan et al. 2017; Wang et al. 2021a, b; López-Adame et al. 2026). Poorly designed embankments or drainage alterations can restrict water flow, leading to stagnation, salinity imbalance, or vegetation die-off (Vulliet et al. 2025).

Coastal bioshield designers often incorporate culverts, tidal gates, or controlled breaches to restore hydrological connectivity. In hybrid designs, such as those implemented in the Mekong Delta and Quang Ngai Province of Vietnam, bamboo and *Melaleuca* fences as well as hard structures (Fig. 6.4) are used to slow wave energy and promote sedimentation, complementing mangrove colonization (Van Cuong et al. 2015; Mai Van et al. 2021). Hydrological modeling can simulate tidal inundation frequency and sediment transport to fine-tune these interventions (Gijsman et al. 2021).

f. ***Implementation Planning—Phasing, Community Involvement, and Cost Estimation***

Implementation of coastal bioshields requires careful phasing to match ecological readiness with social and financial capacity (Piccolo et al. 2024). Initial planting typically begins during favorable seasons—after the monsoon or during low tidal amplitude—to maximize seedling establishment (Friess et al. 2024). Community involvement is crucial at every stage, including site preparation, planting, monitoring, and maintenance. Labor-intensive tasks such as seed collection and nursery management can be organized through local cooperatives, providing employment and building local stewardship. Cost estimation includes expenses for site preparation, planting materials, labor, monitoring equipment, and capacity-building programs (Aerts 2018). Financial models often incorporate ecosystem service valuation, such as carbon sequestration and storm protection, to justify long-term investment (Su et al. 2021).

g. ***Monitoring and Adaptive Management—Establishing Feedback Mechanisms***

Bioshields evolve with environmental dynamics; hence, a static design approach is inadequate. Continuous monitoring and adaptive management are essential for evaluating performance and guiding modifications (Thom 2000; Veettil et al. 2021; Paxton et al. 2024). Key monitoring indicators include vegetation survival rate, canopy cover, sediment accretion rates, biodiversity indices, and hydrological balance (Cadier et al. 2020; Liu et al. 2021; Yu et al. 2021). Remote sensing and field surveys complement each other to provide spatially explicit data on bioshield condition (Veettil et al. 2019b, 2022, 2024).

Adaptive management allows project teams to respond to unexpected outcomes, such as die-off due to salinity spikes or invasive species colonization, by adjusting species composition, density, or hydrological features (Xing et al. 2025). Institutionalizing feedback loops between field monitoring, local committees, and technical experts ensures that lessons learned inform future design iterations (Armstrong et al. 2025).

6.9.2 *Tools and Technologies*

Advancements in geospatial and computational tools have revolutionized bioshield design by enabling predictive modeling, spatial analysis, and real-time monitoring (Veettil et al. 2021). These technologies enhance precision, reduce uncertainty, and optimize decision-making. Geographic Information Systems (GIS) are fundamental to spatial planning (Rehman et al. 2022; Mubeen et al. 2025). GIS integrates topographic, hydrological, and land-use data to delineate potential bioshield zones, assess exposure levels, and identify areas suitable for different vegetation types (Bosso et al. 2025).

Wave attenuation and hydrodynamic simulation models quantify protective performance under various storm and tide scenarios (Vieira et al. 2024). Tools such as SWAN (Simulating Waves Nearshore) and XBeach can model how vegetation density and forest width influence wave dissipation and sediment dynamics (Baron-Hyppolite et al. 2019; Yin et al. 2022). These models guide the design of vegetation structure and layout to maximize coastal defense benefits.

Soil–vegetation interaction models assess nutrient cycling, salinity effects, and root–sediment stability (Peters et al. 2025). These models support species selection and planting density decisions, ensuring that bioshields are ecologically compatible with substrate conditions (Bathmann et al. 2021; Karimi et al. 2022; Yoshikai et al. 2022).

Emerging technologies such as drone-based surveys, LiDAR mapping, and machine learning–based predictive modeling further enhance design efficiency (Wang et al. 2023; Van Alphen et al. 2024; Nawaz et al. 2025). Drones provide high-resolution imagery for vegetation mapping and post-planting monitoring, while LiDAR data can quantify canopy height and structural density (Wang et al. 2023). Machine learning algorithms can analyze long-term climatic and hydrological data to forecast bioshield performance under future sea-level rise scenarios (Nawaz et al. 2025).

6.10 Evaluation of Design Performance

The long-term functionality and sustainability of bioshields depend on rigorous evaluation and continuous improvement based on measurable outcomes. Unlike conventional hard-engineered structures, bioshields are dynamic, living systems whose performance evolves in response to natural processes and human influences (Veettil et al. 2021). Therefore, assessing their effectiveness requires a multidimensional approach that integrates biophysical, ecological, and socioeconomic indicators within a structured monitoring and feedback framework (Murillas-Maza et al. 2023; Duvat et al. 2025). Evaluation is not a one-time exercise; it is an ongoing process that ensures bioshields continue to deliver intended protective, ecological, and social benefits under changing environmental and climatic conditions. This section discusses the core performance indicators for bioshield assessment and outlines an integrated monitoring framework for adaptive management and long-term sustainability.

6.10.1 Performance Indicators

Performance indicators provide quantifiable metrics to evaluate the success of bioshield design and implementation. These indicators should capture the protective efficiency (Duvat et al. 2025), ecological integrity (Zhao et al. 2025), and social acceptance (Paxton et al. 2024) of the bioshield. The following are the key categories of performance measures used in bioshield evaluation.

a. ***Wave Attenuation (%)***

Wave attenuation is the most direct indicator of the protective functions of a bioshield. It quantifies the reduction in wave height or energy as waves pass through vegetated zones. Field measurements (Wei et al. 2024), hydrodynamic modeling (Vinh 2015), and laboratory flume experiments (Peruzzo et al. 2018) are used to determine attenuation rates under various conditions.

Empirical studies, such as Mazda et al. (2006) and Quartel et al. (2007), have shown that mangrove forests can reduce wave heights by 20–50% over 100 m of forest width, depending on species composition, stem density, and water depth. For example, dense stands of *Rhizophora apiculata* in the Mekong Delta demonstrated wave energy reductions exceeding 60% during monsoonal storms (Cuc et al. 2015).

Regular monitoring of wave attenuation through in situ sensors and hydrodynamic models can provide feedback for adjusting vegetation density or expanding forest width. The performance of different zones (seaward, mid, and landward) should also be compared to evaluate their relative contributions to overall protection.

b. ***Sediment Accretion Rate (mm/year)***

Sediment accretion indicates the capacity of a coastal bioshield to stabilize the shoreline and maintain vertical growth in response to sea-level rise (Liu et al. 2021). Measurements are typically conducted using Surface Elevation Tables (SETs), marker horizons, and sediment traps (Callaway et al. 2013).

Accretion rates vary widely depending on sediment supply and vegetation structure. In sediment-rich deltas such as the Sundarbans, mangrove sedimentation rates of 2–18 mm/year have been recorded, effectively counterbalancing the regional sea-level rise of approximately 5 mm/year (Hale et al. 2019). In contrast, low-sediment sandy coasts, such as in Central Vietnam, may require hydrological restoration or sediment augmentation to sustain bioshield growth (Veettil et al. 2021).

Monitoring sediment accretion helps determine whether the bioshield is achieving vertical resilience (Liu et al. 2021). Declining accretion rates can signal hydrological disruption, excessive erosion, or vegetation degradation—prompting management interventions such as channel restoration or replanting.

c. ***Vegetation Survival and Growth (%)***

Vegetation survival rate is one of the most important indicators of bioshield establishment success. It reflects the suitability of site conditions, planting techniques, and species selection. Typically, survival rates above 70–90% within 3 months are considered satisfactory (Hai et al. 2020).

In early stages (0–2 years), survival is heavily influenced by physical stressors such as wave action, sediment burial, or grazing (Sreeranga and Lei 2026). Long-term survival (beyond five years) depends on ecological succession, canopy closure, and nutrient cycling (Lopez Abuchar et al. 2025). Monitoring growth parameters such as height increment, stem diameter, canopy density, and root development provides further insight into ecosystem maturity and resilience (Aye et al. 2024).

Growth performance can also be linked to hydrological and soil parameters. For instance, *Avicennia marina* shows optimal growth under moderate salinity but declines at very high salinity, while *Bruguiera gymnorrhiza* requires relatively stable, lower salinity conditions. Identifying such thresholds is essential for adaptive species management (Basak et al. 2004; Nguyen et al. 2015a, b).

d. ***Biodiversity Indices***

Biodiversity serves as a proxy for ecological health and functional resilience. High species diversity enhances ecosystem stability, nutrient cycling, and resistance to disturbances such as pest outbreaks or storms. For example, the Shannon–Weiner index (H′) is widely used to quantify species diversity, combining richness (number of species) and evenness (distribution of individuals).

$$H' = -\Sigma(p_i \times \ln\ p_i)$$

where p_i is the proportion of individuals belonging to species i in the sample.

An increase in biodiversity over time, including both in flora and fauna, indicates successful habitat creation and ecological integration. Bird abundance, crustacean density, and presence of fish nursery are additional indicators of bioshield ecological performance (Cleland 2011).

Comparing biodiversity between restored and natural reference sites helps evaluate restoration success and functional equivalence (Gerona-Daga and Salmo III 2022). In Vietnam's Ca Mau province, mature mangrove bioshields have demonstrated faunal diversity comparable to natural stands within 10–15 years, suggesting high ecological recovery potential (Tinh et al. 2022).

e. ***Community Satisfaction and Benefit Indicators***

Socioeconomic indicators assess how bioshields contribute to community well-being, local livelihoods, and social cohesion (Ebeler et al. 2025). Metrics include community satisfaction surveys, perceived protection levels, income diversification, and participation in maintenance activities (Ebeler et al. 2025).

In areas like Bangladesh (Hossain et al. 2025) and the Philippines (Valenzuela et al. 2020), communities have reported increased safety, improved fishery resources, and enhanced tourism value following coastal bioshield establishment. Monitoring social outcomes ensures that bioshield projects remain people-centered and equitable.

Community-based monitoring also serves as an early warning system—residents often detect subtle ecological or structural changes before technical teams do (Danielsen et al. 2021). Therefore, integrating community satisfaction with biophysical metrics provides a holistic picture of bioshield performance.

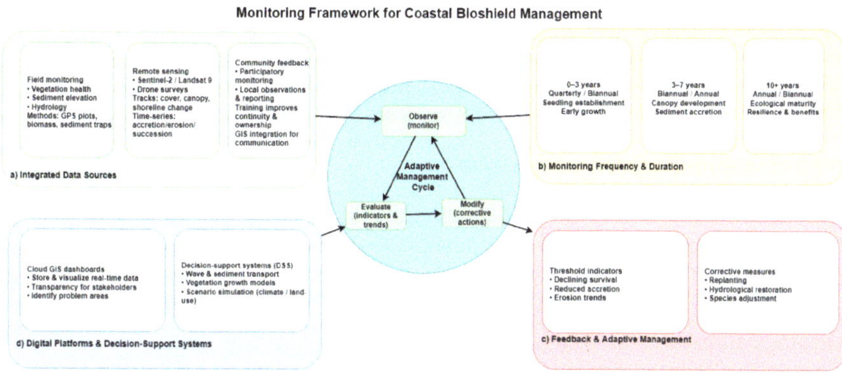

Fig. 6.5 Monitoring framework for coastal bioshield management

6.10.2 Monitoring Framework

An effective monitoring framework (Fig. 6.5) combines scientific measurement with participatory observation to provide continuous, adaptive feedback for bioshield management (Hedge et al. 2017). Because bioshields function as evolving ecosystems, long-term observation is crucial for capturing slow ecological changes, sediment dynamics, and social impacts (Paterson et al. 2018).

a. ***Integrated Data Sources***

Monitoring should integrate field-based data, remote sensing, and community feedback to provide a comprehensive evaluation. Field monitoring provides fine-scale, high-resolution data on vegetation health, sediment elevation, and hydrological parameters. Techniques include GPS-based plot surveys, biomass measurements, and sediment trap analysis. Remote sensing, particularly using satellite imagery (e.g., Sentinel-2, Landsat 9) or drone surveys, enables large-scale temporal tracking of vegetation cover, canopy density, and shoreline change (e.g., Sunny et al. 2022). Time-series analysis helps detect trends in accretion, erosion, and vegetation succession (Veettil et al. 2020). Participatory monitoring involves local stakeholders in routine observation, data collection, and reporting (Burdon et al. 2019). Training community members to record basic environmental indicators enhances data continuity and builds ownership (Danielsen et al. 2021). Integration of these datasets through GIS platforms allows visualization of bioshield evolution over time and facilitates communication among scientists, policymakers, and communities.

b. ***Monitoring Frequency and Duration***

Monitoring should be continuous but structured in phases. Initial post-planting years (0–3 years) require high-frequency monitoring—quarterly or biannually—to assess seedling establishment and early growth (Hai et al. 2020). Intermediate stages (3–7 years) can be monitored biannually or annually, focusing on canopy development and sediment accretion (Gatt et al. 2022). Long-term monitoring (beyond

10 years) may occur annually or biannually to evaluate ecological maturity, resilience, and community benefits (Gatt et al. 2022). Data collection schedules should align with seasonal cycles (e.g., pre- and post-monsoon) to account for variability in hydrodynamic conditions.

c. ***Feedback and Adaptive Management***

The goal of monitoring is to inform adaptive management, which is a cyclical process of observation, evaluation, and modification (Gatt et al. 2022). When indicators fall below threshold levels, such as declining vegetation survival or reduced accretion rates, corrective measures should be implemented. These may include replanting, hydrological restoration, or introducing more suitable species (Gatt et al. 2022).

Adaptive feedback loops involve all stakeholders: local communities, technical experts, and government agencies. Establishing periodic review meetings ensures that monitoring results translate into tangible management decisions (Gatt et al. 2022). Documenting lessons learned across projects builds a knowledge base that strengthens future bioshield design standards.

d. ***Use of Digital Platforms and Decision-Support Systems***

Digital platforms enhance transparency, accessibility, and efficiency in bioshield monitoring. Cloud-based GIS dashboards can store and visualize real-time data, allowing stakeholders to track progress and identify problem areas (Hodbod et al. 2019). Decision-support systems (DSS) that integrate wave modeling, sediment transport data, and vegetation growth models can simulate future performance scenarios under changing climate or land-use conditions (Niswonger et al. 2014).

In Vietnam and Indonesia, pilot programs have begun linking community-based monitoring with mobile applications that allow users to upload photos, GPS coordinates, and qualitative observations (Pratihast et al. 2013; Matondang et al. 2024). These innovations democratize data collection and ensure continuous, decentralized evaluation (Johnson et al. 2021).

6.11 Conclusions

The design of coastal bioshields constitutes an integrative framework that combines ecological science, engineering principles, and social collaboration. As forms of living infrastructure, coastal bioshields differ fundamentally from conventional hard-engineering structures in their capacity to grow, adapt, and co-evolve with coastal processes. Consequently, effective bioshield design necessitates the deliberate integration of ecological integrity, structural performance, and social inclusivity. When planned and implemented holistically, bioshields not only mitigate coastal hazards—such as storm surges, erosion, and tsunamis—but also enhance biodiversity, restore ecosystem services, and support the sustainability of local livelihoods.

Experiences across Asia, particularly in Vietnam, Bangladesh, India, Indonesia, and the Philippines, have underscored the critical importance of context-specific bioshield design informed by local hydrological and geomorphological conditions, as well as meaningful community participation. Shortcomings in early initiatives frequently arose from oversimplified approaches, including the use of monocultures and inappropriate site selection, thereby highlighting the necessity of multidisciplinary collaboration and long-term adaptive management frameworks.

In the context of accelerating climate change, bioshield design must evolve from static, one-off interventions toward dynamic and adaptive systems capable of responding to sea-level rise, sediment redistribution, and increasingly extreme weather events. The integration of empirical models of wave attenuation and sediment accretion with traditional ecological knowledge can enhance both functional performance and social acceptance. Furthermore, embedding bioshields within broader coastal landscape planning, by linking mangroves, dunes, seagrass meadows, and human settlements, strengthens ecological connectivity and overall system resilience.

The future of bioshield design rests on ecosystem-based and community-centered frameworks that integrate contemporary scientific knowledge with indigenous wisdom and participatory governance mechanisms. Such approaches ensure that bioshields operate as sustainable and multifunctional systems, simultaneously safeguarding coastlines, supporting local livelihoods, and preserving the cultural and ecological heritage of coastal communities. In this regard, bioshields exemplify the capacity of nature-based solutions to reconcile human security objectives with environmental stewardship in the context of accelerating global change.

References

Adame MF, Neil D, Wright SF, Lovelock CE (2010) Sedimentation within and among mangrove forests along a gradient of geomorphological settings. Estuar Coast Shelf Sci 86(1):21–30. https://doi.org/10.1016/j.ecss.2009.10.013

Aerts JC (2018) A review of cost estimates for flood adaptation. Water 10(11):1646. https://doi.org/10.3390/w10111646

Ahmed MN, Mohanadhas B (2025) Influence of river vegetation on sediment transport and erosion mitigation: a review. Res Cold Arid Reg. https://doi.org/10.1016/j.rcar.2025.07.001

Ahmed MU, Alam MI, Debnath S, Debrot AO, Rahman MM, Ahsan MN, Verdegem M (2023) The impact of mangroves in small-holder shrimp ponds in south-west Bangladesh on productivity and economic and environmental resilience. Aquaculture 571:739464. https://doi.org/10.1016/j.aquaculture.2023.739464

Airoldi L (2003) The effects of sedimentation on rocky coast assemblages. In: Oceanography and marine biology. CRC Press, pp 169–171

Albers T, Schmitt K (2015) Dyke design, floodplain restoration and mangrove co-management as parts of an area coastal protection strategy for the mud coasts of the Mekong Delta, Vietnam. Wetlands Ecol Manage 23(6):991–1004. https://doi.org/10.1007/s11273-015-9441-3

Alcántara I, Goudie AS (eds) (2010) Geomorphological hazards and disaster prevention. Cambridge University Press, pp 304

An L, Wang J, Xiong N, Wang Y, You J, Li H (2022) Assessment of permeability windbreak forests with different porosities based on laser scanning and computational fluid dynamics. Remote Sens 14(14):3331. https://doi.org/10.3390/rs14143331

Anbleyth-Evans J (2022) Marine ecological democracy: participatory marine planning in indigenous marine areas in Chile. In: Misiune I, Depellegrin D, Egarter Vigl L (eds) Human-nature interactions. Springer, Cham. https://doi.org/10.1007/978-3-031-01980-7_25

Anh HH, Beaulieu A, Da Hanh TM, Tru LC (2024) Economic effects of a controlled mangrove-to-pond coverage ratio policy on mangrove conservation and shrimp farming: a case study in Vietnam using instrumental regression analysis. Trees Forests People 16:100579. https://doi.org/10.1016/j.tfp.2024.100579

Antão LH, Weigel B, Strona G, Hällfors M, Kaarlejärvi E, Dallas T, Opedal ØH, Heliölä J, Henttonen H, Huitu O, Korpimäki E, Kuussaari M, Lehikoinen A, Leinonen R, Lindén A, Merilä P, Pietiäinen H, Pöyry J, Salemaa M, Laine A (2022) Climate change reshuffles northern species within their niches. Nat Clim Chang 12(6):587–592. https://doi.org/10.1038/s41558-022-01381-x

Anthony EJ, Brunier G, Besset M, Goichot M, Dussouillez P, Nguyen VL (2015) Linking rapid erosion of the Mekong River delta to human activities. Sci Rep 5(1):14745. https://doi.org/10.1038/srep14745

Arifanti VB, Basyuni M, Suharti S, Slamet B, Karlina E, Sidik F, Ali HM (2025) Assessing the environmental and socioeconomic impacts of mangrove loss in Indonesia: a synthesis for science-based policy. For Sci Technol 21(4):430–446. https://doi.org/10.1080/21580103.2025.2536595

Armstrong KC, Rakowsky JA, Ferguson Irlanda CE, Kilcullen F, Nestor V, Crowder LB, Ramos J (2025) A framework for just seascape restoration. Npj Ocean Sustain 4(1):4. https://doi.org/10.1038/s44183-025-00106-9

Athikalam PT, Vaideeswaran AK (2021) Vegetation bioshield for coastal protection in South Asia: status and way forward. J Coast Conserv 26:3. https://doi.org/10.1007/s11852-022-00850-x

Aye WN, Tong X, Li J, Tun AW (2024) Assessing the carbon storage potential of a young mangrove plantation in Myanmar. Forests 14(4):824. https://doi.org/10.3390/f14040824

Baaij BM, Kooijman J, Limpens J, Marijnissen RJ, van Loon-Steensma JM (2021) Monitoring impact of salt-marsh vegetation characteristics on sedimentation: an outlook for nature-based flood protection. Wetlands 41(6):76. https://doi.org/10.1007/s13157-021-01467-w

Balke T, Vovides A, Schwarz C, Chmura GL, Ladd C, Basyuni M (2021) Monitoring tidal hydrology in coastal wetlands with the "Mini Buoy": applications for mangrove restoration. Hydrol Earth Syst Sci 25:1229–1244. https://doi.org/10.5194/hess-25-1229-2021

Barbier EB (2017) Marine ecosystem services. Curr Biol 27(11):R507–R510. https://doi.org/10.1016/j.cub.2017.03.020

Baron-Hyppolite C, Lashley CH, Garzon J, Miesse T, Ferreira C, Bricker JD (2019) Comparison of implicit and explicit vegetation representations in SWAN hindcasting wave dissipation by coastal wetlands in Chesapeake bay. Geosciences 9(1):8. https://doi.org/10.3390/geosciences9010008

Basak UC, Gupta N, Rautaray S, Das P (2004) Effects of salinity on the growth of mangrove seedlings. J Trop for Sci 16:437–443

Bathmann J, Peters R, Reef R, Berger U, Walther M, Lovelock CE (2021) Modelling mangrove forest structure and species composition over tidal inundation gradients: the feedback between plant water use and porewater salinity in an arid mangrove ecosystem. Agric For Meteorol 308–309:108547. https://doi.org/10.1016/j.agrformet.2021.108547

Beeston M, Cameron C, Hagger V, Howard J, Lovelock C, Sippo J, Tonneijk F, van Bijsterveldt C, van Eijk P (eds) (2023) Best practice guidelines for mangrove restoration, p 279

Benazir, Triatmadja R, Syamsidik, Nizam, Warniyati (2024). Vegetation-based approached for tsunami risk reduction: insights and challenges. Prog Disaster Sci 23:100352. https://doi.org/10.1016/j.pdisas.2024.100352

Bennett AF (2003) Linkages in the landscape: the role of corridors and connectivity in wildlife conservation. IUCN, Gland, Switzerland and Cambridge, UK. xiv + 254 pp

Best ÜSN, van der Wegen M, Dijkstra J, Reyns J, van Prooijen BC, Roelvink D (2022) Wave attenuation potential, sediment properties and mangrove growth dynamics data over Guyana's

intertidal mudflats: assessing the potential of mangrove restoration works. Earth Syst Sci Data 14:2445–2462. https://doi.org/10.5194/essd-14-2445-2022
Bhanja A, Payra P, Sau I, Makur S, Biswas D, Bhunia T, Sinha N (2025) Diversity of naturally colonized true mangroves and mangrove associates along the bank of Champa Canal, West Bengal, India, with emphasis on different environmental factors. Curr World Environ 20(2):798. https://doi.org/10.12944/CWE.20.2.20
Binh NA, Hauser LT (2025) Mapping mangrove multi-trait functional diversity from satellite observations across dense and fragmented stands using spectral-biophysical derivatives. Sci Rep 15(1):22116. https://doi.org/10.1038/s41598-025-05397-z
Binh TV, Mau LD, Phai VV (2019) Some issues of geomorphology to serve coastal environmental management of Quang Nam province. Vietnam J Mar Sci Technol 19(4A):79–91. https://doi.org/10.15625/1859-3097/19/4A/14599
Binh DV, Kantoush S, Sumi T (2020) Changes to long-term discharge and sediment loads in the Vietnamese Mekong Delta caused by upstream dams. Geomorphology 353:107011. https://doi.org/10.1016/j.geomorph.2019.107011
Bird EC (2008) Coastal geomorphology: an introduction. John Wiley & Sons, pp 436
Blankespoor B, Dasgupta S, Lange GM (2017) Mangroves as a protection from storm surges in a changing climate. Ambio 46(4):478–491. https://doi.org/10.1007/s13280-016-0838-x
Bosso L, Saviano S, Abagnale M, Bellardini D, Bolinesi F, Botte V, Buondonno A, Carotenuto Y, Casotti R, Chiusano ML, Cipolletta F, Conversano F, De Domenico F, Del Gaizo G, Donnarumma V, Furia M, Iudicone D, Kokoszka F, Laface F, Cianelli D (2025) GIS-based integration of marine data for assessment and management of a highly anthropized coastal area. Sci Rep 15(1):16200. https://doi.org/10.1038/s41598-025-00206-z
Breyne J, Dufrêne M, Maréchal K (2021) How integrating 'socio-cultural values' into ecosystem services evaluations can give meaning to value indicators. Ecosyst Serv 49:101278. https://doi.org/10.1016/j.ecoser.2021.101278
Bryan KR, Nardin W, Mullarney JC, Fagherazzi S (2017) The role of cross-shore tidal dynamics in controlling intertidal sediment exchange in mangroves in Cù Lao Dung, Vietnam. Continental Shelf Res 147:128–143. https://doi.org/10.1016/j.csr.2017.06.014
Buenavista D, Purnobasuki H (2023) People and mangroves: biocultural utilization of mangrove forest ecosystem in Southeast Asia. J Mar Island Cult 12(2):95–115. https://doi.org/10.21463/jmic.2023.12.2.07
Bui HH, Nguyen TP, Pham VH, Ho KL (2025) Exploring the interplay of social, economic, and environmental factors on livelihood sustainability in Quang Tri's coastal forest areas. Sustainability 17(17):7661. https://doi.org/10.3390/su17177661
Burdon D, Potts T, McKinley E, Lew S, Shilland R, Gormley K, Thomson S, Forster R (2019) Expanding the role of participatory mapping to assess ecosystem service provision in local coastal environments. Ecosyst Serv 39:101009. https://doi.org/10.1016/j.ecoser.2019.101009
Busoms S, Teres J, Huang XY, Bomblies K, Danku J, Douglas A, Salt DE (2015) Salinity is an agent of divergent selection driving local adaptation of Arabidopsis to coastal habitats. Plant Physiol 168(3):915–929. https://doi.org/10.1104/pp.15.00427
Cadier C, Bayraktarov E, Piccolo R, Adame MF (2020) Indicators of coastal wetlands restoration success: a systematic review. Front Mar Sci 7:600220. https://doi.org/10.3389/fmars.2020.600220
Cahoon DR (2024) Measuring and interpreting the surface and shallow subsurface process influences on coastal wetland elevation: a review. Estuaries and Coasts 47:1708–1734. https://doi.org/10.1007/s12237-024-01332-z
Calado H, Gutierrez D, De Bruyn A (2025) Navigating trade-offs on conservation: the use of participatory mapping in maritime spatial planning. NPJ Ocean Sustain 4(1):8. https://doi.org/10.1038/s44183-025-00109-6
Callaway JC, Cahoon DR, Lynch JC (2013) The surface elevation table–marker horizon method for measuring wetland accretion and elevation dynamics. Methods Biogeochem Wetlands 10:901–917. https://doi.org/10.2136/sssabookser10.c46

Carlos C, Delfino RJ, Juanico DE, David L, Lasco R (2015) Vegetation resistance and regeneration potential of Rhizophora, Sonneratia, and Avicennia in the Typhoon Haiyan-affected mangroves in the Philippines: implications on rehabilitation practices. Clim Disaster Develop J 1(1):1–8. https://doi.org/10.18783/cddj.v001.i01.a01

Carter HN, Schmidt SW, Hirons AC (2015) An international assessment of mangrove management: incorporation in integrated coastal zone management. Diversity 7(2):74–104. https://doi.org/10.3390/d7020074

Charoenlerkthawin W, Bidorn K, Burnett WC, Sopon A, Otarawanna S, Bidorn B (2024) Evaluating the effectiveness of Mangrove rehabilitation: a novel approach for sustainable coastal management. J Environ Manage 365:121673. https://doi.org/10.1016/j.jenvman.2024.121673

Chaudhuri P, Chaudhuri S, Ghosh R (2019) The role of mangroves in coastal and estuarine sedimentary accretion in Southeast Asia. In: Sedimentary processes-examples from Asia, Turkey and Nigeria. IntechOpen

Chávez V, Lithgow D, Losada M, Silva-Casarin R (2021) Coastal green infrastructure to mitigate coastal squeeze. J Infrastruct Preserv Resilience 2(1):7. https://doi.org/10.1186/s43065-021-00026-1

Chen X, Zhang M, Jiang H (2022) Morphological characteristics and hydrological connectivity evaluation of tidal creeks in coastal wetlands. Land 11(10):1707. https://doi.org/10.3390/land11101707

Chen C, Zhang C, Tian B, Wu W, Zhou Y (2023) Mapping intertidal topographic changes in a highly turbid estuary using dense Sentinel-2 time series with deep learning. ISPRS J Photogramm Remote Sens 205:1–16. https://doi.org/10.1016/j.isprsjprs.2023.09.022

Chen Y, Cao Z, Chen Y, Wang L, Shan L, Yu S, Yu K (2025) Evaluation of the stand quality of *Casuarina equisetifolia* in coastal shelterbelts. J Mt Sci 22:1557–1570. https://doi.org/10.1007/s11629-024-9265-9

Cheng H, Chen Y, Peng Y, Wei M, Niu J (2025) Hydraulic traits constrain salinity-dependent niche segregation in mangroves. Plants 14(12):1850. https://doi.org/10.3390/plants14121850

Choudhary B, Dhar V, Pawase AS (2024) Blue carbon and the role of mangroves in carbon sequestration: its mechanisms, estimation, human impacts and conservation strategies for economic incentives. J Sea Res 199:102504. https://doi.org/10.1016/j.seares.2024.102504

Chowdhury A, Naz A, Iyer AS, Bhattacharyya S (2021) Ecosystem based disaster risk reduction at Indian Sundarbans: a lesson learned from AMPHAN supercyclone. IOP Conf Ser: Earth Environ Sci 796(1):012042. https://doi.org/10.1088/1755-1315/796/1/012042

Chowdhury A, Naz A, Sharma SB, Dasgupta R (2023) Changes in salinity, mangrove community ecology, and organic blue carbon stock in response to cyclones at Indian Sundarbans. Life 13(7):1539. https://doi.org/10.3390/life13071539

Ciarletta DJ, Shawler JL, Tenebruso C, Hein CJ, Lorenzo-Trueba J (2019) Reconstructing coastal sediment budgets from beach- and foredune-ridge morphology: a coupled field and modeling approach. J Geophys Res Earth Surf 124(6):1398–1416. https://doi.org/10.1029/2018JF004908

Ciccarelli D, Bona C (2022) Exploring the functional strategies adopted by coastal plants along an ecological gradient using morpho-functional traits. Estuaries Coasts 45:114–129. https://doi.org/10.1007/s12237-021-00945-y

Clarke PJ, Allaway WG (1993) The regeneration niche of the grey mangrove (*Avicennia marina*): effects of salinity, light and sediment factors on establishment, growth and survival in the field. Oecologia 93:548–556. https://doi.org/10.1007/BF00328964

Cleland EE (2011) Biodiversity and ecosystem stability. Nat Educ Knowl 3(10):14

Crosby SC, Hudson DM, Hughes AR, Bartholet A, Burns KT, Donato MK, Susarchick J (2024) Structure and function of restored and natural salt marshes: implications for ecosystem resilience and adaptive potential. Estuaries Coasts 47(6):1561–1578. https://doi.org/10.1007/s12237-024-01395-y

Cuc NTK, Suzuki T, van Steveninck EDDR, Hai H (2015) Modelling the impacts of mangrove vegetation structure on wave dissipation in Ben Tre Province, Vietnam, under different climate

change scenarios. J Coastal Res 31(2):340–347. https://doi.org/10.2112/JCOASTRES-D-12-00271.1
Cutts V, Erftemeijer PLA, Gaffi L, Hagemeijer W, Smith RK, Taylor NG, Sutherland WJ (eds) (2024) Restoration, creation and management of salt marshes and tidal flats: a collation of evidence-based guidance. Report—conservation evidence, wetlands international and the world coastal forum. https://doi.org/10.52201/CGSCOL1/LCNC6109
Da Silva AE, Maracajá KF, Batalhão AC, Silva VF, Borges IM (2025) Ecotourism and co-management: strengthening socio-ecological resilience in local food systems. Sustainability 17(6):2443. https://doi.org/10.3390/su17062443
Dahdouh-Guebas F, Hugé J, Abuchahla GM, Cannicci S, Jayatissa LP, Kairo JG, Kodikara Arachchilage S, Koedam N, Mafaziya Nijamdeen TW, Mukherjee N, Poti M, Prabakaran N, Ratsimbazafy HA, Satyanarayana B, Thavanayagam M, Vande Velde K, Wodehouse D (2021) Reconciling nature, people and policy in the mangrove social-ecological system through the adaptive cycle heuristic. Estuar Coast Shelf Sci 248:106942. https://doi.org/10.1016/j.ecss.2020.106942
Dai E, Zhu J, Wang X, Xi W (2018) Multiple ecosystem services of monoculture and mixed plantations: a case study of the Huitong experimental forest of Southern China. Land Use Policy 79:717–724. https://doi.org/10.1016/j.landusepol.2018.08.014
Damastuti E, Van Wesenbeeck BK, Leemans R, De Groot RS, Silvius MJ (2022) Effectiveness of community-based mangrove management for coastal protection: a case study from Central Java, Indonesia. Ocean Coast Manage 238:106498. https://doi.org/10.1016/j.ocecoaman.2023.106498
Danesh Miah M, Abubokor Siddik M, Yong Shin M (2013) Socio-economic and environmental impacts of casuarina shelterbelt in the Chittagong coast of Bangladesh. For Sci Technol 9(3):156–163. https://doi.org/10.1080/21580103.2013.814592
Danielsen F, Enghoff M, Poulsen MK, Funder M, Jensen PM, Burgess ND (2021) The concept, practice, application, and results of locally based monitoring of the environment. BioScience 71(5):484–502. https://doi.org/10.1093/biosci/biab021
Daro Justine YE, Seenath A (2025) Vegetative nature-based solutions for coastal flood risk management: benefits, challenges, and uncertainties. Ocean Coast Manag 261:107520. https://doi.org/10.1016/j.ocecoaman.2024.107520
Dasgupta S, Islam MS, Huq M, Huque Khan Z, Hasib MR (2019) Quantifying the protective capacity of mangroves from storm surges in coastal Bangladesh. PLoS ONE 14(3):e0214079. https://doi.org/10.1371/journal.pone.0214079
Davidson-Arnott R, Bauer B, Houser C (eds) (2019) Introduction to coastal processes and geomorphology. Cambridge University Press, pp 442
De Zoysa M (2008) Casuarina coastal forest shelterbelts in Hambantota City, Sri Lanka: assessment of impacts. Small-Scale For 7(1):17–27. https://doi.org/10.1007/s11842-008-9038-2
De Zoysa M (2022) Ecotourism development and biodiversity conservation in Sri Lanka: objectives, conflicts and resolutions. Open J Ecol 12:638–666. https://doi.org/10.4236/oje.2022.1210037
De Santana CO, Spealman P, Melo VMM, Gresham D, De Jesus TB, Chinalia FA (2021) Effects of tidal influence on the structure and function of prokaryotic communities in the sediments of a pristine Brazilian mangrove. Biogeosciences 18(7):2259–2273. https://doi.org/10.5194/bg-18-2259-2021
De Sedas A, González Y, Winter K, Lopez OR (2019) Seedling responses to salinity of 26 Neotropical tree species. AoB PLANTS 11(6). https://doi.org/10.1093/aobpla/plz062
Delaney AE, Frangoudes K (2024) Coastal and maritime cultural heritage: from the European Union to East Asia and Latin America. Maritime Stud 23:26. https://doi.org/10.1007/s40152-024-00369-x
Derakhshan S, Dodangeh E, Afzalimehr H, Raeisifar H, Karakouzian M (2022) Effect of vertical variation of submerged vegetation density on the flow resistance. Hydrology 9(11):191. https://doi.org/10.3390/hydrology9110191
Dhawi F (2025) Mastering resilience: avicennia marina's survival in hypersaline arid zones. Front Sustain Food Syst 9:1598548. https://doi.org/10.3389/fsufs.2025.1598548

Doan TV (2012) The role of mangroves in reducing storm waves in in Dai Hop area (Kien Thuy, Hai Phong) (in Vietnamese). Vietnam J Mar Sci Technol 11(1):43–55. https://doi.org/10.15625/1859-3097/11/1/365

Dodman D, Hayward B, Pelling M, Castán Broto V, Chow W, Chu E, Ziervogel G (2023) Cities, settlements and key infrastructure. IPCC 6th assessment report

Dookie S (2025) Mangroves as sacred forests: connecting people and nature. Plant Perspect. https://doi.org/10.3197/whppp.63876246815917

Dulyakasem R, Brooks C, Lehnert SL, Newton P (2026) Community-based natural resource management in coastal communities: the contribution of mangroves to household livelihoods in southern Thailand. Environ Develop 57:101340. https://doi.org/10.1016/j.envdev.2025.101340

Duvat VK, Hatton I, Burban L, Jacobée A, Stahl L (2025) Assessing nature-based coastal defense. Sci Rep 15(1):16798. https://doi.org/10.1038/s41598-025-96744-7

Duy DV, Ty TV, Phat LT, Minh HV, Nam ND, Downes NK, Avtar R, Tanaka H (2025) Simulating the coastal protection performance of breakwaters in the Mekong delta: insights from the western coast of Ca Mau Province, Vietnam. J Mar Sci Eng 13(8):1559. https://doi.org/10.3390/jmse13081559

Dyring M, Hofmann H, Stanton D, Moss P, Froend R (2023) Ecohydrology of coastal aquifers in humid environments and implications of a drying climate. Ecohydrology 16(2):e2491. https://doi.org/10.1002/eco.2491

Ebeler L, Balke T, Laurie E, Le H, Nguyen T, Renaud FG (2025) Local-scale impacts of mangrove restoration and conservation on coastal communities in two Vietnamese deltas: socio-economic and institutional dynamics. Ocean Coast Manag 269:107812. https://doi.org/10.1016/j.ocecoaman.2025.107812

Edwards DP, Cerullo GR (2024) Biodiversity is central for restoration. Curr Biol 34(9):R371–R379. https://doi.org/10.1016/j.cub.2024.02.032

Ellison AM, Felson AJ, Friess DA (2020) Mangrove rehabilitation and restoration as experimental adaptive management. Front Mar Sci 7:536464. https://doi.org/10.3389/fmars.2020.00327

Erasto DA (2021) Assessment of structure and natural regeneration capacity of *Avicennia marina* and *Bruguiera gymnorrhiza* species of mangroves in Mida Creek Kilifi County, Kenya. PhD Thesis, Pwani University, Kenya

Fazlioglu F, Chen L (2020) Introduced non-native mangroves express better growth performance than co-occurring native mangroves. Sci Rep 10(1):3854. https://doi.org/10.1038/s41598-020-60454-z

Feehan CJ, Thomsen MS, Wernberg T, Miles T (2024) Ecosystem damage by increasing tropical cyclones. Commun Earth Environ 5(1):674. https://doi.org/10.1038/s43247-024-01853-2

Feng X, Zhu C, Liu JP, Jia Y (2023) Sediment dynamics in coastal and marine environments: scientific advances. Water 15(7):1404. https://doi.org/10.3390/w15071404

Feng X, Deng Y, Zhong W, Xie Z, Liu H, Li Z, Jia Y, Li X, Chen R, Peng X, Deng Y, Li M, Li M, Guo D (2025) Tracking the expansion of *Sonneratia apetala* and its impact on local mangroves using time-series remote sensing data. Sustainability 17(3):1069. https://doi.org/10.3390/su17031069

Fernández-Pascual E, Pérez-Arcoiza A, Prieto JA, Díaz TE (2017) Environmental filtering drives the shape and breadth of the seed germination niche in coastal plant communities. Ann Bot 119(7):1169–1177

Fisher S, Goff J, Cundy AB, Sear D, McWhorter C (2025) Forested bioshields and tsunami impact mitigation in a Polynesian setting. Nature-Based Solutions 7:100222. https://doi.org/10.1016/j.nbsj.2025.100222

Friess DA, Shribman ZI, Stankovic M, Iram N, Baustian MM, Lewis CJE (2024) Restoring blue carbon ecosystems. Cambridge Prisms: Coastal Futures 2:e9. https://doi.org/10.1017/cft.2024.9

Fu D, Wu X, Hu L, Ma X, Shen C, Shang H, Huang G, He Y, Duan C (2023) Plant traits guide species selection in vegetation restoration for soil and water conservation. Biology 12(4):618. https://doi.org/10.3390/biology12040618

Gairola SU, Bahuguna R, Bhatt SS (2023) Native plant species: a tool for restoration of mined lands. J Soil Sci Plant Nutr 23(2):1438–1448. https://doi.org/10.1007/s42729-023-01181-y

Gao X, Leung F, Wang W (2025) Assessing typhoon impacts on mangroves and post-typhoon recovery patterns to inform projects using mangroves as nature-based solutions. Glob Ecol Conserv 61:e03667. https://doi.org/10.1016/j.gecco.2025.e03667

Garmaeepour R, Alambeigi A, Danehkar A, Shabani AA (2025) Mangrove forest ecosystem services and the social well-being of local communities: unboxing a dilemma. J Nat Conserv 84:126827. https://doi.org/10.1016/j.jnc.2025.126827

Gatt YM, Andradi-Brown DA, Ahmadia GN, Martin PA, Sutherland WJ, Spalding MD, Donnison A, Worthington TA (2022) Quantifying the reporting, coverage and consistency of key indicators in mangrove restoration projects. Front For Glob Change 5:720394. https://doi.org/10.3389/ffgc.2022.720394

Gerona-Daga MEB, Salmo SG III (2022) A systematic review of mangrove restoration studies in Southeast Asia: challenges and opportunities for the United Nation's Decade on Ecosystem Restoration. Front Mar Sci 9:987737. https://doi.org/10.3389/fmars.2022.987737

Giannini TC, Giulietti AM, Harley RM, Viana PL, Jaffe R, Alves R, Pinto CE, O Mota NF, Caldeira CF, Imperatriz-Fonseca VL, Furtini AE, Siqueira JO (2017) Selecting plant species for practical restoration of degraded lands using a multiple-trait approach. Austral Ecol 42(5):510–521. https://doi.org/10.1111/aec.12470

Giesen W, Wulffraat S, Zieren M, Scholten L (eds) (2007) Mangrove guidebook for Southeast Asia. RAP Publication, pp 769

Gijsman R, Horstman EM, Friess DA, Swales A, Wijnberg KM (2021) Nature-based engineering: a review on reducing coastal flood risk with mangroves. Front Mar Sci 8:702412. https://doi.org/10.3389/fmars.2021.702412

Gijsman R, Horstman EM, Swales A, Balke T, Willemsen PWJM, van der Wal D, Wijnberg KM (2024) Biophysical modeling of mangrove seedling establishment and survival across an elevation gradient with forest zones. J Geophys Res: Earth Surf 129(5):e2024JF007664. https://doi.org/10.1029/2024JF007664

Global Invasive Species Database (GISD) (2025) Species profile: *Casuarina equisetifolia.* https://www.iucngisd.org/gisd/speciesname/Casuarina+equisetifolia?utm. Downloaded on 15 Dec 2025

Gomes VJ, Achete FM, Siegle E, De Freitas PP, Lima e Silva WK, Gardunho DC, Fernandes ME, Asp NE (2025) The role of vegetation on sediment transport in a macrotidal fringe mangrove forest. Ocean Coast Manag 261:107527. https://doi.org/10.1016/j.ocecoaman.2024.107527

Gonçalves C, Verdelhos T, Caçador I, Oliveira PJ, Marques D, Veríssimo H (2024) Effectiveness of eco-engineering structures to promote sediment particles retention in estuarine salt marshes. Water 17(5):678. https://doi.org/10.3390/w17050678

Griggs G, Reguero BG (2020) Coastal adaptation to climate change and sea-level rise. Water 13(16):2151. https://doi.org/10.3390/w13162151

Guannel G, Arkema K, Ruggiero P, Verutes G (2016) The power of three: coral reefs, seagrasses and mangroves protect coastal regions and increase their resilience. PLoS ONE 11(7):e0158094. https://doi.org/10.1371/journal.pone.0158094

Guo Y (2022) Hydrodynamics in estuaries and coast: analysis and modeling. Water 14(9):1478. https://doi.org/10.3390/w14091478

Guo Z, Yang X, Wu X, Zou X, Zhang C, Fang H, Xiang H (2021) Optimal design for vegetative windbreaks using 3D numerical simulations. Agric For Meteorol 298–299:108290. https://doi.org/10.1016/j.agrformet.2020.108290

Hai NT, Dell B, Phuong VT, Harper RJ (2020) Towards a more robust approach for the restoration of mangroves in Vietnam. Ann for Sci 77(1):18. https://doi.org/10.1007/s13595-020-0921-0

Hale RP, Wilson CA, Bomer EJ (2019) Seasonal variability of forces controlling sedimentation in the Sundarbans National Forest, Bangladesh. Front Earth Sci 7:439255. https://doi.org/10.3389/feart.2019.00211

Halpin-McCormick A, Sherrill T, Davenport C, Wolkis D, Walsh SK, Barton KE (2025) Sensitivity to elevated salinity in coastal dune plants. Oecologia 207(7):1–12. https://doi.org/10.1007/s00442-025-05759-4

He Z, Sun H, Yu X, Yin Z, Wu M, Zhao L, Hu Z, Peng Y, Lee SY (2021) Monoculture or mixed culture? Relevance of fine root dynamics to carbon sequestration oriented mangrove afforestation and restoration. Front Mar Sci 8:763922. https://doi.org/10.3389/fmars.2021.763922

Hedge P, Molloy F, Sweatman H, Hayes K, Dambacher J, Chandler J, Bax N, Gooch M, Anthony K, Elliot B (2017) An integrated monitoring framework for the Great Barrier Reef World Heritage Area. Mar Policy 77:90–96. https://doi.org/10.1016/j.marpol.2016.12.014

Hegazy I (2024) Strategic environmental assessment for sustainable coastal zone management in Saudi Arabia, aligning with vision 2030. J Umm Al-Qura Univ Eng Archit 15:624–636. https://doi.org/10.1007/s43995-024-00081-1

Hemmerling SA, DeMyers C, Parfait J, Piñero E, Baustian MM, Bregman M, Di Leonardo D, Esposito C, Georgiou IY, Grismore A, Jung H, McMann B, Miner MD (2023) A community-informed transdisciplinary approach to coastal restoration planning: maximizing the social and ecological co-benefits of wetland creation in Port Fourchon, Louisiana, USA. Front Environ Sci 11:1105671. https://doi.org/10.3389/fenvs.2023.1105671

Hernández-Delgado EA (2024) Coastal restoration challenges and strategies for small island developing states in the face of sea level rise and climate change. Coasts 4(2):235–286. https://doi.org/10.3390/coasts4020014

Hochard JP, Barbier EB, Hamilton SE (2021) Mangroves and coastal topography create economic “safe havens” from tropical storms. Sci Rep 11(1):15359. https://doi.org/10.1038/s41598-021-94207-3

Hodbod J, Tebbs E, Chan K, Sharma S (2019) Integrating participatory methods and remote sensing to enhance understanding of ecosystem service dynamics across scales. Land 8(9):132. https://doi.org/10.3390/land8090132

Holland MB, Masuda YJ, Robinson BE (2022) Land tenure security and sustainable development. Palgrave Macmillan Cham, pp 329

Hongwiset S, Rodtassana C, Poungparn S, Umnouysin S, Suchewaboripont V (2022) Synergetic roles of mangrove vegetation on sediment accretion in coastal mangrove plantations in central Thailand. Forests 13(10):1739. https://doi.org/10.3390/f13101739

Hopley CA, Jones BG (2022) Holocene evolution and depositional model of a bayhead delta, Lake Illawarra, Australia. Sedimentology 69(4):1927–1952. https://doi.org/10.1111/sed.12980

Horstman E, Dohmen-Janssen M, Narra P, van den Berg NJ, Siemerink M, Balke T, Hulscher SJ (2012) Wave attenuation in mangrove forests: field data obtained in Trang, Thailand. In: 33rd international conference on coastal engineering, ICCE 2012. Coastal Engineering Research Council. https://doi.org/10.9753/icce.v33.waves.40

Hossain M, Saha C, Saha S (2025) Mangrove restoration in Bangladesh. Open J For 15:35–52. https://doi.org/10.4236/ojf.2025.151003

Houde ALS, Garner SR, Neff BD (2015) Restoring species through reintroductions: strategies for source population selection. Restor Ecol 23(6):746–753. https://doi.org/10.1111/rec.12280

Hu R, Peng S, Guan B, Zhang H, Qu F, Wang X, Wang Z, Yang J, Huang F, Wang G, Han G (2024) Effects of long-term nutrient input on progeny seed nutrient contents, germination and early growth characteristics of typical coastal wetland plants. Plants 14(21):3393. https://doi.org/10.3390/plants14213393

Huang F, Zhang W, Gan X, Huang Y, Tang C (2022) Interactive effects of salinity and light on the growth and survivorship of mangrove *Heritiera littoralis* seedlings. J for Res 27(5):379–386. https://doi.org/10.1080/13416979.2022.2045702

Huang C, Xu Y, Zang R (2024) Low functional redundancy revealed high vulnerability of the subtropical evergreen broadleaved forests to environmental change. Sci Total Environ 935:173307. https://doi.org/10.1016/j.scitotenv.2024.173307

Hülsen S, Dee LE, Kropf CM, Meiler S, Bresch DN (2025) Mangroves and their services are at risk from tropical cyclones and sea level rise under climate change. Commun Earth Environ 6(1):262. https://doi.org/10.1038/s43247-025-02242-z

Huynh LT, Su J, Wang Q, Stringer LC, Switzer AD, Gasparatos A (2024) Meta-analysis indicates better climate adaptation and mitigation performance of hybrid engineering-natural coastal defence measures. Nat Commun 15(1):2870. https://doi.org/10.1038/s41467-024-46970-w

IUCN (2020) Guidance for using the IUCN global standard for nature-based solutions. A user-friendly framework for the verification, design and scaling up of Nature-based Solutions, 1st edn. IUCN, Gland, Switzerland

IUCN (2018) Solutions in focus: sustainable fisheries and aquaculture. Deutsche Gesellschaft für Internationale Zusammenarbeit (GIZ) GmbH. Available at: https://portals.iucn.org/library/sites/library/files/documents/2018-033-En.Pdf

Jackson NL, Nordstrom KF, Feagin RA, Smith WK (2013) Coastal geomorphology and restoration. Geomorphology 199:1–7. https://doi.org/10.1016/j.geomorph.2013.06.027

Jansen LJ, Kalas PP (2022) Improving governance of tenure in policy and practice: agrarian and environmental transition in the Mekong Region and its impacts on sustainability analyzed through the 'Tenure-Scape' approach. Sustainability 15(3):1773. https://doi.org/10.3390/su15031773

Jayson-Quashigah PN, Staneva J, Chen W, Djath B (2025a) Assessing the role of mangroves in shoreline stabilisation through a What-If modelling framework. Discover Appl Sci 7(9):982. https://doi.org/10.1007/s42452-025-07623-9

Jayson-Quashigah PN, Staneva J, Chen W, Djath B, Mahu E, Appeaning Addo K (2025b) Evaluating mangroves as nature-based solutions for coastal protection under current and future sea level rise scenarios. Front Mar Sci 12:1526082. https://doi.org/10.3389/fmars.2025.1526082

Jin B, Geng J, Ding Z, Guo L, Rui Q, Wu J, Peng S, Jin R, Fu X, Pan H, Ding G (2024) Construction and optimization of ecological corridors in coastal cities based on the perspective of "structure-function." Sci Rep 14(1):27945. https://doi.org/10.1038/s41598-024-79433-9

Johnson N, Druckenmiller ML, Danielsen F, Pulsifer PL (2021) The use of digital platforms for community-based monitoring. Bioscience 71(5):452–466. https://doi.org/10.1093/biosci/biaa162

Juanico DE (2022) Does mangrove restoration imply coastal protection? A prospective simulation study. Front Mar Sci 9:968420. https://doi.org/10.3389/fmars.2022.968420

Kamil EA, Takaijudin H, Hashim AM (2021) Mangroves as coastal bio-shield: a review of mangroves performance in wave attenuation. Civil Eng J 7(11):1964–1981. https://doi.org/10.28991/cej-2021-03091772

Kang Y, Lei J, Wang M, Li G, Ding X (2023) Topographic evolution of tidal flats based on remote sensing: an example in Jiangsu coast, Southern Yellow Sea. Front Mar Sci 10:1163302. https://doi.org/10.3389/fmars.2023.1163302

Karimi Z, Abdi E, Deljouei A, Cislaghi A, Shirvany A, Schwarz M, Hales TC (2022) Vegetation-induced soil stabilization in coastal area: An example from a natural mangrove forest. CATENA 216:106410. https://doi.org/10.1016/j.catena.2022.106410

Kathiresan K (2003) How do mangrove forests induce sedimentation? Rev Biol Trop 51(2):355–360

Keith DA, Ferrer-Paris JR, Nicholson E, Kingsford RT (eds) (2020) The IUCN global ecosystem typology 2.0: descriptive profiles for biomes and ecosystem functional groups. IUCN, Gland, Switzerland

Khan WR, Mohamad AL, Ibrahim FH (2025) Biodiversity, health, and ecosystem services of mangroves. Forests 16(8):1278. https://doi.org/10.3390/f16081278

Klain SC, Chan KM (2012) Navigating coastal values: participatory mapping of ecosystem services for spatial planning. Ecol Econ 82:104–113. https://doi.org/10.1016/j.ecolecon.2012.07.008

Krauss KW, Cormier N, Osland MJ, Kirwan ML, Stagg CL, Nestlerode JA, Russell MJ, From AS, Spivak AC, Dantin DD, Harvey JE, Almario AE (2017) Created mangrove wetlands store belowground carbon and surface elevation change enables them to adjust to sea-level rise. Sci Rep 7(1):1030. https://doi.org/10.1038/s41598-017-01224-2

Kremer K, Jonsson B, Dutta T, Tavares MF, Bauhus J (2025) Single- vs mixed-species plantations: a systematic review on the effects on biodiversity. Biol Cons 307:111182. https://doi.org/10.1016/j.biocon.2025.111182

Kropf CM, Vaterlaus L, Bresch DN, Pellissier L (2025) Tropical cyclone risk for global ecosystems in a changing climate. Nat Clim Chang 15(1):92–100. https://doi.org/10.1038/s41558-024-02194-w

Kumbier K, Hughes MG, Rogers K, Woodroffe CD (2021) Inundation characteristics of mangrove and saltmarsh in micro-tidal estuaries. Estuar Coast Shelf Sci 261:107553. https://doi.org/10.1016/j.ecss.2021.107553

Lan Y, Hsu T (2021) Planning and management of coastal buffer zones in Taiwan. Water 13(20):2925. https://doi.org/10.3390/w13202925

Laso Bayas JC, Marohn C, Dercon G, Dewi S, Piepho HP, Joshi L, Van Noordwijk M, Cadisch G (2011) Influence of coastal vegetation on the 2004 tsunami wave impact in west Aceh. Proc Natl Acad Sci 108(46):18612–18617. https://doi.org/10.1073/pnas.1013516108

Laynes-Magaña C, Cetzal-Ix W, López-Castilla HM, Tamayo-Cen I, Martínez-Puc JF, Noguera-Savelli E, Basu SK (2025) Honey production, an economic alternative for coastal areas with mangrove ecosystems: a case study in Sabancuy, Campeche, Southeastern Mexico. J Ethnobiol Ethnomed 21(1):42. https://doi.org/10.1186/s13002-025-00794-0

Le CT, Van Nguyen P, Thu HT, Tran LTD, Nguyen QH (2022) Coastal adaptability to the risk of environmental incidents from the mainland–a case study in the Southeast coastal area. VNUHCM J Earth Sci Environ 6(2):545–555. https://doi.org/10.32508/stdjsee.v6i2.701

Lecerf M, Herr D, Elverum C, Delrieu E, Picourt L (2023) Coastal and marine ecosystems as nature-based solutions in new or updated nationally determined contributions, ocean & climate platform, conservation international, IUCN, rare, the nature conservancy, wetlands international and WWF

Lee JA, Chon J, Ahn C (2014) Planning landscape corridors in ecological infrastructure using least-cost path methods based on the value of ecosystem services. Sustainability 6(11):7564–7585. https://doi.org/10.3390/su6117564

Lewis RR III (2005) Ecological engineering for successful management and restoration of mangrove forests. Ecol Eng 24(4):403–418. https://doi.org/10.1016/j.ecoleng.2004.10.003

Li Y, Zhao D, Yu G, Xie L (2023) Wave height attenuation over a nature-based breakwater of floating emergent vegetation. Sustainability 15(14):10749. https://doi.org/10.3390/su151410749

Li B, Xia Y, Chen X, Wang J, Liu W, Wang Z, Su Z, Ren H (2025a) Enhanced sediment microbial diversity in mangrove forests: indicators of nutrient status in coastal ecosystems. Mar Pollut Bull 211:117421. https://doi.org/10.1016/j.marpolbul.2024.117421

Li T, Wu P, Zuo L, Yang J (2025b) Impacts of tidal creek development on vegetation restoration during ecological water supplement in the Yellow River Delta. J Geog Sci 35(6):1365–1380. https://doi.org/10.1007/s11442-025-2371-y

Li Z, Li Y, Fan B (2025c) Diverse restored mangrove stands enhance carbon storage compared to monospecific plantation: a meta-analysis. For Ecol Manage 578:122446. https://doi.org/10.1016/j.foreco.2024.122446

Lin L, Hao Z, Post CJ, Mikhailova EA (2023) Protection of coastal shelter forests using UAVs: individual tree and tree-height detection in *Casuarina equisetifolia* L. Forests. Forests 14(2):233. https://doi.org/10.3390/f14020233

Liu M, Zhang H, Lin G, Lin H, Tang D (2018) Zonation and directional dynamics of mangrove forests derived from time-series satellite imagery in Mai Po, Hong Kong. Sustainability 10(6):1913. https://doi.org/10.3390/su10061913

Liu Z, Fagherazzi S, Cui B (2021) Success of coastal wetlands restoration is driven by sediment availability. Commun Earth Environ 2(1):44. https://doi.org/10.1038/s43247-021-00117-7

Loch TK, Riechers M (2021) Integrating indigenous and local knowledge in management and research on coastal ecosystems in the Global South: a literature review. Ocean Coast Manag 212:105821. https://doi.org/10.1016/j.ocecoaman.2021.105821

Lopes PF, Constantinescu Ş, Silvano RA, Tatui F, Teampău P, Vaidianu N (2025) Unintended and overlooked consequences of exclusionary marine conservation. ICES J Mar Sci 82(1):fsae190. https://doi.org/10.1093/icesjms/fsae190
Lopez Abuchar LK, Peralta Madriz C, Soto Navarro C, Vinueza Hidalgo G, Whitworth A, Beirne C (2025) Temporal changes in habitat structure and gastropod community assemblage in response to active restoration of a Central American mangrove. Front Mar Sci 12:1563965. https://doi.org/10.3389/fmars.2025.1563965
López-Adame H, Zaragoza-Méndez FÁ, Lara-Domínguez AL, Alvarado-Barrientos MS, Mariño-Tapia I, Sáinz-Hernández E, Pérez-Ceballos R, Zaldívar-Jiménez A, López-Portillo J (2026) Effectiveness of hydrological restoration in the mangrove of a coastal lagoon in the Gulf of Mexico. Ecol Eng 223:107848. https://doi.org/10.1016/j.ecoleng.2025.107848
Ma W, Wang W, Tang C, Chen G, Wang M (2020) Zonation of mangrove flora and fauna in a subtropical estuarine wetland based on surface elevation. Ecol Evol 10(14):7404–7418. https://doi.org/10.1002/ece3.6467
Macintosh DJ, Ashton EC (2002) A review of mangrove biodiversity conservation and management. Centre for Tropical Ecosystems Research, University of Aarhus, Denmark
MacKenzie RA, Foulk PB, Klump JV, Weckerly K, Purbospito J, Murdiyarso D, Nam VN (2016) Sedimentation and belowground carbon accumulation rates in mangrove forests that differ in diversity and land use: a tale of two mangroves. Wetlands Ecol Manage 24(2):245–261. https://doi.org/10.1007/s11273-016-9481-3
Mangora M (2022) Socioeconomic determinants of mangrove exploitation and management in the Pangani River Estuary, Tanzania. Ecol Soc 27:32. https://doi.org/10.5751/ES-13227-270232
Mani M, Nguyen TTL (2025) Boosting Viet Nam's Sustainable Marine Economy (PEIR). World Bank Report, pp 138
Maria TRBDC, Biondi D, Behling A, De Sousa NJ (2021) Evaluation of *Terminalia catappa* street trees: a case study in Itanhaém – São Paulo, Brazil. Urban For Urban Greening 66:127373. https://doi.org/10.1016/j.ufug.2021.127373
Marino M, Baptist MJ, Alkharoubi AI, Nasca S, Cavallaro L, Foti E, Musumeci RE (2025) Nature-based solutions as building blocks for coastal flood risk reduction: a model-based ecosystem service assessment. Sci Rep 15(1):12070. https://doi.org/10.1038/s41598-025-95230-4
Martínez ML, Psuty NP (2004) Coastal dunes. Springer Verlag, Berlin, p 388
Matondang AA, Simanjuntak M, Simangunsong J (2024) Design of digital mobile application for marine and coastal observation in northern part of Sumatra. J Appl Geospat Inform 8(2):146–151
Maun MA (1998) Adaptations of plants to burial in coastal sand dunes. Can J Bot 76(5):713–738. https://doi.org/10.1139/b98-058
Mazda Y, Magi M, Ikeda Y, Kurokawa T, Asano T (2006) Wave reduction in a mangrove forest dominated by *Sonneratia* sp. Wetlands Ecol Manage 14(4):365–378. https://doi.org/10.1007/s11273-005-5388-0
McAfee D, Jones AR, Gillanders BM, Mellin C, Nagelkerken I, Baring MR, Da Silva GM, Tanner JE, Connell SD (2022) Multi-habitat seascape restoration: optimising marine restoration for coastal repair and social benefit. Front Mar Sci 9:910467. https://doi.org/10.3389/fmars.2022.910467
McIvor AL, Möller I, Spencer T, Spalding M (2012) Reduction of wind and swell waves by mangroves. The nature conservancy and wetlands international, p 27. ISSN 2050–7941
McIvor AL, Spencer T, Möller I, Spalding M (2013) The response of mangrove soil surface elevation to sea level rise. Natural coastal protection series: report 3. Cambridge coastal research unit working paper 42. Published by The Nature Conservancy and Wetlands International, p 59. ISSN 2050–7941. http://coastalresilience.org/science/mangroves/surface-elevation-and-sea-level-rise
McLachlan R, Ogston A, Asp N, Fricke A, Nittrouer C, Gomes V (2020) Impacts of tidal-channel connectivity on transport asymmetry and sediment exchange with mangrove forests. Estuar Coast Shelf Sci 233:106524. https://doi.org/10.1016/j.ecss.2019.106524

McQueen A, Suedel B, Copp Franz S, Cardona T, McDermott M, Collins S, Magar V, Herzog J (2025) Application of a nature-based solution adaptive management framework for improving coastal resiliency in a highly urbanized environment. Front Ecol Evol 13:1608372. https://doi.org/10.3389/fevo.2025.1608372

Medina-Calderón JH, Mancera-Pineda JE, Castañeda-Moya E, Rivera-Monroy VH (2021) Hydroperiod and salinity interactions control mangrove root dynamics in a Karstic Oceanic Island in the Caribbean Sea (San Andres, Colombia). Front Mar Sci 7:598132. https://doi.org/10.3389/fmars.2020.598132

Mendez FJ, Losada IJ (2004) An empirical model to estimate the propagation of random breaking and nonbreaking waves over vegetation fields. Coast Eng 51(2):103–118. https://doi.org/10.1016/j.coastaleng.2003.11.003

Méndez-Alonzo R, Moctezuma C, Ordoñez VR, Angeles G, Martínez AJ, López-Portillo J (2015) Root biomechanics in Rhizophora mangle: anatomy, morphology and ecology of mangrove's flying buttresses. Ann Bot 115(5):833–840. https://doi.org/10.1093/aob/mcv002

Mengzhi X, Jixia L, Shixin L, Qianming Z (2025) How biodiversity conservation adapts to climate change: from a cross-spatial scale framework. Front Clim 7:1646318. https://doi.org/10.3389/fclim.2025.1646318

Merven R, Appadoo C, Florens FBV, Iranah P (2023) Dependency on mangroves ecosystem services is modulated by socioeconomic drivers and socio-ecological changes–insights from a small oceanic island. Hum Ecol 51(6):1141–1156. https://doi.org/10.1007/s10745-023-00464-3

Milupi ID, Somers MJ, Ferguson JWH (2017) A review of community-based natural resource management. Appl Ecol Environ Res 15(4):1121–2114. https://doi.org/10.15666/aeer/1504_11211143

Ministry of Agriculture and Rural Development (MARD) (2025) Green Shield protecting the coastline against climate change. https://en.mae.gov.vn/green-shield-protecting-the-coastline-against-climate-change-8852.htm?utm_source=chatgpt.com

Molnar-Tanaka K, Surminski S (2024) Nature-based solutions for flood-management in Asia and the Pacific. OECD Development Centre Working Papers, vol 351, pp 01–49

Mondal BK, Paul AK (2022) Application of participatory rural appraisal and geospatial techniques for analysing the dynamics of mangrove forest and dependent livelihood in Indian Sundarban. In: Sahana M, Areendran G, Raj K (eds) Conservation, management and monitoring of forest resources in India. Springer, Cham. https://doi.org/10.1007/978-3-030-98233-1_16

Monty F, Murti R, Miththapala S, Buyck C (eds) (2017) Ecosystems protecting infrastructure and communities: lessons learned and guidelines for implementation. Gland, Switzerland: IUCN. x + 108pp

Mori N, Chang C, Ono K, Noguchi H, Baba S, Yanagisawa H, Miyagi T, Tsuruta N, Suzuki K, Yamagata S, Hasegawa Y, Tsai Y, Miyashita T, Shimura T (2025) Field experiment study to assess critical wave conditions leading to failure of mangrove *Rhizophora stylosa*. Coast Eng 199:104749. https://doi.org/10.1016/j.coastaleng.2025.104749

Morris RL, Konlechner TM, Ghisalberti M, Swearer SE (2018) From grey to green: efficacy of eco-engineering solutions for nature-based coastal defence. Glob Change Biol 24(5):1827–1842. https://doi.org/10.1111/gcb.14063

Mubeen A, Devanand VB, Ruangpan L, Vojinovic Z, Sanchez Torres A, Plavšić J, Manojlovic N, Paliaga G, Abdullah AF, Leitão JP, Wojcieszak A, Izydorczyk K, Sudar T, Deduš B, Kvesić D, Ikonomov L, Penchev V (2025) A geospatial assessment toolbox for spatial allocation of large-scale nature-based solutions for hydrometeorological risk reduction. Hydrology 12(10):272. https://doi.org/10.3390/hydrology12100272

Mugilan S, Manivasakan S, Baranidharan K, RM J, Ragunath KP, Hemalatha P, Krishnamoorthi S (2024) Mangroves as natural shields: a comprehensive review of their role in mitigating natural disasters and conservation strategies. Asian J Environ Ecol 23:1–10. https://doi.org/10.9734/AJEE/2024/v23i7559

Mukherjee N, Dahdouh-Guebas F, Koedam N, Shanker K (2015) An interdisciplinary framework to evaluate bioshield plantations: Insights from peninsular India. Acta Oecologica 63:91–100. https://doi.org/10.1016/j.actao.2014.01.005

Müller SM, Peisker J, Bieling C, Linnemann K, Reidl K, Schmieder K (2019) The importance of cultural ecosystem services and biodiversity for landscape visitors in the biosphere reserve Swabian Alb (Germany). Sustainability 11(9):2650. https://doi.org/10.3390/su11092650

Mulloy R, Aiken CM, Dwane G, Ellis M, Jackson EL (2025) Scalable mangrove rehabilitation: roots of success for *Rhizophora stylosa* establishment. Ecol Eng 212:107521. https://doi.org/10.1016/j.ecoleng.2025.107521

Murillas-Maza A, Broszeit S, Pouso S, Bueno-Pardo J, Ruiz-Frau A, Terrados J, Jernberg S, Iriondo A, Dolbeth M, Katsanevakis S, Somerfield PJ, Fernandes-Salvador JA (2023) Ecosystem indicators to measure the effectiveness of marine nature-based solutions on society and biodiversity under climate change. Nature-Based Solutions 4:100085. https://doi.org/10.1016/j.nbsj.2023.100085

Nagelkerken I, Blaber S, Bouillon S, Green P, Haywood M, Kirton L, Meynecke J, Pawlik J, Penrose H, Sasekumar A, Somerfield P (2008) The habitat function of mangroves for terrestrial and marine fauna: a review. Aquat Bot 89(2):155–185. https://doi.org/10.1016/j.aquabot.2007.12.007

Narayan S, Beck MW, Reguero BG, Losada IJ, van Wesenbeeck B, Pontee N et al (2016) The effectiveness, costs and coastal protection benefits of natural and nature-based defences. PLoS ONE 11(5):e0154735. https://doi.org/10.1371/journal.pone.0154735

Nawaz U, Anees-ur-Rahaman M, Saeed Z (2025) Survey of deep learning approaches for the monitoring and classification of seagrass. Ocean Sci J 60:19. https://doi.org/10.1007/s12601-025-00213-1

Naz A, Chowdhury A (2022) Eco-engineering and mangrove restoration methods to stabilize earthen embankments and establishing bio-shield against natural disasters: a case study from Sundarban Ramsar Wetland, India. In: Assessing, mapping and modelling of mangrove ecosystem services in the Asia-Pacific region. Springer Nature Singapore, Singapore, pp 183–198. https://doi.org/10.1007/978-981-19-2738-6_10

Nguyen H, Herbohn J, Clendenning J, Lamb D, Dressler W, Vanclay J, Firn J (2015a) What is the available evidence concerning relative performance of different designs of mixed-species plantings for smallholder and community forestry in the tropics? A systematic map protocol. Environ Evidence 4(1):15. https://doi.org/10.1186/s13750-015-0041-8

Nguyen HT, Stanton DE, Schmitz N, Farquhar GD, Ball MC (2015b) Growth responses of the mangrove Avicennia marina to salinity: development and function of shoot hydraulic systems require saline conditions. Ann Bot 115(3):397–407. https://doi.org/10.1093/aob/mcu257

Nguyen TP, Tong VA, Quoi LP, Parnell KE (2016) Mangrove restoration: establishment of a mangrove nursery on acid sulphate soils. J Trop For Sci 28:275–284

Nguyen LT, Hoang HT, Ta HV, Park PS (2020) Comparison of mangrove stand development on accretion and erosion sites in Ca Mau, Vietnam. Forests 11(6):615. https://doi.org/10.3390/f11060615

Nguyen H, Chu L, Harper RJ, Dell B, Hoang H (2022) Mangrove-shrimp farming: a triple-win approach for communities in the Mekong River Delta. Ocean Coast Manag 221:106082. https://doi.org/10.1016/j.ocecoaman.2022.106082

Nguyen LTM, Hoang HT, Choi E, Park PS (2023) Distribution of mangroves with different aerial root morphologies at accretion and erosion sites in Ca Mau Province, Vietnam. Estuar Coast Shelf Sci 287:108324. https://doi.org/10.1016/j.ecss.2023.108324

Nienaber A (2024) Transformative change in coastal biodiversity conservation: a systematic literature review of governance, social-ecological, and cultural pathways. Sustainability 17(24):11186. https://doi.org/10.3390/su172411186

Nishi M, Subramanian SM, Varghese P, Houndonougbo JSH (2025) Synthesis: ecological connectivity in the context of socio-ecological production landscapes and seascapes (SEPLS). In: Ensuring ecological connectivity in socio-ecological production landscapes and seascapes

(SEPLS). Springer Nature Singapore, Singapore, pp 253–275. https://doi.org/10.1007/978-981-95-1474-8_14

Niswonger RG, Allander KK, Jeton AE (2014) Collaborative modelling and integrated decision support system analysis of a developed terminal lake basin. J Hydrol 517:521–537. https://doi.org/10.1016/j.jhydrol.2014.05.043

Pal I, Dhungana G, Baskota A, Udmale P, Gadhawe MA, Doydee P, Nguyen TT, Sophat S (2023) Multi-hazard livelihood security and resilience of lower mekong basin communities. Sustainability 15(11):8469. https://doi.org/10.3390/su15118469

Parida AK, Jha B (2010) Salt tolerance mechanisms in mangroves: a review. Trees 24:199–217. https://doi.org/10.1007/s00468-010-0417-x

Parkinson RW, Juhasz L, Xu J, Fu ZJ (2024) Future shorelines: A living shoreline site selection and design decision support tool that incorporates future conditions induced by sea level rise. Estuaries Coasts 47(8):2641–2654. https://doi.org/10.1007/s12237-024-01425-9

Partani S, Danandeh Mehr A, Jafari A (2024) Enhancing nutrient absorption through the influence of mangrove ecosystem on flow rate and retention time in salt marshes. Sci Total Environ 924:171518. https://doi.org/10.1016/j.scitotenv.2024.171518

Paterson DM, Hope JA, Kenworthy J, Biles CL, Gerbersdorf SU (2018) Form, function and physics: the ecology of biogenic stabilisation. J Soils Sediments 18(10):3044–3054. https://doi.org/10.1007/s11368-018-2005-4

Pattanaik C, Reddy CS, Dhal NK (2008) Phytomedicinal study of coastal sand dune species of Orissa. Indian J Tradit Knowl 7(2):263–268

Paxton AB, Riley TN, Steenrod CL, Puckett BJ, Alemu IJB, Paliotti ST, Silliman BR (2024) Evidence on the performance of nature-based solutions interventions for coastal protection in biogenic, shallow ecosystems: a systematic map. Environ Evid 13(1):28. https://doi.org/10.1186/s13750-024-00350-5

Pérez G, O'Leary BC, Allegri E, Casal G, Cornet CC, De Juan S, Failler P, Fredriksen S, Fonseca C, Furlan E, Gil A, Hawkins JP, Maréchal J, McCarthy T, Roberts CM, Trégarot E, Van der Geest M, Simide R (2024) A conceptual framework to help choose appropriate blue nature-based solutions. J Environ Manage 352:119936. https://doi.org/10.1016/j.jenvman.2023.119936

Pérez-Ceballos R, Zaldívar-Jiménez A, Canales-Delgadillo J, López-Adame H, López-Portillo J, Merino-Ibarra M (2020) Determining hydrological flow paths to enhance restoration in impaired mangrove wetlands. PLoS ONE 15(1):e0227665. https://doi.org/10.1371/journal.pone.0227665

Perri S, Detto M, Porporato A, Molini A (2023) Salinity-induced limits to mangrove canopy height. Glob Ecol Biogeogr 32(9):1561–1574. https://doi.org/10.1111/geb.13720

Perricone V, Mutalipassi M, Mele A, Buono M, Vicinanza D, Contestabile P (2023) Nature-based and bioinspired solutions for coastal protection: an overview among key ecosystems and a promising pathway for new functional and sustainable designs. ICES J Mar Sci 80(5):1218–1239. https://doi.org/10.1093/icesjms/fsad080

Peruzzo P, De Serio F, Defina A, Mossa M (2018) Wave height attenuation and flow resistance due to emergent or near-emergent vegetation. Water 10(4):402. https://doi.org/10.3390/w10040402

Peters R, Reis ÂCA, Mehlig U, Wimmler M, Vollhüter J, Pimple U, Tietjen B, Berger U (2025) Modelling the dynamics of soil moisture and soil water salinity in tropical saltmarshes. Ecol Model 504:111089. https://doi.org/10.1016/j.ecolmodel.2025.111089

Phan H, Reniers A, Ye T, Stive M (2017) Response in the Mekong deltaic coast to its changing sediment sources and sinks. In: Proceedings of coastal dynamics, vol 2017, pp 311–322

Phong NT, Luom TT (2021) Configuration of allocated mangrove areas and protection of mangrove-dominated muddy coasts: knowledge gaps and recommendations. Sustainability 13(11):6258. https://doi.org/10.3390/su13116258

Phong NT, Nuong CT (2023) Thinning, selective harvesting and mangrove protection forests: lessons learned and recommendations from the Vietnamese Mekong Delta. Estuar Coast Shelf Sci 288:108345. https://doi.org/10.1016/j.ecss.2023.108345

Piccolo RL, Brown CJ, Friess DA, Lovelock CE, Brown BM, Buelow C, Saunders MI (2024) Feasibility of habitat restoration under multidisciplinary constraints. Ocean Coast Manag 249:106977. https://doi.org/10.1016/j.ocecoaman.2023.106977
Powell EJ, Tyrrell MC, Milliken A, Tirpak JM, Staudinger MD (2019) A review of coastal management approaches to support the integration of ecological and human community planning for climate change. J Coast Conserv 23(1):1–18. https://doi.org/10.1007/s11852-018-0632-y
Powlen KA, Salerno J, Jones KW, Gavin MC (2023) Identifying socioeconomic and biophysical factors driving forest loss in protected areas. Conserv Biol 37(4):e14058. https://doi.org/10.1111/cobi.14058
Pratihast AK, Herold M, Avitabile V, De Bruin S, Bartholomeus H, Jr. CMS, Ribbe L (2013) Mobile devices for community-based REDD+ monitoring: a case study for central Vietnam. Sensors 13(1):21–38. https://doi.org/10.3390/s130100021
Preston J, Debney A, Gamble C, Hardy MJ, Underwood GJ, Garbutt A, Harley J, Baker R, Dunk RM, Grigg M, Hancock BT, Hendy IW, La Marca EC, Murray J, Pettorelli N, Pittman SJ, Reeves SE, Robertson M, Sturrock AM, Zu Ermgassen PS (2025) Seascape connectivity: evidence, knowledge gaps and implications for temperate coastal ecosystem restoration practice and policy. Npj Ocean Sustain 4(1):33. https://doi.org/10.1038/s44183-025-00128-3
Pretzsch H, Zenner EK (2017) Toward managing mixed-species stands: from parametrization to prescription. For Ecosyst 4:19 (2017). https://doi.org/10.1186/s40663-017-0105-z
Qiu D, Yan J, Ma X, Gao F, Wang F, Wen L, Bai J, Cui B (2019) How vegetation influence the macrobenthos distribution in different saltmarsh zones along coastal topographic gradients. Mar Environ Res 151:104767. https://doi.org/10.1016/j.marenvres.2019.104767
Quadros AF, Helfer V, Nordhaus I, Reuter H, Zimmer M (2021) Functional traits of terrestrial plants in the intertidal: a review on mangrove trees. Biol Bull 241(2):123–139. https://doi.org/10.1086/716510
Quang Bao T (2011) Effect of mangrove forest structures on wave attenuation in coastal Vietnam. Oceanologia 53(3):807–818. https://doi.org/10.5697/oc.53-3.807
Quartel S, Kroon A, Augustinus PGEF, Van Santen P, Tri NH (2007) Wave attenuation in coastal mangroves in the Red River delta, Vietnam. J Asian Earth Sci 29(4):576–584. https://doi.org/10.1016/j.jseaes.2006.05.008
Raju RD, Arockiasamy M (2022) Coastal protection using integration of mangroves with floating barges: an innovative concept. J Mar Sci Eng 10(5):612. https://doi.org/10.3390/jmse10050612
Ravaoarinorotsihoarana LA, Ratefinjanahary I, Aina C, Rakotomahazo C, Glass L, Ranivoarivelo L, Lavitra T (2023) Combining traditional ecological knowledge and scientific observations to support mangrove restoration in Madagascar. Forests 14(7):1368. https://doi.org/10.3390/f14071368
Redelstein R, Dinter T, Hertel D, Leuschner C (2018) Effects of inundation, nutrient availability and plant species diversity on fine root mass and morphology across a saltmarsh flooding gradient. Front Plant Sci 9:330150. https://doi.org/10.3389/fpls.2018.00098
Reed SO, Nghi NV, Minh NA, Lien HTK, Hung TM, Thien NV, Anh NK (2015) Building coastal resilience in Vietnam: an integrated, community-based approach to mangrove management, disaster risk reduction, and climate change adaptation. CARE International in Vietnam, 80 pp
Reef R, Sayers S (2025) Wave attenuation by Australian temperate mangroves. J Mar Sci Eng 13(2):382. https://doi.org/10.3390/jmse13020382
Rehman S, Jahangir S, Azhoni A (2022) GIS based coastal vulnerability assessment and adaptation barriers to coastal regulations in Dakshina Kannada district, India. Reg Stud Mar Sci 55:102509. https://doi.org/10.1016/j.rsma.2022.102509
Risley SC, Britsch ML, Stoll JS, Leslie HM (2025) Mapping local knowledge supports science and stewardship. Ambio. https://doi.org/10.1007/s13280-025-02170-4
Risley SC (2022) Linking local knowledge & community science in support of coastal marine stewardship. MSc Thesis, The University of Maine, USA
Rivers N, Strand M, Fernandes M, Metuge D, Lemahieu A, Nonyane CL, Benkenstein A, Snow B (2023) Pathways to integrate Indigenous and local knowledge in ocean governance processes:

lessons from the Algoa Bay Project, South Africa. Front Mar Sci 9:1084674. https://doi.org/10.3389/fmars.2022.1084674
Robertson AI, Alongi DM (eds) (2013) Tropical mangrove ecosystems, vol 41. American Geophysical Union, USA
Rudianto R, Darmawan V, Isdianto A, Bintoro G (2022) Restoration of coastal ecosystems as an approach to the integrated mangrove ecosystem management and mitigation and adaptation to climate changes in north coast of East Java. J Coast Conserv 26(4):37. https://doi.org/10.1007/s11852-022-00865-4
Saenger P (2002) Mangrove ecology, silviculture and conservation. Springer Science & Business Media, pp 360
Saintilan N, Khan NS, Ashe E, Kelleway JJ, Rogers K, Woodroffe CD, Horton BP (2020) Thresholds of mangrove survival under rapid sea level rise. Science 368:1118–1121. https://doi.org/10.1126/science.aba2656
Sanders CJ, Smoak JM, Naidu AS, Araripe DR, Sanders LM, Patchineelam SR (2010) Mangrove forest sedimentation and its reference to sea level rise, Cananeia, Brazil. Environ Earth Sci 60(6):1291–1301. https://doi.org/10.1007/s12665-009-0269-0
Sarkar P, Banerjee S, Biswas S, Saha S, Pal D, Naskar MK, Srivastava SK, Barman D, Kar G, Mukul SA (2024) Contribution of mangrove ecosystem services to local livelihoods in the Indian Sundarbans. Sustainability 16(16):6804. https://doi.org/10.3390/su16166804
Schmitt K, Albers T, Pham TT, Dinh SC (2013) Site-specific and integrated adaptation to climate change in the coastal mangrove zone of Soc Trang Province, Viet Nam. J Coast Conserv 17(3):545–558. https://doi.org/10.1007/s11852-013-0253-4
Schmitt RJ, Giuliani M, Bizzi S, Kondolf GM, Daily GC, Castelletti A (2021) Strategic basin and delta planning increases the resilience of the Mekong Delta under future uncertainty. Proc Natl Acad Sci 118(36):e2026127118. https://doi.org/10.1073/pnas.2026127118
Schoonees T, Gijón Mancheño A, Scheres B, Bouma TJ, Silva RCCDS, Schlurmann T, Schüttrumpf H (2019) Hard structures for coastal protection, towards greener designs. Estuaries Coasts 42(7):1709–1729. https://doi.org/10.1007/s12237-019-00551-z
Schueler K (2017) Nature-based solutions to enhance coastal resilience. Inter-American Development Bank, Washington, DC. Available at:https://publications.iadb.org/publications/english/document/Naturebased-Solutions-to-Enhance-Coastal-Resilience.pdf
Scott CP, Mach L, Lucas KM, Myers AE (2024) Whose cultural ecosystem service values matter? Exploring power inequities in diverse mangrove communities. Hum Ecol 52(1):81–97. https://doi.org/10.1007/s10745-023-00462-5
Sheaves M (2005) Nature and consequences of biological connectivity in mangrove systems. Mar Ecol Prog Ser 302:293–305. https://doi.org/10.3354/meps302293
Shi B, Cooper JR, Pratolongo PD, Gao S, Bouma TJ, Li G, Wang YP (2017) Erosion and accretion on a mudflat: the importance of very shallow-water effects. J Geophys Res Oceans 122(12):9476–9499. https://doi.org/10.1002/2016JC012316
Siddiqui MR, Ishtiaque A (2025) Roles and nature of public participation in climate change adaptation in coastal South Asia. Discov Environ 3:166. https://doi.org/10.1007/s44274-025-00360-1
Sigren J, Figlus J, Armitage A (2014) Coastal sand dunes and dune vegetation: restoration, erosion, and storm protection. Shore & Beach 82(4):5–12
Sinare H, Coetzer K, Schultz L (2025) Participatory governance for people and nature in multifunctional landscapes—insights from biosphere reserves. Curr Opin Environ Sustain 77:101585. https://doi.org/10.1016/j.cosust.2025.101585
Smith CS, Rudd ME, Gittman RK, Melvin EC, Patterson VS, Renzi JJ, Wellman EH, Silliman BR (2020) Coming to terms with living shorelines: a scoping review of novel restoration strategies for shoreline protection. Front Mar Sci 7:541502. https://doi.org/10.3389/fmars.2020.00434
Sobhani P, Danehkar A, Deljouei A, Marcu MV, Sadeghi SMM (2025) Assessing cultural ecosystem services in the Hara mangrove forests: indicators for sustainable management in a Middle

Eastern coastal marine protected area. Ecol Ind 178:113834. https://doi.org/10.1016/j.ecolind.2025.113834

Son DH, Nhung TT, Phuong TA, Duc NA (2024) Building a WRI-CosMaDSS decision support system for integrated coastal zone management: application to the coastal zone of the Mekong Delta (In Vietnamese). J Hydrometeorol 767:20–31. https://doi.org/10.36335/VNJHM.2024(767).20-31

Sreeranga S, Lei J (2026) Evaluating portable reef for mangrove sapling protection: an experimental study using live mangroves. Coast Eng 204:104903. https://doi.org/10.1016/j.coastaleng.2025.104903

Stori FT, Peres CM, Turra A, Pressey RL (2019) Traditional ecological knowledge supports ecosystem-based management in disturbed coastal marine social-ecological systems. Front Mar Sci 6:452035. https://doi.org/10.3389/fmars.2019.00571

Su J, Friess DA, Gasparatos A (2021) A meta-analysis of the ecological and economic outcomes of mangrove restoration. Nat Commun 12(1):5050. https://doi.org/10.1038/s41467-021-25349-1

Su J, Yin B, Chen L, Gasparatos A (2022) Priority areas for mixed-species mangrove restoration: the suitable species in the right sites. Environ Res Lett 17(6):065001. https://doi.org/10.1088/1748-9326/ac6b48

Sunkur R, Kantamaneni K, Bokhoree C, Ravan S (2023) Mangroves' role in supporting ecosystem-based techniques to reduce disaster risk and adapt to climate change: a review. J Sea Res 196:102449. https://doi.org/10.1016/j.seares.2023.102449

Sunny DS, Islam KA, Mullick MRA, Ellis JT (2022) Performance study of imageries from MODIS, Landsat 8 and Sentinel-2 on measuring shoreline change at a regional scale. Remote Sens Appl Soc Environ 28:100816. https://doi.org/10.1016/j.rsase.2022.100816

Sutton-Grier AE, Gittman RK, Arkema KK, Bennett RO, Benoit J, Blitch S, Burks-Copes KA, Colden A, Dausman A, DeAngelis BM et al (2018) Investing in natural and nature-based infrastructure: building better along our coasts. Sustainability 10(2):523. https://doi.org/10.3390/su10020523

Suzuki T, Zijlema M, Burger B, Meijer MC, Narayan S (2011) Wave dissipation by vegetation with layer schematization in SWAN. Coast Eng 59(1):64–71. https://doi.org/10.1016/j.coastaleng.2011.07.006

Syahid LN, Sakti AD, Ward R, Rosleine D, Windupranata W, Wikantika K (2023) Optimizing the spatial distribution of Southeast Asia mangrove restoration based on zonation, species and carbon projection schemes. Estuar Coast Shelf Sci 293:108477. https://doi.org/10.1016/j.ecss.2023.108477

Syamsiyah N, Sadeli AH, Saidah Z, Noor TI, Widiyanesti S (2025) Community participation in the development of sustainable, environmentally conscious villages in the cirasea sub-watershed, Indonesia. Sustainability 17(11):4871. https://doi.org/10.3390/su17114871

Tada Y, Kashimura T (2009) Proteomic analysis of salt-responsive proteins in the mangrove plant, *Burguiera gymnorhiza*. Plant Cell Physiol 50(3):439–446. https://doi.org/10.1093/pcp/pcp002

Tatongjai S, Kraichak E, Kermanee P (2021) Comparative anatomy and salt management of *Sonneratia caseolaris* (L.) Engl. (Lythraceae) grown in saltwater and freshwater. PeerJ 9:e10962 https://doi.org/10.7717/peerj.10962

Teh SY, Koh HL, Liu PL, Ismail AIM, Lee HL (2009) Analytical and numerical simulation of tsunami mitigation by mangroves in Penang, Malaysia. J Asian Earth Sci 36(1):38–46. https://doi.org/10.1016/j.jseaes.2008.09.007

Thamizoli P, Rengalakshmi R (2022) Sea level change and the livelihood security of coastal communities in Tamil Nadu, Peninsular India. In: Handbook on climate change and disasters. Edward Elgar Publishing, pp 432–452

Thị HPN (2021) The intangible cultural heritage of people living in thua thien hue coastal zone with sustainable development. Hue Univ J Sci: Soc Sci Human 130(6D):155–164. https://doi.org/10.26459/hueunijssh.v130i6D.6215

Thom RM (2000) Adaptive management of coastal ecosystem restoration projects. Ecol Eng 15(3–4):365–372

Thomas E, Jalonen R, Loo J, Boshier D, Gallo L, Cavers S, Bordács S, Smith P, Bozzano M (2014) Genetic considerations in ecosystem restoration using native tree species. For Ecol Manage 333:66–75. https://doi.org/10.1016/j.foreco.2014.07.015

Tien LNH, Dung TX, Hong Phuoc VL (2025) Coastal protection in Cu Lao Dung Mangroves (Soc Trang Province, Vietnam): quantifying wave energy dissipation. J Environ Earth Sci 7(1):514–526. https://doi.org/10.30564/jees.v7i1.7399

Tinh PH, MacKenzie RA, Hung TD, Vinh TV, Ha HT, Lam MH, Hanh NT, Tung NX, Hai PM, Huyen BT (2022) Mangrove restoration in Vietnamese Mekong Delta during 2015–2020: achievements and challenges. Front Mar Sci 9:1043943. https://doi.org/10.3389/fmars.2022.1043943

Tomlinson PB (2016) The botany of mangroves. Cambridge University Press, pp 432

Torabbeigi M, Akbari H, Adibzade M, Abolfathi S (2024) Modeling wave dynamics with coastal vegetation using a smoothed particle hydrodynamics porous flow model. Ocean Eng 311:118756. https://doi.org/10.1016/j.oceaneng.2024.118756

Trujillo AP, Thurman HV (2017) Essentials of oceanography, vol 733. Pearson, Upper Saddle River, NJ

UNEP (2014) The importance of mangroves to people: a call to action. In: van Bochove J, Sullivan E, Nakamura T (eds) United Nations environment programme world conservation monitoring centre, Cambridge, 128 pp

United Nations Framework Convention on Climate Change (UNFCCC) (2022) Intended nationally determined contribution of Viet Nam. Available at: https://unfccc.int/sites/default/files/NDC/2022-06/VIETNAM%27S%20INDC.pdf

Valenzuela BR, Yeo-Chang Y, Park MS, Chun J-N (2020) Local people's participation in mangrove restoration projects and impacts on social capital and livelihood: a case study in the Philippines. Forests 11(5):580. https://doi.org/10.3390/f11050580

Van Alphen R, Rains KC, Rodgers M, Malservisi R, Dixon TH (2024) UAV-based wetland monitoring: multispectral and lidar fusion with random forest classification. Drones 8(3):113. https://doi.org/10.3390/drones8030113

Van Bijsterveldt CE, Van der Wal D, Mancheño AG, Fivash GS, Helmi M, Bouma TJ (2023) Can cheniers protect mangroves along eroding coastlines?—The effect of contrasting foreshore types on mangrove stability. Ecol Eng 187:106863. https://doi.org/10.1016/j.ecoleng.2022.106863

Van Cuong C, Brown S, To HH, Hockings M (2015) Using Melaleuca fences as soft coastal engineering for mangrove restoration in Kien Giang, Vietnam. Ecol Eng 81:256–265. https://doi.org/10.1016/j.ecoleng.2015.04.031

Van Hespen R, Hu Z, Borsje B, De Dominicis M, Friess DA, Jevrejeva S, Kleinhans MG, Maza M, Van Bijsterveldt CE, Van der Stocken T, Van Wesenbeeck B, Xie D, Bouma TJ (2023) Mangrove forests as a nature-based solution for coastal flood protection: biophysical and ecological considerations. Water Sci Eng 16(1):1–13. https://doi.org/10.1016/j.wse.2022.10.004

Van Loon AF, Te Brake B, Van Huijgevoort MHJ, Dijksma R (2016) Hydrological classification, a practical tool for mangrove restoration. PLoS ONE 11(3):e0150302. https://doi.org/10.1371/journal.pone.0150302

Van Mai C, Ngo A, Mai T, Dao HT (2021) Bamboo fences as a nature-based measure for coastal wetland protection in Vietnam. Front Mar Sci 8:756597. https://doi.org/10.3389/fmars.2021.756597

Van Wesenbeeck BK, Van Zelst VT, Antolinez JA, De Boer WP (2025) Quantifying uncertainty in wave attenuation by mangroves to inform coastal green belt policies. Commun Earth Environ 6(1):258. https://doi.org/10.1038/s43247-025-02178-4

Vannucci M (2004) Mangrove management and conservation present and future. United Nations Press, USA

Vasseur L (2020) How ecosystem-based adaptation to climate change can help coastal communities through a participatory approach. Sustainability 13(4):2344. https://doi.org/10.3390/su13042344

Veerkamp C, Ramieri E, Romanovska L, Zandersen M, Förster J, Rogger M, Martinsen L (2021) Assessment frameworks of nature-based solutions for climate change adaptation and disaster risk

reduction. European topic centre on climate change impacts, vulnerability and adaptation (ETC/CCA) Technical Paper 2021/3. https://doi.org/10.25424/cmcc/NBS_assessment_approaches
Veettil BK (2022) Current status of mangrove vegetation in Batticaloa Lagoon, Sri Lanka, using high-resolution satellite imagery. J Coast Conserv 26(6):74. https://doi.org/10.1007/s11852-022-00923-x
Veettil BK, Pereira SFR, Quang NX (2018) Rapidly diminishing mangrove forests in Myanmar (Burma): a review. Hydrobiologia 822(1):19–35. https://doi.org/10.1007/s10750-018-3673-1
Veettil BK, Quang NX, Thu Trang NT (2019a) Changes in mangrove vegetation, aquaculture and paddy cultivation in the Mekong Delta: a study from Ben Tre Province, southern Vietnam. Estuar Coast Shelf Sci 226:106273. https://doi.org/10.1016/j.ecss.2019.106273
Veettil BK, Ward RD, Quang NX, Trang NTT, Giang TH (2019b) Mangroves of Vietnam: historical development, current state of research and future threats. Estuar Coast Shelf Sci 218:212–236. https://doi.org/10.1016/j.ecss.2018.12.021
Veettil BK, Costi J, Marques WC, Tran X, Quang NX, Van DD, Hoai PN (2020) Coastal environmental changes in Southeast Asia: a study from Quang Nam Province, Central Vietnam. Reg Stud Mar Sci 39:101420. https://doi.org/10.1016/j.rsma.2020.101420
Veettil BK, Ward RD, Dung NTK, Van DD, Quang NX, Hoai PN, Hoang N (2021) The use of bioshields for coastal protection in Vietnam: current status and potential. Reg Stud Mar Sci 47:101945. https://doi.org/10.1016/j.rsma.2021.101945
Veettil BK, Quang NX, Dung NTK, Bao DQ, Xuan-Linh T, Nhat-Duc H (2022) Mangrove forest degradation and shoreline erosion in the Mekong delta: a case study from Ben Tre province. Vietnam J Sci Technol 60(5):825–836. https://doi.org/10.15625/2525-2518/16555
Veettil BK, Wickramasinghe D, Amarakoon V (2023) Mangrove forests in Sri Lanka: an updated review on distribution, diversity, current state of research and future perspectives. Reg Stud Mar Sci 62:102932. https://doi.org/10.1016/j.rsma.2023.102932
Veettil BK, Tran X, Lopes MS (2024) Mangrove vegetation changes and shoreline erosion along the central coastline of Vietnam: a study from Quang Ngai province. Environ Develop 52:101065. https://doi.org/10.1016/j.envdev.2024.101065
Vieira B, Pinho J, Barros J, Antunes do Carmo J (2024) Optimizing coastal protection: nature-based engineering for longitudinal drift reversal and erosion reduction. Ocean Coast Manage 256:107288. https://doi.org/10.1016/j.ocecoaman.2024.107288
Vinh VD (2015) Applying numerical model to assess role of mangrove forest in wave attenuation in Hai Phong coastal area. Vietnam J Mar Sci Technol 15(1):67–76. https://doi.org/10.15625/1859-3097/15/1/6082
Vinh VD, Quan NV (2015) Characteristics of hydrodynamics and flushing time in Nai Lagoon (Ninh Thuan)—results based on Delft3D. Vietnam J Mar Sci Technol 15(3):250–256. https://doi.org/10.15625/1859-3097/15/3/7220
Vulliet C, Koci J, Sheaves M, Waltham N (2025) Contrasting effects in tidal inundation under varying sea levels on the ecological structure and functions of tropical marsh ecosystems. Sci Total Environ 980:179500. https://doi.org/10.1016/j.scitotenv.2025.179500
Walters BB, Rönnbäck P, Kovacs JM, Crona B, Hussain SA, Badola R, Dahdouh-Guebas F (2008) Ethnobiology, socio-economics and management of mangrove forests: a review. Aquat Bot 89(2):220–236. https://doi.org/10.1016/j.aquabot.2008.02.009
Wambugu PW, Nyamongo DO, Kirwa EC (2023) Role of seed banks in supporting ecosystem and biodiversity conservation and restoration. Diversity 15(8):896. https://doi.org/10.3390/d15080896
Wang WF (2024) Eutrophication mechanisms and their impacts on coastal marine ecosystems. Int J Mar Sci 14(4):285–294. https://doi.org/10.5376/ijms.2024.14.0032
Wang L, Mu M, Li X, Lin P, Wang W (2011) Differentiation between true mangroves and mangrove associates based on leaf traits and salt contents. J Plant Ecol 4(4):292–301. https://doi.org/10.1093/jpe/rtq008

Wang C, Jian S, Ren H, Yan J, Liu N (2021a) A web-based software platform for restoration-oriented species selection based on plant functional traits. Front Ecol Evol 9:570454. https://doi.org/10.3389/fevo.2021.570454
Wang F, Cheng P, Chen N, Kuo Y (2021b) Tidal driven nutrient exchange between mangroves and estuary reveals a dynamic source-sink pattern. Chemosphere 270:128665. https://doi.org/10.1016/j.chemosphere.2020.128665
Wang C, Morgan GR, Morris JT (2023) Drone lidar deep learning for fine-scale bare earth surface and 3D marsh mapping in intertidal estuaries. Sustainability 15(22):15823. https://doi.org/10.3390/su152215823
Wang J, Patruno L, Chen Z, Yang Q, Tamura Y (2024) Windbreak effectiveness of single and double-arranged shelterbelts: a parametric study using large eddy simulation. Forests 15(10):1760. https://doi.org/10.3390/f15101760
Wang J, Wu Y, Xing D, Zhang Q, Quispe Puma MR, Chen Q (2025) Dynamic traits of intracellular water and salt based on electrophysiological measurements during adaptations of three mangrove species under salinity stresses. Horticulturae 11(3):309. https://doi.org/10.3390/horticulturae11030309
Wang H, Hu Z (2023) Modeling wave attenuation by vegetation with accompanying currents in SWAN. Acta Oceanol Sin 42:63–76 (2023). https://doi.org/10.1007/s13131-023-2199-1
Wei X, Mo W, Xiong L, Hu X, Cheng H (2024) Field investigation of wave attenuation in a mangrove forest dominated by *Avicennia marina* (Forsk.) Viern. Plants 14(1):135. https://doi.org/10.3390/plants14010135
Williams SL, Sur C, Abbott JM, Limbong SR (2017) Species richness accelerates marine ecosystem restoration in the Coral Triangle. Proc Natl Acad Sci 114(45):11986–11991. https://doi.org/10.1073/pnas.1707962114
Winterwerp H, van Wesenbeeck B, van Dalfsen J, Tonneijck F, Astra A, Verschure S, Van Eijk P (2014) A sustainable solution for massive coastal erosion in Central Java. Wetlands International, Netherlands, p 45
Wu T, Barrett J (2022) Coastal land use management methodologies under pressure from climate change and population growth. Environmental Management 70:827–839. https://doi.org/10.1007/s00267-022-01705-9
Xie W, He Q, Wang X, Guo L, Zhang K (2018) Role of mudflat-creek sediment exchanges in intertidal sedimentary processes. J Hydrol 567:351–360. https://doi.org/10.1016/j.jhydrol.2018.10.027
Xie K, Xu Y, Lu X (2025) Effects of mixed plantations on soil microbial diversity and ecosystem multifunctionality across China: a meta-analysis. Front For Glob Change 8:1694630. https://doi.org/10.3389/ffgc.2025.1694630
Xie D, Schwarz C, Kleinhans MG, Zhou Z (2022) Implications of coastal conditions and sea-level rise on mangrove vulnerability: a bio-morphodynamic modeling study. J Geophys Res: Earth Surf 127(3):e2021JF006301. https://doi.org/10.1029/2021JF006301
Xing X, Qi Q, Zhang D, Wu H, Xin Y, Du W, Tong S, An Y, Cui G, Zhang M, Liu G (2025) The critical role of adaptive management in wetland restoration: evidence from long-term monitoring in Northeast China. Ecol Ind 178:113966. https://doi.org/10.1016/j.ecolind.2025.113966
Xing F, Wang YP, Jia J (2022) Hydrodynamics and sediment transport patterns on intertidal flats along middle Jiangsu coast. Anthropocene Coasts 5:12. https://doi.org/10.1007/s44218-022-00012-4
Xiong W, Li J, Liu B (2024) Coastal wetland conservation and urban sustainable development synergy pathway research: insights from Qingdao and Weihai for Qinhuangdao. Sustainability 17(21):9902. https://doi.org/10.3390/su17219902
Yan H, Shi B, Gao F (2025) Wetland ecological restoration and geomorphological evolution: a hydrodynamic-sediment-vegetation coupled modeling study. J Mar Sci Eng 13(7):1326. https://doi.org/10.3390/jmse13071326

Yang Z, Tang J, Shen Y (2017) Numerical study for vegetation effects on coastal wave propagation by using nonlinear Boussinesq model. Appl Ocean Res 70:32–40. https://doi.org/10.1016/j.apor.2017.09.001

Yang Y, Liu L, Shi P, Zhao M, Dai J, Lyu Y, Liu Y (2019) Converging effects of shrubs on shadow dune formation and sand trapping. J Geophys Res Earth Surf 124(7):1835–1853. https://doi.org/10.1029/2018JF004695

Yang Q, He C, Xu Y, Ye Y, Guo H, Zhang S, Li F, Liu K (2022) Coastal wetlands play an important role in the ecological security pattern of the coastal zone. Land 11(12):2317. https://doi.org/10.3390/land11122317

Yin K, Xu S, Gong S, Chen J, Wang Y, Li M (2022) Modeling wave attenuation by submerged flexible vegetation with XBeach phase-averaged model. Ocean Eng 257:111646. https://doi.org/10.1016/j.oceaneng.2022.111646

Yoshikai M, Nakamura T, Suwa R, Sharma S, Rollon R, Yasuoka J, Egawa R, Nadaoka K (2022) Predicting mangrove forest dynamics across a soil salinity gradient using an individual-based vegetation model linked with plant hydraulics. Biogeosciences 19:1813–1832. https://doi.org/10.5194/bg-19-1813-2022

Yoshikai M, Nakamura T, Herrera EC, Suwa R, Rollon R, Ray R, Furukawa K, Nadaoka K (2023) Representing the impact of Rhizophora mangroves on flow in a hydrodynamic model (COAWST_rh v1.0): the importance of three-dimensional root system structures. Geosci Model Dev 16:5847–5863. https://doi.org/10.5194/gmd-16-5847-2023

Yu S, Cui B, Xie T, Wang Q, Yan J, Ning Z (2021) Research progress and development trend of coastal wetland restoration in greater bay areas. Watershed Ecol Environ 4:177–187. https://doi.org/10.1016/j.wsee.2022.11.004

Zainol Z, Peris AD, Akhir MF, Rahim NH, Satyanarayana B (2022) Mangrove propagule dispersal in a shallow and narrow coastal lagoon: a simulation-based assessment of the Setiu Wetlands, Malaysia. Forests 13(9):1525. https://doi.org/10.3390/f13091525

Zeng Y, Friess DA, Sarira TV, Siman K, Koh LP (2021) Global potential and limits of mangrove blue carbon for climate change mitigation. Curr Biol 31(8):1737-1743.e3. https://doi.org/10.1016/j.cub.2021.01.070

Zhang K, Liu H, Li Y, Xu H, Shen J, Rhome J, Smith TJ (2012) The role of mangroves in attenuating storm surges. Estuar Coast Shelf Sci 102–103:11–23. https://doi.org/10.1016/j.ecss.2012.02.021

Zhang X, Chua VP, Cheong H (2015) Hydrodynamics in mangrove prop roots and their physical properties. J Hydro-Environ Res 9(2):281–294. https://doi.org/10.1016/j.jher.2014.07.010

Zhang Y, Xin K, Sheng N, Xie Z, Liao B (2021a) The regenerative capacity of eight mangrove species based on propagule traits in Dongzhai Harbor, Hainan Province, China. Glob Ecol Conserv 30:e01788. https://doi.org/10.1016/j.gecco.2021.e01788

Zhang Z, Huang B, Tan C, Cheng X (2021b) A study on the drag coefficient in wave attenuation by vegetation. Hydrol Earth Syst Sci 25:4825–4834. https://doi.org/10.5194/hess-25-4825-2021

Zhang X, Fang C, Wang Y, Lou X, Su Y, Huang D (2022a) Review of effects of dam construction on the ecosystems of river estuary and nearby marine areas. Sustainability 14(10):5974. https://doi.org/10.3390/su14105974

Zhang X, Lin P, Chen X (2022b) Coastal protection by planted mangrove forest during Typhoon Mangkhut. J Mar Sci Eng 10(9):1288. https://doi.org/10.3390/jmse10091288

Zhang R, Chen Y, Lei J, Zhou X, Yao P, Stive MJ (2023) Experimental investigation of wave attenuation by mangrove forests with submerged canopies. Coast Eng 186:104403. https://doi.org/10.1016/j.coastaleng.2023.104403

Zhang J, Jiang Z, Li D, Zhou P (2025a) Mangrove vertical soil accretion and potential risk—resilience assessment of sea-level rise in the Beilun Estuary and Guangxi Coastal Zone, China. Sustainability 17(18):8099. https://doi.org/10.3390/su17188099

Zhang L, Luo J, Wang H, Liu C, Chen H, Zhang M, Wei D, Xing Y, Fang S, Sun X (2025b) Reforestation potential for integrated mangrove aquaculture target at realizing a synergy of ecological and economic benefits. J Environ Manage 386:125741. https://doi.org/10.1016/j.jenvman.2025.125741

Zhao B, Liu Y, Xu W, Liu Y, Sun J, Wang L (2019) Morphological characteristics of tidal creeks in the central coastal region of Jiangsu, China, using LiDAR. Remote Sens 11(20):2426. https://doi.org/10.3390/rs11202426

Zhao L, Fan X, Xiao S (2025) Remote-sensing indicators and methods for coastal-ecosystem health assessment: a review of progress, challenges, and future directions. Water 17(13):1971. https://doi.org/10.3390/w17131971

Zoccarato C, Minderhoud PS, Teatini P (2018) The role of sedimentation and natural compaction in a prograding delta: insights from the mega Mekong delta, Vietnam. Sci Rep 8(1):11437. https://doi.org/10.1038/s41598-018-29734-7

Zu Ermgassen PS, Worthington TA, Gair JR, Garnett EE, Mukherjee N, Nagelkerken I, Abrantes K, Acosta A, Araujo AR, Baker R, Barnett A, Beitl CM, Benzeev R, Brookes J, A G, Ching Chong V, Connolly RM, Diele K, Dwyer PG, Spalding MD (2025) Mangroves support an estimated annual abundance of over 700 billion juvenile fish and invertebrates. Commun Earth Environ 6(1):299. https://doi.org/10.1038/s43247-025-02229-w

Chapter 7
Techniques for Establishment and Restoration of Coastal Bioshields

Abstract The establishment and restoration of coastal bioshields require scientifically informed techniques that integrate ecological, hydrological, and social dimensions to ensure long-term functionality and sustainability. This chapter provides a comprehensive overview of contemporary and traditional methods used to establish and rehabilitate coastal vegetation such as mangroves, salt marshes, and dune systems. It discusses the principles of site assessment, species selection, planting design, and post-restoration management, emphasizing the importance of aligning restoration strategies with local geomorphological and tidal conditions. The chapter also explores nursery propagation, direct planting, natural regeneration, and assisted colonization as key approaches for vegetation establishment. Emphasis is placed on adaptive restoration frameworks that account for climate variability, sea-level rise, and anthropogenic pressures. The integration of remote sensing and GIS technologies for site monitoring and success evaluation is highlighted as a major advancement in bioshield restoration science. Several case studies from Asia and other tropical regions illustrate best practices, lessons learned, and community-based approaches that enhance project success. Moreover, this chapter underscores the value of participatory management and local stewardship in maintaining restored bioshields over time. By combining technical expertise with ecosystem-based principles, this chapter provides a roadmap for implementing effective, scalable, and resilient bioshield restoration projects that contribute to both coastal protection and ecological recovery.

Keywords Adaptive management · Coastal vegetation establishment · Community-based restoration · Ecological restoration · Mangrove rehabilitation · Remote sensing and monitoring

7.1 Introduction

Coastal bioshields, including natural vegetative barriers such as mangroves, coastal forests, and dune vegetation, play a vital role in protecting shorelines from erosion, storm surges, and tidal inundation while simultaneously supporting biodiversity,

B. Kozhikkodan Veettil, *The Potential of Bioshields for Coastal Protection*, Advances in Natural and Technological Hazards Research 55,
https://doi.org/10.1007/978-3-032-22888-8_7

carbon sequestration, and local livelihoods (Veettil et al. 2021). Unlike conventional engineering structures, bioshields operate as dynamic ecological systems, whose effectiveness depends on the interplay of ecological, hydrological, geomorphological, and socio-economic factors (Feagin et al. 2010; Mukherjee et al., 2015; Marino et al. 2025). Their establishment and restoration thus require careful, site-specific planning and long-term management, particularly in regions where climate change, land-use pressures, and human activities threaten coastal resilience (Veettil et al. 2021).

Across Asia, countries including Vietnam, Bangladesh, India, Indonesia, and the Philippines have increasingly invested in bioshield initiatives as core components of nature-based coastal defense strategies (Menéndez et al. 2020; Gerona-Daga and Salmo III 2022; Molnar-Tanaka and Surminski 2024). In these areas, densely populated coastlines are highly vulnerable to tropical cyclones, storm surges, and saltwater intrusion, making the restoration of mangroves and other coastal vegetation both an ecological and socio-economic priority (Menéndez et al. 2020; Veettil et al. 2021). Experience from these contexts demonstrates that technical success is closely linked to strategic site selection, appropriate species choice, planting methodologies, and ongoing maintenance (Syahid et al. 2023; Forrester et al. 2024). Achieving functional and resilient bioshields requires aligning ecological requirements with local hydrodynamic and sedimentary conditions, often necessitating detailed site assessments, hydrological modeling, and adaptive planning (Liu et al. 2021; Chen et al. 2024; Yan et al. 2025).

Equally important is the involvement of local communities in restoration activities. Participatory approaches that engage communities in propagule collection, planting, and monitoring enhance both ecological outcomes and social ownership (Damastuti and de Groot 2019; Dinda et al. 2019). Local stewardship contributes to higher survival rates, improved species diversity, and sustained ecosystem service provision (Sathiyamoorthy and Sakurai 2024). Moreover, integrating community knowledge with scientific planning fosters culturally appropriate interventions and ensures that restoration projects deliver tangible benefits to coastal livelihoods, including fisheries, aquaculture, and eco-tourism opportunities (Sathiyamoorthy and Sakurai 2024).

The long-term sustainability of bioshields also depends on adaptive management (Thom 2000). Continuous monitoring, iterative learning, and flexible responses to changing environmental conditions enable restored systems to maintain functionality under increasing climatic variability (Thom 2000). Innovations such as hybrid engineering–ecological approaches, assisted regeneration, and controlled hydrological interventions have further enhanced the resilience and effectiveness of bioshield projects (Schmitt et al. 2013; Huynh et al. 2024).

This chapter provides an overview of technical methodologies and best practices for the establishment and restoration of coastal bioshields. Drawing on lessons from successful Asian case studies, it explores species selection, site preparation, planting techniques, and post-establishment management strategies. It also addresses challenges such as salinity stress, hydrological fluctuations, and anthropogenic pressures, highlighting innovative solutions and adaptive frameworks that enhance long-term success. By framing bioshield restoration as an integrated ecological engineering process rather than a simple tree-planting exercise, the chapter emphasizes the need for combining scientific rigor, community participation, and flexible management to achieve resilient, multifunctional coastal landscapes. The insights presented here are intended to guide researchers, practitioners, and policymakers in designing and implementing bioshield interventions that are ecologically robust, socially inclusive, and adaptable to future environmental and socio-economic challenges.

7.2 Site Selection and Environmental Assessment

Before the installation of coastal bioshields, assessments of species suitability, environmental conditions, and soil parameters are essential. Figure 7.1 illustrates the various steps involved in the pre-assessment of coastal bioshield sites.

Fig. 7.1 Overview of the steps involved in assessing environmental and pedological conditions before installing a coastal bioshield

7.2.1 *Importance of Site-Specific Planning*

The success of coastal bioshields fundamentally depends on rigorous, site-specific planning that integrates geomorphological, hydrological, and ecological characteristics (Schmitt et al. 2013; Liu et al. 2021; Paxton et al. 2024; Huang et al. 2025). Coastal environments are highly heterogeneous, with variations in tidal amplitude, sediment composition, salinity gradients, and freshwater inflow creating distinct microhabitats (Veettil et al. 2021; Dalkıran and Zünbülgil-Ünsal 2023). These factors directly influence species suitability, growth performance, and survival, making careful site selection the cornerstone of restoration success (Sahana et al. 2022; Jaffé et al. 2025). For example, mangrove species generally thrive under specific intertidal conditions (including moderate salinity, suitable inundation, and fine sediments), and mangrove roots help stabilize sediment and promote accretion, contributing to structural and ecological resilience (Hongwiset et al. 2022; Zhang et al. 2025).

In contrast, coastal shelterbelts, dune vegetation, and other salt-tolerant plant communities occupy elevated sandy substrates, experiencing minimal inundation but higher exposure to wind and salt spray (Maun 2009; Du and Hesp 2020). These species are less tolerant of prolonged waterlogging but play an important role in wind buffering, sand stabilization, and early-stage coastline defense (Maun 2009). Consequently, pre-restoration assessments must carefully delineate elevation gradients, tidal zones, and microhabitats to ensure species-to-site compatibility (Sahana et al. 2022; Cutts et al. 2024; Piccolo et al. 2024). Failure to match species with appropriate environmental conditions often results in low survival, reduced growth, and compromised coastal protection (Veettil et al. 2021).

7.2.2 *Environmental and Soil Parameters*

A comprehensive understanding of environmental and soil characteristics is essential for informed species selection and effective planting strategies (Gorman et al. 2022; Fu et al. 2023). Key parameters typically include:

a. ***Soil texture and organic matter***: Soil structure determines water retention, nutrient availability, and root penetration. Fine-textured, organic-rich sediments promote higher survival rates in mangrove species, whereas sandy, low-organic substrates favor dune vegetation (Maun 2009).
b. ***Salinity and pH***: Salinity tolerance varies among species and is a critical determinant of establishment success, particularly in areas with fluctuating tidal regimes or high evaporation rates (Ye et al. 2005).
c. ***Sediment accretion and erosion rates***: Active sedimentation supports natural stabilization and root anchorage, whereas erosion-prone sites require supplementary engineering or pre-stabilization measures (Lovelock et al. 2015).

d. ***Hydrodynamic regime***: Wave height, current velocity, and tidal amplitude influence stress levels, sediment transport, and species zonation (Mazda et al. 2006).
e. ***Freshwater input and groundwater salinity***: These factors shape ecological niches and influence the physiological stress of plant communities (Adame et al. 2021).

For instance, in the Mekong Delta, Vietnam, mangrove restoration achieved optimal success in areas with high sedimentation rates annually, enabling roots to anchor effectively and stabilize soil (Tinh et al. 2022). Conversely, highly eroded or compacted sites exhibited poor survival, emphasizing the necessity of thorough pre-planting environmental assessments (Fagherazzi et al. 2017). Similar findings have been reported in Bangladesh's Sundarbans, where hydrological modification and salinity intrusion strongly influenced species composition and restoration outcomes (Islam and Gnauck 2007; Rahman 2020).

7.2.3 Use of Remote Sensing and GIS Tools

Technological innovations, particularly remote sensing, UAV mapping, and GIS modeling, have revolutionized site selection and environmental assessment for coastal bioshields (Veettil et al. 2021). Medium-to-high-resolution satellite imagery, such as PlanetScope, Sentinel-2, and Landsat 8, allows the identification of degraded zones, sedimentation trends, land-use conflicts, and tidal inundation patterns at regional scales (Veettil et al. 2021, 2024). GIS-based suitability modeling enables practitioners to prioritize restoration sites, optimize resource allocation, and reduce the risk of planting failures (Lan et al. 2013).

In Gujarat (India) (Upadhyay et al. 2015) and North Java (Indonesia) (Arjasakusuma et al. 2021), remote sensing-derived models are employed to integrate sediment dynamics, elevation, salinity, and land tenure data into restoration planning, thereby improving long-term success rates. UAVs provide detailed microtopographic mapping and facilitate monitoring of early growth stages, erosion hotspots, and water-logging patterns, supporting adaptive management interventions. When combined with on-ground environmental surveys, these tools provide a holistic, data-driven approach to site selection, ensuring that restoration efforts are both ecologically appropriate and operationally efficient.

By integrating ecological assessment, hydrological understanding, and technological tools, bioshield projects can achieve higher survival rates, accelerated growth, and enhanced coastal protection (Schmitt et al. 2013). Site-specific planning thus forms the backbone of successful restoration, linking science, technology, and local environmental knowledge to build resilient, multifunctional coastal landscapes (Guerry et al. 2022; Paxton et al. 2024).

7.3 Species Selection and Planting Design

Thoughtful species selection combined with strategic planting design enhances the structural integrity, ecological function, and long-term resilience of coastal bioshields, ensuring that they effectively serve as living barriers against coastal hazards while supporting diverse ecosystem services.

7.3.1 Principles of Species Selection

Species selection is a foundational determinant of both the protective effectiveness and ecological integrity of coastal bioshields. Appropriate species must be native or locally adapted, ecologically compatible with site conditions, and resilient to variations in tidal inundation, salinity, and climatic extremes (Sloey et al. 2025). While exotic or non-native species may offer rapid initial establishment, their use often poses long-term risks, including altered ecosystem processes, reduced biodiversity, and increased vulnerability to pests and diseases (Wang et al. 2013; Cordero et al. 2023). Consequently, bioshield design should prioritize indigenous species that contribute to functional resilience and ecological stability.

In mangrove-based bioshields, species selection is commonly structured along the intertidal gradient to reflect natural zonation patterns (Ellison et al. 2000). Pioneer species such as *Avicennia marina* and *Sonneratia alba* dominate seaward zones due to their high tolerance to salinity, wave action, and frequent inundation (Triest et al. 2021). In the middle intertidal zone, species such as *Rhizophora apiculata* and *Bruguiera gymnorrhiza* are favored for their robust prop root and knee root systems, which enhance sediment trapping and wave energy dissipation (Krauss et al. 2014). Landward zones, characterized by lower salinity and less frequent flooding, support species including *Nypa fruticans*, *Heritiera littoralis*, and *Xylocarpus granatum*, which contribute to canopy complexity and ecological connectivity with terrestrial systems (Allen et al. 2003; Wang et al. 2011; Nwobi et al. 2020).

In non-mangrove bioshields such as coastal forests and dune systems, species selection focuses on wind resistance, root stabilization, and salt tolerance (Maun 2009; Lee et al. 2020). Commonly used species include *Casuarina equisetifolia*, *Pandanus odorifer*, *Thespesia populnea*, and the creeping vine *Ipomoea pes-caprae*, which stabilizes sand surfaces and reduces wind-driven erosion (Chan and Baba 2009; Rodrigues et al. 2011; Veettil et al. 2021). Mixed-species planting designs are strongly preferred over monocultures, as they enhance structural diversity, improve resistance to pests and diseases, and increase adaptive capacity under changing environmental conditions (Liu et al. 2018).

7.3.2 Nursery Propagation Techniques

The availability of healthy, well-adapted planting material is essential for successful bioshield establishment. Mangrove nurseries are typically established near restoration sites to minimize transportation stress and enhance acclimatization (Fig. 7.2). Propagules are collected from mature, healthy stands and sorted based on size, vigor, and the absence of disease (Amenoudji et al. 2024). Preconditioning is often applied to enhance salinity tolerance and improve early survival following transplantation (Kodikara et al. 2017a).

Standard nursery practices include raising propagules in polybags filled with a mixture of local soil and organic compost, maintaining partial shade and adequate freshwater flushing during early growth stages, and gradually exposing seedlings to saline conditions through a hardening process before out-planting (Clarke and Johns 2002). These practices strengthen root development and physiological tolerance to field conditions.

For non-mangrove bioshields, such as casuarina shelterbelts or dune vegetation, nursery techniques often involve seed pre-treatment (e.g., soaking or scarification), controlled irrigation, and root pruning to promote fibrous root systems and drought resistance. In Vietnam, the Forestry Science Institute has developed improved nursery protocols that have increased seedling survival by up to 30% compared to conventional practices, underscoring the importance of standardized propagation techniques (VAFS 2025).

Fig. 7.2 Mangrove (*Rhizophora apiculata*) nursery established near Can Gio Mangrove reserve, Southern Vietnam, and the succession of vegetation in the area. *Photographs* Bijeesh Kozhikkodan Veettil

7.3.3 Planting Patterns and Density

Planting design must balance rapid establishment with long-term ecological resilience (Ferreira et al. 2023). High-density planting (e.g., 5,000–10,000 propagules per hectare) can accelerate canopy closure and sediment stabilization in highly dynamic environments, but excessive density may suppress natural regeneration and reduce species diversity (Rohal et al. 2025). Alternatively, cluster or patch planting designs allow natural recruitment to occur between planted areas, enhancing structural heterogeneity and ecosystem adaptability (Pistorius et al. 2025). An ideal example of patch planting of mangroves can be seen in Bau Ca Cai mangrove forest in Quang Ngai province in Central Vietnam (Fig. 7.3).

Species-specific spacing guidelines are commonly applied in mangrove bioshields, with recommended distances of approximately 1.5 × 1.5 m for *Rhizophora* spp., 2 × 2 m for *Avicennia* spp., and 2.5 × 2.5 m for *Sonneratia* spp. (Hassan et al. 2018; Yoshikai et al. 2022). Spatial alignment is also critical: planting rows oriented perpendicular to prevailing wave or wind direction improves wave attenuation and energy dissipation (Wei et al. 2025). In coastal dune systems, staggered or triangular planting configurations maximize wind reduction and stabilization (Charbonneau et al. 2021).

Fig. 7.3 Satellite imagery showing the Bau Ca Cai mangrove plantation site (Sentinel-2 image acquired in 2021) in Quang Ngai Province, Vietnam, and mangroves planted in clusters/patches. *Photographs* Bijeesh Kozhikkodan Veettil

7.4 Hydrological Engineering and Site Preparation

The integration of hydrological engineering and site preparation is a foundational step in bioshield restoration (Twomey et al. 2024). By restoring natural water flow, reconnecting tidal channels, and stabilizing substrates with low-cost, biodegradable materials, practitioners create an environment that allows vegetation to establish, grow, and perform both protective and ecological functions (Rey et al. 2019; Twomey et al. 2024). These preparatory interventions not only increase survival rates but also reduce the need for intensive management, demonstrating that thoughtful engineering aligned with ecological principles is key to resilient coastal bioshields (Liu et al. 2021; Rudianto et al. 2022).

7.4.1 Restoring Natural Hydrology

A critical determinant of successful coastal bioshield restoration is the restoration and management of natural hydrological processes (Liu et al. 2021; Twomey et al. 2024). Many failed projects across Asia have been linked not to poor planting techniques, but to inadequate consideration of tidal dynamics and water exchange (Van Loon et al. 2016; Sidik et al. 2025). Bioshields, particularly mangroves and intertidal vegetation, rely on regular tidal flushing for nutrient delivery, sediment deposition, and root aeration (Pérez-Ceballos et al. 2020; López-Adame et al. 2026). Disruption of natural hydrological connectivity—through blocked tidal creeks, embankments, or intensive aquaculture infrastructure such as shrimp ponds—can severely compromise establishment and growth (Canales-Delgadillo et al. 2019; Ferreira et al. 2023).

Hydrological restoration, therefore, is often more crucial than the act of planting itself. In Bangladesh, the Coastal Embankment Rehabilitation Program has implemented Tidal River Management (TRM) to reinstate natural sedimentation patterns before mangrove establishment (Adnan et al. 2020; Islam et al. 2022). By strategically breaching embankments, TRM facilitates controlled flooding, sediment redistribution, and soil fertility improvement, creating optimal conditions for subsequent planting (Adnan et al. 2020; Hafiz et al. 2023). Similarly, in Indonesia's Demak district, reopening blocked canals and restoring tidal channels enabled the natural regeneration of mangroves, reducing the need for large-scale planting and minimizing intervention costs (van der Lelij et al. 2021; Tonneijck et al. 2022). These examples underscore that understanding and managing site-specific hydrodynamics are fundamental to successful bioshield restoration, as they ensure the functional resilience of the ecosystem from the outset.

7.4.2 Site Stabilization and Protection

Before planting, coastal sites frequently require stabilization to prevent erosion, promote sediment retention, and create suitable microhabitats for seedlings (Gonçalves et al. 2025). Various low-cost and environmentally compatible interventions are widely employed to achieve these objectives:

a. ***Temporary bamboo fencing or brushwood barriers***: These reduce wave energy and wind exposure, protecting young seedlings and promoting sediment accumulation (Mai Van et al. 2021).
b. ***Coir logs and sediment traps***: Placed along eroding shorelines, these structures encourage natural sediment deposition, increase substrate elevation, and facilitate seedling anchorage (Schoell et al. 2023).
c. ***Mulching or grass cover in dune systems***: This approach stabilizes sand, reduces wind-driven erosion, and maintains moisture for establishing vegetation (Tereszkiewicz et al. 2025).

The use of biodegradable and locally available materials is advantageous, as these interventions integrate seamlessly with natural processes, decompose over time, and avoid long-term ecological disruption. For instance, in Vietnam's Mekong Delta, temporary bamboo wave barriers effectively increased sediment deposition rates, creating favorable conditions for mangrove seedling establishment in several areas (Mai Van et al. 2021). Similarly, in Indonesia (van der Lelij et al. 2021; Tonneijck et al. 2022), bamboo or brushwood barriers combined with minor topographic modifications significantly enhanced early survival rates of coastal vegetation by mitigating hydrodynamic stress and reducing sediment loss.

Effective site preparation also includes careful leveling, removal of debris, and minor contouring to guide tidal flows and prevent waterlogging or excessive erosion (Jarzemsky et al. 2013; Della Bella et al. 2024). Integrating hydrological restoration with site stabilization ensures that newly planted or naturally regenerating vegetation can withstand extreme events such as cyclones, storm surges, and high tides, enhancing long-term ecosystem resilience (Turner and Lewis 1996; Liu et al. 2021).

7.5 Assisted Natural Regeneration (ANR)

7.5.1 Concept and Application

Assisted Natural Regeneration (ANR) is an increasingly recognized approach in coastal bioshield restoration that emphasizes the recovery of ecosystems through the protection and enhancement of natural regenerative processes (Wilson et al. 2022; Oluwajuwon et al. 2024). Rather than relying solely on large-scale planting, ANR capitalizes on existing seed banks, remnant vegetation, and viable rootstocks to facilitate ecosystem recovery with minimal external inputs (Wilson et al. 2022).

The approach involves reducing anthropogenic pressures and removing barriers to natural regeneration while applying targeted interventions to accelerate ecological succession and improve vegetation establishment (Wilson et al. 2022; Oluwajuwon et al. 2024).

ANR is particularly well-suited to coastal environments where natural propagule supply remains functional and hydrological conditions are conducive to regeneration (Wilson et al. 2022). Core interventions typically include the protection of degraded sites from further disturbance, such as grazing, fuelwood collection, or unregulated access, alongside low-intensity management actions that enhance seedling survival (Oluwajuwon et al. 2024). These may involve selective weeding and mulching to reduce competition, fencing or community-based guarding to exclude livestock and human interference, and minor hydrological enhancements such as small channel excavations to improve tidal flushing and sediment delivery (Burbridge and Hallin 2002). Where natural recruitment is uneven or insufficient, supplementary planting using locally adapted species may be applied in a targeted manner to fill gaps and reinforce ecosystem structure (Burbridge and Hallin 2002).

In the Philippines, ANR has been widely adopted under the National Greening Program as a cost-effective strategy for large-scale mangrove rehabilitation. By prioritizing site protection and natural recovery processes, the program has enabled the restoration of extensive mangrove areas despite limited financial resources (Lozano et al. 2025). Similar applications in parts of Indonesia and Vietnam have demonstrated that ANR can accelerate vegetation recovery while maintaining ecological integrity, particularly when combined with community participation and adaptive management frameworks (e.g., Kusumadewi et al. 2022).

7.5.2 *Benefits and Limitations*

One of the primary advantages of ANR is its cost efficiency (Oluwajuwon et al. 2024). Compared with conventional planting-based restoration programs, ANR can reduce implementation costs by approximately >30%, largely by minimizing nursery production, labor-intensive planting, and post-planting maintenance (Qiu et al. 2025). In addition to economic benefits, ANR enhances genetic diversity and ecological resilience by relying on naturally dispersed propagules that are well adapted to local environmental conditions (Miles et al. 2010; Ugah et al. 2025). This often results in more heterogeneous and structurally complex vegetation, which improves long-term stability and ecosystem functioning (Edwards and Cerullo 2024).

However, the applicability of ANR is context-dependent: the approach is most effective in areas where remnant vegetation, viable seed sources, and relatively intact hydrological processes persist (FAO 2015; Oluwajuwon et al. 2024). In severely degraded landscapes, such as sites with compacted soils, chronic erosion, altered hydrology, or complete loss of propagule sources, natural regeneration may be insufficient or excessively slow (Zivec et al. 2021; Oluwajuwon et al. 2024). Under such conditions, ANR must be complemented by active interventions, including soil rehabilitation, hydrological restoration, and strategic planning (FAO 2015).

Consequently, ANR should be viewed not as a universal substitute for planting, but as a flexible and adaptive restoration strategy (FAO 2015). When applied in appropriate contexts and integrated with site protection, hydrological management, and community stewardship, ANR offers a scientifically robust, economically efficient, and ecologically sound pathway for restoring coastal bioshields (Oluwajuwon et al. 2024).

7.6 Post-planting Maintenance and Monitoring

Post-planting maintenance and monitoring are not ancillary activities but integral components of bioshield restoration. Sustained investment in early maintenance, coupled with systematic monitoring and adaptive management, ensures that bioshields evolve from planted stands into resilient, self-sustaining coastal ecosystems capable of delivering long-term protective and ecological benefits.

7.6.1 Early Maintenance

The initial two to three years following planting represent a critical phase in the establishment of coastal bioshields, during which seedling mortality is typically the highest (Kodikara et al. 2017b; Qiu et al. 2021). During this period, active maintenance is essential to ensure survival, structural development, and functional integration of vegetation into the coastal landscape (Statton et al. 2017; Farrer et al. 2022). Early maintenance activities are designed to mitigate biophysical stressors and anthropogenic pressures that commonly undermine restoration success (Sloey et al. 2025).

Key maintenance interventions include replanting in areas where survival rates are too low, which helps maintain adequate stand density and continuity of the bioshield (Oluwajuwon et al. 2024; Thorhaug et al. 2025). Regular removal of algal mats, debris, and excessive sediment accumulation is also necessary, as these factors can suppress seedlings, restrict gas exchange, and impede early growth (Zakaria et al. 2025). In addition, effective protection measures must be implemented to prevent damage from grazing, trampling, and illegal harvesting, all of which can substantially reduce survival rates in unprotected sites (Oluwajuwon et al. 2024).

Maintaining hydrological connectivity during this early phase is equally important (Canales-Delgadillo et al. 2019; Ferreira et al. 2023). Blocked drainage channels, altered tidal flows, or prolonged waterlogging can stress seedlings and inhibit root development (Wu et al. 2021; Wang et al. 2022). Periodic inspection and minor corrective interventions, such as clearing obstructions or adjusting microtopography, can help preserve natural tidal exchange and sediment dynamics.

Evidence across Asia indicates that community-based maintenance systems are among the most effective mechanisms for sustaining early-stage bioshield establishment (e.g., Dulyakasem et al. 2026). For example, in Xuan Thuy National Park in Vietnam, local cooperatives are assigned responsibility for designated bioshield zones, undertaking routine maintenance activities under the supervision of local authorities (KimDung et al. 2016). These arrangements are supported by government incentives and benefit-sharing mechanisms, which enhance local ownership, accountability, and long-term stewardship (Veettil et al. 2021).

7.6.2 *Monitoring Protocols*

Systematic and well-structured monitoring is fundamental to evaluating restoration performance and informing adaptive management. Monitoring protocols should be designed to track both ecological outcomes and functional performance of bioshields over time (Cadier et al. 2020). Standard indicators commonly employed in coastal restoration programs include seedling survival rates, height growth and canopy development, sediment accretion and erosion trends, and indicators of biodiversity recovery such as species richness and faunal return (Cadier et al. 2020; Dang 2024; Tran et al. 2025; Zhao et al. 2025).

Monitoring frequency is typically highest during the early establishment phase, with quarterly or biannual assessments recommended in the first three years, followed by annual or biennial monitoring as the bioshield matures (England et al. 2021). These temporal frameworks allow practitioners to detect early signs of failure, identify emerging stressors, and implement corrective measures promptly (Peters et al. 2014; England et al. 2021).

Recent advances in remote sensing, geographic information systems, and unmanned aerial vehicle (UAV) technologies have substantially enhanced monitoring capacity at both local and landscape scales (Dang 2024; Christofi et al. 2025; Zhao et al. 2025). High-resolution drone imagery can be used to detect canopy gaps, assess spatial patterns of survival, and estimate biomass growth with greater efficiency than ground-based surveys alone (Christofi et al. 2025; Gracia et al. 2025). When combined with field-based measurements and community monitoring, these tools provide a robust, multi-scale monitoring framework that supports evidence-based decision-making (Christofi et al. 2025).

7.7 Innovative and Emerging Techniques

Various innovative and emerging techniques, such as ecological engineering and hybrid approaches, advanced propagation technologies, molecular genetics, and the use of drones and artificial intelligence, signal a shift toward more adaptive, data-driven, and ecologically integrated approaches to coastal bioshield restoration. While many of these methods remain in experimental or early implementation stages, their continued refinement and contextual application hold significant promise for improving resilience and sustainability in future coastal defense strategies.

7.7.1 Ecological Engineering and Hybrid Approaches

Recent advances in coastal restoration have increasingly emphasized the integration of ecological processes with engineered structures, giving rise to hybrid coastal protection systems that combine bioshields with low-impact engineering solutions (Powell et al. 2019; Huynh et al. 2024). These ecological engineering approaches are designed to enhance coastal defense while simultaneously promoting habitat creation, sediment stabilization, and biodiversity conservation (Powell et al. 2019). By working with natural dynamics rather than against them, hybrid systems offer adaptive and resilient alternatives to conventional hard infrastructure (Adelesi and Ajani 2023).

Examples of such approaches include bio-dykes, which consist of earthen, bamboo, or brushwood embankments reinforced with mangrove or salt-tolerant vegetation (Mai Van et al. 2021; Alferink et al. 2023). These structures reduce wave energy and facilitate sediment accretion, creating favorable conditions for vegetation establishment (Dao et al. 2018). Living shorelines represent another widely adopted hybrid approach, combining coastal vegetation with oyster reefs, coir barriers, or rock sills to stabilize shorelines while maintaining ecological connectivity (Warnell et al. 2023; Chan et al. 2024). In addition, eco-concrete structures—engineered with textured surfaces, cavities, and bio-receptive materials—are increasingly being used to support natural colonization by vegetation and marine organisms (Teong et al. 2024).

Empirical evidence from Indonesia's northern Java coast demonstrates the effectiveness of hybrid systems. There, brushwood dams combined with mangrove planting have successfully promoted stable sedimentation and achieved high vegetation survival rates, even in highly eroding environments (Winterwerp et al. 2020; Gijón Mancheño et al. 2024). Such outcomes highlight the potential of ecological engineering to enhance both the protective and ecological functions of coastal bioshields (Gijsman et al. 2021).

7.7.2 Propagation Technology and Genetic Considerations

Innovations in propagation techniques and genetic science are also reshaping bioshield restoration practices (Srinivasa Rao et al. 1998). Conventional nursery-based propagation remains effective for many species; however, some ecologically important mangroves and coastal plants are difficult to propagate using traditional methods. Biotechnological approaches, including tissue culture and micropropagation, are now being explored to overcome these limitations, particularly for rare or slow-growing species such as *Excoecaria agallocha* and *Bruguiera hainesii* (Arumugam and Panneerselvam 2012; Hoang et al. 2025).

In parallel, advances in molecular genetics have improved the capacity to assess and ensure genetic suitability in restoration efforts. DNA-based provenance testing allows practitioners to select propagules that are genetically adapted to local environmental conditions, thereby enhancing resilience to salinity stress, temperature extremes, and diseases (Natarajan et al. 2021). In Vietnam, research institutions are actively investigating salt-tolerant mangrove genotypes for restoration in high-salinity deltaic zones, reflecting the growing importance of climate-adaptive genetic selection in coastal bioshield design (e.g., Hoang et al. 2025; Hong Nguyen et al. 2025).

7.7.3 Use of Drones and AI in Restoration Planning

Emerging digital technologies, particularly unmanned aerial vehicles (UAVs) and artificial intelligence (AI), are opening new frontiers in bioshield planning, implementation, and monitoring. Drone-based applications range from high-resolution site assessment and change detection to experimental seed and propagule dispersal in remote or inaccessible tidal flats (Ayub et al. 2021). Pilot programs in Panama and the UAE have demonstrated that the aerial dispersal of mangrove propagules can significantly increase operational efficiency while reducing labor costs and physical risks associated with manual planting (Flying Labs 2021; Aramcoworld 2023).

Artificial intelligence and machine learning algorithms are being increasingly used to analyze large environmental datasets, including those related to salinity, elevation, hydrodynamics, and sedimentation patterns (Luan and Cai 2023). These tools enable the prediction of survival probabilities and growth trajectories under varying environmental conditions, supporting evidence-based site selection and adaptive management decisions (Pimenow et al. 2025). When integrated with remote sensing and field data, AI-driven models offer powerful decision-support systems that enhance the precision, scalability, and long-term effectiveness of bioshield restoration initiatives (Veettil et al. 2021; Pimenow et al. 2025; Yu et al. 2025).

7.8 Case Examples from Asia

Experiences from across Asia provide valuable empirical evidence on the effectiveness of coastal bioshields when ecological principles, appropriate engineering, and community participation are effectively integrated. The following examples illustrate diverse approaches to bioshield establishment and restoration under varying environmental, institutional, and socio-economic contexts.

7.8.1 Vietnam: Mekong Delta Bioshield Belt

Since the early 1990s, Vietnam has implemented large-scale mangrove restoration programs across the Mekong Delta as part of its national coastal protection strategy (Veettil et al. 2019, 2021; Tinh et al. 2022). By 2020, more than 150,000 ha of coastal bioshields had been established, primarily composed of *Rhizophora apiculata* in seaward zones and *Avicennia alba* in more landward or dynamic sedimentary environments (Tinh et al. 2022). These species were selected for their complementary rooting structures, tolerance to tidal inundation, and capacity to enhance sediment accretion (Veettil et al. 2019).

Empirical assessments indicate that mangrove bioshields can reduce coastal erosion by up to 60%, while also attenuating wave energy and stabilizing shorelines (Jayson-Quashigah et al. 2025). Beyond their protective function, the mangrove belts in Vietnam have generated significant livelihood co-benefits, particularly through sustainable crab, shrimp, and clam harvesting (To et al. 2017). Co-management arrangements involving local households and cooperatives have further strengthened stewardship and long-term maintenance, demonstrating the value of integrating ecological restoration with socio-economic incentives (Veettil et al. 2019, 2021).

7.8.2 India: Casuarina Shelterbelts in Tamil Nadu

In the aftermath of the 2004 Indian Ocean tsunami, the Government of Tamil Nadu initiated extensive coastal shelterbelt programs aimed at enhancing protection against cyclones, storm surges, and salt sprays (Gonsalves and Mohan 2011). Several hundred kilometers of *Casuarina equisetifolia* shelterbelts were planted along the southeastern coastline, often in combination with dune stabilization measures (Gonsalves and Mohan 2011).

The rapid growth rate, wind resistance, and salt tolerance of *Casuarina* made it particularly suitable for immediate coastal protection (Gonsalves and Mohan 2011; Zhang et al. 2024; Li et al. 2025). Subsequent cyclonic events demonstrated that these shelterbelts effectively reduced wind speed and minimized damage to inland agricultural areas and settlements (Gonsalves and Mohan 2011). This case highlights the importance of species selection based on functional traits and underscores the role of fast-growing coastal trees in bioshield formation where mangroves are ecologically unsuitable.

7.8.3 Bangladesh: Coastal Afforestation Program

Bangladesh's Coastal Greenbelt Project, initiated in the early 1980s, represents one of the world's largest coastal afforestation initiatives. To date, mangrove bioshields have been established across approximately 200,000 ha of newly accreted coastal land (UDD 2023). The program primarily employs *Sonneratia apetala* for rapid establishment, often supplemented by *Avicennia officinalis* to enhance structural diversity and ecological stability (UDD 2023).

A defining feature of the program is its strong emphasis on community participation. Local communities are actively involved in planting, maintenance, and benefit-sharing arrangements, which have improved both project sustainability and rural employment opportunities (Ahammad et al. 2013; Nandy et al. 2013). Evaluations indicate that these bioshields have significantly enhanced cyclone resilience, reduced loss of life and property, and contributed to poverty alleviation in vulnerable coastal regions (Ahammad et al. 2013).

7.8.4 Philippines: Community-Based Mangrove Restoration

In the Philippines, particularly in provinces such as Bohol and Palawan, community-managed mangrove restoration initiatives have emphasized assisted natural regeneration, local stewardship, and adaptive management (Pulhin et al. 2017). These projects often rely on traditional ecological knowledge to guide site selection and species choice, minimizing planting in unsuitable areas and prioritizing hydrological restoration (Marquez and Olavides 2024).

Reported survival rates frequently exceed 80%, reflecting the effectiveness of participatory governance and locally adapted restoration practices (Pulhin et al. 2017; Buncag et al. 2019). These experiences demonstrate that community-based approaches, when supported by appropriate technical guidance and institutional frameworks, can deliver high ecological performance while strengthening social cohesion and long-term bioshield sustainability.

7.9 Conclusions

The establishment and restoration of coastal bioshields require an integrated approach that is both scientifically robust and socially inclusive. Empirical evidence from diverse Asian contexts indicates that bioshields achieve optimal performance when ecological processes, hydrological dynamics, and socio-institutional factors are addressed in a coordinated and holistic manner, rather than through isolated or purely plantation-based interventions. Successful and sustainable outcomes consistently depend on the alignment of restoration strategies with site-specific environmental conditions, existing institutional capacities, and the needs, knowledge, and participation of local communities.

Lessons derived from experiences across Asia underscore several key principles for effective bioshield restoration. First, rigorous site-specific ecological assessments are essential for informing the selection of appropriate species, propagation methods, and management strategies. Restoration initiatives that neglect key biophysical factors—such as geomorphology, hydrological regimes, and salinity gradients—are unlikely to attain long-term stability or functional integrity. Second, the restoration of natural hydrological connectivity, including the re-establishment of tidal exchange and sediment transport processes, is frequently a prerequisite for successful vegetation recovery and sustained ecosystem resilience. In many contexts, hydrological rehabilitation is more critical to restoration success than planting density or species selection in isolation.

Third, community participation and local stewardship emerge as critical determinants of the long-term sustainability of coastal bioshields. Participatory planning processes, equitable benefit-sharing mechanisms, and community-based maintenance and monitoring systems not only enhance survival and growth rates but also cultivate a sense of ownership and responsibility among local stakeholders. This social engagement reduces the risk of degradation and supports sustained management over time. Fourth, hybrid and innovative approaches that integrate ecological restoration with low-impact engineering solutions, biotechnological interventions, and digital tools can significantly strengthen system resilience, particularly in highly dynamic or severely degraded coastal environments. By complementing natural processes, such approaches contribute to substrate stabilization, accelerate ecosystem recovery, and enhance adaptive capacity in the face of climatic variability and extremes.

Finally, continuous monitoring and adaptive management are essential for sustaining ecological stability and protective effectiveness over time. Systematic and long-term monitoring facilitates the early detection of emerging stressors, supports evidence-based decision-making, and enables restoration strategies to be adjusted in response to evolving environmental dynamics and socio-economic conditions.

Taken together, these insights indicate that bioshield establishment in Asia has progressed from isolated, short-term tree-planting initiatives toward integrated frameworks for coastal ecosystem recovery. By synthesizing traditional ecological knowledge with advances in restoration science and emerging technologies, bioshields can deliver durable and multifunctional benefits that extend beyond shoreline protection to encompass biodiversity conservation, climate change mitigation, and the support of coastal livelihoods. Sustained technical innovation, effective institutional coordination, and strong community ownership will remain fundamental to transforming these living systems into resilient assets capable of safeguarding both coastal populations and natural ecosystems in the face of ongoing and future environmental change.

References

Adame M, Reef R, Santini N, Najera E, Turschwell M, Hayes M, Masque P, Lovelock C (2021) Mangroves in arid regions: ecology, threats, and opportunities. Estuar Coast Shelf Sci 248:106796. https://doi.org/10.1016/j.ecss.2020.106796

Adelesi O, Ajani OL (2023) Resilience by design: engineering climate-adaptive infrastructure for coastal protection and flood mitigation. Int J Sci Res Arch 10(02):1473–1491. https://doi.org/10.30574/ijsra.2023.10.2.1104

Adnan MSG, Talchabhadel R, Nakagawa H, Hall JW (2020) The potential of Tidal River Management for flood alleviation in South Western Bangladesh. Sci Total Environ 731:138747. https://doi.org/10.1016/j.scitotenv.2020.138747

Ahammad R, Nandy P, Husnain P (2013) Unlocking ecosystem based adaptation opportunities in coastal Bangladesh. J Coast Conserv 17(4):833–840. https://doi.org/10.1007/s11852-013-0284-x

Alferink M, Gijon Mancheno A, Suzuki T, Reniers AJHM (2023) Optimising the wave attenuation of bamboo fences using the numerical wave model SWASH. J Coast Hydraul Struct 3:27. https://doi.org/10.48438/jchs.2023.0027

Allen JA, Krauss KW, Hauff RD (2003) Factors limiting the intertidal distribution of the mangrove species *Xylocarpus granatum*. Oecologia 135(1):110–121. https://doi.org/10.1007/s00442-002-1167-2

Amenoudji CI, Sanogo S, Djogli KR, Adanve J-F, Sodedji KFA (2024) Effect of salinity and substrate on the emergence and growth of propagules of the mangrove species *Rhizophora racemosa* in the Sassandra-Dagbego Ramsar Complex, Côte d'Ivoire. Ann ResRev Biol 39(7):21–31. https://doi.org/10.9734/arrb/2024/v39i72095

Aramcoworld (2023) https://www.aramcoworld.com/articles/2023/drones-mangroves-and-carbon-superpowers. Accessed on 18 Dec 2025

Arjasakusuma S, Kusuma SS, Saringatin S, Wicaksono P, Mutaqin BW, Rafif R (2021) Shoreline dynamics in East Java Province, Indonesia, from 2000 to 2019 using multi-sensor remote sensing data. Land 10(2):100. https://doi.org/10.3390/land10020100

Arumugam M, Panneerselvam R (2012) Micropropagation and Phenolic exudation protocol for *Excoecaria agallocha*-an important mangrove. Asian Pac J Trop Biomed 2(2):S1096–S1101. https://doi.org/10.1016/S2221-1691(12)60368-2

Ayub AS, Nugroho F, Nurul Suci AN, Anggoro A (2021) Utilization of unmanned aerial vehicle (UAV) technology for mapping mangrove ecosystem. J Sylva Indonesiana 4:70–77. https://doi.org/10.32734/jsi.v4i02.6149

Buncag MJJ, Esguerra WM, Asnawi DL (2019) Community-Based mangrove rehabilitation: the case of Calatagan mangrove park- marine protected area, Batangas, Philippines. Int J Sci Manage Stud i6:95–102. https://doi.org/10.51386/25815946/ijsms-v2i6p111

Burbridge PR, Hallin DC (2002) Rehabilitation of coastal wetland forests degraded through their conversion to shrimp farms. In: Holland MM, Warren ML, Stanturf JA (eds) Proceedings of a conference on sustainability of wetlands and water resources, May 23–25, Oxford, Mississippi, pp 20–29

Cadier C, Bayraktarov E, Piccolo R, Adame MF (2020) Indicators of coastal wetlands restoration success: a systematic review. Front Mar Sci 7:600220. https://doi.org/10.3389/fmars.2020.600220

Canales-Delgadillo JC, Perez-Ceballos R, Zaldivar-Jimenez MA, Merino-Ibarra M, Cardoza G, Cardoso-Mohedano JG (2019) The effect of mangrove restoration on avian assemblages of a coastal lagoon in southern Mexico. PeerJ 7:e7493. https://doi.org/10.7717/peerj.7493

Chan HT, Baba S (2009) Manual on guidelines for rehabilitation of coastal forests damaged by natural hazards in the Asia-Pacific Region. International Society for Mangrove Ecosystems (ISME) and International Tropical Timber Organization (ITTO), 66 pp

Chan SCY, Swearer SE, Morris RL (2024) Mangrove cover and extent of protection influence lateral erosion control at hybrid mangrove living shorelines. Estuar Coasts 47:1517–1530. https://doi.org/10.1007/s12237-024-01391-2

Charbonneau BR, Dohner SM, Wnek JP, Barber D, Zarnetske P, Casper BB (2021) Vegetation effects on coastal foredune initiation: Wind tunnel experiments and field validation for three dune-building plants. Geomorphology 378:107594. https://doi.org/10.1016/j.geomorph.2021.107594

Chen Z, Luo F, Zhou G, Zhu F, Wu H, Li R, Zhang C (2024) Hydrodynamic modeling study of nature-based hybrid coastal defense strategy applied in salt marsh restoration. Estuar Coast Shelf Sci 298:108666. https://doi.org/10.1016/j.ecss.2024.108666

Christofi D, Mettas C, Evagorou E, Stylianou N, Eliades M, Theocharidis C, Chatzipavlis A, Hasiotis T, Hadjimitsis D (2025) A review of open remote sensing data with GIS, AI, and UAV support for shoreline detection and coastal erosion monitoring. Appl Sci 15(9):4771. https://doi.org/10.3390/app15094771

Clarke A, Johns L (2002) Mangrove nurseries: construction, propagation and planting: fisheries guidelines, Department of Primary Industries, Queensland, Fish Habitat Guideline FHG 004, 32 pp

Cordero S, Gálvez F, Fontúrbel FE (2023) Ecological impacts of exotic species on native seed dispersal systems: a systematic review. Plants 12(2):261. https://doi.org/10.3390/plants12020261

Cutts V, Erftemeijer PLA, Gaffi L, Hagemeijer W, Smith RK, Taylor NG, Sutherland WJ (eds) (2024) Restoration, creation and management of salt marshes and tidal flats: a collation of evidence-based guidance. Report of Conservation Evidence, Wetlands International and the World Coastal Forum. https://doi.org/10.52201/CGSCOL1/LCNC6109

Dalkıran N, Zünbülgil-Ünsal B (2023) Environmental heterogeneity and salinity gradient impacted the alpha and beta diversities of diatom assemblages in a coastal delta wetland. Water 15(19):3414. https://doi.org/10.3390/w15193414

Damastuti E, de Groot R (2019) Participatory ecosystem service mapping to enhance community-based mangrove rehabilitation and management in Demak, Indonesia. Reg Environ Change 19:65–78. https://doi.org/10.1007/s10113-018-1378-7

Dang KB (2024) Application of artificial intelligence for monitoring shoreline changes in the central coast of Vietnam. VNU J Sci: Earth Environ Sci 40(1):82–97. https://doi.org/10.25073/2588-1094/vnuees.5012

Dao T, Stive MJ, Hofland B, Mai T (2018) Wave damping due to wooden fences along mangrove coasts. J Coastal Res 34(6):1317–1327. https://doi.org/10.2112/JCOASTRES-D-18-00015.1

Della Bella A, Del Vecchio S, Fantinato E, Buffa G (2024) Coastal dune restoration: a checklist approach to site selection. Land 13(2):135. https://doi.org/10.3390/land13020135

Dinda N, Suadi S, Sahubawa L (2019) Rehabilitation of mangrove ecosystem through community-based project and the current economic value: a case study of rehabilitation project in Lubuk Kertang Village, Langkat Regency, North Sumatera. Jurnal Perikanan Universitas Gadjah Mada 20(2):71–78. https://doi.org/10.22146/jfs.34067

Du J, Hesp PA (2020) Salt spray distribution and its impact on vegetation zonation on coastal dunes: a review. Estuaries Coasts 43(8):1885–1907. https://doi.org/10.1007/s12237-020-00820-2

Dulyakasem R, Brooks C, Lehnert SL, Newton P (2026) Community-based natural resource management in coastal communities: the contribution of mangroves to household livelihoods in southern Thailand. Environ Develop 57:101340. https://doi.org/10.1016/j.envdev.2025.101340

Edwards DP, Cerullo GR (2024) Biodiversity is central for restoration. Curr Biol 34(9):R371–R379. https://doi.org/10.1016/j.cub.2024.02.032

Ellison AM, Mukherjee BB, Karim A (2000) Testing patterns of zonation in mangroves: scale dependence and environmental correlates in the Sundarbans of Bangladesh. J Ecol 88(5):813–824. https://doi.org/10.1046/j.1365-2745.2000.00500.x

England J, Angelopoulos N, Cooksley S, Dodd J, Gill A, Gilvear D, Johnson M, Naura M, Tree A, Wheeldon J, Wilkes MA (2021) Best practices for monitoring and assessing the ecological response to river restoration. Water 13(23):3352. https://doi.org/10.3390/w13233352

Fagherazzi S, Bryan KR, Nardin W (2017) Buried alive or washed away: the challenging life of mangroves in the Mekong Delta. Oceanography 30(3):48–59. https://doi.org/10.5670/oceanog.2017.313

FAO (2015) Global guidelines for the restoration of degraded forests and landscapes in drylands: building resilience and benefiting livelihoods. Berrahmouni N, Regato P, Parfondry M (eds.) Forestry Paper No. 175. Rome, Food and Agriculture Organization of the United Nations

Farrer EC, Van Bael SA, Clay K, Smith MK (2022) Plant-microbial symbioses in coastal systems: their ecological importance and role in coastal restoration. Estuaries Coasts 45(7):1805–1822. https://doi.org/10.1007/s12237-022-01052-2

Feagin RA, Mukherjee N, Shanker K, Baird AH, Cinner J, Kerr AM, Koedam N, Sridhar A, Arthur R, Jayatissa LP, Seen DL, Menon M, Rodriguez S, Shamsuddoha M, Dahdouh-Guebas F (2010) Shelter from the storm? Use and misuse of coastal vegetation bioshields for managing natural disasters. Conserv Lett 3(1):1–11. https://doi.org/10.1111/j.1755-263X.2009.00087.x

Ferreira AC, De Lacerda LD, Rodrigues JVM, Bezerra LEA (2023) New contributions to mangrove rehabilitation/restoration protocols and practices. Wetlands Ecol Manage 31(1):89–114. https://doi.org/10.1007/s11273-022-09903-2

Flying Labs (2021) A new ally in aerial seeding and mangrove restoration. https://flyinglabs.org/blog/a-new-ally-in-aerial-seeding-and-mangrove-restoration. Accessed on 18 Dec 2025

Forrester J, Leonardi N, Cooper JR, Kumar P (2024) Seagrass as a nature-based solution for coastal protection. Ecol Eng 206:107316. https://doi.org/10.1016/j.ecoleng.2024.107316

Fu D, Wu X, Hu L, Ma X, Shen C, Shang H, Huang G, He Y, Duan C (2023) Plant traits guide species selection in vegetation restoration for soil and water conservation. Biology 12(4):618. https://doi.org/10.3390/biology12040618

Gerona-Daga MEB, Salmo SG III (2022) A systematic review of mangrove restoration studies in Southeast Asia: challenges and opportunities for the United Nation's decade on ecosystem restoration. Front Mar Sci 9:987737. https://doi.org/10.3389/fmars.2022.987737

Gijón Mancheño A, Maulana B, Reniers AJ, Tas SA, Wilms T, Rejeki S, Ariyati RW, Widowati LL (2024) Monitoring pilot study of temporary permeable structures for mangrove restoration. Water 17(4):558. https://doi.org/10.3390/w17040558

Gijsman R, Horstman EM, Friess DA, Swales A, Wijnberg KM (2021) Nature-based engineering: a review on reducing coastal flood risk with mangroves. Front Mar Sci 8:702412. https://doi.org/10.3389/fmars.2021.702412

Gonçalves C, Verdelhos T, Caçador I, Oliveira PJ, Marques D, Veríssimo H (2025) Effectiveness of eco-engineering structures to promote sediment particles retention in estuarine salt marshes. Water 17(5):678. https://doi.org/10.3390/w17050678

Gonsalves J, Mohan P (eds) (2011) Strengthening resilience in post-disaster situations: stories, experience and lessons from South Asia. IDRC

Gorman D, Vanderklift MA, Lafratta A (2022) Quantitative analysis of methodological and environmental influences on survival of planted mangroves in restoration and afforestation. Forests 13(3):404. https://doi.org/10.3390/f13030404

Gracia V, Dietrich MM, Sierra JP, Valero F, Espanya A, Mösso C (2025) Using UAVs to monitor the evolution of restored coastal dunes. Remote Sens 17(19):3263. https://doi.org/10.3390/rs17193263

Guerry AD, Silver J, Beagle J, Wyatt K, Arkema K, Lowe J, Hamel P, Griffin R, Wolny S, Plane E, Griswold M, Papendick H, Sharma J (2022) Protection and restoration of coastal habitats yield multiple benefits for urban residents as sea levels rise. Npj Urban Sustain 2(1):13. https://doi.org/10.1038/s42949-022-00056-y

Hafiz N, Biswas S, Mondal MS, Islam MA, Khan MS (2023) Variations in water and deposited sediment qualities in the tidal river basins of bangladesh and their implications for TRM success. Sustainability 15(18):13855. https://doi.org/10.3390/su151813855

Hassan MK, Jintana V, Kuittinen S, Pappinen A (2018) Management practices and aboveground biomass production patterns of *Rhizophora apiculata* plantation: study from a mangrove area in Samut Songkram Province, Thailand. Biores 13(4):7826–7850. https://doi.org/10.15376/biores.13.4.7826-7850

Hoang TTT, Pham VD, Pham MP, Dang NH, Le XD, Vu DD (2025) Development of chloroplast microsatellite markers for Bruguiera hainesii C.G.Rogers (Rhizophoraceae) in the Con Dao National Park, Ba Ria—Vung Tau province, Vietnam. J Trop Sci Eng 37:21–34. https://doi.org/10.58334/vrtc.jtst.n37.02

Hong Nguyen N, Nguyen TD, Thi Tran GH, Quoc Nguyen H, Pham NB, Hoang Chu H, Do PT (2025) Genetic diversity of the tall-stilt mangrove (*Rhizophora apiculata* Blume) in Southern Central Coast of Vietnam using SSR markers. Acad J Biol 47(1):137–150. https://doi.org/10.15625/2615-9023/21433

Hongwiset S, Rodtassana C, Poungparn S, Umnouysin S, Suchewaboripont V (2022) Synergetic roles of mangrove vegetation on sediment accretion in coastal mangrove plantations in central Thailand. Forests 13(10):1739. https://doi.org/10.3390/f13101739

Huang R, Tian Z, Fan D, Ye Q, Liu Q, Kong M, Wang Y, Lyu J, Sun L (2025) Nature's double defense: how mangroves and intertidal topography shape coastal flood mitigation. Ocean Coast Manag 272:108004. https://doi.org/10.1016/j.ocecoaman.2025.108004

Huynh LT, Su J, Wang Q, Stringer LC, Switzer AD, Gasparatos A (2024) Meta-analysis indicates better climate adaptation and mitigation performance of hybrid engineering-natural coastal defence measures. Nat Commun 15(1):2870. https://doi.org/10.1038/s41467-024-46970-w

Islam SN, Gnauck A (2007) Effects of salinity intrusion in mangrove wetlands ecosystems in the Sundarbans: an alternative approach for sustainable management. Wetlands: Monitoring Modelling and Management. Taylor & Francis, Leiden, pp 315–322

Islam MF, Schot PP, Dekker SC, Griffioen J, Middelkoop H (2022) Physical controls and a priori estimation of raising land surface elevation across the southwestern Bangladesh delta using tidal river management. Hydrol Earth Syst Sci 26:903–921. https://doi.org/10.5194/hess-26-903-2022

Jaffé R, Paul-Gorsline C, McDermott M, Fluharty S, Al-Shaikh I, Skeat SL, Abdulwahab UA, Nelis L, Jaffe BD (2025) Using habitat suitability modeling to integrate ecosystem-based approaches for mangrove restoration site selection. Ecosphere 16(3):e70222. https://doi.org/10.1002/ecs2.70222

Jarzemsky RD, Burchell MR, Evans RO (2013) The impact of manipulating surface topography on the hydrologic restoration of a forested coastal wetland. Ecol Eng 58:35–43. https://doi.org/10.1016/j.ecoleng.2013.06.002

Jayson-Quashigah PN, Staneva J, Chen W, Djath B (2025) Assessing the role of mangroves in shoreline stabilisation through a What-If modelling framework. Discover Appl Sci 7(9):982. https://doi.org/10.1007/s42452-025-07623-9

KimDung N, Bush SR, Mol AP (2016) The Vietnamese state and administrative co-management of nature reserves. Sustainability 8(3):292. https://doi.org/10.3390/su8030292

Kodikara KAS, Jayatissa LP, Huxham M, Dahdouh-Guebas F, Koedam N (2017a) The effects of salinity on growth and survival of mangrove seedlings changes with age. Acta Botanica Brasilica 32(01):37–46

Kodikara KAS, Mukherjee N, Jayatissa LP, Dahdouh-Guebas F, Koedam N (2017b) Have mangrove restoration projects worked? An in-depth study in Sri Lanka. Restor Ecol 25(5):705–716. https://doi.org/10.1111/rec.12492

Krauss KW, McKee KL, Lovelock CE, Cahoon DR, Saintilan N, Reef R, Chen L (2014) How mangrove forests adjust to rising sea level. New Phytol 202(1):19–34. https://doi.org/10.1111/nph.12605

Kusumadewi SD, Purnomo H, Nadhira S, Putrizulfan R (2024) Systematic review on the implementation of mangrove community-based restoration in Indonesia and beyond. IOP Conf Ser: Earth Environ Sci 1315:012052. https://doi.org/10.1088/1755-1315/1315/1/012052

Lan PT, Son TS, Gunasekara K, Nhan NT, Hien LP (2013) Application of remote sensing and GIS technology for monitoring coastal changes in estuary area of the Red river system, Vietnam. J Korean Soc Survey Geodesy, Photogrammetry Cartogr 31(6–2):529–538. https://doi.org/10.7848/ksgpc.2013.31.6-2.529

Lee JT, Yen LZ, Chu MY, Lin YS, Chang CC, Lin RS, Lee MJ (2020) Growth characteristics and anti-wind erosion ability of three tropical foredune pioneer species for sand dune stabilization. Sustainability 12(8):3353. https://doi.org/10.3390/su12083353

Li LL, Jiang C, Chen C, Chen YQ, Lin H, Xie A, Lin Y, Yuan F (2025) Response of stem shape to long-term wind stress in coastal *Casuarina equisetifolia* forest. For Res 38(3):181–189. https://doi.org/10.12403/j.1001-1498.20240312

Liu CLC, Kuchma O, Krutovsky KV (2018) Mixed-species versus monocultures in plantation forestry: development, benefits, ecosystem services and perspectives for the future. Glob Ecol Conserv 15:e00419. https://doi.org/10.1016/j.gecco.2018.e00419

Liu Z, Fagherazzi S, Cui B (2021) Success of coastal wetlands restoration is driven by sediment availability. Commun Earth Environ 2(1):44. https://doi.org/10.1038/s43247-021-00117-7

López-Adame H, Zaragoza-Méndez FÁ, Lara-Domínguez AL, Alvarado-Barrientos MS, Mariño-Tapia I, Sáinz-Hernández E, Pérez-Ceballos R, Zaldívar-Jiménez A, López-Portillo J (2026) Effectiveness of hydrological restoration in the mangrove of a coastal lagoon in the Gulf of Mexico. Ecol Eng 223:107848. https://doi.org/10.1016/j.ecoleng.2025.107848

Lovelock CE, Cahoon DR, Friess DA, Guntenspergen GR, Krauss KW, Reef R, Rogers K, Saunders ML, Sidik F, Swales A, Saintilan N, Thuyen LX, Triet T (2015) The vulnerability of Indo-Pacific mangrove forests to sea-level rise. Nature 526(7574):559–563. https://doi.org/10.1038/nature15538

Lozano AL, Cortado JMG, Cosep PA II, Javier JAL (2025) National greening program in the Philippines: a thematic review of policy frameworks, implementation practices and measured outcomes. Int J Environ Clim Change 15(7):125–139

Luan H, Cai Z (2023) Introduction to artificial intelligence and machine learning in environmental science. Environ Sci: Adv 2(9):1149–1150. https://doi.org/10.1039/D3VA90026F

Marino M, Nasca S, Alkharoubi AI, Cavallaro L, Foti E, Musumeci RE (2025) Efficacy of nature-based solutions for coastal protection under a changing climate: a modelling approach. Coast Eng 198:104700. https://doi.org/10.1016/j.coastaleng.2025.104700

Marquez GPB, Olavides RD (2024) Integrating science-based and local ecological knowledge: a case study of mangrove restoration and rehabilitation projects in the Philippines. Disasters 48:e12630. https://doi.org/10.1111/disa.12630

Maun MA (2009) The biology of coastal sand dunes. Oxford University Press, 280 pp

Mazda Y, Magi M, Ikeda Y, Kurokawa T, Asano T (2006) Wave reduction in a mangrove forest dominated by Sonneratia sp. Wetlands Ecol Manage 14(4):365–378. https://doi.org/10.1007/s11273-005-5388-0

Menéndez P, Losada IJ, Narayan S, Beck MW (2020) The global flood protection benefits of mangroves. Sci Rep 10(1):4404. https://doi.org/10.1038/s41598-020-61136-6

Miles L, Kapos V, Dunning E (2010) Ecosystem services from new and restored forests: tool development. Multiple benefits series 5. Prepared on behalf of the UN-REDD Programme. UNEP World Conservation Monitoring Centre, Cambridge, UK

Molnar-Tanaka K, Surminski S (2024) Nature-based solutions for flood-management in Asia and the Pacific. OECD development Centre working papers, vol 351, pp 01–49

Mukherjee N, Dahdouh-Guebas F, Koedam N, Shanker K (2015) An interdisciplinary framework to evaluate bioshield plantations: insights from peninsular India. Acta Oecologica 63:91–100. https://doi.org/10.1016/j.actao.2014.01.005

Nandy P, Ahammad R, Alam M, Islam A (2013) Coastal ecosystem based adaptation: Bangladesh experience. In: Climate change adaptation actions in Bangladesh. Springer Japan, Tokyo, pp 277–303. https://doi.org/10.1007/978-4-431-54249-0_15

Natarajan P, Murugesan AK, Govindan G, Gopalakrishnan A, Kumar R, Duraisamy P, Balaji R, Shyamli PS, Parida AK, Parani M (2021) A reference-grade genome identifies salt-tolerance genes from the salt-secreting mangrove species *Avicennia marina*. Commun Biol 4(1):851. https://doi.org/10.1038/s42003-021-02384-8

Nwobi C, Williams M, Mitchard ET (2020) Rapid mangrove forest loss and nipa palm (*Nypa fruticans*) expansion in the Niger Delta, 2007–2017. Remote Sens 12(14):2344. https://doi.org/10.3390/rs12142344

Oluwajuwon TV, Chazdon RL, Ota L, Gregorio N, Herbohn J (2024) Bibliometric and literature synthesis on assisted natural regeneration: an evidence base for forest and landscape restoration in the tropics. Front for Glob Change 7:1412075. https://doi.org/10.3389/ffgc.2024.1412075

Paxton AB, Riley TN, Steenrod CL, Puckett BJ, Alemu IJB, Paliotti ST, Silliman BR (2024) Evidence on the performance of nature-based solutions interventions for coastal protection in biogenic, shallow ecosystems: a systematic map. Environ Evidence 13(1):28. https://doi.org/10.1186/s13750-024-00350-5

Pérez-Ceballos R, Zaldívar-Jiménez A, Canales-Delgadillo J, López-Adame H, López-Portillo J, Merino-Ibarra M (2020) Determining hydrological flow paths to enhance restoration in impaired mangrove wetlands. PLoS ONE 15(1):e0227665. https://doi.org/10.1371/journal.pone.0227665

Peters RJ, Duda JJ, Pess GR, Zimmerman M, Crain P, Hughes Z,... Warheit K (2014) Guidelines for monitoring and adaptively managing restoration of Chinook Salmon (*Oncorhynchus tshawytscha*) and steelhead (*O. mykiss*) on the Elwha River. In joint federal interagency conference, 111 pp

Piccolo RL, Brown CJ, Friess DA, Lovelock CE, Brown BM, Buelow C, Saunders MI (2024) Feasibility of habitat restoration under multidisciplinary constraints. Ocean Coast Manag 249:106977. https://doi.org/10.1016/j.ocecoaman.2023.106977

Pimenow S, Pimenowa O, Prus P, Niklas A (2025) The impact of artificial intelligence on the sustainability of regional ecosystems: current challenges and future prospects. Sustainability 17(11):4795. https://doi.org/10.3390/su17114795

Pistorius T, Hoang HDT, Dinh TT, Hung LT, Roth M, Duy TK (2025) Restoration of degraded coastal forests in Vietnam's North Central coast. In: Handbook of nature-based solutions to mitigation and adaptation to climate change. Springer Nature Switzerland, Cham, pp 1101–1121. https://doi.org/10.1007/978-3-031-34967-6_20

Powell EJ, Tyrrell MC, Milliken A, Tirpak JM, Staudinger MD (2019) A review of coastal management approaches to support the integration of ecological and human community planning for climate change. J Coast Conserv 23(1):1–18. https://doi.org/10.1007/s11852-018-0632-y

Pulhin JM, Gevaña DT, Pulhin FB (2017) Community-based mangrove management in the Philippines: experience and challenges in the context of changing climate. In: DasGupta R, Shaw R (eds) Participatory mangrove management in a changing climate. Disaster Risk Reduction. Springer, Tokyo. https://doi.org/10.1007/978-4-431-56481-2_16

Qiu D, Ma X, Yan J, Shao D, Bai J, Cui B (2021) Biogeomorphological processes and structures facilitate seedling establishment and distribution of annual plants: implications for coastal restoration. Sci Total Environ 756:143842. https://doi.org/10.1016/j.scitotenv.2020.143842

Qiu L, Liu F, Zhao X, Tong M, He H, Fortin J, Huang L (2025) Assessing assisted natural regeneration as a cost-efficient mitigation for climate change and biodiversity loss in China. Earth's Future 13(3):e2024EF005257. https://doi.org/10.1029/2024EF005257

Rahman MM (2020) Impact of increased salinity on the plant community of the Sundarbans Mangrove of Bangladesh. Commun Ecol 21(3):273–284. https://doi.org/10.1007/s42974-020-00028-1

Rey F, Bifulco C, Bischetti G, Bourrier F, De Cesare G, Florineth F, Graf F, Marden M, Mickovski S, Phillips C, Peklo K, Poesen J, Polster D, Preti F, Rauch H, Raymond P, Sangalli P, Tardio G, Stokes A (2019) Soil and water bioengineering: practice and research needs for reconciling natural hazard control and ecological restoration. Sci Total Environ 648:1210–1218. https://doi.org/10.1016/j.scitotenv.2018.08.217

Rodrigues RS, Mascarenhas A, Jagtap TG (2011) An evaluation of flora from coastal sand dunes of India: rationale for conservation and management. Ocean Coast Manag 54(2):181–188. https://doi.org/10.1016/j.ocecoaman.2010.11.005

Rohal CB, Yu X, Crawford JM, Montgomery O, Reynolds LK, Adams CR (2025) Sediment stability is optimized by manipulating planting design during coastal marsh establishment. Sci Rep 15(1):19854. https://doi.org/10.1038/s41598-025-04281-0

Rudianto R, Darmawan V, Isdianto A, Bintoro G (2022) Restoration of coastal ecosystems as an approach to the integrated mangrove ecosystem management and mitigation and adaptation to climate changes in north coast of East Java. J Coast Conserv 26(4):37. https://doi.org/10.1007/s11852-022-00865-4

Sahana M, Areendran G, Sajjad H (2022) Assessment of suitable habitat of mangrove species for prioritizing restoration in coastal ecosystem of Sundarban Biosphere Reserve, India. Sci Rep 12(1):20997. https://doi.org/10.1038/s41598-022-24953-5

Sathiyamoorthy S, Sakurai T (2024) Effectiveness of community participation in Mangrove restoration: the evidence from northern Sri Lanka. Environ Econ Policy Stud 26:759–779. https://doi.org/10.1007/s10018-024-00397-1

Schmitt K, Albers T, Pham TT, Dinh SC (2013) Site-specific and integrated adaptation to climate change in the coastal mangrove zone of Soc Trang Province, Viet Nam. J Coast Conserv 17(3):545–558. https://doi.org/10.1007/s11852-013-0253-4

Schoell M, Ayvazian S, Cobb D, Grunden D, Chintala M, Gerber-Williams A, Rocha K (2023) Exploring the use of living shorelines for stabilization and nutrient mitigation in New England. Ecol Restor 41(2–3):84–98. https://doi.org/10.3368/er.41.2-3.84

Sidik F, Basyuni M, Pradisty NA, Hanintyo R, Widagti N, Priyono B, Amelia R, Macklin P, Surana N, Slamet B (2025) Hydrological suitability for mangrove species in restored mangroves in abandoned ponds. Aquat Bot 199:103873. https://doi.org/10.1016/j.aquabot.2025.103873

Sloey TM, Hildebrandt S, Morris RL, Bilskie MV, Bland A, Bushek D, Bilkovic DM (2025) A blueprint to greener shorelines: advancing the effectiveness, sustainability, and widespread adoption of coastal nature-based solutions through transdisciplinary research. Estuaries Coasts 48(5):133. https://doi.org/10.1007/s12237-025-01571-8

Srinivasa Rao C, Eganathan P, Anand A, Balakrishna P, Reddy TP (1998) Protocol for in vitro propagation of *Excoecaria agallocha* L., a medicinally important mangrove species. Plant Cell Rep 17(11):861–865. https://doi.org/10.1007/s002990050498

Statton J, Montoya LR, Orth RJ, Dixon KW, Kendrick GA (2017) Identifying critical recruitment bottlenecks limiting seedling establishment in a degraded seagrass ecosystem. Sci Rep 7(1):14786. https://doi.org/10.1038/s41598-017-13833-y

Syahid LN, Sakti AD, Ward R, Rosleine D, Windupranata W, Wikantika K (2023) Optimizing the spatial distribution of Southeast Asia mangrove restoration based on zonation, species and carbon projection schemes. Estuar Coast Shelf Sci 293:108477. https://doi.org/10.1016/j.ecss.2023.108477

Teong TK, Hartanto RS, Ledet J, Du H, Gupta S, Jiang HY, Kua HW, Pang SD, Wang S, Todd PA (2024) Marine bioreceptivity among green concretes. Ecol Eng 204:107284. https://doi.org/10.1016/j.ecoleng.2024.107284

Tereszkiewicz PA, Provost LA, Everett C (2025) Coastal sand dunes: a review of management strategies for dune stabilization. Technical Note ERDC/TN EWN-25-5. U.S. Army Engineer Research and Development Center

Thom RM (2000) Adaptive management of coastal ecosystem restoration projects. Ecol Eng 15(3–4):365–372. https://doi.org/10.1016/S0925-8574(00)00086-0

Thorhaug A, Verduin JJ, Kiswara W, Prathep A, Huang X, Gallagher JB, Yap T, Dorward SE, Schwarz A, Berlyn GP (2025) Seagrass restoration in the greater Southeast Asia region: techniques, species, survival and comparisons among investigations. Front Mar Sci 12:1505222. https://doi.org/10.3389/fmars.2025.1505222

Tinh PH, MacKenzie RA, Hung TD, Vinh TV, Ha HT, Lam MH, Hanh NT, Tung NX, Hai PM, Huyen BT (2022) Mangrove restoration in Vietnamese Mekong Delta during 2015–2020: achievements and challenges. Front Mar Sci 9:1043943. https://doi.org/10.3389/fmars.2022.1043943

To LP, Nguyen HT, Roel B, Lam TS, Dang KN (2017) Sustainability and livelihood opportunities for mangrove-shrimp farmers in the coastal areas of SocTrang province. CTU J Innov Sustain Develop 5:18–29. https://doi.org/10.22144/ctu.jen.2017.003
Tonneijck F, Van der Goot F, Pearce F (2022) Building with Nature in Indonesia. Restoring an eroding coastline and inspiring action at scale. Wetlands International and Ecoshape Foundation. February 2022
Tran TV, Reef R, Zhu X, Mohan M (2025) Spectral analysis for monitoring mangrove restoration: a case study in the Vietnamese Southern Coastline. Cambridge Prisms: Coastal Futures 3:e8. https://doi.org/10.1017/cft.2025.5
Triest L, Del Socorro A, Gado VJ, Mazo AM, Sierens T (2021) Avicennia genetic diversity and fine-scaled structure influenced by coastal proximity of mangrove fragments. Front Mar Sci 8:643982. https://doi.org/10.3389/fmars.2021.643982
Turner RE, Lewis RR (1996) Hydrologic restoration of coastal wetlands. Wetlands Ecol Manage 4:65–72. https://doi.org/10.1007/BF01876229
Twomey AJ, Nunez K, Carr JA, Crooks S, Friess DA, Glamore W, Orr M, Reef R, Rogers K, Waltham NJ, Lovelock CE (2024) Planning hydrological restoration of coastal wetlands: key model considerations and solutions. Sci Total Environ 915:169881. https://doi.org/10.1016/j.scitotenv.2024.169881
Ugah EPTA, Ndubuisi OG, FNisafetyE FISPON (2025) Gauging the effectiveness of assisted natural regeneration and active restoration in enhancing biodiversity and ecosystem services in degraded tropical forest. Int J Innov Sci Eng Technol Res 13(2):131–145. https://doi.org/10.5281/zenodo.15608063
Upadhyay R, Joshi N, Sampat AC, Verma AK, Patel A, Singh V, Kalubarme MH (2015) Mangrove restoration and regeneration monitoring in Gulf of Kachchh, Gujarat State, India, using remote sensing and geo-informatics. Int J Geosci 6(4):299–310. https://doi.org/10.4236/ijg.2015.64023
Urban Development Directorate (UDD) (2023) Payra-Kuakata comprehensive plan focusing on eco-tourism: regional plan: 2021–2041. Ministry of Housing and Public Works, Government of the People's Republic of Bangladesh. https://ceip-bwdb.gov.bd/report/Regional%20plan/Regional%20Plan.pdf
van der Lelij AC, van Rees F, Pardo MdL, Weseenbeeck v (2021) Ecological mangrove restoration with permeable structures: monitoring report 1220476–000-ZKS-0010. Deltares, The Netherlands
Van Loon AF, Te Brake B, Van Huijgevoort MHJ, Dijksma R (2016) Hydrological classification, a practical tool for mangrove restoration. PLoS ONE 11(3):e0150302. https://doi.org/10.1371/journal.pone.0150302
Van Mai C, Ngo A, Mai T, Dao HT (2021) Bamboo fences as a nature-based measure for coastal wetland protection in Vietnam. Front Mar Sci 8:756597. https://doi.org/10.3389/fmars.2021.756597
Veettil BK, Ward RD, Quang NX, Trang NTT, Giang TH (2019) Mangroves of Vietnam: historical development, current state of research and future threats. Estuar Coast Shelf Sci 218:212–236. https://doi.org/10.1016/j.ecss.2018.12.021
Veettil BK, Ward RD, Dung NTK, Van DD, Quang NX, Hoai PN, Hoang N (2021) The use of bioshields for coastal protection in Vietnam: current status and potential. Reg Stud Mar Sci 47:101945. https://doi.org/10.1016/j.rsma.2021.101945
Veettil BK, Tran X, Lopes MS (2024) Mangrove vegetation changes and shoreline erosion along the central coastline of Vietnam: a study from Quang Ngai province. Environ Develop 52:101065. https://doi.org/10.1016/j.envdev.2024.101065
Vietnam Academy of Forest Sciences (VAFS) (2025) Technical procedures for planting casuarina forests. https://vafs.gov.vn/vn/quy-trinh-ky-thuat-trong-rung-phi-lao/. Accessed on: 02 Feb 2025
Wang L, Mu M, Li X, Lin P, Wang W (2011) Differentiation between true mangroves and mangrove associates based on leaf traits and salt contents. J Plant Ecol 4(4):292–301. https://doi.org/10.1093/jpe/rtq008

Wang X, Wang Y, Wang Y (2013) Use of exotic species during ecological restoration can produce effects that resemble vegetation invasions and other unintended consequences. Ecol Eng 52:247–251. https://doi.org/10.1016/j.ecoleng.2012.11.007

Wang X, Zou Y, Zhu T, Guan B, Yang J, Yu J (2022) The effects of hydrological connectivity blocking on Suaeda salsa development in the Yellow River Delta, China. Front Ecol Evol 10:1005677. https://doi.org/10.3389/fevo.2022.1005677

Warnell K, Mason S, Siegle A, Merritt M, Olander L (2023) Fact sheet: living shoreline creation. NBS Roadmap Project. Nicholas Institute for Energy, Environment & Sustainability, Duke University, Durham, NC. www.nicholasinstitute.duke.edu/roadmap

Wei X, Mo W, Xiong L, Hu X, Cheng H (2025) Field investigation of wave attenuation in a mangrove forest dominated by *Avicennia marina* (Forsk.) Viern. Plants 14(1):135. https://doi.org/10.3390/plants14010135

Wilson SJ, Smith R, Chazdon R, Durst P, Metzel R, Sprenkle-Hippolite S (2022) Assisted natural regeneration. A guide for restoring tropical forests. Conservation International, Arlington, Virginia

Winterwerp JC, Albers T, Anthony EJ, Friess DA, Mancheño AG, Moseley K, Muhari A, Naipal S, Noordermeer J, Oost A, Saengsupavanich C, Tas SA, Tonneijck FH, Wilms T, Van Bijsterveldt C, Van Eijk P, Van Lavieren E, Van Wesenbeeck BK (2020) Managing erosion of mangrove-mud coasts with permeable dams—lessons learned. Ecol Eng 158:106078. https://doi.org/10.1016/j.ecoleng.2020.106078

Wu Y, Zhang Y, Dai L, Xie L, Zhao S, Liu Y, Zhang Z (2021) Hydrological connectivity improves soil nutrients and root architecture at the soil profile scale in a wetland ecosystem. Sci Total Environ 762:143162. https://doi.org/10.1016/j.scitotenv.2020.143162

Yan H, Shi B, Gao F (2025) Wetland ecological restoration and geomorphological evolution: a hydrodynamic-sediment-vegetation coupled modeling study. J Mar Sci Eng 13(7):1326. https://doi.org/10.3390/jmse13071326

Ye Y, Tam NFY, Lu C, Wong Y (2005) Effects of salinity on germination, seedling growth and physiology of three salt-secreting mangrove species. Aquat Bot 83(3):193–205. https://doi.org/10.1016/j.aquabot.2005.06.006

Yoshikai M, Nakamura T, Bautista DM, Herrera EC, Baloloy A, Suwa R, Basina R, Primavera-Tirol YH, Blanco AC, Nadaoka K (2022) Field measurement and prediction of drag in a planted Rhizophora mangrove forest. J Geophys Res: Oceans 127(11):e2021JC018320. https://doi.org/10.1029/2021JC018320

Yu Z, Bu C, Li Y (2025) Machine learning for ecological analysis. Chem Eng J 507:160780. https://doi.org/10.1016/j.cej.2025.160780

Zakaria MH, Syed NN, Ramaiya SD, Bujang JS (2025) Seagrass assisted recovery and long-term monitoring in the Sungai Pulai estuary, Johor, Malaysia. Front Conserv Sci 6:1646399. https://doi.org/10.3389/fcosc.2025.1646399

Zhang R, Liu L, Yang Y, Xu S, He Z, Tracy ME, Guo Z (2025) Seawater tides shaped mangrove sediment a bacterial community drastically distinct from supratidal land soil. BMC Microbiol 25(1):742. https://doi.org/10.1186/s12866-025-04486-3

Zhang S, Wang G, Yu W, Wei L, Gao C, Li D, et al. (2024) Multi-omics analyses reveal the mechanisms underlying the responses of *Casuarina equisetifolia* ssp. incana to seawater atomization and encroachment stress. BMC Plant Biol 24(1):854. https://doi.org/10.1186/s12870-024-05561-z

Zhao L, Fan X, Xiao S (2025) Remote-sensing indicators and methods for coastal-ecosystem health assessment: a review of progress, challenges, and future directions. Water 17(13):1971. https://doi.org/10.3390/w17131971

Zivec P, Balcombe S, McBroom J, Sheldon F, Capon SJ (2021) Patterns and drivers of natural regeneration on old-fields in semi-arid floodplain ecosystems. Agr Ecosyst Environ 316:107466. https://doi.org/10.1016/j.agee.2021.107466

Chapter 8
Management Challenges and Long-Term Sustainability of Coastal Bioshields

Abstract Effective management of bioshields requires a long-term, adaptive approach that accounts for ecological complexity, socio-economic pressures, and climate uncertainty. This chapter examines the major challenges associated with maintaining bioshield functionality and sustainability over time, focusing on governance gaps, ecological degradation, institutional fragmentation, and competing land-use priorities. It highlights how inadequate monitoring, poor stakeholder coordination, and short-term project planning often undermine the long-term resilience of coastal protection systems. The discussion integrates ecological perspectives—such as species turnover, sediment dynamics, and habitat connectivity—with socio-political dimensions, including policy enforcement, financing mechanisms, and community participation. Particular attention is given to the drivers of bioshield decline, including pollution, coastal development, aquaculture expansion, and unsustainable resource extraction. The chapter underscores the need for adaptive management frameworks that incorporate continuous learning, scientific feedback, and stakeholder engagement. It also explores financial sustainability models, such as payment for ecosystem services (PES) and blue carbon initiatives, that can incentivize long-term bioshield conservation. Drawing from regional experiences in Asia and other vulnerable coastal regions, this chapter synthesizes practical recommendations for integrating bioshield management into broader coastal zone management and climate adaptation policies. Ultimately, it emphasizes that ensuring bioshield longevity depends on harmonizing ecological integrity, institutional support, and community stewardship under a unified sustainability vision.

Keywords Adaptive management · Coastal zone governance · Community stewardship · Ecosystem degradation · Payment for ecosystem services (PES) · Sustainability frameworks

B. Kozhikkodan Veettil, *The Potential of Bioshields for Coastal Protection*, Advances in Natural and Technological Hazards Research 55,
https://doi.org/10.1007/978-3-032-22888-8_8

8.1 Introduction

The management and long-term sustainability of coastal bioshields, which are nature-based protective systems comprising mangroves, casuarina belts, beach vegetation, and coastal dune complexes, constitute a central challenge within contemporary ecosystem-based approaches to coastal protection (Veettil et al. 2021; Perricone et al. 2023; Forrester et al. 2024; White et al. 2024). Over the past two decades, bioshields have gained increasing recognition as multifunctional forms of green infrastructure that deliver a wide range of ecological, economic, and social benefits (Veettil et al. 2021; Jordan and Fröhle 2022). These include wave attenuation, shoreline stabilization, biodiversity conservation, carbon sequestration, and the enhancement of coastal livelihoods (Veettil et al. 2021; Perricone et al. 2023; Paxton et al. 2024; Daro Justine and Seenath 2025). However, the capacity of bioshields to provide sustained protection is not guaranteed by their establishment alone (Veettil et al. 2021; Paxton et al. 2024; Sloey et al. 2025). Their long-term effectiveness depends on dynamic interactions among biophysical processes, governance arrangements, financial resources, and active engagement of coastal communities (Craig and Ruhl 2010; Paxton et al. 2024; Dulyakasem et al. 2026).

In much of Asia, the management of bioshields is further complicated by rapid population growth, accelerated industrialization, and extensive coastal land-use transformation (Veettil et al. 2021). The expansion of aquaculture, urban development, tourism infrastructure, and port construction is exerting increasing pressure on coastal ecosystems, often undermining the integrity of bioshield systems (Veettil et al. 2020, 2021). At the same time, climate change has intensified coastal hazards through sea-level rise, stronger storm events, salinity intrusion, and altered sediment dynamics (Xu et al. 2024). Under these conditions, bioshields must be understood not as static barriers, but as living, adaptive systems whose resilience depends on continuous management, ecological feedback, and institutional support. This reality challenges conventional environmental management models that emphasize short-term project cycles and sectorally fragmented decision-making.

Long-term sustainability of bioshields extends well beyond initial planting or restoration activities. It encompasses the maintenance of ecosystem structure and function, the capacity to respond to environmental variability and climate stressors, and the establishment of enduring social, technical, and financial support mechanisms (Paxton et al. 2024; Sloey et al. 2025). Experiences from Vietnam and other Southeast Asian countries illustrate how early bioshield initiatives were frequently compromised by inappropriate species-site matching, limited post-plantation maintenance, and weak coordination between government agencies and local stakeholders (Veettil et al. 2021; Gerona-Daga and Salmo 2022). Comparable patterns have been observed in South Asia, where large-scale coastal afforestation programs in India and Bangladesh have experienced degradation over time due to inconsistent monitoring, institutional fragmentation, and conflicts over land tenure and resource use

(Iftekhar and Islam 2004; Thivakaran 2017). These experiences underscore a persistent gap between the conceptual appeal of bioshields and the practical challenges of sustaining them over decades.

Despite the growing policy endorsement of bioshields as cost-effective and environmentally sound alternatives to hard engineering structures, their long-term operational sustainability remains insufficiently addressed. Many initiatives continue to prioritize area-based planting targets over ecological performance, social legitimacy, and adaptive capacity (Paxton et al. 2024). As a result, bioshields may fail to mature into functional protective systems or may deteriorate under cumulative environmental and anthropogenic pressures (Sloey et al. 2025).

Against this backdrop, this chapter examines the key management challenges and sustainability determinants influencing the long-term performance of coastal bioshields in Asia. It critically reviews ecological, institutional, socio-economic, and technical constraints, while highlighting the importance of adaptive management, participatory governance, and stable financing mechanisms. By synthesizing regional experiences and lessons learned, the chapter aims to identify strategic pathways through which bioshields can be maintained as resilient, self-sustaining systems capable of safeguarding coastal communities in the face of accelerating climate change and development pressures.

8.2 Ecological and Environmental Challenges

The ecological integrity of bioshields determines their ability to deliver long-term protective functions. However, in many Asian contexts, bioshields face multiple environmental and biological constraints that threaten their sustainability (Veettil et al. 2021).

8.2.1 Species Selection and Ecological Suitability

Inappropriate species selection is one of the most recurrent management issues. Many bioshield projects in Asia, particularly in Vietnam, India, and the Philippines, have relied on monocultures of fast-growing species such as *Casuarina equisetifolia* or *Rhizophora apiculata* (Primavera and Esteban 2008; Das and Sandhu 2014; Hai et al. 2020). Although these species establish quickly and provide short-term protection, monocultures reduce ecological resilience and limit biodiversity (Le et al. 2021; Li et al. 2025). For example, *Casuarina* plantations are susceptible to pest outbreaks and storm damage due to shallow root systems (Willis 2017; Wang et al. 2025), while dense *Rhizophora* stands can impede natural sedimentation processes critical for mangrove expansion (Yoshikai et al. 2023).

Long-term ecological sustainability requires mixed-species planting strategies that reflect local environmental gradients such as salinity, tidal exposure, and soil texture (Chen et al. 2022; Yoshikai et al. 2022; Li et al. 2025). Studies conducted at both global scales (Li et al. 2025) and in southern Vietnam's Mekong Delta (Nguyen et al. 2020) demonstrate that mixed-species mangrove belts exhibit higher survival rates, greater sediment accretion, and improved resistance to storms compared with single-species plantations.

8.2.2 Natural Disturbances and Climate Stressors

Climate-related stressors, such as rising sea levels, temperature shifts, increasing salinity, and changing storm patterns, pose severe risks to bioshields (Veettil et al. 2020, 2021). Sea-level rise can lead to saltwater intrusion and the landward retreat of coastal vegetation zones (Di Nitto et al. 2014; Xie et al. 2022). In low-lying regions such as the Ganges–Brahmaputra and Mekong deltas, bioshields are increasingly vulnerable to coastal squeeze, a process in which urban and hard infrastructure constrain or prevent their natural landward migration (Phan et al. 2015; Dang et al. 2022; Rahman et al. 2022). Extreme weather events, such as Super Typhoon Haiyan (2013) in the Philippines, have demonstrated that even mature bioshields can be uprooted or severely damaged when storm intensity exceeds historical norms (Villamayor et al. 2016; Takagi 2025).

Adaptation to such disturbances requires dynamic management strategies, including allowing space for natural regeneration, maintaining ecological corridors, and incorporating sediment management practices (Schmitt et al. 2013; Trung and Thuy 2019). For example, in Vietnam's Bac Lieu province (Fig. 8.1), sediment nourishment and controlled tidal exchange have been used to enhance mangrove regeneration behind dykes, promoting long-term bioshield recovery (Schmitt et al. 2013).

8.2.3 Pollution and Anthropogenic Stressors

Industrial discharges, agricultural runoffs, and plastic pollution significantly degrade coastal bioshield ecosystems (Sabdaningsih et al. 2023; Chan et al. 2025a, b; Deakin et al. 2025; Neelavannan et al. 2025). Pollutants alter sediment chemistry, reduce soil oxygenation, and inhibit seedling growth. In aquaculture-intensive regions like the Mekong and Red River deltas, nutrient loading from shrimp ponds has led to eutrophication, algal blooms, and increased susceptibility to disease in mangrove ecosystems (Oanh et al. 2014; Le et al. 2022; Dung et al. 2025). Addressing these threats requires stricter enforcement of environmental regulations and the integration of pollution control into bioshield management plans. Figure 8.2 illustrates pollution in various vegetated coastal areas in Vietnam.

Fig. 8.1 Sediment nourishment techniques using melaleuca fencing and hard structures in Nha Mat, Bac Lieu, Mekong Delta (Vietnam). *Photographs* Tran Thi Nhu Phuong

Fig. 8.2 Plastic pollution in mangrove areas in Bac Lieu and algal blooms in abandoned fish/shrimp ponds in Ben Tre province (Vietnam). *Photo* Bijeesh Kozhikkodan Veettil

8.2.4 Invasive Species and Ecological Imbalance

Invasive species such as *Acrostichum aureum* (mangrove fern) and *Nypa fruticans* (nipa palm) have disrupted the composition and structure of several mangrove ecosystems (Moreno-Domínguez et al. 2016; Numbere 2019). These species can dominate degraded areas and outcompete native mangroves, leading to reduced ecosystem services and altered hydrodynamics (Augusthy et al. 2024). It is worth noting that

Nypa fruticans is not considered as an invasive species in the Asian region. An effective bioshield management, therefore, requires continuous ecological monitoring, early detection of invasive species, and periodic thinning or enrichment planting to maintain species diversity and functional balance (Tran et al. 2025).

8.3 Institutional and Governance Challenges

Beyond ecological concerns, institutional weaknesses represent some of the most persistent barriers to bioshield sustainability in Asia. Fragmented governance structures, overlapping mandates, and inconsistent policies often undermine coordination among agencies responsible for forestry, coastal protection, and disaster management (Beresnev et al. 2016; Pham et al. 2022; Muhammad et al. 2025).

8.3.1 Fragmented Institutional Responsibilities

In most Asian countries, coastal bioshields fall under the jurisdiction of multiple ministries or departments. In Vietnam, for example, mangrove management involves the Ministry of Agriculture and Rural Development (MARD), the Ministry of Natural Resources and Environment (MONRE), and provincial governments, each with distinct mandates and budgets (Veettil et al. 2019, 2021). This fragmentation results in inconsistent management, duplication of efforts, and a lack of accountability. Similar institutional overlaps are evident in India, where coastal vegetation management falls between the Ministry of Environment, Forest and Climate Change (MoEFCC) and state-level coastal zone authorities (Madhanagopal et al. 2022).

Effective coordination mechanisms, such as inter-ministerial task forces or integrated coastal zone management (ICZM) committees, are essential to overcome these challenges (Chou et al. 2021). Bangladesh's Coastal Zone Policy (2005) and the Philippines' Integrated Coastal Management (ICM) framework under the PEMSEA initiative offer useful examples of coordinated institutional arrangements that enhance bioshield governance (Chua et al. 2006; Iftekhar 2006).

8.3.2 Policy Inconsistency and Regulatory Gaps

In many cases, national policies on forestry, land use, and disaster risk reduction are not harmonized, resulting in conflicting objectives. For example, aquaculture expansion policies may contradict mangrove conservation laws, leading to unauthorized clearing of bioshield areas (Veettil et al. 2019). In Indonesia, despite national mangrove rehabilitation targets, weak enforcement of coastal land-use zoning allows continued conversion for shrimp farming (Mursyid et al. 2021; Muenzel et al.

2025). Similarly, in Vietnam, inconsistencies between the Forestry Law and the Land Law create uncertainty about land tenure and benefit-sharing arrangements for communities managing coastal forests (Veettil et al. 2019).

Policy coherence and legal clarity are therefore critical. Countries like India have begun to address this through the Coastal Regulation Zone (CRZ) Notification (2019), which specifies buffer zones for mangroves and mandates environmental impact assessments for coastal development projects (Sakthivel and Khan 2024). Strengthening legal frameworks to explicitly recognize bioshields as protected natural infrastructure can enhance their long-term sustainability.

8.3.3 Limited Institutional Capacity and Resources

Even where appropriate policies exist, limited technical and financial capacity within local institutions often constrain effective management. Field-level forest officers or coastal managers may lack training in ecosystem-based adaptation or remote sensing-based monitoring (Kanan and Giupponi 2024). Budget constraints further limit routine maintenance, replanting, and community engagement. Development agencies and NGOs frequently fill these gaps, but their project-based funding models often end after short timeframes, leaving bioshields without sustained support (MFF Vietnam 2015). Building institutional capacity through continuous training, adequate staffing, and long-term financing is thus essential for sustainability.

8.4 Socioeconomic and Land-Use Pressures

The socio-economic dynamics of coastal regions exert a significant influence over the management and survival of bioshields (Fig. 8.3). High population density, competing land demands, and shifting livelihoods create continuous tension between conservation objectives and development pressures (Hagger et al. 2022; Tran et al. 2024).

8.4.1 Land-Use Conversion and Encroachment

In many parts of Asia, bioshield areas have been converted into aquaculture ponds, salt pans, and settlements (figures with these-ninh thuan etc.). The rapid expansion of shrimp farming in the Mekong Delta, for instance, resulted in the loss of over half of the mangrove cover between 1980 and 2000 (Veettil et al. 2019, 2021). While reforestation efforts have since been launched, illegal encroachments remain common (Veettil et al. 2019). Similarly, in several coastal areas of Bangladesh,

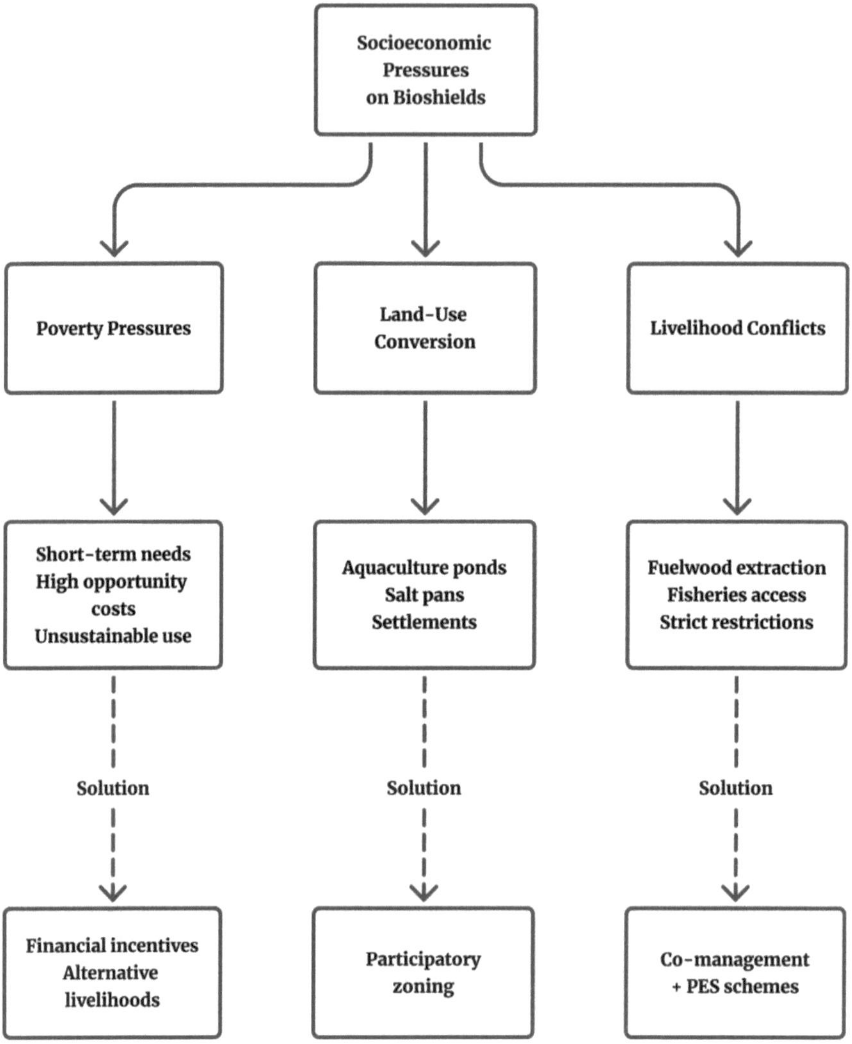

Fig. 8.3 Various socioeconomic pressures on coastal vegetation bioshields and solutions to overcome each

mangrove buffer zones have been encroached upon for rice cultivation and fishponds, undermining the sustainability of the Sundarbans bioshield system (Hoq 2006).

Strengthening land-use planning and tenure systems is therefore fundamental. Participatory zoning approaches, where communities and authorities jointly delineate bioshield protection areas, have proven effective in several Vietnamese provinces, balancing livelihood needs with conservation priorities (Cuong et al. 2016; Hue et al. 2024).

8.4.2 Livelihood Dependence and Community Conflicts

Coastal communities often depend on bioshield ecosystems for fuelwood, fisheries, and non-timber forest products, and unsustainable extraction can degrade vegetation and compromise protective functions (Debrot et al. 2020; Veettil et al. 2023). Conversely, strict protection policies that restrict access can provoke local resistance. Successful management thus depends on participatory models that link bioshield protection with livelihood benefits, such as eco-tourism, aqua-silviculture, or PES (Payment for Ecosystem Services) schemes. In Vietnam, for example, co-management arrangements between forest rangers and local households have improved compliance and vegetation health while generating supplemental income (Swan 2010).

8.4.3 Poverty and Short-Term Economic Pressures

Poverty remains a key driver of bioshield degradation. In marginalized coastal communities, immediate economic needs often take precedence over long-term sustainability considerations. The opportunity cost of conserving bioshields can be perceived as high when land could otherwise be used for lucrative shrimp farming or infrastructure (Kwan et al. 2025). Introducing financial incentives, microcredit schemes, or alternative livelihoods can alleviate such pressures (Lan et al. 2022; Karpowicz et al. 2024). Linking bioshields to tangible socio-economic benefits, such as disaster insurance discounts or carbon revenue sharing, can also improve long-term stewardship.

8.5 Technical and Monitoring Constraints

Robust technical and monitoring systems are fundamental to evaluating bioshield performance and ensuring its long-term sustainability. Effective monitoring enables managers to assess ecological conditions, detect emerging threats, and implement timely corrective measures (Tremblay et al. 2023). Nevertheless, across much of Asia, bioshield initiatives continue to be constrained by weak technical capacity, limited baseline information, and the absence of systematic feedback mechanisms (Gerona-Daga and Salmo 2022). These shortcomings undermine evidence-based decision-making and reduce the ability of bioshields to function as resilient, adaptive forms of coastal protection.

8.5.1 Lack of Long-Term Monitoring Programs

A persistent limitation of many bioshield projects is their reliance on short-term, project-based funding, which typically prioritizes establishment targets over sustained monitoring (Friess et al. 2022; Beeston et al. 2023). As a result, monitoring efforts often cease once external funding ends, leaving critical gaps in knowledge regarding long-term ecological trajectories (Baker and Gittman 2025). Processes such as vegetation succession, sediment accretion or erosion, and changes in species composition are rarely tracked over sufficient temporal scales (Gatt et al. 2022). Without longitudinal datasets, it becomes difficult to distinguish between natural variability and structural decline, and management interventions are frequently based on assumptions rather than empirical evidence.

In Vietnam, recent initiatives to integrate long-term, satellite-based mangrove monitoring through partnerships between the Ministry of Agriculture and Environment (MAE) and national research institutions represent a significant advance (MAE 2025a, b). These programs may provide continuous data on canopy conditions and spatial change, demonstrating how institutionalized monitoring frameworks can support more informed and proactive management. Such approaches offer valuable lessons for other countries seeking to move beyond short-term assessment models.

8.5.2 Limited Use of Remote Sensing and Geospatial Tools

Advances in remote sensing and geospatial technologies, including freely available satellite platforms such as Landsat and Sentinel-2, as well as unmanned aerial vehicles (UAVs), have greatly expanded the potential for cost-effective, large-scale bioshield monitoring. These tools can generate high-resolution information on vegetation cover, shoreline dynamics, canopy density, and disturbance impacts, often with greater spatial and temporal consistency than ground-based surveys alone (Veettil et al. 2021). Despite these advantages, their application in bioshield management remains uneven and limited in many coastal areas globally.

Key barriers include insufficient technical expertise within local agencies, a lack of appropriate hardware and software infrastructure, and weak integration of geospatial data into routine management processes (Avtar et al. 2020; Christofi et al. 2024; Chan et al. 2025a, b). Targeted capacity-building programs that train local authorities to use open-access satellite data and basic geographic information systems (GIS) platforms could substantially enhance monitoring efficiency (Chan et al. 2025a, b). Moreover, GIS-based decision support systems can assist in optimizing planting designs, identifying erosion-prone areas, and prioritizing zones requiring reinforcement or restoration (Ozkiper et al. 2024).

8.5.3 *Insufficient Feedback Loops and Adaptive Management*

Beyond data collection, a critical weakness in many bioshield programs lies in the absence of effective feedback loops linking monitoring outcomes to management actions (NASEM 2017). Management systems often follow a linear logic that emphasizes implementation outputs rather than continuous learning and adjustment. In contrast, adaptive management frameworks require iterative processes in which monitoring results directly inform decisions such as replanting, species substitution, or hydrological modification (Williams and Brown 2014).

Establishing formal mechanisms for data sharing and coordination among research institutions, government agencies, and local communities is essential to operationalizing adaptive management (Onley et al. 2025). Transparent communication of monitoring results can strengthen accountability, encourage local stewardship, and enable more responsive interventions (Beeston et al. 2023; Karcher et al. 2024). Without such feedback mechanisms, technical monitoring risks becoming an isolated exercise, disconnected from the practical realities of sustaining bioshields under changing environmental and socio-economic conditions.

8.6 Adaptive Management and Restoration Strategies

Given the ecological complexity and socio-economic dynamics surrounding coastal bioshields, adaptive management has emerged as an essential strategy for ensuring long-term viability (Williams and Brown 2014; Birgé et al. 2016). Unlike static management models, adaptive management emphasizes learning-by-doing, iterative decision-making, and the integration of monitoring feedback into planning processes (Williams and Brown 2014).

8.6.1 *Principles of Adaptive Management*

Adaptive management for bioshields involves continuous assessment of ecological conditions, stakeholder feedback, and environmental stressors to inform corrective actions (Thom 2000). It recognizes that bioshield ecosystems are dynamic systems influenced by tidal regimes, sediment fluxes, and human interventions (Thom 2000). The core principles include:

a. Iterative Planning: Adjusting management plans based on observed outcomes and updated data.
b. Flexibility: Allowing shifts in management priorities or species composition in response to environmental change.
c. Participatory Decision-Making: Involving communities, scientists, and policy-makers in the co-production of knowledge.

d. Integration Across Scales: Coordinating actions at local, provincial, and national levels.

These principles can be applied effectively in the Integrated Coastal Zone Management (ICZM) frameworks, where adaptive feedback from local monitoring has guided reforestation density, species mix, and buffer zone design (Plaoky 2024).

8.6.2 *Restoration-Based Approaches*

In areas where bioshields have been degraded, restoration strategies are critical for rebuilding ecological structure and function. Two dominant approaches are used in Asia: ecological restoration and rehabilitation.

Ecological restoration seeks to re-establish native vegetation and natural hydrology (Veettil et al. 2019). For example, the Can Gio Mangrove Biosphere Reserve in southern Vietnam restored over 30,000 ha of degraded mangroves through hydrological reconnection and natural regeneration (Fig. 8.4).

Rehabilitation focuses on recovering protective functions, even if full biodiversity restoration is not possible. This is seen in India's Odisha coast, where casuarina belts are maintained primarily for storm buffering (Das and Sandhu 2014).

Recent advancements in assisted natural regeneration (ANR), where local communities protect and nurture natural seedlings rather than replanting, have proven cost-effective and ecologically sustainable (Hagger et al. 2022; Oluwajuwon et al. 2025). This approach is widely practiced in Indonesia's North Kalimantan region (Budi et al. 2022).

Fig. 8.4 Restored mangrove areas (*Nypa fruticans, Rhizophora apiculata*) in Can Gio Mangrove Biosphere Reserve, Southern Vietnam, through hydrological reconnection and natural regeneration. *Photographs* Bijeesh Kozhikkodan Veettil

8.6.3 Integrating Climate Adaptation into Bioshield Management

Climate change adaptation must be embedded into bioshield design and management. This involves identifying future climate risks (e.g., sea-level rise, changing precipitation patterns) and modifying management to accommodate them. For example, in Vietnam's Ca Mau province, bioshield belts are now designed with retreat zones to allow inland migration of mangroves as sea levels rise (MAE 2025c). Similarly, in the Philippines, climate-resilient mangrove species and mixed planting are promoted in areas frequently hit by typhoons (Juanico 2022; Takagi 2025). Integrating climate modeling into bioshield planning can also guide decisions about species choice, buffer width, and hydrological design, enhancing long-term resilience.

8.7 Financing and Long-Term Maintenance Mechanisms

The financial sustainability of bioshields is one of the most critical and often neglected dimensions of management. Many projects fail not because of ecological or technical deficiencies but because of the absence of consistent funding mechanisms for maintenance, monitoring, and community engagement (Hai et al. 2020; Stewart-Sinclair et al. 2020).

8.7.1 Public Sector Financing and Budgetary Limitations

Most Asian countries rely heavily on government budgets and donor-funded projects to establish bioshields. However, once external funding ends, maintenance often lapses. This discontinuity has been observed in several provinces in Vietnam and Indonesia, where initial donor projects achieved impressive reforestation outcomes but lacked post-project care (Nguyen et al. 2015; Damastuti et al. 2022). Governments must therefore institutionalize bioshield maintenance within recurrent public budgets rather than treating them as short-term development projects (Gerona-Daga and Salmo 2022).

Fiscal decentralization can also empower local authorities to allocate funds for bioshield upkeep (Wever et al. 2012). For example, India's coastal states have introduced state-level disaster mitigation funds that can be used for bioshield maintenance as part of broader disaster risk reduction (DRR) investments (DMD 2025).

8.7.2 Payment for Ecosystem Services (PES) and Community Incentives

PES mechanisms provide a promising pathway for sustained financing by compensating communities that manage and protect bioshields. Vietnam's Payment for Forest Environmental Services (PFES) program, which was initially applied to upland forests, is now being piloted for coastal mangrove systems in provinces like Ca Mau (Pham et al. 2019). Households receive annual payments for maintaining vegetation cover and preventing encroachment, fostering a sense of ownership and accountability.

In the Philippines, community-based forestry management agreements (CBFMAs) provide local cooperatives with usufruct rights to bioshield areas, allowing them to derive controlled economic benefits (e.g., honey collection, eco-tourism) while maintaining protective vegetation (Dugan and Pulhin 2007).

8.7.3 Blue Carbon and Climate Finance Mechanisms

Coastal bioshields, particularly mangrove systems, serve as vital carbon sinks (Veettil et al. 2024, 2025). The growing recognition of blue carbon—carbon stored in coastal and marine ecosystems—has opened new funding avenues. Under international mechanisms such as the Green Climate Fund (GCF) and REDD + (Reducing Emissions from Deforestation and Forest Degradation), countries can receive payments for verified carbon sequestration (Cui and Huang 2017; Kelly 2017).

Indonesia and Vietnam have initiated pilot blue carbon projects that monetize the carbon storage of mangrove bioshields (Vu et al. 2023; Zaid et al. 2025; Veettil et al. 2025). By linking carbon finance with local community benefits, these initiatives may promote both environmental and economic sustainability. However, technical challenges related to carbon measurement and verification remain significant barriers to widespread adoption.

8.7.4 Public–Private Partnerships (PPPs) and Insurance Incentives

Private-sector engagement can enhance the durability of bioshield programs. Tourism operators, aquaculture businesses, and insurance companies have direct interests in coastal stability. PPP models, where businesses co-invest in bioshield maintenance in exchange for risk reduction or green branding, are emerging in Asia (Morita and Matsumoto 2021). Moreover, the insurance industry increasingly recognizes bioshields as natural risk buffers, potentially lowering premiums for properties protected by natural vegetation—thus providing an economic incentive for their preservation (Kousky 2022).

8.8 Sustainability Frameworks and Indicators

To achieve measurable and enduring outcomes, bioshield management must be guided by clear sustainability frameworks and performance indicators (Ioannou et al. 2025). These frameworks provide the basis for assessing ecological health, socio-economic benefits, and institutional effectiveness over time (Karnauskaitė et al. 2019; Olander et al. 2024).

8.8.1 Ecological Sustainability Indicators

Ecological indicators assess the biological and physical condition of bioshields (Schmitt et al. 2013; Schaduw et al. 2024). Common indicators include:

a. Vegetation Survival Rate (%): Proportion of planted or natural seedlings surviving after 3–5 years (Schmitt et al. 2013).
b. Canopy Cover and Density (m^2/ha): Reflecting ecosystem maturity and carbon storage potential (Schaduw et al. 2024).
c. Sediment Accretion Rate (mm/year): Indicative of shoreline stability (Danh et al. 2025).
d. Biodiversity Index: Presence of associated flora and fauna, indicating ecological integrity (Schaduw et al. 2024).

For instance, in the Mekong Delta, mangrove belts showing sediment accretion rates of 3–5 mm/year have demonstrated resilience against moderate storm surges (Dang et al. 2022). These measurable indicators allow for quantitative evaluation of long-term ecological performance.

8.8.2 Socioeconomic and Institutional Indicators

Socioeconomic sustainability is equally critical, as bioshields thrive where local communities perceive tangible benefits (Ebeler et al. 2025; Dulyakasem et al. 2026). Indicators include:

a. Community Participation Rate: Percentage of local households actively involved in bioshield management (Dulyakasem et al. 2026).
b. Household Income Change: Change in income derived from bioshield-related livelihoods (e.g., eco-tourism, fisheries) (Ebeler et al. 2025).

c. Conflict Resolution Mechanisms: Presence of functioning systems to address land-use disputes (Ebeler et al. 2025).
d. Policy Integration Index (hypothetical): Degree to which bioshields are embedded in national or regional development plans.

For example, the Philippines' ICM program provides an instructive model, where local sustainability indicators (e.g., income diversification, stakeholder participation) are used to assess progress toward coastal resilience goals (White et al., 2005; Cervania et al. 2018).

8.8.3 Integrated Sustainability Assessment Frameworks

Integrated frameworks combine ecological, economic, and governance dimensions into holistic assessments (Ioannou et al. 2025). One such model is the Ecosystem Approach to Coastal Management (EACM), developed under UNEP and PEMSEA, which evaluates ecological health, economic viability, and institutional capacity simultaneously (Chua 2013; Pazoto et al. 2021). In Vietnam, adaptation of this framework under the national Sustainable Coastal Forest Development Program (2015–2025) has improved long-term bioshield monitoring by linking forest performance indicators with socio-economic outcomes (Prime Minister's Decision No. 120/QĐ-TTg, 2015). Embedding bioshields within national Natural Capital Accounting systems can also institutionalize their value in public policy, ensuring that long-term ecosystem services are reflected in economic planning.

8.9 Conclusions

The sustainability of coastal bioshields in Asia is contingent upon a delicate and interdependent balance among ecological integrity, institutional coordination, financial continuity, and meaningful community participation. Evidence synthesized from experiences in Vietnam, India, Bangladesh, Indonesia, and the Philippines demonstrates that bioshields can function as effective long-term protective systems while delivering substantial socio-economic co-benefits when they are managed through adaptive, inclusive, and well-coordinated frameworks. Nevertheless, this sustainability remains inherently fragile in contexts where governance fragmentation, discontinuous funding, and competing socio-economic pressures persist.

Several overarching insights emerge from the analysis presented in this chapter. First, ecological resilience is fundamentally linked to diversity and adaptive management. Bioshields designed with mixed-species assemblages, aligned with site-specific ecological gradients and supported by ongoing management interventions, exhibit greater structural stability and functional longevity than monoculture plantations. Such diversity enhances resistance to pests, disease, and climatic stressors, while allowing bioshields to adjust dynamically to changing environmental conditions.

Second, governance coherence plays a decisive role in long-term success. Fragmentation among forestry, coastal zone management, fisheries, and disaster risk reduction institutions frequently undermines bioshield effectiveness. Harmonizing policy frameworks, clarifying mandates, and strengthening inter-agency coordination are essential to reduce institutional conflict and ensure that bioshields are integrated into broader coastal planning and risk management strategies.

Third, sustained local participation emerges as a critical determinant of stewardship and performance. Where communities are actively involved in planning, management, and monitoring—and where benefit-sharing mechanisms are transparent and equitable—bioshields are more likely to be protected from encroachment, degradation, and neglect. Community-managed systems consistently demonstrate higher compliance, stronger social legitimacy, and improved ecological outcomes.

Fourth, sustainable financing constitutes the cornerstone of long-term bioshield viability. Innovative mechanisms such as payments for ecosystem services, carbon finance, and PPPs offer pathways to move beyond short-term, donor-dependent projects. When effectively designed, these mechanisms can provide predictable funding streams that support maintenance, monitoring, and adaptive management over decadal timescales.

Finally, monitoring and performance indicators must be institutionalized rather than treated as ancillary project components. Continuous assessment using integrated ecological and socio-economic indicators is essential to track bioshield condition, inform management decisions, and ensure effectiveness under evolving climatic and developmental pressures.

In a nutshell, the long-term sustainability of bioshields cannot be understood solely as an ecological challenge. It is equally a question of governance quality, social equity, and long-term vision. Integrating adaptive management principles with secure financing arrangements and strong community engagement offers the most reliable pathway toward durable, nature-based coastal protection systems capable of safeguarding Asia's vulnerable shorelines for generations to come.

References

Augusthy S, Nizam A, Kumar A (2024) The diversity, drivers, consequences and management of plant invasions in the mangrove ecosystems. Sci Total Environ 945:173851. https://doi.org/10.1016/j.scitotenv.2024.173851

Avtar R, Komolafe AA, Kouser A, Singh D, Yunus AP, Dou J, Kumar P, Gupta RD, Johnson BA, Thu Minh HV, Aggarwal AK, Kurniawan TA (2020) Assessing sustainable development prospects through remote sensing: a review. Remote Sens Appl: Soc Environ 20:100402. https://doi.org/10.1016/j.rsase.2020.100402

Baker R, Gittman RK (2025) Co-funding robust monitoring with living shoreline construction is critical for maximizing beneficial outcomes. Estuaries Coasts 48:5. https://doi.org/10.1007/s12237-024-01433-9

Beeston M, Cameron C, Hagger V, Howard J, Lovelock C, Sippo J, Tonneijk F, van Bijsterveldt C, van Eijk P (eds) (2023) Best practice guidelines for mangrove restoration. Global Mangrove Alliance, pp 141

Beresnev N, Phung T, Broadhead J (2016) Mangrove-related policy and institutional frameworks in Pakistan. Thailand and Viet Nam. Food and agriculture organization of the United Nations, regional office for Asia and the pacific. International Union for Conservation of Nature (IUCN) Bangkok, 2016

Birgé HE, Allen CR, Garmestani AS, Pope KL (2016) Adaptive management for ecosystem services. J Environ Manage 183:343–352. https://doi.org/10.1016/j.jenvman.2016.07.054

Budi G, Putro HA, Sasongko BT (2022) Application of Assisted Natural Regeneration (ANR) to enhance forest biodiversity in West Kalimantan, Indonesia. FAO

Cervania AB, Perdio AC, Llave DS, Zapanta AB (2018) Bataan coastal resource management programs: environmental, socio-economic, and implementation issues from stakeholders' views. Asia Pac J Multidiscip Res 6(1):46–58

Chan HCM, Owers CJ, Fuller S, Hayward MW, Moverley D, Griffin AS (2025a) Capacity and capability of remote sensing to inform invasive plant species management in the Pacific Islands region. Conserv Biol 39(1):e14344. https://doi.org/10.1111/cobi.14344

Chan MW, Ibrahim M, Qasim M, Siddiqui MH, Alamri S, Siddiqui PJ, Gulzar S, Azeem M, Handlovic TT, Wahab MF, Alfagham AT, Rasheed M (2025b) Pollution indices of brackish water and sediments of major mangrove sites along Arabian Sea coast of Pakistan. Sci Rep 15(1):31026. https://doi.org/10.1038/s41598-025-01712-w

Chen S, Sun Y, Tang K, Zhang F, Ding W, Wang A (2022) Distribution characteristics and restoration application of vegetation in Chengcun bay surrounding areas of Yangjiang City. Int J Environ Res Public Health 19(16):10399. https://doi.org/10.3390/ijerph191610399

Chou LM, Chua TE, Bonga D (2021) "Integrated coastal management" enhances coastal resilience to climate change—The East Asia experience. In: Climate change science. Elsevier, pp 59–79. https://doi.org/10.1016/B978-0-12-823767-0.00004-5

Christofi D, Mettas C, Evagorou E, Stylianou N, Eliades M, Theocharidis C, Chatzipavlis A, Hasiotis T, Hadjimitsis D (2024) A review of open remote sensing data with GIS, AI, and UAV support for shoreline detection and coastal erosion monitoring. Appl Sci 15(9):4771. https://doi.org/10.3390/app15094771

Chua TE (2013) Coastal and ocean governance in the seas of East Asia: PEMSEA's experience. Coast Manag 41(2):99–119. https://doi.org/10.1080/08920753.2013.768517

Chua TE, Bonga D, Bermas-Atrigenio N (2006) Dynamics of integrated coastal management: PEMSEA's experience. Coast Manag 34(3):303–322. https://doi.org/10.1080/08920750600686695

Craig RK, Ruhl JB (2010) Governing for sustainable coasts: complexity, climate change, and coastal ecosystem protection. Sustainability 2(5):1361–1388. https://doi.org/10.3390/su2051361

Cui L, Huang Y (2017) Exploring the schemes for green climate fund financing: international lessons. World Dev 101:173–187. https://doi.org/10.1016/j.worlddev.2017.08.009

Cuong CV, Dart P, Dudley N, Hockings M (2016) Factors influencing successful implementation of biosphere reserves in Vietnam: challenges, opportunities and lessons learnt. Environ Sci Policy 67:16–26. https://doi.org/10.1016/j.envsci.2016.10.002

Damastuti E, De Groot R, Debrot AO, Silvius MJ (2022) Effectiveness of community-based mangrove management for biodiversity conservation: a case study from Central Java, Indonesia. Trees, For People 7:100202. https://doi.org/10.1016/j.tfp.2022.100202

Dang ATN, Reid M, Kumar L (2022) Assessing potential impacts of sea level rise on mangrove ecosystems in the Mekong Delta, Vietnam. Reg Environ Change 22:70. https://doi.org/10.1007/s10113-022-01925-z

Danh M, Thai LT, Thi Hong Dang N, Manh NNB (2025) Building sustainable indicators for mangrove forests based on a survey to determine forest erosion in the Southwestern Sea of Vietnam. J Environ Earth Sci 7(8):85–100. https://doi.org/10.30564/jees.v7i8.10762

Daro Justine YE, Seenath A (2025) Vegetative nature-based solutions for coastal flood risk management: benefits, challenges, and uncertainties. Ocean Coast Manag 261:107520. https://doi.org/10.1016/j.ocecoaman.2024.107520

Das S, Sandhu H (2014) Role of exotic vegetation in coastal protection: an investigation into the ecosystem services of Casuarina in Odisha. Econ Polit Weekly, pp 42–50

Deakin K, Porter A, Osorio Baquero A, Lewis C (2025) Plastic pollution in mangrove ecosystems: a global meta-analysis. Mar Pollut Bull 218:118165. https://doi.org/10.1016/j.marpolbul.2025.118165

Debrot AO, Veldhuizen A, K, SW, Klapwijk CJ, Islam MN, Alam MI, Ahsan MN, Ahmed MU, Hasan SR, Fadilah R, Noor YR, Pribadi R, Rejeki S, Damastuti E, Koopmanschap E, Reinhard S, Verburg C, Poelman M (2020) Non-timber forest product livelihood-focused interventions in support of mangrove restoration: a call to action. Forests 11(11):1224. https://doi.org/10.3390/f11111224

Di Nitto D, Neukermans G, Koedam N, Defever H, Pattyn F, Kairo JG, Dahdouh-Guebas F (2014) Mangroves facing climate change: landward migration potential in response to projected scenarios of sea level rise. Biogeosciences 11:857–871. https://doi.org/10.5194/bg-11-857-2014

Disaster Management Division (DMD) (2025) Fiscal decentralization can also empower local authorities to allocate funds for bioshield upkeep. Ministry of Home Affairs, Government of India. https://ndmindia.mha.gov.in/ndmi/response-fund.

Dugan P, Pulhin JM (2007) Forest harvesting in community based forest management in the Philippines: simple tools versus complex procedures. RAP Publication, 2007, No. No.09, 38–46 ref. 12 Conference title: A cut for the poor. Proceedings of the international conference on Managing forests for poverty reduction: capturing opportunities for forest harvesting and wood processing for the benefit of the poor, Ho Chi Minh City, Vietnam, 3–6 October 2006

Dulyakasem R, Brooks C, Lehnert SL, Newton P (2026) Community-based natural resource management in coastal communities: the contribution of mangroves to household livelihoods in southern Thailand. Environ Develop 57:101340. https://doi.org/10.1016/j.envdev.2025.101340

Dung LV, Tue NT, Dang Quy T, Nhuan MT (2025) Ammonium, nitrate, and phosphate in coastal waters of red river biosphere reserve, Vietnam. Environ Nat Resourc J 23(1):55–64. https://doi.org/10.32526/ennrj/23/20240128

Ebeler L, Balke T, Laurie E, Le H, Nguyen T, Renaud FG (2025) Local-scale impacts of mangrove restoration and conservation on coastal communities in two Vietnamese deltas: socio-economic and institutional dynamics. Ocean Coast Manag 269:107812. https://doi.org/10.1016/j.ocecoaman.2025.107812

Forrester J, Leonardi N, Cooper JR, Kumar P (2024) Seagrass as a nature-based solution for coastal protection. Ecol Eng 206:107316. https://doi.org/10.1016/j.ecoleng.2024.107316

Friess DA, Gatt YM, Ahmad R, Brown BM, Sidik F, Wodehouse D (2022) Achieving ambitious mangrove restoration targets will need a transdisciplinary and evidence-informed approach. One Earth 5(5):456–460. https://doi.org/10.1016/j.oneear.2022.04.013

Gatt YM, Andradi-Brown DA, Ahmadia GN, Martin PA, Sutherland WJ, Spalding MD, Donnison A, Worthington TA (2022) Quantifying the reporting, coverage and consistency of key indicators in mangrove restoration projects. Front For Glob Change 5:720394. https://doi.org/10.3389/ffgc.2022.720394

Gerona-Daga MEB, Salmo SG III (2022) A systematic review of mangrove restoration studies in Southeast Asia: challenges and opportunities for the United Nation's decade on ecosystem restoration. Front Mar Sci 9:987737. https://doi.org/10.3389/fmars.2022.987737

Hagger V, Worthington TA, Lovelock CE, Adame MF, Amano T, Brown BM, Friess DA, Landis E, Mumby PJ, Morrison TH, R K, Wilson KA, Zganjar C, Saunders MI (2022) Drivers of global mangrove loss and gain in social-ecological systems. Nat Commun 13(1):6373. https://doi.org/10.1038/s41467-022-33962-x

Hai NT, Dell B, Phuong VT, Harper RJ (2020) Towards a more robust approach for the restoration of mangroves in Vietnam. Ann for Sci 77(1):18. https://doi.org/10.1007/s13595-020-0921-0

Hoq ME (2006) An analysis of fisheries exploitation and management practices in Sundarbans mangrove ecosystem, Bangladesh. Ocean Coast Manag 50(5–6):411–427. https://doi.org/10.1016/j.ocecoaman.2006.11.001

Hue HTT, Quan PZ, Lap KQ (2024) Review of biodiversity assessment and resource management orientation for sustainable rural development in Lang Son province, Vietnam. Environ Res Commun 6(9):095019. https://doi.org/10.1088/2515-7620/ad7928

Iftekhar MS (2006) Conservation and management of the Bangladesh coastal ecosystem: overview of an integrated approach. Nat Res Forum 30(3):230–237. https://doi.org/10.1111/j.1477-8947.2006.00111.x

Iftekhar MS, Islam MR (2004) Managing mangroves in Bangladesh: a strategy analysis. J Coast Conserv 10:139–146. https://doi.org/10.1652/1400-0350(2004)010[0139:MMIBAS]2.0.CO;2

Ioannou A, Bataka E, Kokosis N, Kofinas D, Billinis C, Laspidou C (2025) A holistic one health assessment framework for coastal areas. Sustainability 17(21):9359. https://doi.org/10.3390/su17219359

FröhleJordan P, Fröhle P (2022) Bridging the gap between coastal engineering and nature conservation?. J Coast Conserv 26:4. https://doi.org/10.1007/s11852-021-00848-x

Juanico DE (2022) Does mangrove restoration imply coastal protection? A prospective simulation study. Front Mar Sci 9:968420. https://doi.org/10.3389/fmars.2022.968420

Kanan AH, Giupponi C (2024) Coastal socio-ecological systems adapting to climate change: a global overview. Sustainability 16(22):10000. https://doi.org/10.3390/su162210000

Karcher DB, Tuohy P, Cooke SJ, Cvitanovic C (2024) Knowledge exchange at the interface of marine science and policy: a review of progress and research needs. Ocean Coast Manag 253:107137. https://doi.org/10.1016/j.ocecoaman.2024.107137

Karnauskaitė D, Schernewski G, Støttrup JG, Kataržytė M (2019) Indicator-based sustainability assessment tool to support coastal and marine management. Sustainability 11(11):3175. https://doi.org/10.3390/su11113175

Karpowicz DA, Mohan M, Watt MS, Montenegro JF, King SA, Selvam PP, Nithyanandan M, Robyn B, Ali T, Abdullah MM, Doaemo W, Ewane EB (2024) Mangrove-based carbon market projects: 15 considerations for engaging and supporting local communities. Diversity 16(9):574. https://doi.org/10.3390/d16090574

Kelly AC (2017) Improving REDD+ (reducing emissions from deforestation and forest degradation) programs (doctoral dissertation). University of Washington, USA

Kousky C (2022) Insurance-sector tools to combat biodiversity loss. Science 377:714–716. https://doi.org/10.1126/science.abo7282

Kwan V, Friess DA, Sarira TV, Zeng Y (2025) Permanence risks limit blue carbon financing strategies to safeguard Southeast Asian mangroves. Commun Earth Environ 6(1):57. https://doi.org/10.1038/s43247-025-02035-4

Lan BT, Phuong TT, Dat TT, Truong DD (2022) Payment for urban mangrove forest conservation in Vietnam: a community case study of Can Gio Biosphere Reserve, Ho Chi Ming City. Sustainability 15(13):10299. https://doi.org/10.3390/su151310299

Le HD, Smith C, Herbohn J, Nguyen H (2021) A comparison of growth, structure and diversity of mixed species and monoculture reforestation systems in the Philippines. J Sustain For 40(4):401–430. https://doi.org/10.1080/10549811.2020.1767145

Le ND, Nguyen TH, Duong TT, Rochelle-Newall E, Hoang TTH, Vu TH, Pham TMH, Nguyen TAH, Dinh LM, Phung TXB, Nguyen TD, Pham TMH, Dao VH, Le TPQ (2022) Risk of eutrophication in the seawater of the coastal Red River aquaculture zone (Thai Binh province, Vietnam). Reg Stud Mar Sci 55:102587. https://doi.org/10.1016/j.rsma.2022.102587

Li Z, Li Y, Fan B (2025) Diverse restored mangrove stands enhance carbon storage compared to monospecific plantation: a meta-analysis. For Ecol Manage 578:122446. https://doi.org/10.1016/j.foreco.2024.122446

Madhanagopal D, Samuel L, Gandhi K (2022) Localized policy responses for climate change and developmental complications: discussions from three coastal regions of India. In: Climate change and risk in south and Southeast Asia. Routledge, pp 236–265

MFF Vietnam (2015) Mangroves for the Future Phase III—National Strategic Action Plan (2015–2018). IUCN, Gland, Switzerland, 44pp

Ministry of Agriculture and Environment (MAE) (2025a) Satellites monitor the health of mangrove forests (in Vietnamese). https://mae.gov.vn/ve-tinh-theo-doi-suc-khoe-rung-ngap-man-18323.htm. Accessed on 18 Dec 2025

Ministry of Agriculture and Environment (MAE) (2025b) Starting 2026, Vietnam to boost remote sensing applications in agriculture and environmental monitoring. https://en.mae.gov.vn/starting-2026-vietnam-to-boost-remote-sensing-applications-in-agriculture-and-environmental-monitoring-8856.htm. Accessed on 18 Dec 2025

Ministry of Agriculture and Environment (MAE) (2025c) IUCN partners to restore coastal mangrove forests in Ca Mau. https://en.mae.gov.vn/iucn-partners-to-restore-coastal-mangrove-forests-in-ca-mau-8964.htm. Accessed on 18 Dec 2025

Moreno-Domínguez R, Cascales-Miñana B, Ferrer J, Diez JB (2016) Acrostichum, a Pioneering fern of floodplain areas from the Late Oligocene Sariñena Formation of the Iberian Peninsula. PLoS ONE 11(9):e0162334. https://doi.org/10.1371/journal.pone.0162334

Morita K, Matsumoto KI (2021) Governance challenges for implementing nature-based solutions in the Asian region. Polit Govern 9(4):102–113

Muenzel D, Anggoro AW, Bulan DE, Nurfadilah Y, Pratama R, Beger M (2025) Optimising site selection for ecosystem approaches to shrimp aquaculture in mangrove systems. Aquacult Int 33(7):632. https://doi.org/10.1007/s10499-025-02338-5

Muhammad F, Marsum M, Putro SP, Hadi M, Murningsih M, Hidayat JW, Wardoyo S (2025) Vulnerability analysis and sustainable rehabilitation strategies for mangrove ecosystems in Bedono coastal area, Central Java. BMC Ecol Evol 25(1):86. https://doi.org/10.1186/s12862-025-02406-7

Mursyid H, Daulay MH, Pratama AA, Laraswati D, Novita N, Malik A, Maryudi A (2021) Governance issues related to the management and conservation of mangrove ecosystems to support climate change mitigation actions in Indonesia. For Policy Econ 133:102622. https://doi.org/10.1016/j.forpol.2021.102622

National Academies of Sciences, Engineering, and Medicine (NASEM) (2017) Effective monitoring to evaluate ecological restoration in the Gulf of Mexico. The National Academies Press, Washington, DC. https://doi.org/10.17226/23476

Neelavannan K, Pulikkoden A, Jinoy G, Meleppura R, Panickan P, Rajaneesh KM, Dixit S, Bawazeer A, Reshi O, Ponnambalam M, Joydas T, Maneja RH, Albert J, Alhems L (2025) Geospatial distribution and anthropogenic litter impact on coastal mangrove ecosystems from the Saudi Arabia coast of the Gulf. Sci Rep 15(1):15964. https://doi.org/10.1038/s41598-025-98136-3

Nguyen T, Van Tam N, Quoi LP, Parnell KE (2015) Community perspectives on an internationally funded mangrove restoration project: Kien Giang province, Vietnam. Ocean Coast Manage 119:146–154. https://doi.org/10.1016/j.ocecoaman.2015.10.008

Nguyen LT, Hoang HT, Ta HV, Park PS (2020) Comparison of mangrove stand development on accretion and erosion sites in Ca Mau, Vietnam. Forests 11(6):615. https://doi.org/10.3390/f11060615

Numbere AO (2019) Impact of invasive Nypa Palm (*Nypa Fruticans*) on mangroves in coastal areas of the Niger Delta Region, Nigeria. In: Makowski C, Finkl C (eds) Impacts of invasive species on coastal environments. Coastal Research Library, vol 29. Springer, Cham. https://doi.org/10.1007/978-3-319-91382-7_13

Oanh DTH, Giang HT, Lien NTK (2014) Relationship between shrimp health and phytoplankton population fluctuation in intensive white leg shrimp (*Litopenaeus vannamei*) ponds. Can Tho Univ J Sci 159–168

Olander L, Warnell K, Mason S, Karasik R, Tallis H, Coffey K, Hale C, Shepard C, Wowk K, Yoskowitz D (2024) Advancing consistent socio-economic monitoring of coastal ecosystem restoration through collaborative metric development. Commun Earth Environ 5(1):537. https://doi.org/10.1038/s43247-024-01705-z

Oluwajuwon TV, Chazdon RL, Gregorio N, Ota L, Tripoli R, Pasa A, Herbohn J (2025) Assisted natural regeneration for tropical forest and landscape restoration in the Philippines: Implementation, motivations, challenges and future directions. Trees, Forests and People 20:100896. https://doi.org/10.1016/j.tfp.2025.100896

Onley IR, Cassey P, McGeoch MA (2025) Biodiversity data sharing platforms are vital for the management and prevention of biological invasions. Biodivers Conserv 34:2247–2257. https://doi.org/10.1007/s10531-025-03058-1

Ozkiper O, Allegri E, Bianconi A, Pham HV, Furlan E, Simide R, Van der Geest M, Critto A (2024) A GIS-MCDA approach to map environmental suitability of Posidonia oceanica meadows as blue nature-based solutions in the Mediterranean eco-region. Sci Total Environ 955:176803. https://doi.org/10.1016/j.scitotenv.2024.176803

Paxton AB, Riley TN, Steenrod CL, Puckett BJ, Alemu IJB, Paliotti ST, Silliman BR (2024) Evidence on the performance of nature-based solutions interventions for coastal protection in biogenic, shallow ecosystems: a systematic map. Environ Evidence 13(1):28. https://doi.org/10.1186/s13750-024-00350-5

Pazoto CE, Silva EP, Andrade LAB, Favero JMD, Alô CFS, Duarte MR (2021) Ocean literacy, formal education, and governance: a diagnosis of Brazilian school curricula as a strategy to guide actions during the Ocean Decade and beyond. Ocean Coast Res 69:e21041. https://doi.org/10.1590/2675-2824069.21008cep

Perricone V, Mutalipassi M, Mele A, Buono M, Vicinanza D, Contestabile P (2023) Nature-based and bioinspired solutions for coastal protection: an overview among key ecosystems and a promising pathway for new functional and sustainable designs. ICES J Mar Sci 80(5):1218–1239. https://doi.org/10.1093/icesjms/fsad080

Pham TT, Vu TP, Hoang TL, Dao TL, Nguyen DT, Pham DC, Dao LH, Nguyen VT, Hoang NV (2022) The effectiveness of financial incentives for addressing mangrove loss in Northern Vietnam. Front For Glob Change 4:709073. https://doi.org/10.3389/ffgc.2021.709073

Pham TT, Vu TP, Pham DC, Dao LHT, Nguyen VT, Hoang NVH, Hoang TL, Dao TLC, Nguyen DT (2019) Opportunities and challenges for mangrove management in Vietnam: lessons learned from Thai Binh, Quang Ninh and Thanh Hoa provinces. Occasional Paper 197. Bogor, Indonesia: CIFOR

Phan LK, de Vries JSVT, Stive MJ (2015) Coastal mangrove squeeze in the Mekong Delta. J Coastal Res 31(2):233–243. https://doi.org/10.2112/JCOASTRES-D-14-00049.1

Plaoky I (2024) Integrated coastal zone management: a comprehensive approach to sustainable development in coastal areas. J Coast Zone Manag 27:617. https://doi.org/10.35248/2473-3350.24.27.617

Primavera JH, Esteban JMA (2008) A review of mangrove rehabilitation in the Philippines: successes, failures and future prospects. Wetlands Ecol Manage 16:345–358. https://doi.org/10.1007/s11273-008-9101-y

Rahman MM, Haque A, Nicholls RJ, Darby SE, Urmi MT, Dustegir MM, Dunn FE, Tahsin A, Razzaque S, Horsburgh K, Haque MA (2022) Sustainability of the coastal zone of the Ganges-Brahmaputra-Meghna delta under climatic and anthropogenic stresses. Sci Total Environ 829:154547. https://doi.org/10.1016/j.scitotenv.2022.154547

Sabdaningsih A, Adyasari D, Suryanti S, Febrianto S, Eshananda Y (2023) Environmental legacy of aquaculture and industrial activities in mangrove ecosystems. J Sea Res 196:102454. https://doi.org/10.1016/j.seares.2023.102454

Sakthivel M, Khan N (2024) Protection of the Indian Coastal Ecosystem through Coastal Regulation Zone (CRZ) Notifications*: an Analysis. In: The Routledge companion to Indian ethics. Routledge India, pp 211–222

Schaduw JNW, Tallei TE, Sumilat DA (2024) Mangrove health index, community structure and canopy cover in small islands of Bunaken National Park, Indonesia: insights into dominant mangrove species and overall mangrove condition. Trop Life Sci Res 35(2):187. https://doi.org/10.21315/tlsr2024.35.2.9

Schmitt K, Albers T, Pham TT, Dinh SC (2013) Site-specific and integrated adaptation to climate change in the coastal mangrove zone of Soc Trang Province, Viet Nam. J Coast Conserv 17(3):545–558. https://doi.org/10.1007/s11852-013-0253-4

Sloey TM, Hildebrandt S, Morris RL, Bilskie MV, Bland A, Bushek D, Bilkovic DM (2025) A blueprint to greener shorelines: advancing the effectiveness, sustainability, and widespread adoption of coastal nature-based solutions through transdisciplinary research. Estuaries Coasts 48(5):133. https://doi.org/10.1007/s12237-025-01571-8

Stewart-Sinclair PJ, Purandare J, Bayraktarov E, Waltham N, Reeves S, Statton J, Lovelock CE (2020) Blue restoration–building confidence and overcoming barriers. Front Mar Sci 7:541700. https://doi.org/10.3389/fmars.2020.541700

Swan S (2010) Co-management: concepts and practices in Viet Nam. Co-management/Shared Governance of Natural Resources and Protected Areas in Viet Nam, pp 27–44

Takagi H (2025) Mangrove forest recovery a decade after Super Typhoon Haiyan. J Environ Manage 395:127840. https://doi.org/10.1016/j.jenvman.2025.127840

Thivakaran GA (2017) Mangrove restoration: an overview of coastal afforestation in India. In: Prusty B, Chandra R, Azeez P (eds) Wetland science. Springer, New Delhi. https://doi.org/10.1007/978-81-322-3715-0_26

Thom RM (2000) Adaptive management of coastal ecosystem restoration projects. Ecol Eng 15(3–4):365–372. https://doi.org/10.1016/S0925-8574(00)00086-0

Tran TV, Reef R, Zhu X (2024) Long-term changes of mangrove distribution and its response to anthropogenic impacts in the Vietnamese Southern Coastal Region. J Environ Manage 370:122658. https://doi.org/10.1016/j.jenvman.2024.122658

Tran TV, Reef R, Zhu X, Mohan M (2025) Spectral analysis for monitoring mangrove restoration: a case study in the Vietnamese Southern Coastline. Cambridge Prisms: Coastal Futures 3:e8. https://doi.org/10.1017/cft.2025.5

Tremblay LA, Chariton AA, Li M, Zhang Y, Horiguchi T, Ellis JI (2023) Monitoring the health of coastal environments in the pacific region—a review. Toxics 11(3):277. https://doi.org/10.3390/toxics11030277

Trung TC, Thuy DV (2019) Effectiveness of the co-management model of mangrove forests and natural resources for the coastal areas of the Mekong Delta (in Vietnamese). J Water Resourc Sci Technol 56:58–66

Veettil BK, Puri V (2025) Blue carbon ecosystems in Vietnam: a review. Ocean Coast Manag 267:107711. https://doi.org/10.1016/j.ocecoaman.2025.107711

Veettil BK, Ward RD, Quang NX, Trang NTT, Giang TH (2019) Mangroves of Vietnam: historical development, current state of research and future threats. Estuar Coast Shelf Sci 218:212–236. https://doi.org/10.1016/j.ecss.2018.12.021

Veettil BK, Costi J, Marques WC, Tran X, Quang NX, Van DD, Hoai PN (2020) Coastal environmental changes in Southeast Asia: a study from Quang Nam Province, Central Vietnam. Reg Stud Mar Sci 39:101420. https://doi.org/10.1016/j.rsma.2020.101420

Veettil BK, Ward RD, Dung NTK, Van DD, Quang NX, Hoai PN, Hoang N (2021) The use of bioshields for coastal protection in Vietnam: current status and potential. Reg Stud Mar Sci 47:101945. https://doi.org/10.1016/j.rsma.2021.101945

Veettil BK, Wickramasinghe D, Amarakoon V (2023) Mangrove forests in Sri Lanka: an updated review on distribution, diversity, current state of research and future perspectives. Reg Stud Mar Sci 62:102932. https://doi.org/10.1016/j.rsma.2023.102932

Veettil BK, Puri V, Wickramasinghe D, Ward RD, Stankovic M, Udagedara S, Madawala Y (2024) Blue carbon ecosystems in Sri Lanka: a review. Estuar Coast Shelf Sci 306:108907. https://doi.org/10.1016/j.ecss.2024.108907

Villamayor BMR, Rollon RN, Samson MS, Albano GMG, Primavera JH (2016) Impact of Haiyan on Philippine mangroves: implications to the fate of the widespread monospecific Rhizophora

plantations against strong typhoons. Ocean Coast Manag 132:1–14. https://doi.org/10.1016/j.ocecoaman.2016.07.011
Vu TP, Pham TT, Tran NMH, Nguyen TTA, Nguyen TVA, Tang TKH (2023) Blue carbon market in Vietnam: potential and challenges for future development. Occasional Paper 8. Center for International Forestry Research (CIFOR), Bogor, Indonesia; and World Agroforestry (ICRAF), Nairobi, Kenya
Wang X, Shi H, Wu X, Chen X, Liu S, Yu Q, Zhang H, Zhou X (2025) Microbial diversity in coastal *Casuarina equisetifolia* forest and its potential in counteracting bacterial wilt infections. Pest Manag Sci 81(8):4621–4634. https://doi.org/10.1002/ps.8822
Wever L, Glaser M, Gorris P, Ferrol-Schulte D (2012) Decentralization and participation in integrated coastal management: policy lessons from Brazil and Indonesia. Ocean Coast Manag 66:63–72. https://doi.org/10.1016/j.ocecoaman.2012.05.001
White A, Christie P, D'Agnes H, Lowry K, Milne N (2004) Designing ICM projects for sustainability: lessons from the Philippines and Indonesia. Ocean Coast Manag 48(3–6):271–296. https://doi.org/10.1016/j.ocecoaman.2005.04.007
White AE, Cohn N, Davis EH, Hein CJ, Zinnert JC (2024) Coastal dune management affects above and belowground biotic characteristics. Sci Rep 14(1):22688. https://doi.org/10.1038/s41598-024-73312-z
Williams BK, Brown ED (2014) Adaptive management: from more talk to real action. Environ Manage 53:465–479. https://doi.org/10.1007/s00267-013-0205-7
Willis KJ (2017) State of the world's plants 2017. Royal Botanic Gardens, Kew, London: England
Xie D, Schwarz C, Kleinhans MG, Zhou Z (2022) Implications of coastal conditions and sea-level rise on mangrove vulnerability: a bio-morphodynamic modeling study. J Geophys Res: Earth Surf 127(3):e2021JF006301. https://doi.org/10.1029/2021JF006301
Xu H, Wang C, Ge M, Sardans J, Peñuelas J, Tong C, Wang W (2024) Salinity increases under sea level rise strengthens the chemical protection of SOC in subtropical tidal marshes. Sci Total Environ 954:176512. https://doi.org/10.1016/j.scitotenv.2024.176512
Yoshikai M, Nakamura T, Suwa R, Sharma S, Rollon R, Yasuoka J, Egawa R, Nadaoka K (2022) Predicting mangrove forest dynamics across a soil salinity gradient using an individual-based vegetation model linked with plant hydraulics. Biogeosciences 19:1813–1832. https://doi.org/10.5194/bg-19-1813-2022
Yoshikai M, Nakamura T, Herrera EC, Suwa R, Rollon R, Ray R, Nadaoka K (2023) Representing the impact of Rhizophora mangroves on flow and sediment transport in a hydrodynamic model (COAWST_rh v1. 0): the importance of three-dimensional root system structures. Geosci Model Dev 16:5847–5863. https://doi.org/10.5194/egusphere-2022-1350
Zaid M, Ricky R, Sedera RM (2025) Blue carbon regulations and implementation in several countries: lessons for Indonesia. J Law, Environ Justice 3(1):30–78

Part IV
Socioeconomic and Policy Dimensions of Coastal Bioshields

Chapter 9
Ecosystem Services and Socioeconomic Benefits of Bioshields

Abstract Coastal bioshields provide a wide range of ecosystem services that extend far beyond coastal protection, delivering critical ecological, economic, and social benefits to coastal communities. This chapter explores the multiple dimensions of ecosystem services associated with coastal vegetation systems such as mangroves, salt marshes, seagrass meadows, and dune forests. It categorizes these services into provisioning, regulating, supporting, and cultural functions, emphasizing their interconnected role in sustaining livelihoods, biodiversity, and climate regulation. Key regulating services—such as carbon sequestration, shoreline stabilization, and water filtration—are analyzed alongside provisioning services, including fisheries enhancement, timber and non-timber products, and habitat provision for marine species. The chapter also evaluates the socioeconomic benefits derived from bioshields, focusing on income diversification, food security, and disaster risk reduction in vulnerable coastal regions. Empirical studies from Asia and other tropical zones illustrate how the economic valuation of these services informs policy design, conservation financing, and community-based management strategies. Furthermore, the chapter addresses trade-offs and equity considerations, highlighting how ecosystem service distribution can influence social resilience and inclusivity. Integrating ecological economics with social science perspectives, this chapter demonstrates that recognizing and quantifying the full spectrum of bioshield benefits is essential for evidence-based policymaking and sustainable coastal development. It concludes by advocating for the inclusion of ecosystem service valuation in national accounting systems to strengthen the case for nature-based solutions in coastal resilience planning.

Keywords Coastal livelihoods · Carbon sequestration · Disaster risk reduction · Ecosystem services valuation · Nature-based solutions · Socioeconomic benefits

B. Kozhikkodan Veettil, *The Potential of Bioshields for Coastal Protection*, Advances in Natural and Technological Hazards Research 55,
https://doi.org/10.1007/978-3-032-22888-8_9

9.1 Introduction

Economic valuation provides a critical framework for understanding the tangible and intangible benefits of coastal bioshields, including natural vegetative barriers such as mangroves, coastal forests, and dune vegetation that mitigate coastal hazards (Barbier 2016, 2020; Hernández-Blanco et al. 2021; Brander et al. 2024). While their ecological functions have long been recognized, quantifying their economic value remains crucial for integrating them into national investment and policy frameworks (Hanley et al. 2015; Hernández-Blanco et al. 2021; Brander et al. 2024). Governments and donors increasingly demand financial justification for nature-based solutions (NbS), particularly in regions such as Asia, where coastal zones are densely populated and economically vital (Swann et al. 2021).

Asia is home to more than 60% of the world's population living within 100 km of the coast, and the region experiences frequent cyclones, storm surges, and sea-level rise impacts (Brecht et al. 2012). The economic losses from coastal hazards are staggering, with annual damages exceeding USD 200 billion (Monioudi et al. 2025). In this context, bioshields offer low-cost, multi-functional alternatives or complements to engineered structures such as seawalls and breakwaters (Paxton et al. 2024). Their ability to attenuate wave energy, trap sediments, and stabilize shorelines yields direct protection benefits, while their role in carbon sequestration, fisheries enhancement, and ecotourism provides long-term economic co-benefits (Veettil et al. 2021; Sunkur et al. 2023; Weaver and Stehno 2024).

However, despite growing empirical evidence, bioshields are often undervalued in cost–benefit analyses (CBA) due to methodological challenges, limited data, and the undervaluation of ecosystem services (Himes-Cornell et al. 2018; Gargaran et al. 2024; Lopez-Rivas and Cardenas 2024; Chelli et al. 2025). This chapter aims to synthesize approaches to economic valuation of bioshields in Asia, analyze cost–benefit evidence from key countries, and identify policy implications for scaling up investment in nature-based coastal protection.

9.2 Conceptual Framework of Economic Valuation

Economic valuation seeks to express the total value of bioshields in monetary terms, encompassing both use values (direct and indirect) and non-use values (existence, option, and bequest) (Ranasinghe and Kallesoe 2006; Somda and Awaïss 2013; Milon and Alvarez 2019; Maldonado et al. 2020). In the context of bioshields, this translates into a comprehensive framework (Table 9.1) encompassing the following:

a. ***Direct Use Values***: Tangible products such as timber, fishery resources, honey, and tourism revenues derived from bioshield ecosystems (Mehvar et al. 2018).
b. ***Indirect Use Values***: Ecosystem services such as flood mitigation, storm buffering, sediment retention, and groundwater recharge (Pandeya et al. 2016).

Table 9.1 Conceptual framework for economic evaluation of coastal bioshields

Framework/ value category	Definition	Key components/ indicators	Typical valuation methods	Illustrative examples/ evidence
Direct use values	Values derived from the direct extraction or consumption of bioshield resources	Timber, fuelwood, fisheries, honey, and eco-tourism revenues	Market price method, production function approach	Fishery and timber products from mangrove bioshields (Mehvar et al. 2018)
Indirect use values	Benefits obtained from ecosystem functions that regulate environmental processes	Storm buffering, flood mitigation, sediment retention, groundwater recharge	Avoided damage cost, replacement cost, benefit transfer	Regulating services often exceed direct outputs in Asian bioshields (Brander et al. 2012; Getzner and Islam 2020)
Non-use values	Values independent of current or future direct use of bioshields	Existence, bequest, and option values; cultural and spiritual significance	Contingent valuation, choice experiments	Cultural and intrinsic values of coastal forests (Mehvar et al. 2018)
Total Economic Value (TEV)	Aggregate monetary value of all use and non-use benefits provided by bioshields	Provisioning, regulating, cultural, and supporting services	Integrated TEV accounting, ecosystem service valuation	Mangrove TEV in southern Vietnam ~US$3,000/ha/yr (Quoc Vo et al. 2015)
Cost–benefit analysis	Comparison of discounted benefits and costs of bioshield interventions over time	Planting & maintenance costs, opportunity costs, avoided damages, ecosystem service benefits	Net Present Value (NPV), Benefit–Cost Ratio (BCR)	BCRs of 3.2–7.5 for bioshields in Vietnam, India, and Bangladesh (ADB 2021a)

c. ***Non-Use Values***: Intrinsic ecological, cultural, and spiritual values of bioshields, often captured through contingent valuation or choice modeling (Mehvar et al. 2018).

9.2.1 Total Economic Value (TEV) Framework

The Total Economic Value (TEV) framework provides the most widely accepted approach for evaluating bioshields (Milon and Alvarez 2019). It aggregates all categories of value to represent the overall contribution of an ecosystem to human welfare. In the case of mangroves and coastal forests, TEV captures provisioning services (e.g., wood, food), regulating services (e.g., wave attenuation, carbon sequestration), cultural services (e.g., recreation, heritage), and supporting services (e.g., nutrient cycling) (BOBLME 2014; Wan Ab Rashid et al. 2024).

In Asia, empirical applications of the TEV framework have demonstrated that the indirect regulating services of bioshields often exceed their direct economic outputs (Brander et al. 2012; Getzner and Islam 2020; Gargaran et al. 2024). For instance, the ecosystem value of mangroves in southern Vietnam is estimated at ~US $3,000/ha/yr, which is considerable when compared to the value of timber and fish products (Quoc Vo et al. 2015).

9.2.2 Cost–Benefit Analysis (CBA) Framework

Cost–Benefit Analysis provides a structured method to compare the monetary benefits of coastal bioshields with their establishment, maintenance, and opportunity costs (Chelli et al. 2025). The benefit–cost ratio (BCR) and net present value (NPV) are commonly used indicators (OECD 2018; van Zanten et al. 2023). The general formula for bioshield CBA is:

$$NPV = \sum_{t=0}^{n} \frac{B_t - C_t}{(1+r)^t}$$

where B_t represents annual benefits, C_t represents annual costs, r is the discount rate, and n is the analysis period.

CBAs in Asia typically consider:

a. Initial planting and maintenance costs (labor, seedlings, fencing) (Aerts 2018)
b. Opportunity costs (forgone aquaculture or land use) (Van Zanten et al. 2025)
c. Avoided damage costs (reduction in flood and storm losses) (Menéndez et al. 2020; Van Zanten et al. 2023)
d. Ecosystem service benefits (carbon credits, fisheries productivity, etc.) (Van Zanten et al. 2023; Zhang et al. 2025)

A multi-country analysis by the Asian Development Bank (ADB 2021a) found that bioshields in coastal protection projects in Vietnam, India, and Bangladesh yielded benefit–cost ratios ranging from 3.2 to 7.5, depending on local conditions and discount rates. These values are substantially higher than most hard-engineered alternatives over a 20-year lifespan (Huynh et al. 2024).

9.3 Economic Valuation Methods Applied to Coastal Bioshields

Economic valuation of coastal bioshields employs a range of methods (Table 9.2) tailored to the type of benefits being assessed (Freeman et al. 2014; Milon and Alvarez 2019). These can be broadly grouped into market-based valuation, replacement cost method, and avoided damage cost approaches (Mehvar et al. 2018; Pisani et al. 2020).

9.3.1 Market-Based Valuation

Market-based valuation quantifies services that have direct market prices. Examples include:

a. Timber and non-timber products: In mangrove bioshields, fuelwood and honey can be valued using local market rates (World Bank 2022a; Kandulu et al. 2025).
b. Carbon sequestration: Estimated through carbon storage and sequestration rates (tC/ha/year) multiplied by prevailing carbon market prices. For instance, mangroves in the Mekong Delta store an average of 250–300 tC/ha, equating to roughly USD 10,000–12,000/ha under voluntary carbon market rates (Hilmi et al. 2023; Vu et al. 2023).
c. Tourism revenue: Calculated using entry fees or average tourist expenditures associated with ecotourism in bioshield sites (Nobi et al. 2021; Hilmi et al. 2023).

While straightforward, this approach captures only a fraction of bioshields' total economic value, as most ecosystem services are not traded in markets.

9.3.2 Replacement Cost Method

The replacement cost method estimates the value of bioshields by calculating the cost of replacing their protective function with human-made alternatives (e.g., seawalls, dikes, or breakwaters) (Barbier 2015, 2016). Studies in India and the Philippines have shown that the cost of constructing seawalls equivalent to mangrove protection is about 3–6 times compared to mangrove restoration costs (Baig et al. 2016; Narayan et al. 2016; Barbier 2020).

In Vietnam, coastal bioshields in Nam Dinh Province were found to reduce wave heights by 60–70% (Pham et al. 2020), equivalent to a concrete seawall valued at USD 1.2 million/km. Similar studies were conducted in Central Vietnam (Quang Bao 2011) and at the global scale (Van Wesenbeeck et al. 2025). Thus, replacement cost methods make a strong economic argument for bioshields as cost-effective natural defenses (Narayan et al. 2016).

Table 9.2 Economic valuation methods used for coastal bioshields

Valuation method	Type of economic value captured	Key ecosystem service assessed	Typical metrics/units	Illustrative applications and findings	Strengths and limitations
Market-based evaluation	Direct use values with observable prices	Timber and non-timber forest products; carbon sequestration; tourism and recreation	Market prices (USD/unit); tC/ha/year × carbon price; tourist expenditure (USD/year)	Mangrove fuelwood and honey valued at local market rates; Mekong Delta mangroves storing 250–300 tC/ha valued at USD 10,000–12,000/ha; ecotourism revenues from bioshield sites	Simple and transparent; data often readily available; underestimates total value by excluding non-market services
Replacement cost method	Indirect use values (protective function)	Coastal protection; wave attenuation and erosion control	Cost of artificial structures (USD/km or USD/ha)	Seawalls cost 3–6 times more than mangrove restoration in India and the Philippines; mangroves in Nam Dinh (Vietnam) are equivalent to seawalls worth ~USD 1.2 million/km	Intuitive for policymakers; strong infrastructure comparison; assumes artificial structures provide equivalent protection
Avoided damage cost method	Indirect use values (risk reduction)	Flood mitigation; storm surge buffering; protection of lives and assets	Avoided losses (USD/household or USD/event)	Post–Typhoon Haiyan villages with mangroves avoided USD 1,000–3,000 in household damages; lower mortality and infrastructure loss in protected areas	Directly linked to disaster risk reduction; data-intensive and event-specific; results sensitive to hazard magnitude

9.3.3 *Avoided Damage Cost Method*

This approach calculates the economic losses avoided due to bioshield presence, such as reduced flooding, crop damage, and infrastructure loss (Barbier 2016). Following Typhoon Haiyan (2013), a comparative study in the Philippines showed that villages with mangrove belts suffered considerably lower property damage and fewer casualties than unprotected areas (Seriño et al. 2017). The avoided damage was valued at USD 1,000–3,000 per household, depending on location and asset exposure (Losada et al. 2017; Seriño et al. 2017).

9.3.4 *Contingent Valuation and Choice Modelling*

Stated preference methods capture non-market and non-use values, including cultural, aesthetic, and existence values (Do and Frör 2021). Surveys ask respondents about their willingness to pay (WTP) for bioshield restoration or protection (e.g., Nguyen et al. 2022a, b). For instance, in northern Vietnam, households reported an average WTP of USD 5–9 per month to support mangrove restoration programs (Pham et al. 2018). In India's Tamil Nadu coast, contingent valuation revealed that households valued bioshields not only for storm protection but also for fishery enhancement and shoreline aesthetics (Debroy et al. 2014).

9.3.5 *Benefit Transfer*

When local valuation data are unavailable, benefit transfer applies results from other studies in similar ecological and socio-economic contexts (Himes-Cornell et al. 2018). For example, storm protection values for mangroves in Southeast Asia often fall in the range of several thousand USD per hectare per year, and such values from places like the Philippines have been transferred in regional policy assessments for Vietnam and Indonesia instead of site-specific estimates (Malik et al. 2015; Getzner and Islam 2020). While cost-effective, this approach requires careful adjustment for context differences such as income levels, coastal geomorphology, and hazard frequency (Himes-Cornell et al. 2018).

9.4 Tangible and Intangible Economic Benefits of Bioshields

Bioshields deliver a spectrum of benefits (Fig. 9.1) that can be categorized into tangible (market-based) and intangible (non-market) economic values (Barbier et al. 2009; Koetse et al. 2015; Romanazzi et al. 2023; Lopez-Rivas et al. 2024). Several studies consistently show that even conservative valuations of coastal bioshield services yield benefit–cost ratios above 3:1 over a 20–30-year period (Barbier 2012; Su et al. 2021; Van Zanten et al. 2023; Huynh et al. 2024; Zhang et al. 2025).

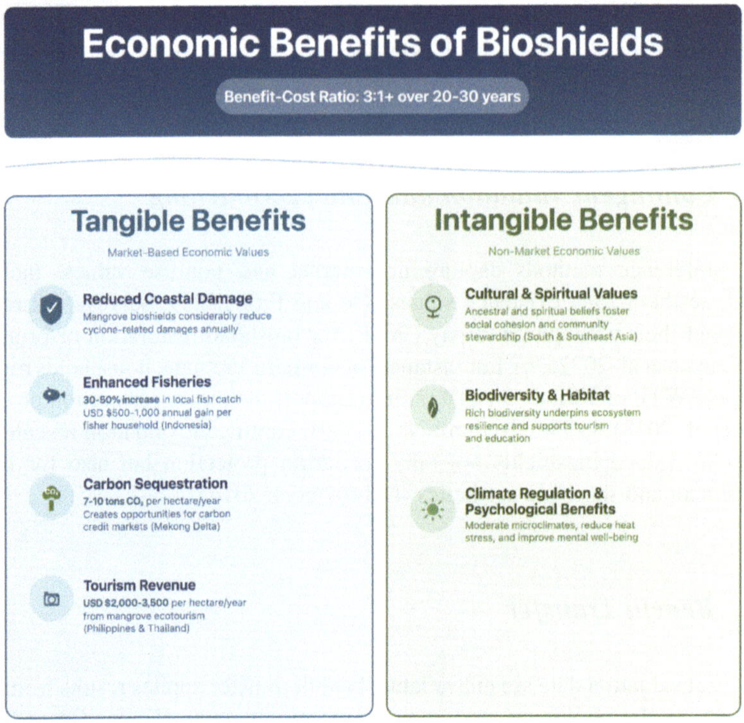

Fig. 9.1 Economic benefits of bioshields

9.4.1 Tangible Benefits

Tangible benefits of coastal bioshields include:

a. ***Reduced Coastal Damage***: Empirical evidence from Bangladesh, India, and Vietnam indicates that mangrove bioshields can considerably reduce cyclone-related damages annually (e.g., Menéndez et al. 2020).
b. ***Enhanced Fisheries Productivity***: Mangrove ecosystems serve as nurseries for shrimp and fish species. In Indonesia, restored mangrove areas increased local fish catch by 30–50%, equivalent to an annual gain of USD 500–1,000 per fisher household (Syawaludin and Kurniawan 2024; Zu Ermgassen et al. 2025).
c. ***Carbon Sequestration***: In the Mekong Delta, bioshields sequester 7–10 tons of CO_2/ha per year, creating opportunities for participation in carbon credit markets (Hilmi et al. 2023; Asilo and Yap-Dejeto 2025).
d. ***Tourism Revenue***: In the Philippines and Thailand, mangrove ecotourism generates up to USD 2,000–3,500/ha annually in visitor spending and entrance fees (Nobi et al. 2021; Malik et al. 2025).

9.4.2 *Intangible Benefits*

Intangible benefits of coastal bioshields include:

a. ***Cultural and Spiritual Values***: Many coastal communities in South and Southeast Asia associate mangroves and coastal forests with ancestral and spiritual beliefs. These values contribute to social cohesion and community stewardship, although they are often excluded from formal valuation (Buenavista and Purnobasuki 2023; Scott et al. 2024).
b. ***Biodiversity and Habitat Support***: Coastal bioshields sustain rich biodiversity, which underpins ecosystem resilience and supports tourism and education (Ruslan et al. 2022; Lovelock et al. 2025).
c. ***Climate Regulation and Psychological Benefits***: Coastal greenbelts moderate microclimates, reduce heat stress, and improve mental well-being among local residents—an increasingly recognized but rarely monetized benefit (Ke et al. 2022; Scott et al. 2024).

9.5 Country Case Studies from Asia

To illustrate the application of valuation methods and the diversity of outcomes, this section reviews key examples from Vietnam, India, Bangladesh, Indonesia, and the Philippines—countries that collectively represent leading experiences with bioshield economics in Asia.

9.5.1 *Vietnam*

Vietnam has implemented numerous bioshield projects along the Mekong and Red River Deltas since the 1990s, primarily through the Vietnam Red Cross Coastal Afforestation Project and MARD's Coastal Protection Forest Program (Hai et al. 2020; Veettil et al. 2021; Pham et al. 2022). Several comprehensive economic analyses have found that mangrove rehabilitation yields high benefit–cost ratios (above 1) and an NPV of USD 1,200/ha per year over 25 years (Hoang et al. 1998; Su et al. 2021). Figure 9.2 shows a few direct benefits from mangrove forests in Vietnam, such as food, timber (non-destructive), and tourism activities.

Key benefits of coastal bioshields in Vietnam included:

a. Avoided damage from typhoons and coastal erosion (~USD 700/ha/year) (NBI 2025)
b. Carbon sequestration (~USD 250/ha/year) (World Bank 2022b; UNDP 2024)
c. Fishery enhancement and local livelihood (a significant portion out of the total ecosystem value ~USD 3000/ha/year) (Vo 2013; Veettil et al. 2021)

Fig. 9.2 Mangrove-related benefits in Vietnam: food (water coconut—nipa palm), tourism activities (clockwise), timber. *Photographs* Bijeesh Kozhikkodan Veettil

Costs were dominated by seedling establishment and maintenance, estimated at USD 350–500/ha (Su et al. 2021; Simpson et al. 2025). When ecosystem co-benefits were included, the economic returns significantly outperformed traditional grey infrastructure investments (Oanh et al. 2020; NBI 2025).

9.5.2 India

Following the 2004 tsunami, India has launched large-scale bioshield programs along its coastlines (Mukherjee et al. 2015). Valuation studies have revealed that mangrove buffers reduced flood damage by up to 80% during subsequent cyclones (Das and Vincent 2009; Menéndez et al. 2020). Economic analysis using replacement cost and avoided damage methods estimated benefit–cost ratios ranging from 4.0 to 6.8, depending on vegetation density and exposure (Su et al. 2021). Additionally, households adjacent to bioshields experienced fewer livelihood disruptions and faster recovery times after storms, indicating broader socio-economic benefits (Ramesh et al. 2022).

9.5.3 Bangladesh

The Bangladesh Coastal Greenbelt Project (initiated in 1995) represents one of Asia's most comprehensive bioshield programs (Dey et al. 2021). Over 200,000 has of mangroves have been planted along the Bay of Bengal (Mahmood et al. 2023). Valuation by Barua (2020) reported total annual benefits of USD 840/ha by all forests in Bangladesh, and this value is high (USD 2176 ha^{-1} yr^{-1}) in Sundarbans mangrove forests, mainly from avoided cyclone damages, fuelwood collection, and carbon sequestration. With establishment costs averaging USD 150–200/ha, the benefit–cost ratio exceeded 3.5, making it a highly efficient adaptation measure (Salem and Mercer 2012; Golub and Golub 2016).

9.5.4 *Indonesia*

Indonesia's mangrove rehabilitation under the National Mangrove Rehabilitation Program (2021–2024) aims to restore 600,000 ha of degraded coastal forests (Jompa and Murdiyarso 2023). Preliminary economic modeling by the Ministry of Environment and Forestry indicates a potential annual benefit of USD 2 billion nationwide, primarily through fisheries enhancement, shoreline stabilization, and carbon markets (Van Zanten et al. 2021, 2025). Local studies in Java and Kalimantan show benefit–cost ratios between 4.5 and 7.2, depending on mangrove density and tourism potential (Van Zanten et al. 2021).

9.5.5 *The Philippines*

The Philippines has pioneered community-based mangrove management since the 1980s (Pulhin et al. 2017; Song et al. 2021). A cost–benefit study following Typhoon Haiyan found that communities with well-managed bioshields suffered less property loss and fewer casualties compared to those without vegetation buffers (Losada et al. 2017; Takagi 2025). Valuations of mangrove restoration in Leyte and Panay Islands showed benefit–cost ratios of 6.0–9.5, with major benefits arising from avoided damage, carbon storage, and eco-tourism revenues (Su et al. 2021).

9.6 Comparative Analysis of Coastal Bioshield Economics Across Asia

The comparative economic evidence from Asia reveals several consistent trends despite differences in geography, socio-economic conditions, and institutional capacity (Menéndez et al. 2020; Su et al. 2021). These patterns highlight both the strengths and the variability of bioshield performance in economic terms (e.g., Weiss 2009; Malik et al. 2015).

9.6.1 Regional Patterns in Cost and Benefit Structures

Across the five focal countries—Vietnam, India, Bangladesh, Indonesia, and the Philippines—coastal bioshields exhibit similar cost structures but different benefit magnitudes due to variations in hazard exposure, ecological productivity, and management practices (Su et al. 2021). Establishment costs typically range from USD 300 to 800/ha, while maintenance costs average USD 50–100/ha annually (Su et al. 2021; Van Zanten et al. 2025).

Benefits, however, are more variable. The avoided damage value tends to be the largest component, accounting for 50–70% of total benefits (Menéndez et al. 2020). In regions with high cyclone frequency, such as coastal Bangladesh or the Bay of Bengal, avoided damage benefits can reach USD 1,000/ha per year (Sarker et al. 2020). In contrast, on more stable coastlines such as southern Vietnam, carbon sequestration and fishery enhancement contribute a larger share of the total economic value (Getzner and Islam 2020; Van Zanten et al. 2025).

9.6.2 Economic Efficiency Relative to Engineered Structures

Comparative cost–benefit analyses show that bioshields often outperform hard-engineering structures over long time horizons (Huynh et al. 2024). For example, a seawall may have a shorter payback period but significantly higher capital and maintenance costs. Bioshields, while slower to mature, offer sustainable, self-repairing systems that generate co-benefits (Narayan et al. 2016; Meraj and Hashimoto 2025). Table 9.3 summarizes benefit–cost ratios of various coastal protection structures, including bioshields, synthesized from available literature.

The above table underscores the economic efficiency of bioshields. Even when discounting co-benefits, bioshields maintain higher BCRs than grey infrastructure, especially when evaluated over 25–40 years. When ecosystem service co-benefits, such as carbon credits, fisheries income, and tourism, are included, the total BCR often exceeds 7:1, a strong justification for investment.

Table 9.3 Summary of the benefit–cost ratio of various coastline protection structures

Protection type	Capital cost (USD/km)	Maintenance (USD/km)	Average lifespan (years)	Benefit–cost ratio (BCR)
Seawalls	1.5–3 million	50,000–100,00	20–25	0.8–1.5
Breakwaters	2–4 million	60,000–120,000	20–25	1–2
Bioshield (mangroves/ coastal forests)	0.1–0.3 million	5,000–10,000	25–40	3–7.5

9.6.3 *Role of Hazard Context and Ecosystem Type*

Bioshield effectiveness and value vary by hazard type and ecosystem. Mangrove bioshields perform best against storm surges and wave attenuation (Spalding et al. 2014). *Casuarina* and *Pandanus* belts (Kayum et al. 2022; Lin et al. 2023) are effective against wind and sand drift, whereas dune vegetation systems stabilize beaches (Johnston et al. 2023). The economic return on investment correlates with the hazard profile:

a. In cyclone-prone coasts (e.g., Bay of Bengal), avoided damage benefits dominate (e.g., Sarker et al. 2020).
b. In erosion-prone deltas (e.g., Mekong), shoreline stabilization and carbon value prevail (e.g., Vo 2013).
c. In small island settings (e.g., the Philippines, Indonesia), tourism and habitat services are key (e.g., Malik et al. 2025).

The adaptive multifunctionality of bioshields ensures that they remain economically viable even as hazard conditions evolve due to climate change (Hülsen et al. 2025).

9.7 Methodological and Practical Challenges

Despite substantial progress in quantifying the economic value of bioshields, methodological and practical challenges remain pervasive in the Asian context. These challenges stem from data limitations and geographic gaps (Gargaran et al. 2024), ecosystem complexity and valuation methodology challenges (Himes-Cornell et al. 2018; De La Cruz 2021), and governance constraints and framework limitations (Thompson et al. 2017; De Valck et al. 2023).

9.7.1 *Valuing Non-market and Co-benefits*

While avoided damage and carbon sequestration can be quantified with reasonable precision, other benefits, such as biodiversity support, cultural heritage, or mental health, are rarely included (Himes-Cornell et al. 2018). The absence of robust valuation methods for these non-market benefits leads to the underestimation of bioshields' true economic potential (Van Schoubroeck et al. 2024). Moreover, economic models often fail to account for the synergistic benefits among services, such as how enhanced fishery productivity reinforces food security and livelihood resilience (Milon and Alvarez 2019).

9.7.2 Temporal and Spatial Variability

The performance of bioshields changes over time as vegetation matures, ecosystems stabilize, and climate patterns shift (Su et al. 2021). Most economic assessments use static models, which may not capture dynamic ecological processes (Barbier 2016). Similarly, spatial heterogeneity, such as differences in mangrove species composition, tidal range, and geomorphology, makes benefit transfer difficult without proper local calibration (Getzner and Islam 2020; Worthington et al. 2020; Korb et al. 2025).

9.7.3 Opportunity Costs and Land-Use Conflicts

Bioshields often compete with aquaculture, agriculture, and infrastructure for limited coastal land (Veettil et al. 2021; Athikalam and Vaideeswaran 2022). Opportunity costs can be high, particularly in densely populated deltas like the Mekong (Primavera 2005). Accurately quantifying these trade-offs is critical for realistic cost–benefit assessments (Berchoux et al. 2023). Additionally, informal or unregulated land tenure systems complicate the allocation of economic benefits among stakeholders (Edwards and Anderson 1984; Berchoux et al. 2023).

9.7.4 Data Limitations and Uncertainty

Reliable data on storm damage, fishery productivity, and ecosystem carbon fluxes are scarce in many Asian developing countries. Consequently, analysts rely on proxy data or international literature, which introduces uncertainty (Drupp et al. 2025). Furthermore, the choice of discount rates significantly influences economic outcomes; higher rates (e.g., 10%) can undervalue long-term ecosystem benefits, while lower rates (e.g., 3%) make bioshields appear more economically favorable (Farber and Hemmersbaugh 1993).

9.7.5 Integration with Disaster Risk Reduction (DRR) and Climate Adaptation Frameworks

Another key challenge is aligning coastal bioshield valuation with existing DRR and climate adaptation policy frameworks (Favero and Hinkel 2024). Many governments still prioritize short-term engineering solutions, and the economic case for coastal bioshields is often excluded from national investment plans due to a lack of standardized valuation tools or institutional familiarity with ecosystem accounting methods

(Dang et al. 2021; De Valck et al. 2023; Favero and Hinkel 2024; Vollmer et al. 2024).

9.8 Policy and Investment Implications

Translating valuation results into actionable policy is essential for scaling up bioshield adoption in many developing countries (Pascual et al. 2023). The following implications emerge from empirical and comparative evidence.

9.8.1 *Mainstreaming Bioshield Valuation into Policy Frameworks*

Governments should incorporate bioshield valuation into national cost–benefit guidelines and environmental impact assessment (EIA) processes (Fairbrass et al. 2020; Walz et al. 2021). For example, Vietnam's 2020 Law on Environmental Protection (Law No. 72/2020/QH14, effective from January 2022) now recognizes ecosystem-based adaptation as an eligible measure for state support, allowing bioshield investments to be justified using economic valuation data. Similarly, India's National Disaster Management Authority (NDMA) includes ecosystem services in resilience planning metrics (Paudel et al. 2024). Institutionalizing valuation standards can ensure coastal bioshields are considered alongside grey infrastructure in public investment decisions (Fairbrass et al. 2020; Walz et al. 2021).

9.8.2 *Financing Mechanisms and Incentives*

Economic valuation supports innovative financing for bioshields, including:

a. ***Payment for Ecosystem Services (PES)***: Vietnam's pilot PES schemes (e.g., Ca Mau province) compensate communities for maintaining coastal protection forests (e.g., Nguyen et al. 2022a, b), even though the institutional priorities and local values often cause contradictions and compromises in PES policy (McElwee et al. 2019).
b. ***Carbon Credit Markets***: With proper monitoring and verification, mangrove bioshields can generate blue carbon credits, attracting international finance (Li and Liu 2025; Perera et al. 2025; Swails et al. 2025).
c. ***Public–Private Partnerships (PPPs)***: Insurance and tourism sectors can co-finance bioshields to reduce risk exposure and enhance local attractiveness (Marchal et al. 2019; Biasin et al. 2024).

d. ***Green and Blue Bonds***: Regional development banks, such as ADB, are exploring green and blue bonds to fund large-scale coastal ecosystem restoration (ADB 2021b; Kılıç 2024; Moses and Tafadzwa 2025).

Such financial instruments hinge on credible valuation data to demonstrate return on investment and risk mitigation potential.

9.8.3 Integrating Bioshields into Coastal Zone Planning

Economic valuation evidence can inform Integrated Coastal Zone Management (ICZM) and spatial planning frameworks (Burke et al. 2020; van Zanten et al. 2025). In Indonesia (Mustofa et al. 2025; van Zanten et al. 2025) and the Philippines (Menéndez et al. 2018), bioshields are now included in coastal land-use plans, supported by cost–benefit analyses showing long-term savings compared to seawall maintenance (Van Zels et al. 2021; Huynh et al. 2024). Incorporating valuation results into GIS-based planning tools enables policymakers to identify priority restoration areas that balance both ecological suitability and economic efficiency (van Zanten et al. 2025).

9.8.4 Capacity Building and Knowledge Sharing

To institutionalize bioshield valuation, capacity development is required at multiple levels:

a. Technical capacity for economic modeling and ecosystem accounting within government agencies and universities.
b. Community-level training in data collection, participatory monitoring, and local valuation initiatives (Pagdee and Kawasaki 2021).
c. Regional cooperation, such as through ASEAN, SAARC, and ADB platforms, to exchange case studies and harmonize valuation methodologies.
d. Strengthening human and institutional capacity ensures valuation becomes an integral part of climate adaptation investment planning.

9.8.5 Synergies with the Blue Economy and SDGs

The economic rationale for bioshields aligns with the broader Blue Economy and Sustainable Development Goals (SDGs), particularly SDG 13 (Climate Action), SDG 14 (Life Below Water), and SDG 15 (Life on Land) (Pakseresht et al. 2025). By providing quantifiable economic evidence, bioshield valuation strengthens the case for integrated coastal ecosystem management as both an environmental and development strategy (Lopez-Rivas and Cardenas 2024).

9.9 Conclusions

The economic valuation and cost–benefit analysis of bioshields reveal a compelling case for their inclusion in national and regional coastal protection strategies across Asia. Empirical studies consistently demonstrate that bioshields not only deliver substantial avoided damage and livelihood benefits but also outperform traditional infrastructure in long-term economic efficiency.

Key insights include:

a. Confirming that bioshields are cost-effective investments.
b. Multi-Functional Co-Benefits: Beyond protection, bioshields enhance fisheries, store carbon, support biodiversity, and generate tourism income.
c. Policy Relevance: Quantitative valuation enables governments to integrate bioshields into fiscal planning, insurance mechanisms, and climate adaptation programs.
d. Methodological Gaps: Improved data collection, dynamic modeling, and valuation of non-market benefits are essential to refine estimates.
e. Equity and Participation: Ensuring that local communities equitably share economic benefits, which enhances project sustainability and long-term stewardship.

As climate risks intensify, the economic case for bioshields will strengthen further. Integrating robust valuation into national decision-making not only elevates the visibility of nature-based solutions but also ensures the long-term security of coastal livelihoods and ecosystems. The future of coastal protection in Asia will depend increasingly on how effectively nations can translate these economic insights into large-scale, inclusive, and adaptive bioshield programs.

References

Aerts JC (2018) A review of cost estimates for flood adaptation. Water 10(11):1646. https://doi.org/10.3390/w10111646

Asian Development Bank (ADB) (2021a) Asian development outlook 2021—financing green and inclusive recovery. Asian Development Bank, 414pp. https://doi.org/10.22617/FLS210163-3

Asian Development Bank (ADB) (2021b) Green and blue bond framework. Asian Development Bank Treasury Department Funding Division, Manila, Philippines

Asilo HAM, Yap-Dejeto LG (2025) Understanding blue carbon management strategies: quantitative assessment of blue carbon stock among three mangrove stands in Eastern Visayas, Philippines, W. Pacific Border. Carbon Res 4:63 (2025). https://doi.org/10.1007/s44246-025-00216-6

Athikalam PT, Vaideeswaran AK (2022) Vegetation bioshield for coastal protection in South Asia: status and way forward. J Coast Conserv 26(1):3. https://doi.org/10.1007/s11852-022-00850-x

Baig SP, Rizvi A, Josella M, Palanca-Tan R (2016) Cost and benefits of ecosystem based adaptation: the case of the Philippines. IUCN, Gland, Switzerland, viii, 32

Barbier EB (2012) Progress and challenges in valuing coastal and marine ecosystem services. Rev Environ Econ Policy 6:1–19. https://doi.org/10.1093/reep/rer017

Barbier EB (2015) Valuing the storm protection service of estuarine and coastal ecosystems. Ecosyst Serv 11:32–38. https://doi.org/10.1016/j.ecoser.2014.06.010
Barbier EB (2016) The protective service of mangrove ecosystems: a review of valuation methods. Mar Pollut Bull 109(2):676–681. https://doi.org/10.1016/j.marpolbul.2016.01.033
Barbier EB (2020) Estuarine and coastal ecosystems as defense against flood damages: an economic perspective. Front Clim 2:594254. https://doi.org/10.3389/fclim.2020.594254
Barbier EB, Baumgärtner S, Chopra K, Costello C, Duraiappah A, Hassan R, Perrings C (2009) The valuation of ecosystem services. In: Biodiversity, ecosystem functioning, and human wellbeing: an ecological and economic perspective, pp 248–262
Barua SK, Boscolo M, Animon I (2020) Valuing forest-based ecosystem services in Bangladesh: implications for research and policies. Ecosyst Serv 42:101069. https://doi.org/10.1016/j.ecoser.2020.101069
Berchoux T, Hutton CW, Hensengerth O, Voepel HE, Tri VP, Vu PT, Hung NN, Parsons D, Darby SE (2023) Effect of planning policies on land use dynamics and livelihood opportunities under global environmental change: evidence from the Mekong Delta. Land Use Policy 131:106752. https://doi.org/10.1016/j.landusepol.2023.106752
Biasin A, Toxopeus H, Pettenella D, Polzin F, Masiero M (2024) Financing urban Nature-based Solutions (NBS): a literature review from the perspective of funders. Nature-Based Solutions 6:100195. https://doi.org/10.1016/j.nbsj.2024.100195
BOBLME (2014) Assessing, demonstrating and capturing the economic value of marine & coastal ecosystem services in the Bay of Bengal Large Marine Ecosystem BOBLME-2014-Socioec-02
Brander ML, Wagtendonk JA, Hussain SS, McVittie A, Verburg PH, De Groot RS, Van der Ploeg S (2012) Ecosystem service values for mangroves in Southeast Asia: a meta-analysis and value transfer application. Ecosyst Serv 1(1):62–69. https://doi.org/10.1016/j.ecoser.2012.06.003
Brander L, De Groot R, Schägner J, Guisado-Goñi V, Van 't Hoff V, Solomonides S, McVittie A, Eppink F, Sposato M, Do L, Ghermandi A, Sinclair M, Thomas R (2024) Economic values for ecosystem services: a global synthesis and way forward. Ecosyst Serv 66:101606. https://doi.org/10.1016/j.ecoser.2024.101606
Brecht H, Dasgupta S, Laplante B, Murray S, Wheeler D (2012) Sea-level rise and storm surges. J Environ Develop 21(1):120–138. https://doi.org/10.1177/1070496511433601
Buenavista D, Purnobasuki H (2023) People and mangroves: biocultural utilization of mangrove forest ecosystem in Southeast Asia. J Mar Island Cult 12(2):95–115. https://doi.org/10.21463/jmic10.21463/jmic.2023.12.210.21463/jmic.2023.12.2.07
Burke L, Larsen G, Lau W, Kushner B, Hori T (2020) Climate-resilient integrated coastal zone management performance indicators. Inter-American Development Bank, 48pp. https://doi.org/10.18235/0002164
Chelli A, Brander L, Geneletti D (2025) Cost-benefit analysis of urban nature-based solutions: a systematic review of approaches and scales with a focus on benefit valuation. Ecosyst Serv 71:101684. https://doi.org/10.1016/j.ecoser.2024.101684
Dang AN, Jackson BM, Benavidez R, Tomscha SA (2021) Review of ecosystem service assessments: pathways for policy integration in Southeast Asia. Ecosyst Serv 49:101266. https://doi.org/10.1016/j.ecoser.2021.101266
Das S, Vincent JR (2009) Mangroves protected villages and reduced death toll during Indian super cyclone. Proc Natl Acad Sci 106(18):7357–7360. https://doi.org/10.1073/pnas.0810440106
De La Cruz PMC (2021) The knowledge status of coastal and marine ecosystem services—challenges, limitations and lessons learned from the application of the ecosystem services approach in management. Front Mar Sci 8:684770. https://doi.org/10.3389/fmars.2021.684770
De Valck J, Jarvis D, Coggan A, Schirru E, Pert P, Graham V, Newlands M (2023) Valuing ecosystem services in complex coastal settings: an extended ecosystem accounting framework for improved decision-making. Mar Policy 155:105761. https://doi.org/10.1016/j.marpol.2023.105761
Debroy P, Jayaraman R, Krishnan M, Criddle KR (2014) Importance of mangrove conservation and valuation to women and men-a case study of Pichavaram mangroves in India. Asian Fish Sci 27S:79–90

Dey T, Rajib MGM, Islam MA (2021) People's perceptions about the socio-economic and environmental impact of Coastal Green Belt in Bangladesh. Indonesian J Soc Environ Issues (IJSEI) 2(2):155–162
Do HH, Frör O (2021) River ecosystem resilience: applying the contingent valuation method in Vietnam. Sustainability 14(19):12029. https://doi.org/10.3390/su141912029
Drupp MA, Turk ZM, Groom B, Heckenhahn J (2025) Global evidence on the income elasticity of willingness to pay, relative price changes and public natural capital values. Environ Resource Econ. https://doi.org/10.1007/s10640-025-01042-5
Edwards SF, Anderson GD (1984) Land use conflicts in the coastal zone: an approach for the analysis of the opportunity costs of protecting coastal resources. J Northeastern Agricult Econ Council 13(1):73–81. https://doi.org/10.1017/S0163548400004118
Fairbrass A, Mace G, Ekins P, Milligan B (2020) The natural capital indicator framework (NCIF) for improved national natural capital reporting. Ecosyst Serv 46:101198. https://doi.org/10.1016/j.ecoser.2020.101198
Farber DA, Hemmersbaugh PA (1993) The shadow of the future: discount rates, later generations, and the environment. Vand L Rev 46:267
Favero F, Hinkel J (2024) Key innovations in financing nature-based solutions for coastal adaptation. Climate 12(4):53. https://doi.org/10.3390/cli12040053
Freeman III AM, Herriges JA, Kling CL (2014) The measurement of environmental and resource values: theory and methods. Routledge, 478pp
Gargaran JPS, Capuno RDY, Novicio VFP, Petiluna CC, Catalba CET, Paras DAP, Garces JJC (2024) Mangrove ecosystem in Asia: review and synthesis of ecosystem services and economic valuation methods. Environ Exp Biol 22(2):59–70. https://doi.org/10.22364/eeb.22.06
Getzner M, Islam MS (2020) Ecosystem services of mangrove forests: results of a meta-analysis of economic values. Int J Environ Res Public Health 17(16):5830. https://doi.org/10.3390/ijerph17165830
Golub A, Golub ES (2016) Cost-benefit analysis of adaptation strategy in Bangladesh-Bangladesh Priorities, 74pp. Creative Commons Attribution CC BY 4.0
Hai NT, Dell B, Phuong VT, Harper RJ (2020) Towards a more robust approach for the restoration of mangroves in Vietnam. Ann for Sci 77(1):18. https://doi.org/10.1007/s13595-020-0921-0
Hanley N, Hynes S, Patterson D, Jobstvogt N (2015) Economic valuation of marine and coastal ecosystems: is it currently fit for purpose? J Ocean Coast Econ 2:1. https://doi.org/10.15351/2373-8456.1014
Hernández-Blanco M, Costanza R, Cifuentes-Jara M (2021) Economic valuation of the ecosystem services provided by the mangroves of the Gulf of Nicoya using a hybrid methodology. Ecosyst Serv 49:101258. https://doi.org/10.1016/j.ecoser.2021.101258
Hilmi N, Benitez Carranco MB, Broussard D, Mathew M, Djoundourian S, Cassotta S, Ferrier-Pagès C (2023) Tropical blue carbon: solutions and perspectives for valuations of carbon sequestration. Front Clim 5:1169663. https://doi.org/10.3389/fclim.2023.1169663
Himes-Cornell A, Grose SO, Pendleton L (2018) Mangrove ecosystem service values and methodological approaches to valuation: where do we stand? Front Mar Sci 5:409061. https://doi.org/10.3389/fmars.2018.00376
Hoang TN, Adger W, Kelly P (1998) Natural resource management in mitigating climate impacts: the example of mangrove restoration in Vietnam. Glob Environ Chang 8(1):49–61. https://doi.org/10.1016/S0959-3780(97)00023-X
Hülsen S, Dee LE, Kropf CM, Meiler S, Bresch DN (2025) Mangroves and their services are at risk from tropical cyclones and sea level rise under climate change. Commun Earth Environ 6(1):262. https://doi.org/10.1038/s43247-025-02242-z
Huynh LT, Su J, Wang Q, Stringer LC, Switzer AD, Gasparatos A (2024) Meta-analysis indicates better climate adaptation and mitigation performance of hybrid engineering-natural coastal defence measures. Nat Commun 15(1):2870. https://doi.org/10.1038/s41467-024-46970-w

Johnston KK, Dugan JE, Hubbard DM, Emery KA, Grubbs MW (2023) Using dune restoration on an urban beach as a coastal resilience approach. Front Mar Sci 10:1187488. https://doi.org/10.3389/fmars.2023.1187488

Jompa J, Murdiyarso D (2023) Coastal area rehabilitation for climate change adaptation: the key role of mangroves in Nationally Determined Contributions. CIFOR. https://www.cifor-icraf.org/knowledge/publication/8956

Kandulu J, Wheeler S, Pham TT (2025) Carbon pricing: Approaches, lessons and recommendations for blue carbon in Vietnam. Working Paper 48. CIFOR, Bogor, Indonesia; ICRAF, Nairobi, Kenya. https://doi.org/10.17528/cifor-icraf/009371

Kayum S, Shimatani Y, Minagawa T (2022) Evaluation of Pandanus trees as a means of eco-DRR against storm surge wave on Saint Martin's Island, Bangladesh. Water 14(11):1781. https://doi.org/10.3390/w14111781

Ke G, Utama IK, Wagner T, Sweetman AK, Arshad A, Nath TK, Neoh JY, Muchamad LS, Suroso DS (2022) Influence of mangrove forests on subjective and psychological wellbeing of coastal communities: case studies in Malaysia and Indonesia. Front Public Health 10:898276. https://doi.org/10.3389/fpubh.2022.898276

Kılıç AO (2024) Seychelles blue bond: indebting ecological restructuring of fisheries. Mar Policy 163:106144. https://doi.org/10.1016/j.marpol.2024.106144

Koetse MJ, Brouwer R, Van Beukering PJ (2015) Economic valuation methods for ecosystem services. In: Ecosystem services: from concept to practice, pp 108–131

Korb CC, Guasselli LA, Hasenack H, Belloli TF, Cunha CS (2025) Temporal and spatial variability of hydrogeomorphological attributes in coastal wetlands—Lagoa do Peixe national park, Brazil. Coasts 5(3):23. https://doi.org/10.3390/coasts5030023

Li P, Liu D (2025) How blue carbon financing can sustain blue carbon ecosystems protection and restoration: a proposed conceptual framework for the blue carbon financing mechanism. Ocean Coast Manag 265:107644. https://doi.org/10.1016/j.ocecoaman.2025.107644

Lin L, Hao Z, Post CJ, Mikhailova EA (2023) Protection of coastal shelter forests using UAVs: Individual tree and tree-height detection in *Casuarina equisetifolia* L. Forests. Forests 14(2):233. https://doi.org/10.3390/f14020233

Lopez-Rivas JD, Cardenas J (2024) What is the economic value of coastal and marine ecosystem services? A systematic literature review. Mar Policy 161:106033. https://doi.org/10.1016/j.marpol.2024.106033

Losada IJ, Beck M, Menéndez P, Espejo A, Torres S, Díaz-Simal P, Fernández F, Abad S, Ripoll N, García J, Narayan S, Trespalacios D (2017) Valuation of the coastal protection services of mangroves in the Philippines. World Bank, Washington, DC, USA. Technical report. http://hdl.handle.net/10986/27666 License: CC BY 3.0 IGO

Lovelock CE, Hagger V, Feller IC, Amir AA, Machava António V, Owuor MA, Friess DA (2025) Mangrove biodiversity and ecosystem services. Nat Rev Biodivers 1(12):772–787. https://doi.org/10.1038/s44358-025-00103-3

Mahmood R, Zhang L, Li G (2023) Assessing effectiveness of nature-based solution with big earth data: 60 years mangrove plantation program in Bangladesh coast. Ecol Process 12:11. https://doi.org/10.1186/s13717-023-00419-y

Maldonado JH, Sánchez RDPM, Morales MEV, Henao JPH, Tarazona YG, Compeán RG, Schling M (2020) Innovation in economic analysis and evaluation approaches for coastal protection and restoration investments in the Caribbean. Inter-American development bank environment, rural development and risk management division. Technical Note No. IDB-TN-01861

Malik A, Fensholt R, Mertz O (2015) Economic valuation of mangroves for comparison with commercial aquaculture in South Sulawesi, Indonesia. Forests 6(9):3028–3044. https://doi.org/10.3390/f6093028

Malik A, Ichsan Ali M, Rasyid Jalil A, Mannan A, Musyawarah R (2025) Promoting sustainable mangrove tourism through payments for ecosystem services: insights from Tongke-Tongke Village, South Sulawesi, Indonesia. Reg Sustain 6(2):100213. https://doi.org/10.1016/j.regsus.2025.100213

Marchal R, Piton G, Dartée KW, Pengal P, Matthews JH, Tacnet J, Graveline N, Altamirano MA, Joyce J, Nanu F, Groza I, Peña K, Cokan B, Burke S, Moncoulon D (2019) The (Re)Insurance industry's roles in the integration of nature-based solutions for prevention in disaster risk reduction—insights from a European survey. Sustainability 11(22):6212. https://doi.org/10.3390/su11226212

McElwee P, Huber B, Vân Nguyễn TH (2019) Hybrid outcomes of payments for ecosystem services policies in Vietnam: between theory and practice. Dev Chang 51(1):253–280. https://doi.org/10.1111/dech.12548

Mehvar S, Filatova T, Dastgheib A, Ranasinghe R (2018) Quantifying economic value of coastal ecosystem services: a review. J Mar Sci Eng 6(1):5. https://doi.org/10.3390/jmse6010005

Menéndez P, Losada IJ, Beck MW, Torres-Ortega S, Espejo A, Narayan S, Díaz-Simal P, Lange G (2018) Valuing the protection services of mangroves at national scale: the Philippines. Ecosyst Serv 34:24–36. https://doi.org/10.1016/j.ecoser.2018.09.005

Menéndez P, Losada IJ, Narayan S, Beck MW (2020) The global flood protection benefits of mangroves. Sci Rep 10(1):4404. https://doi.org/10.1038/s41598-020-61136-6

Meraj G, Hashimoto S (2025) Bridging the adaptation finance gap: the role of nature-based solutions for climate resilience. Sustain Sci 20:1093–1107. https://doi.org/10.1007/s11625-025-01655-1

Milon JW, Alvarez S (2019) The elusive quest for valuation of coastal and marine ecosystem services. Water 11(7):1518. https://doi.org/10.3390/w11071518

Monioudi IN, Vousdoukas MI, Giardino A, Stocchino A, Mentaschi L, Feyen L (2025) Impacts of sea level rise and adaptation across Asia and the Pacific. Sci Rep 15(1):35742. https://doi.org/10.1038/s41598-025-11517-6

Moses M, Tafadzwa J (2025) Innovative blue finance investment model for sustainable ocean economies for resilient coastal development. Indiana J Econ Bus Manage 5(2):74–85

Mukherjee N, Dahdouh-Guebas F, Koedam N, Shanker K (2015) An interdisciplinary framework to evaluate bioshield plantations: insights from peninsular India. Acta Oecologica 63:91–100. https://doi.org/10.1016/j.actao.2014.01.005

Mustofa I, Mistoro NH, Suharyanto HHR, Hasanah AU, Prawitasari DA (2025) Nature-based solutions for climate-resilient urban landscapes: implementation in the new capital city of Indonesia. IOP conference series: earth and environmental science, vol 1447, p 012019. https://doi.org/10.1088/1755-1315/1447/1/012019

Narayan S, Beck MW, Reguero BG, Losada IJ, van Wesenbeeck B, Pontee N et al (2016) The effectiveness, costs and coastal protection benefits of natural and nature-based defences. PLoS ONE 11(5):e0154735. https://doi.org/10.1371/journal.pone.0154735

Nature-Based Infrastructure NBI (2025) Sustainable Asset Valuation (SAVi) of Mangroves in Vietnam: an economic valuation of mangroves for flood protection and erosion control in the DEEP C Industrial Zone. International Institute for Sustainable Development, 48pp

Nguyen TH, Dell B, Harper RJ (2022a) Assessment of the feasibility of applying payment for forest ecosystem services in Vietnamese mangrove forests. APN Sci Bull 12(1):184–189. https://doi.org/10.30852/sb.2022.2016

Nguyen TV, Simioni M, Vo HT (2022b) Valuing mangrove conservation attributes in Red River Delta, Vietnam: a choice experiment approach. Mar Resour Econ 37(3):349–368

Nobi MN, Sarker AHM, Nath B, Røskaft E, Suza M, Kvinta P (2021) Economic valuation of tourism of the Sundarban Mangroves, Bangladesh. J Ecol Nat Environ 13(4):100–109. https://doi.org/10.5897/JENE2021.0910

OECD (2018) Cost-benefit analysis and the environment: further developments and policy use. OECD Publishing, Paris. https://doi.org/10.1787/9789264085169-en

Oanh PT, Tamura M, Kumano N, Nguyen QV (2020) Cost-benefit analysis of mixing gray and green infrastructures to adapt to sea level rise in the vietnamese mekong river delta. Sustainability 12(24):10356. https://doi.org/10.3390/su122410356

Pagdee A, Kawasaki J (2021) The importance of community perceptions and capacity building in payment for ecosystems services: a case study at Phu Kao, Thailand. Ecosyst Serv 47:101224. https://doi.org/10.1016/j.ecoser.2020.101224

Pakseresht A, Kermani A, Decker-Lange C (2025) Towards a sustainable and circular blue bioeconomy: a scoping review. Technol Forecast Soc Chang 216:124157. https://doi.org/10.1016/j.techfore.2025.124157

Pandeya B, Buytaert W, Zulkafli Z, Karpouzoglou T, Mao F, Hannah D (2016) A comparative analysis of ecosystem services valuation approaches for application at the local scale and in data scarce regions. Ecosyst Serv 22:250–259. https://doi.org/10.1016/j.ecoser.2016.10.015

Pascual U, Balvanera P, Anderson CB, Christie M, Martin A, Raymond CM, Termansen M, Vatn A, Athayde S, Baptiste B, Barton DN, Jacobs S, Kelemen E, Kumar R, Lazos E, Mwampamba TH, Nakangu B, Subramanian SM, Van Noordwijk M, Zent E (2023) Diverse values of nature for sustainability. Nature 620(7975):813–823. https://doi.org/10.1038/s41586-023-06406-9

Paudel PK, Parajuli S, Bohara M, Kibria MG, Abedin MA, Sinha R (2024) Mainstreaming ecosystem-based approaches into disaster risk reduction policies: a comparative study of Nepal, India, and Bangladesh. Policy Des Pract 7(3):324–342. https://doi.org/10.1080/25741292.2024.2368919

Paxton AB, Riley TN, Steenrod CL, Puckett BJ, I BA, Paliotti ST, Adler AM, Exar L, T McLean JE, Kelley J, Zhang YS, Smith CS, Gittman RK, Silliman BR (2024) Evidence on the performance of nature-based solutions interventions for coastal protection in biogenic, shallow ecosystems: a systematic map. Environ Evidence 13:28. https://doi.org/10.1186/s13750-024-00350-5

Perera NS, P Costa MD, Macreadie PI, Wartman M (2025) Trends in market-based blue carbon projects. Sustain Develop 33(3):3377–3402. https://doi.org/10.1002/sd.3293

Pham TD, Kaida N, Yoshino K, Nguyen XH, Nguyen HT, Bui DT (2018) Willingness to pay for mangrove restoration in the context of climate change in the Cat Ba biosphere reserve, Vietnam. Ocean Coast Manag 163:269–277. https://doi.org/10.1016/j.ocecoaman.2018.07.005

Pham HT, Nguyen QC, Tran TT (2020) The role of mangrove forest in coastal wave attenuation: case study of Hau Loc, Vietnam. Tạp Chí Khoa Học Kỹ Thuật Thủy Lợi Và Môi Trường 72:67–74

Pham TT, Vu TP, Hoang TL, Dao TL, Nguyen DT, Pham DC, Dao LH, Nguyen VT, Hoang NV (2022) The effectiveness of financial incentives for addressing mangrove loss in Northern Vietnam. Front For Glob Change 4:709073. https://doi.org/10.3389/ffgc.2021.709073

Pisani D, Pazienza P, Perrino EV, Caporale D, De Lucia C (2020) The economic valuation of ecosystem services of biodiversity components in protected areas: a review for a framework of analysis for the Gargano National Park. Sustainability 13(21):11726. https://doi.org/10.3390/su132111726

Primavera J (2005) Overcoming the impacts of aquaculture on the coastal zone. Ocean Coast Manag 49(9–10):531–545. https://doi.org/10.1016/j.ocecoaman.2006.06.018

Pulhin JM, Gevaña DT, Pulhin FB (2017) Community-based mangrove management in the Philippines: experience and challenges in the context of changing climate. In: DasGupta R, Shaw R (eds) Participatory mangrove management in a changing climate. disaster risk reduction. Springer, Tokyo. https://doi.org/10.1007/978-4-431-56481-2_16

Quang Bao T (2011) Effect of mangrove forest structures on wave attenuation in coastal Vietnam. Oceanologia 53(3):807–818. https://doi.org/10.5697/oc.53-3.807

Quoc Vo T, Kuenzer C, Oppelt N (2015) How remote sensing supports mangrove ecosystem service valuation: a case study in Ca Mau province, Vietnam. Ecosyst Serv 14:67–75. https://doi.org/10.1016/j.ecoser.2015.04.007

Ramesh DA, Karthi N, Infantina A, Priya P (2022) Economic assessment mangrove ecosystems' goods services of India. J Soc Rev Develop 1(2):42–51

Ranasinghe T, Kallesoe M (2006) Valuation, rehabilitation and conservation of mangroves in tsunami affected areas of Hambantota, Sri Lanka: economic valuation of Tsunami affected mangroves—the world conservation union, ecosystems and livelihoods group Asia

Romanazzi GR, Koto R, De Boni A, Ottomano Palmisano G, Cioffi M, Roma R (2023) Cultural ecosystem services: a review of methods and tools for economic evaluation. Environ Sustain Indic 20:100304. https://doi.org/10.1016/j.indic.2023.100304

Ruslan NFN, Goh HC, Hattam C, Edwards-Jones A, Moh HH (2022) Mangrove ecosystem services: contribution to the well-being of the coastal communities in Klang Islands. Mar Policy 144:105222. https://doi.org/10.1016/j.marpol.2022.105222

Salem ME, Mercer DE (2012) The economic value of mangroves: a meta-analysis. Sustainability 4(3):359–383. https://doi.org/10.3390/su4030359

Sarker AR, Nobi MN, Roskaft E, Chivers DJ, Suza M (2020) Value of the storm-protection function of Sundarban mangroves in Bangladesh. J Sustain Develop 13(3):128–137. https://doi.org/10.5539/jsd.v13n3p128

Scott CP, Mach L, Lucas KM, Myers AE (2024) Whose cultural ecosystem service values matter? Exploring power inequities in diverse mangrove communities. Hum Ecol 52:81–97. https://doi.org/10.1007/s10745-023-00462-5

Seriño MN, Ureta JC, Baldesco J, Galvez KJ, Predo C, Seriño EK (2017) Valuing the protection service provided by mangroves in typhoon-hit areas in the Philippines. EEPSEA Research Report No. 2017-RR19. Economy and Environment Program for Southeast Asia, Laguna, Philippines

Simpson S, Smart LS, Landis E, Van Laere S, Kibria AS, Albot O, Beeston M, Lovelock CE, McDonald RI, Moyer RP, Worthington TA, Maxwell TL, Spalding M (2025) The Blue Carbon Cost Tool—understanding market potential and investment requirements for high-quality coastal wetland projects. Front Mar Sci 12:1622255. https://doi.org/10.3389/fmars.2025.1622255

Somda J, Awaïss A (2013) Economic valuation of ecological functions and services of natural ecosystems: guide on the use of simple methods. IUCN, Ouagadougou, Burkina Faso, p 32

Song AM, Dressler WH, Satizábal P, Fabinyi M (2021) From conversion to conservation to carbon: the changing policy discourse on mangrove governance and use in the Philippines. J Rural Stud 82:184–195. https://doi.org/10.1016/j.jrurstud.2021.01.008

Spalding M, McIvor A, Tonneijck FH, Tol S, van Eijk P (2014) Mangroves for coastal defence. Guidelines for coastal managers & policy makers. Published by Wetlands International and The Nature Conservancy, 42 p

Su J, Friess DA, Gasparatos A (2021) A meta-analysis of the ecological and economic outcomes of mangrove restoration. Nat Commun 12(1):5050. https://doi.org/10.1038/s41467-021-25349-1

Sunkur R, Kantamaneni K, Bokhoree C, Ravan S (2023) Mangroves' role in supporting ecosystem-based techniques to reduce disaster risk and adapt to climate change: a review. J Sea Res 196:102449. https://doi.org/10.1016/j.seares.2023.102449

Swails E, Tuan VQ, Pham TT (2025) Vietnam blue carbon opportunity assessment. Working Paper 47. CIFOR, Bogor, Indonesia; ICRAF, Nairobi, Kenya. https://doi.org/10.17528/cifor-icraf/009370

Swann S, Blandford L, Cheng S, Cook J, Miller A, Barr R (2021) Public international funding of nature-based solutions for adaptation: a landscape assessment. Working Paper. World Resources Institute, Washington, DC. Available online at: https://doi.org/10.46830/wriwp.20.00065

Syawaludin M, Kurniawan R (2024) Mangrove restoration and fisheries productivity: a bibliometric and systematic review with SDG 15 (Life on Land) impact. Assyfa J Farm Agric 2(1):27–39. https://doi.org/10.61650/ajfa.v2i1.863

Takagi H (2025) Mangrove forest recovery a decade after Super Typhoon Haiyan. J Environ Manage 395:127840. https://doi.org/10.1016/j.jenvman.2025.127840

Thompson BS, Primavera JH, Friess DA (2017) Governance and implementation challenges for mangrove forest Payments for Ecosystem Services (PES): empirical evidence from the Philippines. Ecosyst Serv 23:146–155. https://doi.org/10.1016/j.ecoser.2016.12.007

UNDP (2024) Mangroves, the lifeline of Viet Nam's coasts: Stewards of climate, biodiversity, and communities. https://www.undp.org/vietnam/stories/mangroves-lifeline-viet-nams-coasts-stewards-climate-biodiversity-and-communities. Accessed on 9 Nov 2025

Van Schoubroeck S, Anougmar S, Finizola e Silva M, Ala-Harja V, Statzu V, Everaert G, Watt L, Barboza FR, Compernolle T (2024) Valuation of ecosystem services in marine protected areas: a comprehensive review of methods and needed developments. Ecosyst Serv 70:101678. https://doi.org/10.1016/j.ecoser.2024.101678

Van Wesenbeeck BK, Van Zelst VT, Antolinez JA, De Boer WP (2025) Quantifying uncertainty in wave attenuation by mangroves to inform coastal green belt policies. Commun Earth Environ 6(1):258. https://doi.org/10.1038/s43247-025-02178-4

Van Zanten B, Brander LM, Castaneda J, Herrera D, Kaczan D, Brown BM, Smith IA, Jongman B (2025) Integrated spatial cost-benefit analysis of large-scale mangrove conservation and restoration in Indonesia. Front Environ Sci 13:1459034. https://doi.org/10.3389/fenvs.2025.1459034

Van Zelst VT, Dijkstra JT, Van Wesenbeeck BK, Eilander D, Morris EP, Winsemius HC, Ward PJ, De Vries MB (2021) Cutting the costs of coastal protection by integrating vegetation in flood defences. Nat Commun 12(1):6533. https://doi.org/10.1038/s41467-021-26887-4

Van Zanten BT, Brander LM, Gutierrez Torres D, Uyttendaele GYP, Herrera Garcia LD, Patrama D, Kaczan DJ (2021) The economics of large-scale mangrove conservation and restoration in Indonesia. https://thedocs.worldbank.org/en/doc/89fd7ff87561a4a913ed3371278e7933-0070062022/related/The-Economics-of-Large-scale-Mangrove-Conservation-andRestoration-in-Indonesia-Brief-Note-Interactive-220516.pdf.

van Zanten BT, Goizueta GG, Brander L, Reguero BG, Griffin R, Macleod KK, Alves A, Midgley A, Herrera LD, Jongman B (2023) Assessing the benefits and costs of nature-based solutions for climate resilience: a guideline for project developers. World Bank, Washington, DC. License: Creative Commons Attribution CC BY 3.0 IGO

Veettil BK, Ward RD, Dung NTK, Van DD, Quang NX, Hoai PN, Hoang N (2021) The use of bioshields for coastal protection in Vietnam: current status and potential. Reg Stud Mar Sci 47:101945. https://doi.org/10.1016/j.rsma.2021.101945

Vo Q (2013) Valuation of mangrove ecosystems along the coast of the Mekong Delta in Vietnam an approach combining socio-economic and remote sensing methods, Doctoral dissertation, Kiel University, Kiel, Germany

Vollmer A, Ceolotto S, Farrell N (2024) Nature-based solutions for climate adaptation: review of barriers to adoption and guidelines for policymakers, ESRI Working Paper, No. 794, The Economic and Social Research Institute (ESRI), Dublin

Vu TP, Pham TT, Tran NMH, Nguyen TTA, Nguyen TVA, Tang TKH (2023) Blue carbon market in Vietnam: potential and challenges for future development. Occasional Paper 8. Center for International Forestry Research (CIFOR), Bogor, Indonesia; and World Agroforestry (ICRAF), Nairobi, Kenya

Walz Y, Janzen S, Narvaez L, Ortiz-Vargas A, Woelki J, Doswald N, Sebesvari Z (2021) Disaster-related losses of ecosystems and their services. Why and how do losses matter for disaster risk reduction? Int J Disaster Risk Reduction 63:102425. https://doi.org/10.1016/j.ijdrr.2021.102425

Wan Ab Rashid WNA, Mohamad J, Hamzah MA (2024) Total economic value of marine ecosystem service in Malaysia: a review. Built Environ J 21(2):108–126. https://doi.org/10.24191/bej.v21i2.938

Weaver RJ, Stehno AL (2024) Mangroves as coastal protection for restoring low-energy waterfront property. J Mar Sci Eng 12(3):470. https://doi.org/10.3390/jmse12030470

Weiss J (2009) The economics of climate change in Southeast Asia: a regional review. Asian Development Bank. http://hdl.handle.net/11540/179.

World Bank (2022a) Accelerating clean, green, and climate-resilient growth in Vietnam: a country environmental analysis. International Bank for Reconstruction and Development/The World Bank, 160pp

World Bank (2022b) What is it worth?: testing a practical approach to assess and value natural assets in coastal areas of Vietnam's Quang Ninh Province and Tam Giang—Cau Hai Wetland in Thua Thien Hue Province. © World Bank. http://hdl.handle.net/10986/37695 License: CC BY 3.0 IGO

Worthington TA, Zu Ermgassen PS, Friess DA, Krauss KW, Lovelock CE, Thorley J, Tingey R, Woodroffe CD, Bunting P, Cormier N, Lagomasino D, Lucas R, Murray NJ, Sutherland WJ, Spalding M (2020) A global biophysical typology of mangroves and its relevance for ecosystem structure and deforestation. Sci Rep 10(1):14652. https://doi.org/10.1038/s41598-020-71194-5

Zhang J, Lu Z, Zhou J, Qin G, Bai Y, Sanders CJ, Macreadie PI, Yuan J, Huang X, Wang F (2025) Getting the best of carbon bang for mangrove restoration buck. Nat Commun 16(1):1297. https://doi.org/10.1038/s41467-025-56587-2

Zu Ermgassen PS, Worthington TA, Gair JR, Garnett EE, Mukherjee N, Nagelkerken I, Abrantes K, Acosta A, Araujo AR, Baker R, Barnett A, Beitl CM, Benzeev R, Brookes JAG, Ching Chong V, Connolly RM, Diele K, Dwyer PG, Spalding MD (2025) Mangroves support an estimated annual abundance of over 700 billion juvenile fish and invertebrates. Commun Earth Environ 6(1):299. https://doi.org/10.1038/s43247-025-02229-w

Chapter 10
Governance, Policy, and Regulatory Frameworks of Coastal Bioshields

Abstract The effectiveness and sustainability of bioshields as nature-based solutions for coastal protection depend heavily on the strength and coherence of governance, policy, and legal frameworks. This chapter examines various institutional architectures, regulatory instruments, and policy mechanisms that shape the management and implementation of bioshield initiatives across different governance levels. It analyzes the evolution of national and regional policies promoting coastal ecosystem conservation, disaster risk reduction, and climate adaptation, highlighting both enabling conditions and persistent governance gaps. Special attention is given to the integration of bioshields within Integrated Coastal Zone Management (ICZM) frameworks and international conventions such as the Ramsar Convention, the Convention on Biological Diversity (CBD), and the Paris Agreement. The chapter explores challenges related to cross-sectoral coordination, enforcement, and funding, which often hinder effective policy execution. It presents comparative case studies from Asian nations—such as Vietnam, India, and Indonesia—illustrating diverse governance models and policy innovations that link environmental protection with socioeconomic development. Furthermore, the role of decentralization, community co-management, and legal recognition of ecosystem services is discussed to enhance local ownership and compliance. By synthesizing legal and institutional perspectives, the chapter argues that adaptive, inclusive, and science-informed governance is vital to scaling up bioshield initiatives. It concludes with policy recommendations aimed at strengthening legal frameworks to ensure long-term coastal resilience through ecosystem-based approaches.

Keywords Coastal policy frameworks · Community co-management · Ecosystem-based adaptation · Environmental governance · Integrated Coastal Zone Management (ICZM) · Legal instruments

B. Kozhikkodan Veettil, *The Potential of Bioshields for Coastal Protection*, Advances in Natural and Technological Hazards Research 55,
https://doi.org/10.1007/978-3-032-22888-8_10

10.1 Introduction

The governance, policy, and legal frameworks surrounding coastal bioshields form the institutional backbone that determines their success, sustainability, and scalability (Rahman et al. 2018; Biermann et al. 2010). While ecological and technical knowledge are critical for bioshield design and implementation, these interventions are ultimately shaped by the policy environments within which they operate (Elrick-Barr et al. 2024). Governance defines who makes decisions, how resources are allocated, and how accountability is maintained (Iza 2021). In contrast, legal frameworks provide enforceable rules that protect bioshield zones from degradation, ensure stakeholder participation, and align national practices with international commitments (Vasseur 2020; Bilawal Khaskheli et al. 2025).

In recent decades, global recognition of the importance of nature-based solutions (NbS) in climate adaptation and disaster risk reduction has driven the inclusion of bioshields within environmental and coastal management policies (Cohen-Shacham et al. 2016; Bisaro and Meyer 2022; Martin et al. 2025). International agreements such as the Paris Agreement (2015), the Convention on Biological Diversity (CBD), and the Sendai Framework for Disaster Risk Reduction (2015–2030) have explicitly encouraged ecosystem-based adaptation and the restoration of coastal ecosystems (OECD 2020). At the same time, regional organizations and national governments have developed diverse policy instruments to translate these global commitments into actionable strategies (Toledo et al. 2025).

However, despite this growing policy attention, many bioshield projects remain hindered by fragmented governance, overlapping institutional mandates, and inadequate legal protection (Chamberland-Fontaine et al. 2022). For example, in many coastal nations, mangroves—the primary bioshield vegetation—fall under multiple agencies (forestry, fisheries, and environment), resulting in jurisdictional conflicts (Veettil et al. 2019; Foggia et al. 2025). Furthermore, inadequate enforcement of land-use laws often exposes bioshield zones to aquaculture expansion, urban encroachment, and pollution (Romañach et al. 2018; Velástegui-Montoya et al. 2025).

This chapter examines the global and regional governance systems that support bioshields, reviews key legal instruments, analyzes country-specific examples, and identifies challenges and policy innovations that can strengthen institutional foundations for bioshield development and maintenance.

10.2 Global Governance and Policy Context

The global governance architecture for bioshields (Fig. 10.1) emerges at the intersection of environmental conservation, climate change adaptation, and disaster risk reduction (Lo 2016). While no single global treaty explicitly governs "bioshields," multiple international instruments collectively provide a comprehensive policy environment for their protection and expansion (Lo 2016; Seddon et al. 2020).

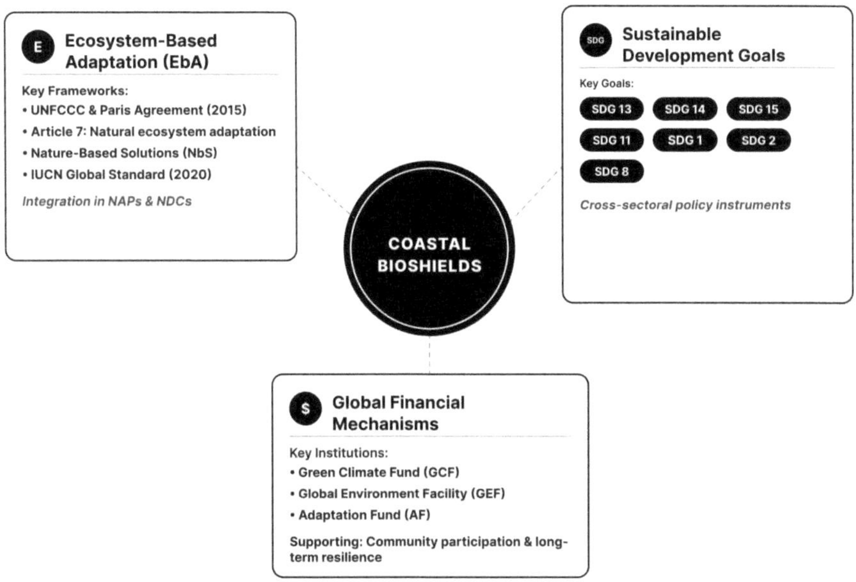

Fig. 10.1 Global context of coastal bioshield governance

10.2.1 Ecosystem-Based Adaptation and Nature-Based Solutions

The recognition of Ecosystem-Based Adaptation (EbA) as a formal climate adaptation strategy has been pivotal in advancing bioshields within policy discourse (Seddon et al. 2020; Johnson et al. 2022). The UN Framework Convention on Climate Change (UNFCCC) and its subsidiary agreements, particularly the Paris Agreement (2015), have acknowledged ecosystems as critical assets for climate resilience (Seddon et al. 2020). Article 7 of the Paris Agreement emphasizes the role of natural ecosystems in adaptation planning and implementation (Paris Agreement 2015). Consequently, bioshields have been integrated into National Adaptation Plans (NAPs) and Nationally Determined Contributions (NDCs) of many coastal nations, including Vietnam, India, and Indonesia (Seddon et al. 2020; Veettil et al. 2021).

Similarly, the UN Environment Programme (UNEP) and the International Union for Conservation of Nature (IUCN) promote Nature-Based Solutions (NbS) as a unifying concept that includes bioshields (Dunlop et al. 2024). These solutions aim to deliver multiple co-benefits, such as biodiversity conservation, climate change mitigation, and socio-economic improvement, while reducing disaster risks (Corgo et al. 2024). The IUCN Global Standard for NbS (2020) provides a policy framework for designing and evaluating bioshields that meet ecological integrity and governance criteria (IUCN 2020; Le Gouvello et al. 2023).

10.2.2 Coastal Protection Within the Sustainable Development Goals

The UN Sustainable Development Goals (SDGs) provide an overarching agenda that supports bioshields through integrated targets (Sever et al. 2025). SDG 13 (Climate Action), SDG 14 (Life Below Water), and SDG 15 (Life on Land) call for ecosystem-based management and restoration of coastal and marine ecosystems (Pakseresht et al. 2025). Additionally, SDG 11 (Sustainable Cities and Communities) and SDG 1 (No Poverty) underscore the social and economic dimensions of bioshield development, particularly their roles in protecting vulnerable coastal populations.

For instance, bioshield projects that protect agricultural lands from saline intrusion directly support SDG 2 (Zero Hunger), while those that generate sustainable livelihoods through ecotourism or mangrove-based aquaculture align with SDG 8 (Decent Work and Economic Growth). Thus, bioshields represent cross-sectoral policy instruments that advance multiple global sustainability goals simultaneously (Bimrah et al. 2022).

10.2.3 Global Financial Mechanisms

The successful implementation of bioshield initiatives often depends on international financial mechanisms. Institutions such as the Green Climate Fund (GCF), Global Environment Facility (GEF), and Adaptation Fund (AF) provide financial support for ecosystem-based adaptation projects, including coastal bioshields. For example, the GCF has funded mangrove restoration and coastal protection initiatives in Bangladesh and Fiji, emphasizing community participation and long-term resilience (Majumdar and Matsui 2025; Independent Evaluation Unit 2025). These funding mechanisms encourage developing countries to integrate bioshields into national adaptation strategies and provide technical assistance for scaling up best practices.

10.3 International Legal Instruments Supporting Bioshields

Several international legal frameworks indirectly or directly support the conservation and management of ecosystems that function as bioshields. These agreements provide the normative basis for countries to legislate and implement bioshield-related policies at national and local levels (Trouwborst et al. 2017; Ivanova 2020; Ekardt et al. 2023). Together, these instruments (e.g., Convention on Biological Diversity—CBD, Ramsar Convention on Wetlands, UN Convention on the Law of the Sea—UNCLOS, Sendai Framework for Disaster Risk Reduction 2015–2030, and Paris

Agreement and Nationally Determined Contributions—NDCs) form a comprehensive legal architecture that, while fragmented, provides overlapping mandates and incentives for bioshield development and protection (Biermann et al. 2009).

10.3.1 Convention on Biological Diversity (CBD)

The CBD (UN 2011) serves as one of the most important international frameworks for coastal protection by bioshields. Its objectives (conservation, sustainable use, and equitable benefit-sharing) directly support ecosystem-based approaches to coastal management. The Aichi Biodiversity Targets (2011–2020) and the Kunming–Montreal Global Biodiversity Framework (2022–2030) call for the restoration of degraded ecosystems, including mangroves, wetlands, and coastal forests, which function as bioshields (Carr et al. 2020; Fu et al. 2024). Target 11 of the Aichi Biodiversity Targets were 17% for terrestrial/inland waters and 10% for coastal and marine areas by 2020 as part of conserving areas of importance (Juffe-Bignoli et al. 2016). Target 3 of the Kunming–Montreal Global Biodiversity Framework emphasizes the protection of at least 30% of coastal and marine areas by 2030 ("30 by 30" goal), offering legal impetus for bioshield conservation (Shen et al. 2023).

10.3.2 Ramsar Convention on Wetlands

The Ramsar Convention (1971) provides a legal mechanism for the protection of wetlands of international importance. Many bioshield habitats—mangroves, salt marshes, and coastal lagoons—fall under this definition (Ramsar 2016). The convention's wise use principle encourages sustainable management of these ecosystems, balancing conservation and human use (Ramsar 2016). Over 2,400 sites worldwide are designated as Ramsar wetlands, many of which function as natural bioshields (Bera 2022). For example, Vietnam's Xuan Thuy and Ca Mau Ramsar sites have mangrove-based ecosystems that serve as buffers against storm surges and coastal erosion (Veettil et al. 2021).

10.3.3 UN Convention on the Law of the Sea (UNCLOS)

The UNCLOS (1982) provides the overarching legal framework for coastal and marine governance. Although it does not explicitly mention bioshields, its provisions on coastal zone management, pollution prevention, and habitat protection create enabling conditions for bioshield implementation (IUCN 1981). Article 194 mandates that states take all measures necessary to prevent, reduce, and control pollution of the marine environment, including through the protection of ecosystems. Coastal bioshields contribute directly to fulfilling this obligation by filtering sediments, absorbing pollutants, and stabilizing coastlines (Zhang et al. 2023).

10.3.4 Sendai Framework for Disaster Risk Reduction (2015–2030)

The Sendai Framework emphasizes the integration of ecosystem-based and community-led approaches in disaster risk management (Pearson and Pelling 2015). It recognizes that healthy ecosystems, including mangroves and coastal vegetation, serve as natural barriers against hazards such as cyclones, tsunamis, and flooding. The framework encourages states to invest in "green infrastructure" alongside engineered defenses, reinforcing the legitimacy of bioshields within disaster risk governance (Raymond et al. 2017; UNDRR 2020).

10.3.5 Paris Agreement and Nationally Determined Contributions (NDCs)

The Paris Agreement (2015) under the UNFCCC establishes a legal foundation for ecosystem-based adaptation, including bioshields. Many signatory countries have incorporated coastal ecosystem restoration into their NDCs (Lecerf et al. 2023). For instance, Vietnam's updated NDC (2022) highlights mangrove restoration as a key measure for both adaptation and mitigation (Buckingham and Hanson 2015). Similarly, Indonesia, Bangladesh, and the Philippines have committed to scaling up coastal vegetation as part of their climate strategies.

10.4 Regional Policy Frameworks and Cooperation

Regional organizations have played an instrumental role in translating global legal principles into regionally relevant coastal governance frameworks (Rochette et al. 2015; Adewumi 2021). These initiatives often reflect the specific ecological, cultural, and socio-economic contexts of their member states. The regional frameworks demonstrate that effective bioshield governance benefits from multi-country cooperation, which facilitates shared learning, harmonized policy approaches, and pooled funding opportunities.

10.4.1 Asia–Pacific

The Association of Southeast Asian Nations (ASEAN) has taken significant steps toward regional cooperation in coastal protection. The ASEAN Working Group on Coastal and Marine Environment (AWGCME) promotes information exchange, joint research, and capacity-building for ecosystem-based adaptation (Sudirman et al. 2023). The European Union and ASEAN participate in mangrove restoration and nature-based coastal defenses in the ASEAN region (EU 2025). Furthermore, the Regional Programme on Mangroves for the Future (MFF), jointly implemented by IUCN and UNDP, has successfully supported bioshield projects in Vietnam, Thailand, Indonesia, and Sri Lanka (e.g., Ekaratne and Vidanage 2013; MFF Vietnam 2015).

10.4.2 South Asia

The South Asian Association for Regional Cooperation (SAARC) launched the Climate Change Action Plan (2008), emphasizing ecosystem-based coastal adaptation and the sharing of traditional knowledge (SAARC 2008). Countries such as India, Bangladesh, and Sri Lanka cooperate through regional workshops and joint monitoring programs to restore mangrove belts in cyclone-prone areas (Mohanty et al. 2022).

10.4.3 Africa

In Africa, the African Ministerial Conference on the Environment (AMCEN) promotes integrated coastal zone management (ICZM) and blue economy policies that include bioshields (Karani et al. 2022; March et al. 2024a). The West African Coastal Areas Management Program (WACA), supported by the World Bank, collaborates with countries such as Senegal, Benin, and Ghana to enhance governance, community engagement, and legal protection for mangrove ecosystems (Almar et al. 2023).

10.4.4 Europe

Although bioshields are less common in temperate regions, European frameworks, such as the EU Strategy on Adaptation to Climate Change (2021) and the EU Biodiversity Strategy for 2030, recognize coastal ecosystem restoration as a form of green infrastructure (Hermoso et al. 2022). The EU policies and coastal communities generally support the use of salt marshes and dunes as natural flood barriers (Apine and Stojanovic 2025; Kampa et al. 2025).

10.4.5 Latin America and the Caribbean

Regional cooperation under the Caribbean Community Climate Change Centre (CCCCC) and Latin American platforms for mangrove management (e.g., UNESCO-MangRes Project) promotes bioshields through integrated coastal zone management (Barragán Muñoz 2020; Rosenberg 2020; Grimm et al. 2023). Countries such as Mexico, Cuba, and Colombia have adopted regionally harmonized policies to strengthen ecosystem resilience and disaster preparedness (Barragán Muñoz 2020; Lucatello and Alcántara-Ayala 2023).

10.5 National Policy Case Studies

10.5.1 Vietnam

Vietnam represents one of the most progressive examples of bioshield governance in Southeast Asia (McElwee et al. 2016; Veettil et al. 2021). The Law on Environmental Protection in 2020 (Law No. 72/2020/QH14) and the National Strategy on Climate Change (2021–2030) in 2020 (Decision No. 1055/QD-TTg) explicitly promote ecosystem-based adaptation, with mangrove restoration serving as a central component (Hai et al. 2020). The Vietnam Red Cross Coastal Afforestation Programme, initiated in the late 1990s, has successfully integrated community participation into national climate adaptation strategies (DeWit et al. 2022).

Furthermore, Vietnam's National Adaptation Plan (NAP 2022) emphasizes bioshield expansion along vulnerable deltas such as the Mekong and Red River (Nguyen et al. 2021). Provincial authorities are empowered to designate coastal protection forests as legally protected zones under the Law on Forest Protection and Development in 2017 (Law on Forestry No. 16/2017/QH14). Despite these advances, enforcement remains uneven due to overlapping mandates between the Ministry of Agriculture and Rural Development (MARD) and the Ministry of Natural Resources and Environment (MONRE) (Pham et al. 2019; Veettil et al. 2019, 2021). Strengthening inter-ministerial coordination and integrating bioshield monitoring into national environmental databases remain policy priorities.

10.5.2 India

India's policy framework for bioshields has evolved significantly since the 2004 Indian Ocean tsunami (Veettil et al. 2020; Poti et al. 2025). The Coastal Regulation Zone (CRZ) Notification (2019) identifies mangroves as ecologically sensitive areas that must be preserved and restored (Thomas 2025). Additionally, India's National Action Plan on Climate Change (Government of India 2008) recognizes bioshields as a key measure under its National Mission for a Green India and National Mission on Sustainable Agriculture (Jha 2012; Gupta et al. 2021). These policies have been complemented by community-based projects led by NGOs, such as the M.S. Swaminathan Research Foundation's mangrove restoration in Pichavaram, demonstrating effective integration between national policy and grassroots-level action (Selvam and Thamizoli 2021).

10.5.3 Bangladesh

Bangladesh's Coastal Greenbelt Project, initiated in 1995, represents one of the earliest national-level bioshield programs (Dey et al. 2021). Managed jointly by the Bangladesh Forest Department and local cooperatives, the project has planted thousands of hectares of mangroves and coastal forests along the Bay of Bengal (Alam et al. 2022; Ashrafuzzaman 2025). The Bangladesh Climate Change Strategy and Action Plan (BCCSAP) in 2009 identifies coastal afforestation as a top adaptation priority (Islam et al. 2013). Moreover, the National Forest Policy (2019) institutionalizes community participation, ensuring that local groups receive benefits from bioshield maintenance (Islam and Rahman 2015).

10.5.4 Indonesia

Indonesia's governance framework integrates bioshields within its broader Blue Economy (Sidik et al. 2022) and National Mangrove Rehabilitation Program (Sasmito et al. 2023). The government aims to restore 600,000 ha of mangrove forests by 2024, combining central coordination with local community engagement (Sasmito et al. 2023). The Village Fund (Dana Desa) mechanism enables decentralized financing for community-based coastal rehabilitation, illustrating how fiscal devolution supports bioshield governance (Watts et al. 2019).

10.5.5 The Philippines

The Philippines institutionalized bioshields under its National Greening Program (2011–2028) (Executive Order No. 26) and Integrated Coastal Management (ICM) framework (H. No 7767, The Congress of the Philippines). The Philippine Climate Change Act 2009 (Republic Act 9729) and Disaster Risk Reduction and Management Act 2010 (Republic Act 10,121) legally mandate the use of ecosystem-based approaches in national adaptation and risk reduction strategies. Local governments, empowered under the Local Government Code 1991 (Republic Act No. 7160), play a central role in bioshield management, integrating traditional ecological knowledge into municipal coastal resource plans (Maliao et al. 2025).

10.6 Institutional Arrangements and Multi-level Governance

Effective governance of bioshields requires coordination among multiple institutional levels—from international agreements to local community management (Pittman and Armitage 2016; Elrick-Barr et al. 2024). The success or failure of bioshield initiatives is often determined not by ecological feasibility but by the institutional coherence of the system in which they operate.

10.6.1 Vertical Integration

Multi-level environmental governance refers to the distribution of authorities across various administrative levels internationally, nationally, provincially, and locally (Yi et al. 2018). Coastal bioshields such as mangroves, being cross-sectoral interventions, demand strong vertical integration (Walker et al. 2022; Quevedo et al. 2025). For instance, in Vietnam, mangrove management falls under the jurisdiction of both national ministries and provincial People's Committees, requiring alignment between top-down directives and bottom-up implementation (Orchard et al. 2015). Similarly, India's mangrove protection programs involve cooperation between central agencies, such as the Ministry of Environment, Forest and Climate Change (MoEFCC), and local self-government bodies (D'Souza and Ishwar 2012).

In many developing countries, weak vertical coordination results in fragmented responsibilities and duplication of efforts (Larson and Soto 2008). Decentralization reforms—while improving local participation—can sometimes lead to the devolution of responsibilities without adequate financial or technical support (Ribot et al. 2006). Therefore, capacity-building and fiscal decentralization are crucial for ensuring that local authorities can effectively manage and monitor bioshields (Barr et al. 2006; OECD 2021).

10.6.2 Horizontal Coordination

Horizontal coordination refers to collaboration across different sectors and ministries (Cash et al. 2006). Since bioshields intersect with forestry, agriculture, fisheries, and disaster management, institutional overlap is common (Berkes 2009). For example, in Indonesia, both the Ministry of Environment and Forestry and the Ministry of Marine Affairs and Fisheries oversee mangrove governance, sometimes resulting in policy inconsistencies (Mursyid et al. 2021; Arifanti et al. 2022; Firdasari 2022). Similarly, in Bangladesh, coordination between the Forest Department and the Water Development Board has been a persistent challenge (Begum et al. 2021; Khan and Giessen 2021).

To address such issues, inter-ministerial committees, task forces, and integrated coastal management (ICM) frameworks have been established in several countries. The Philippines' National Coastal Zone Management Council and Vietnam's National Steering Committee on Climate Change serve as examples of horizontal coordination mechanisms that facilitate cross-sectoral dialogue and planning.

10.6.3 Role of Local Governments and Communities

Local governments are often the most immediate actors responsible for bioshield maintenance and enforcement of land-use regulations. Empowering them through legal mandates and financial mechanisms enhances long-term sustainability. Community-based forest management (CBFM) models, as seen in the Philippines (Lasco and Pulhin 2006) and Vietnam (Decision No 2157/QD-TTg, 2021), demonstrate that local stewardship can significantly improve the survival and expansion of bioshields.

In Vietnam, for instance, local cooperatives and women's unions are directly involved in mangrove planting, patrolling, and benefit-sharing under provincial afforestation programs (Pham et al. 2022). These arrangements integrate traditional ecological knowledge with modern management practices, leading to better ecological outcomes and social acceptance (Ebeler et al. 2025).

10.7 Legal Challenges in Bioshield Implementation

Despite growing policy support, numerous legal and administrative challenges (Table 10.1) hinder the full realization of coastal bioshields' potential across the globe. These challenges include ambiguity in property and tenure rights, inadequate legal enforcement and monitoring, conflicts between economic development and conservation laws, and limited recognition of customary laws and traditional rights.

Legal pluralism—acknowledging both statutory and customary systems—offers a pathway to more inclusive bioshield governance.

10.7.1 Ambiguity in Property and Tenure Rights

Ambiguous land and resource tenure often undermines bioshield governance (Rotich et al. 2016; Bell-James et al. 2023a). In many coastal areas, the ownership of mangrove forests and intertidal zones remains unclear or contested among government agencies, private landowners, and local communities (Miller and Tonoto 2023). This uncertainty discourages long-term investment in restoration and sometimes leads to encroachment or illegal conversion for aquaculture (Song et al. 2021; Bell-James et al. 2023a).

Legal reforms that clarify ownership and usage rights are crucial. For example, Bangladesh's Social Forestry Rules (2010) (Jashimuddin and Inoue 2012) and Vietnam's Land Law in 2013 (Law No. 45/2013/QH13) provide legal bases for community co-management, allowing local groups to benefit from sustainable resource use while ensuring conservation obligations.

10.7.2 Inadequate Legal Enforcement and Monitoring

Even where laws exist, enforcement remains weak due to limited institutional capacity, corruption, or competing development priorities (Bisson et al. 2003; Yuza et al. 2023). In countries like Indonesia and Myanmar, illegal cutting of mangroves continues despite national bans (Veettil et al. 2018; Onrizal et al. 2019). Similarly, coastal zoning regulations in India's CRZ Notification are frequently violated by tourism and aquaculture interests (Sakthivel and Khan 2024).

Establishing dedicated coastal enforcement agencies and integrating remote sensing technologies for monitoring can enhance compliance (Veettil et al. 2020, 2021). The use of satellite-based mangrove mapping in Vietnam (Veettil et al. 2022) and the Philippines (Pillodar et al. 2023) has shown promise for detecting illegal clearing and assessing restoration progress.

10.7.3 Conflicts Between Economic Development and Conservation Laws

Rapid coastal development—ports, resorts, aquaculture, and industrial zones—often conflicts with bioshield conservation goals (Veettil et al. 2021). Legal frameworks that prioritize short-term economic growth can inadvertently undermine long-term

Table 10.1 Legal challenges affecting bioshield implementation and governance

Legal challenge	Key issues	Implications for bioshield implementation	Illustrative examples/case studies	Potential legal and policy responses
Ambiguity in property and tenure rights	Unclear ownership of mangroves and intertidal zones; overlapping claims by state agencies, private actors, and local communities	Discourages long-term investment in restoration; facilitates illegal conversion and encroachment; weakens accountability and stewardship	Mangrove tenure conflicts in Southeast Asia (Song et al. 2021); Bangladesh Social Forestry Rules (2010) enabling co-management; Vietnam Land Law (2013) recognizing community use rights	Legal clarification of tenure and usufruct rights; community-based and co-management frameworks; secure, long-term tenure arrangements linked to conservation obligations
Inadequate legal enforcement and monitoring	Limited institutional capacity; corruption and weak compliance mechanisms; conflicting development priorities	Continued illegal mangrove cutting and degradation; ineffective coastal zoning and conservation laws; loss of trust in regulatory institutions	Illegal mangrove clearing in Indonesia and Myanmar; violations of India's Coastal Regulation Zone (CRZ) Notification; satellite-based monitoring in Vietnam and the Philippines	Establishment of dedicated coastal enforcement agencies; integration of remote sensing and GIS-based monitoring; transparent reporting and penalty mechanisms
Conflicts between economic development and conservation laws	Priority given to ports, tourism, aquaculture, and SEZs; environmental safeguards overridden by development policies	Bioshield degradation and fragmentation; reduced long-term coastal resilience; increased vulnerability to erosion and storm surges	Mangrove clearance under SEZ policies in Myanmar and Southeast Asia; ecosystem valuation in the Philippines and Vietnam supporting conservation	Mandatory EIA and SEA integration in coastal planning; ecosystem service valuation in cost–benefit analyses; harmonization of development and environmental legislation

(continued)

Table 10.1 (continued)

Legal challenge	Key issues	Implications for bioshield implementation	Illustrative examples/case studies	Potential legal and policy responses
Limited recognition of customary law and traditional rights	Marginalization of indigenous and customary governance systems; lack of formal legal recognition	Reduced local legitimacy and compliance; loss of traditional ecological knowledge; weak community engagement in bioshield management	Customary marine tenure (tabu, raui) in Pacific Islands; traditional mangrove stewardship practices in small island states	Legal pluralism integrating statutory and customary law; formal conventional tenure recognition and management systems; participatory governance and co-designed legal frameworks

ecological resilience. For example, special economic zone (SEZ) policies may override environmental protection laws, enabling mangrove clearance for industrial purposes (Aung et al. 2022).

Balancing economic and ecological objectives requires the integration of environmental impact assessments (EIA), strategic environmental assessments (SEA), and ecosystem valuation into coastal development planning. Countries such as the Philippines (Losada et al. 2017; Rosales et al. 2025) and Vietnam (Oanh et al. 2020) have begun incorporating ecosystem service valuation into cost–benefit analyses to justify bioshield investments.

10.7.4 Limited Recognition of Customary Law and Traditional Rights

In many coastal communities, traditional systems of land tenure and natural resource management predate modern statutory laws (Fairbairn 1991; Govan et al. 2009). However, these customary practices are often ignored or marginalized in formal legal systems (Cinner and Aswani 2007). Recognizing and integrating customary law can strengthen local legitimacy and enforcement (Cinner and Aswani 2007). For instance, in the Pacific Islands, customary marine tenure systems (locally known as tabu or raui) effectively regulate mangrove use and protection (Love 2021).

10.8 Future Policy Directions and Institutional Innovations

The evolving nature of climate change, sea-level rise, and coastal urbanization calls for adaptive and forward-looking policy approaches (Griggs and Reguero 2020). The future of bioshield governance depends on innovations that combine legal robustness, institutional flexibility, and community participation.

10.8.1 Integrating Bioshields into National Climate and Development Plans

To ensure long-term sustainability, bioshields should be mainstreamed into national climate strategies, land-use plans, and disaster management frameworks. This requires incorporating bioshield indicators into Nationally Determined Contributions (NDCs) and Sustainable Development Plans (SDPs). For example, Vietnam's updated NDC (2022) includes specific targets for mangrove expansion (Vu et al. 2023), while India's National Mission for a Green India promotes bioshields as key carbon sinks (Datta 2016). Integration across ministries, such as environment, agriculture, transport, and housing, is necessary to prevent conflicts and optimize land-use efficiency (UNDP 2019).

10.8.2 Strengthening Legal Protection for Coastal Ecosystems

Legal reforms should elevate the status of bioshields as critical climate adaptation infrastructure (Bell-James 2019). This can be achieved through explicit recognition in environmental and land-use legislation (Bell-James et al. 2023b). For instance, enacting a Coastal Ecosystem Protection Act (Iza and Cordoba-Munoz 2023) or including bioshields within national disaster management laws could ensure continuous legal protection (Covele et al. 2024).

Furthermore, adopting ecosystem-based management principles within coastal regulation zones (CRZs) can improve resilience (Borges et al. 2017; Powell et al. 2019). Legal instruments should mandate the inclusion of bioshield buffers in all new coastal infrastructure projects (Nilsson and Bohman 2015; Iza 2021).

10.8.3 Institutionalizing Community Co-management

Empirical evidence consistently demonstrates that bioshields are more successful when local communities are co-managers rather than passive beneficiaries

(Voorberg 2020; Gnansounou et al. 2022). Governments should therefore establish legal mechanisms for co-management agreements that delineate roles, responsibilities, and benefit-sharing arrangements (De Oliveira et al. 2024; Rölfer et al. 2024).

Decentralized funding mechanisms—such as Vietnam's Payment for Forest Environmental Services (PFES) (Duong and De Groot 2020) and Indonesia's Village Fund (Dana Desa) (Watts et al. 2019)—should be expanded to include coastal bioshield maintenance. Legal mandates for gender inclusion and indigenous participation further enhance equity and local ownership (Ty et al. 2023).

10.8.4 Leveraging Technology for Governance and Transparency

Emerging technologies such as remote sensing, GIS, drones, and blockchain offer tools for transparent and accountable coastal bioshield governance (Bhambri and Kautish 2024). Real-time satellite monitoring can track changes in coastal bioshield coverage (Veettil et al. 2021), while blockchain-based land registries can secure community tenure rights (Daniel and Ifejika Speranza 2020).

Digital dashboards integrating ecological, social, and financial data can provide decision-makers with dynamic insights for adaptive management (Benson et al. 2021). These tools also improve compliance reporting for international frameworks such as the Paris Agreement and the Convention on Biological Diversity.

10.8.5 Financing Mechanisms and Public–Private Partnerships

Sustainable financing remains a core challenge. Governments should expand the use of innovative financing instruments such as blue bonds (March et al. 2024b), carbon credits (Farahmand et al. 2025), and ecosystem service payments (Meraj and Hashimoto 2025). Public–private partnerships (PPPs) can mobilize additional resources for large-scale bioshield restoration (Pascal et al. 2021). For instance, corporate social responsibility (CSR) programs in the Philippines have supported mangrove rehabilitation as part of climate adaptation portfolios (Camacho et al. 2020). Integrating coastal bioshields into climate finance mechanisms ensures long-term funding and institutional stability.

10.9 Conclusions

Governance, policy, and legal frameworks play a decisive role in determining whether bioshields remain small, project-based interventions or evolve into institutionalized components of national climate strategies. The global shift towards nature-based solutions has brought bioshields to the forefront of environmental policy discussions, but translating this recognition into effective action requires coherent governance systems, enforceable legal frameworks, and inclusive institutions.

The analysis in this chapter reveals several key insights. First, international legal instruments such as the CBD, Ramsar Convention, and Paris Agreement provide a strong normative foundation for bioshield promotion, but implementation gaps persist at national levels. Second, institutional fragmentation and weak enforcement continue to undermine progress, especially in developing countries where overlapping mandates hinder coordination. Third, integrating community participation, recognizing customary rights, and decentralizing decision-making are essential for ensuring legitimacy and sustainability.

Moving forward, adaptive governance models that combine statutory law with local knowledge, leverage technology for transparency, and embed bioshields within broader development planning are needed. Legal recognition of bioshields as critical infrastructure for disaster risk reduction and climate resilience will provide the policy momentum required to mainstream them into national and regional frameworks.

Ultimately, effective governance of bioshields represents a test of our collective ability to harmonize ecological integrity, social equity, and economic development in the face of accelerating coastal risks.

References

Adewumi IJ (2021) Exploring the nexus and utilities between regional and global ocean governance architecture. Front Mar Sci 8:645557. https://doi.org/10.3389/fmars.2021.645557

Alam S, Jahan S, Noor F (2022) The surface water system, flood and water resources management of Bangladesh. Bangladesh Geosciences and Resources Potential, pp 467–546

Almar R, Stieglitz T, Addo KA, Ba K, Ondoa GA, Bergsma EW, Arino O (2023) Coastal zone changes in West Africa: challenges and opportunities for satellite earth observations. Surv Geophys 44(1):249–275. https://doi.org/10.1007/s10712-022-09721-4

Apine E, Stojanovic T (2025) Community acceptance of nature-based solutions for coastal flood risk management. People Nat 8(1):21–36. https://doi.org/10.1002/pan3.70182

Arifanti VB, Sidik F, Mulyanto B, Susilowati A, Wahyuni T, Yuniarti N, Aminah A, Suita E, Karlina E, Suharti S, Turjaman M, Hidayat A, Rachmat HH, Imanuddin R, Yeny I, Darwiati W, Sari N, Hakim SS, Slamet WY, Novita N (2022) Challenges and strategies for sustainable mangrove management in Indonesia: a review. Forests 13(5):695. https://doi.org/10.3390/f13050695

Ashrafuzzaman M (2025) Mangrove is the facto nature-based solutions to tackle the climate change around the globe. In: Nagy GJ, Ayal DY (eds) Handbook of nature-based solutions to mitigation and adaptation to climate change. Springer, Cham. https://doi.org/10.1007/978-3-031-34967-6_61

Aung TS, Overland I, Vakulchuk R, Xie Y (2022) The environmental burdens of special economic zones on the coastal and marine environment: a remote sensing assessment in Myanmar. Remote Sens Appl Soc Environ 8:100809. https://doi.org/10.1016/j.rsase.2022.100809

Barr CM, Resosudarmo IAP, Dermawan A, McCarthy J, Moeliono M, Setiono B (eds) (2006) Decentralization of forest administration in Indonesia: implications for forest sustainability, economic development, and community livelihoods. CIFOR, Bogor

Barragán Muñoz JM (2020) Progress of coastal management in Latin America and the Caribbean. Ocean Coast Manag 184:105009. https://doi.org/10.1016/j.ocecoaman.2019.105009

Begum F, Lobry de Bruyn L, Kristiansen P, Islam MA (2021) Institutionalising co-management activities for conservation of forest resources: evidence from the Sundarban mangrove forest management of Bangladesh. J Environ Manage 298:113504. https://doi.org/10.1016/j.jenvman.2021.113504

Bell-James J (2019) Integrating the ecosystem services paradigm into environmental law: a mechanism to protect mangrove ecosystems? J Environ Law 31(2):291–314. https://doi.org/10.1093/jel/eqz010

Bell-James J, Foster R, Lovelock CE (2023b) Identifying priorities for reform to integrate coastal wetland ecosystem services into law and policy. Environ Sci Policy 142:164–172. https://doi.org/10.1016/j.envsci.2023.02.013

Bell-James J, Fitzsimons JA, Lovelock CE (2023a) Land tenure, ownership and use as barriers to coastal wetland restoration projects in Australia: recommendations and solutions. Environ Manage 72:179–189. https://doi.org/10.1007/s00267-023-01817-w

Benson A, Murray T, Canonico G, Montes E, Muller-Karger FE, Kavanaugh MT, Dewitt LM (2021) Data management and interactive visualizations for the evolving marine biodiversity observation network. Oceanography 34(2):130–141. https://doi.org/10.5670/oceanog.2021.220

Bera B (2022) Ramsar Wetlands in India: its type and importance. Environment and sustainable development problems, prospects and mitigation. Bharti Publications, New Delhi, India, pp 163–203

Berkes F (2009) Evolution of co-management: role of knowledge generation, bridging organizations and social learning. J Environ Manage 90(5):1692–1702. https://doi.org/10.1016/j.jenvman.2008.12.001

Bhambri P, Kautish S (2024) Technological advancements in promoting ecosystem health. In: Leal Filho W, Kautish S, Wall T, Rewhorn S, Paul SK (eds) Digital technologies to implement the UN sustainable development goals. World Sustainability Series. Springer, Cham. https://doi.org/10.1007/978-3-031-68427-2_21

Biermann F, Betsill MM, Gupta J, Kanie N, Lebel L, Liverman D, Schroeder H, Siebenhüner B, Zondervan R (2010) Earth system governance: a research framework. Int Environ Agreements: Politics, Law Econ 10(4):277–298. https://doi.org/10.1007/s10784-010-9137-3

Biermann F, Betsill M, Gupta J, Kanie N, Lebel L, Liverman D, Schroeder H, Siebenhüner B (2009) Earth system governance: people, places and the planet. Science and implementation plan of the earth system governance project. IHDP: The Earth System Governance Project, 148pp

Bilawal Khaskheli M, Li Y, Zhao Y, Shumin W (2025) The legal and socioeconomic challenges for environmental protection in marine policy and the silk route maritime. Front Mar Sci 12:1597932. https://doi.org/10.3389/fmars.2025.1597932

Bimrah K, Dasgupta R, Hashimoto S, Saizen I, Dhyani S (2022) Ecosystem services of mangroves: a systematic review and synthesis of contemporary scientific literature. Sustainability 14(19):12051. https://doi.org/10.3390/su141912051

Bisaro A, Meyer K (2022) Integrating nature-based Solutions into policies for climate change adaptation and disaster risk reduction. IUCN, Gland, Switzerland

Bisson J, Guiang ES, Walpole P, Tolentino D (2003) Better governance critical to reversing forest degradation trends in Southeast Asia. Paper presented at XII Forestry Congress, Quebec City, Canada. http://www.fao.org/DOCREP/ARTICLE/WFC/XII/0837-A4.HTM

Borges R, Ferreira AC, Lacerda LD (2017) Systematic planning and ecosystem-based management as strategies to reconcile mangrove conservation with resource use. Front Mar Sci 4:276476. https://doi.org/10.3389/fmars.2017.00353

Buckingham K, Hanson C (2015) The restoration diagnostic. Case example: restoration of mangrove forests in Vietnam. World Resources Institute, Washington DC, USA, 10pp

Camacho L, Gevaña D, Sabino L, Ruzol C, Garcia J, Camacho A, Oo TN, Maung AC, Saxena KG, Liang L, Yiu E, Takeuchi K (2020) Sustainable mangrove rehabilitation: lessons and insights from community-based management in the Philippines and Myanmar. APN Sci Bull 10(1). https://doi.org/10.30852/sb.2020.983

Carr H, Abas M, Boutahar L, Caretti ON, Chan WY, Chapman AS, Zivian A (2020) The Aichi biodiversity targets: achievements for marine conservation and priorities beyond 2020. PeerJ 8:e9743. https://doi.org/10.7717/peerj.9743

Cash DW, Adger W, Berkes F, Garden P, Lebel L, Olsson P, Pritchard L, Young O (2006) Scale and cross-scale dynamics: governance and information in a multilevel world. Ecol Soc 11(2):8

Chamberland-Fontaine S, Heckadon-Moreno S, Hickey GM (2022) Tangled roots and murky waters: piecing together Panama's mangrove policy puzzle. Front for Glob Change 5:818722. https://doi.org/10.3389/ffgc.2022.818722

Cinner JE, Aswani S (2007) Integrating customary management into marine conservation. Biol Cons 140(3–4):201–216. https://doi.org/10.1016/j.biocon.2007.08.008

Cohen-Shacham E, Walters G, Janzen C, Maginnis S (eds) (2016) Nature-based solutions to address global societal challenges. IUCN, Gland, Switzerland, xiii + 97pp

Corgo J, Cruz SS, Conceição P (2024) Nature-based solutions in spatial planning and policies for climate change adaptation: a literature review. Ambio 53:1599–1617. https://doi.org/10.1007/s13280-024-02052-1

Covele AA, Van Niekerk D, Cilliers D (2024) Statutory and policy-based eco-disaster risk reduction in SADC member states. Jàmbá: J Disaster Risk Stud 16(2):a1799. https://doi.org/10.4102/jamba.v16i2.1799

D'Souza N, Ishwar NM (2012) Coastal sustainability: learnings from MFF (India) projects, IUCN India Country Office, New Delhi, 32pp

Daniel D, Ifejika Speranza C (2020) The role of blockchain in documenting land users' rights: the canonical case of farmers in the vernacular land market. Front Blockchain 3:509605. https://doi.org/10.3389/fbloc.2020.00019

Datta S (2016) Continuing the forest conservation debate: rhetoric and reality of green India mission. Econ Pol Wkly 51:49–55

De Oliveira M, Morrison T, O'Brien KR, Lovelock CE (2024) Governance of coastal wetlands: beyond the community conservation paradigm. Ocean Coast Manag 255:107253. https://doi.org/10.1016/j.ocecoaman.2024.107253

DeWit W, Burns ET, Guinchard J-C, Pardo MF (2022) Working with nature to protect people: how nature-based solutions reduce climate change and weather-related disasters. Published in June 2022 by IFRC—International Federation of Red Cross and Red Crescent Societies and WWF—World Wide Fund For Nature, 43pp

Dey T, Rajib MGM, Islam MA (2021) People's perceptions about the socio-economic and environmental impact of coastal green belt in Bangladesh. Indonesian J Soc Environ Issues (IJSEI) 2(2):155–162. https://doi.org/10.47540/ijsei.v2i2.280

Dunlop T, Khojasteh D, Glamore W, Haghani M, Rizzi D, Greve P, Felder S (2024) The evolution and future of research on nature-based Solutions to address societal challenges. Commun Earth Environ 5(1):132. https://doi.org/10.1038/s43247-024-01308-8

Duong NT, De Groot WT (2020) The impact of payment for forest environmental services (PFES) on community-level forest management in Vietnam. Forest Policy Econ 113:102135. https://doi.org/10.1016/j.forpol.2020.102135

Ebeler L, Balke T, Laurie E, Le H, Nguyen T, Renaud FG (2025) Local-scale impacts of mangrove restoration and conservation on coastal communities in two Vietnamese deltas: socio-economic

and institutional dynamics. Ocean Coast Manag 269:107812. https://doi.org/10.1016/j.ocecoaman.2025.107812
Ekaratne K, Vidanage SP (2013) Small grants, large gains: lessons from MFF small grant facility projects in Sri Lanka (2011–2013). IUCN Sri Lanka Country Office, Colombo, viii + 115pp
Ekardt F, Günther P, Hagemann K, Garske B, Heyl K, Weyland R (2023) Legally binding and ambitious biodiversity protection under the CBD, the global biodiversity framework, and human rights law. Environ Sci Eur 35(1):80. https://doi.org/10.1186/s12302-023-00786-5
Elrick-Barr CE, Thomsen DC, Smith TF (2024) Governance innovations in the coastal zone: towards social-ecological resilience. Environ Sci Policy 153:103687. https://doi.org/10.1016/j.envsci.2024.103687
EU (2025) EU and ASEAN participate in mangroves restoration effort for a sustainable future. Delegation of the European Union to the Association of Southeast Asian Nations (ASEAN). https://www.eeas.europa.eu/EU-ASEAN_Mangrove_en. Accessed on 09 Jan 2026
Fairbairn TI (1991) Customary marine tenure in the South Pacific region and implications for giant clam mariculture. Research Reports and Papers in Economics of Giant Clam Mariculture 206541, University of Queensland, School of Economics. https://doi.org/10.22004/ag.econ.206541
Farahmand S, Hilmi N, Duarte CM (2025) The rise and flows of blue carbon credits advance global climate and biodiversity goals. Npj Ocean Sustain 4(1):39. https://doi.org/10.1038/s44183-025-00141-6
Firdasari MF (2022) Multi-level governance in practice: investigating mangrove forest governance in Indonesia. Doctoral dissertation, University of Birmingham, United Kingdom
Foggia C, Attorre F, Bourgeois C, Candeago E, Malatesta L (2025) Shields of the shore: mangrove ecosystem shifts and climate vulnerability in Mozambique. Front for Glob Change 8:1648754. https://doi.org/10.3389/ffgc.2025.1648754
Fu C, Steckbauer A, Mann H, Duarte CM (2024) Achieving the Kunming-Montreal global biodiversity targets for blue carbon ecosystems. Nat Rev Earth Environ 5(7):538–552. https://doi.org/10.1038/s43017-024-00566-6
Gnansounou SC, Sagoe AA, Mattah PAD, Salako KV, Aheto DW, Glèlè Kakaï R (2022) The co-management approach has positive impacts on mangrove conservation: evidence from the mono transboundary biosphere reserve (Togo-Benin), West Africa. Wetlands Ecol Manage 30(6):1245–1259. https://doi.org/10.1007/s11273-022-09894-0
Govan H, TawakeA, Tabunakawai K, Jenkins A, Lasgorceix A, Techera E, Tafea H, Kinch J, Feehely J, Ifopo P, Hills R, Alefaio S, Meo S, Troniak S, Malimali, George S, Tauaefa T, Obed T (2009) Community conserved areas: a review of status & needs in Melanesia and Polynesia. ICCA regional review for CENESTA/TILCEPA/TGER/IUCN/GEF-SGP
Government of India (2008) National action plan on climate change. Prime Minister's Council on Climate Change, 56pp
Griggs G, Reguero BG (2020) Coastal adaptation to climate change and sea-level rise. Water 13(16):2151. https://doi.org/10.3390/w13162151
Grimm KE, Archibald JL, Axelsson EP, Grady KC (2023) Follow the money: understanding the Latin America and Caribbean mangrove restoration funding landscape to assist organizations and funders in improved social-ecological outcomes. Conserv Sci Pract 5(5):e12815. https://doi.org/10.1111/csp2.12815
Gupta N, Pradhan S, Jain A, Patel N (2021) Sustainable agriculture in India 2021: what we know and how to scale up. Council on Energy, Environment and Water, New Delhi
Hai N, Dell B, Phuong V, Harper RJ (2020) Towards a more robust approach for the restoration of mangroves in Vietnam. Ann for Sci 77:18. https://doi.org/10.1007/s13595-020-0921-0
Hermoso V, Carvalho S, Giakoumi S, Goldsborough D, Katsanevakis S, Leontiou S, Markantonatou V, Rumes B, Vogiatzakis I, Yates K (2022) The EU biodiversity strategy for 2030: opportunities and challenges on the path towards biodiversity recovery. Environ Sci Policy 127:263–271. https://doi.org/10.1016/j.envsci.2021.10.028

Independent Evaluation Unit (2025). Independent evaluation of the GCF's result area "Health and Wellbeing, and Food and Water Security". Country case study report: Fiji (January). Independent Evaluation Unit, Green Climate Fund, Songdo, South Korea

Islam A, Shaw R, Mallick F (2013) Bangladesh climate change strategy and action plans. In: Climate change adaptation actions in Bangladesh. Springer Japan, Tokyo, pp 107–118. https://doi.org/10.1007/978-4-431-54249-0_7

Islam SA, Rahman MM (2015) Coastal afforestation in Bangladesh to combat climate change induced hazards. J Sci Technol Environ Inform 2(1):13–25

IUCN (2020) Global standard for nature-based solutions. A user-friendly framework for the verification, design and scaling up of NbS, 1st edn. IUCN, Gland, Switzerland

IUCN (1981) The environmental law of the sea. In: Johnston DM (ed). International Union for Conservation of Nature and Natural Resources. Gland, Switzerland, 420pp

Ivanova M (2020) Implementation of multilateral environmental agreements: rationale and design of the environmental conventions index. Sustainability 12(17):7098. https://doi.org/10.3390/su12177098

Iza A (ed) (2021) Governance for ecosystem-based adaptation. IUCN Environmental Policy and Law Paper, No. 89. IUCN, Bonn, Germany

Iza A, Cordoba-Munoz R (2023) Governance for ecosystem-based adaptation. CLIMA—The evaluation tool. IUCN, Gland, Switzerland

Jashimuddin M, Inoue M (2012) Community forestry for sustainable forest management: experiences from Bangladesh and policy recommendations. Formath 11:133–166

Jha S (2012) The Green India Mission (GIM): a roadmap for neoliberal exploitation in forest. Indian J Polit Sci 73:385–398

Johnson BA, Kumar P, Okano N, Dasgupta R, Shivakoti BR (2022) Nature-based solutions for climate change adaptation: a systematic review of systematic reviews. Nature-Based Solutions 2:100042. https://doi.org/10.1016/j.nbsj.2022.100042

Juffe-Bignoli D, Harrison I, Butchart SH, Flitcroft R, Hermoso V, Jonas H, Lukasiewicz A, Thieme M, Turak E, Bingham H, Dalton J, Darwall W, Deguignet M, Dudley N, Gardner R, Higgins J, Kumar R, Linke S, Milton GR, Smith KG (2016) Achieving Aichi biodiversity target 11 to improve the performance of protected areas and conserve freshwater biodiversity. Aquat Conserv Mar Freshwat Ecosyst 26:133–151. https://doi.org/10.1002/aqc.2638

Kampa E, Elkina E, Bueb B, Otero Villanueva MD (2025) Restoring European coastal wetlands for climate and biodiversity: do EU policies and international agreements support restoration? Sustainability 17(21):9469. https://doi.org/10.3390/su17219469

Karani P, Failler P, Gilau AM, Ndende M, Diop ST (2022) Africa blue economy strategies integrated in planning to achieve sustainable development at national and regional economic communities (RECs). J Sustain Res e220011. https://doi.org/10.20900/jsr20220011

Khan MFA, Giessen L (2021) Exceptional bureaucratic rivalry in mangrove forest policy: explanations from the Sundarbans, Bangladesh. Ocean Coast Manag 203:105510. https://doi.org/10.1016/j.ocecoaman.2020.105510

Larson AM, Soto F (2008) Decentralization of natural resource governance regimes. Annu Rev Environ Resour 33:213–239. https://doi.org/10.1146/annurev.environ.33.020607.095522

Lasco RD, Pulhin JM (2006) Environmental impacts of community-based forest management in the Philippines. Int J Environ Sustain Develop 5(1):46–56. https://doi.org/10.1504/IJESD.2006.008682

Le Gouvello R, Herr D, Spadone A, Simard F, Brugere C (2023) The IUCN global standard for nature-based Solutions™ as a tool for enhancing the sustainable development of marine aquaculture. Front Mar Sci 10:1146637. https://doi.org/10.3389/fmars.2023.1146637

Lecerf M, Herr D, Elverum C, Delrieu E, Picourt L (2023) Coastal and marine ecosystems as nature based solutions in new or updated Nationally Determined Contributions, Ocean & Climate Platform, Conservation International, IUCN, Rare, The Nature Conservancy, Wetlands International and WWF

Lo V (2016) Synthesis report on experiences with ecosystem-based approaches to climate change adaptation and disaster risk reduction. Technical Series No.85. Secretariat of the Convention on Biological Diversity, Montreal, p 106

Losada IJ, Beck M, Menendez P, Espejo A, Torres S, Diaz-Simal P, Fernandez F, Abad S, Ripoll N, Garcia J, Narayan S, Trespalacios D, Quiroz A (2017) Valuing protective services of mangroves in the Philippines: technical report. World Bank. http://hdl.handle.net/10986/27666. License: CC BY 3.0 IGO

Love MW (2021) Tabu, MPA and community: nomenclature and the political ecology of marine conservation in Vanuatu. J Polit Ecol 28(1):91–110. https://doi.org/10.2458/jpe.2301

Lucatello S, Alcántara-Ayala I (2023) Addressing the interplay of the Sendai Framework with sustainable development goals in Latin America and the Caribbean: moving forward or going backwards? Disaster Prevent Manage Int J 32(1):206–233. https://doi.org/10.1108/DPM-07-2022-0152

Majumdar MS, Matsui K (2025) Understanding the contribution of the green climate fund (GCF) in Mangrove Forest conservation: a case study on Sundarbans Mangrove Forest, Bangladesh. Sustainability 17(8):3583. https://doi.org/10.3390/su17083583

Maliao RJ, Cahilig RC, Cahilig RR, Jaspe BT (2025) Climate change, traditional ecological knowledge, and riverine biodiversity conservation: a case in Aklan, Central Philippines. Environ Dev Sustain 27(2):4745–4767. https://doi.org/10.1007/s10668-023-04096-x

March A, Bennett M, Germishuizen M, Evans T, Failler P (2024a) The status of blue economy development in Africa. Mar Policy 165:106205. https://doi.org/10.1016/j.marpol.2024.106205

March A, Evans T, Laing S, Raguain J (2024b) Evaluating the world's first sovereign blue bond: lessons for operationalising blue finance. Commodities 3(2):151–167. https://doi.org/10.3390/commodities3020010

Martin JG, Scolobig A, Linnerooth-Bayer J, Irshaid J, Aguilera Rodriguez JJ, Fresolone-Caparrós A, Oen A (2025) The nature-based solution implementation gap: a review of nature-based solution governance barriers and enablers. J Environ Manage 388:126007. https://doi.org/10.1016/j.jenvman.2025.126007

McElwee P, Thi Nguyen VH, Nguyen DV, Tran NH, Le HV, Nghiem TP, Thi Vu HD (2016) Using REDD+ policy to facilitate climate adaptation at the local level: synergies and challenges in Vietnam. Forests 8(1):11. https://doi.org/10.3390/f8010011

Meraj G, Hashimoto S (2025) Bridging the adaptation finance gap: the role of nature-based solutions for climate resilience. Sustain Sci 20:1093–1107. https://doi.org/10.1007/s11625-025-01655-1

MFF Viet Nam (2015) Mangroves for the future phase III—national strategic action plan (2015–2018). IUCN, Gland, Switzerland, p 44

Miller MA, Tonoto P (2023) Leveraging plural valuations of mangroves for climate interventions in Indonesia. Sustain Sci 18(3):1533–1547. https://doi.org/10.1007/s11625-023-01297-1

Mohanty A, Dubey A, Singh RB (2022) Cyclonic disasters and resilience. Springer, Singapore, 211pp. https://doi.org/10.1007/978-981-19-1215-3

Mursyid H, Daulay MH, Pratama AA, Laraswati D, Novita N, Malik A, Maryudi A (2021) Governance issues related to the management and conservation of mangrove ecosystems to support climate change mitigation actions in Indonesia. Forest Policy Econ 133:102622. https://doi.org/10.1016/j.forpol.2021.102622

Nguyen TA, Huynh LHT, Pannier E, Nguyen A, Tessier O, et al (2021) Climate change adaptation policies in Viet Nam: from national perspective to local practices. Climate change in Viet Nam, impacts and adaptation: a COP26 assessment report of the GEMMES Viet Nam project, pp 501–532. ffhal-03987275f

Nilsson AK, Bohman B (2015) Legal prerequisites for ecosystem-based management in the Baltic Sea area: the example of eutrophication. Ambio 44:370–380. https://doi.org/10.1007/s13280-015-0656-6

Oanh PT, Tamura M, Kumano N, Nguyen QV (2020) Cost-benefit analysis of mixing gray and green infrastructures to adapt to sea level rise in the Vietnamese Mekong River Delta. Sustainability 12(24):10356. https://doi.org/10.3390/su122410356

OECD (2020) Common ground between the Paris agreement and the Sendai framework: climate change adaptation and disaster risk reduction. OECD Publishing, Paris. https://doi.org/10.1787/3edc8d09-en

OECD (2021) Delivering quality education and health care to all: preparing regions for demographic change, OECD Rural Studies, OECD Publishing, Paris. https://doi.org/10.1787/83025c02-en

Onrizal O, Amelia R, Amri K, Sulistiyono N, Mansor M (2019) Stand structure and diversity of restored mangroves at abandoned pond in eastern coast of North Sumatra. IOP Conf Ser: Earth Environ Sci 305:012050. https://doi.org/10.1088/1755-1315/305/1/012050

Orchard SE, Stringer LC, Quinn CH (2015) Environmental entitlements: institutional influence on mangrove social-ecological systems in Northern Vietnam. Resources 4(4):903–938. https://doi.org/10.3390/resources4040903

Pakseresht A, Kermani A, Decker-Lange C (2025) Towards a sustainable and circular blue bioeconomy: a scoping review. Technol Forecast Soc Chang 216:124157. https://doi.org/10.1016/j.techfore.2025.124157

Paris Agreement (2015) Report of the conference of the parties to the United Nations framework convention on climate change. https://unfccc.int/files/essential_background/convention/application/pdf/english_paris_agreement.pdf. Accessed 2 Feb 2025

Pascal N, Brathwaite A, Bladon A, Claudet J, Clua E (2021) Impact investment in marine conservation. Ecosyst Serv 48:101248. https://doi.org/10.1016/j.ecoser.2021.101248

Pearson L, Pelling M (2015) The UN Sendai framework for disaster risk reduction 2015–2030: negotiation process and prospects for science and practice. J Extreme Events 2(01):1571001. https://doi.org/10.1142/S2345737615710013

Pham TT, Vu TP, Hoang TL, Dao TL, Nguyen DT, Pham DC, Dao LH, Nguyen VT, Hoang NV (2022) The effectiveness of financial incentives for addressing mangrove loss in Northern Vietnam. Front for Glob Change 4:709073. https://doi.org/10.3389/ffgc.2021.709073

Pham TT, Vu TP, Pham DC, Dao LHT, Nguyen VT, Hoang NVH, Hoang TL, Dao TLC, Nguyen DT (2019) Opportunities and challenges for mangrove management in Vietnam: lessons learned from Thai Binh, Quang Ninh and Thanh Hoa provinces. Occasional Paper 197. CIFOR, Bogor, Indonesia

Pillodar F, Suson P, Aguilos M, Amparado R (2023) Mangrove resource mapping using remote sensing in the philippines: a systematic review and meta-analysis. Forests 14(6):1080. https://doi.org/10.3390/f14061080

Pittman J, Armitage D (2016) Governance across the land-sea interface: a systematic review. Environ Sci Policy 64:9–17. https://doi.org/10.1016/j.envsci.2016.05.022

Poti M, Singh AR, Shanker K, Koedam N, Prabakaran N, Dahdouh-Guebas F, Hugé J (2025) Mangrove governance, its challenges, and responses to the 2004 earthquake and tsunami in the Andaman islands, India: a stakeholder perspective. Ocean Coast Manag 269:107786. https://doi.org/10.1016/j.ocecoaman.2025.107786

Powell EJ, Tyrrell MC, Milliken A, Tirpak JM, Staudinger MD (2019) A review of coastal management approaches to support the integration of ecological and human community planning for climate change. J Coast Conserv 23:1–18. https://doi.org/10.1007/s11852-018-0632-y

Quevedo JMD, Miller MA, Gevaña DT, Marks D, Friess DA, Tonoto P, Taylor D (2025) Perceptions of coastal vegetated ecosystems: a systematic review across geographical and sectoral dimensions. Ambio. https://doi.org/10.1007/s13280-025-02193-x

Rahman HM, Sherren K, Van Proosdij D (2018) Institutional innovation for nature-based coastal adaptation: lessons from salt marsh restoration in Nova Scotia, Canada. Sustainability 11(23):6735. https://doi.org/10.3390/su11236735

Ramsar (2016) An introduction to the Ramsar convention on wetlands, 7th ed. (previously The Ramsar Convention Manual). Ramsar Convention Secretariat, Gland, Switzerland

Raymond CM, Berry P, Breil M, Nita MR, Kabisch N, de Bel M, Enzi V, Frantzeskaki N, Geneletti D, Cardinaletti M, Lovinger L, Basnou C, Monteiro A, Robrecht H, Sgrigna G, Munari L, Calfapietra C (2017) An impact evaluation framework to support planning and evaluation of nature-based solutions projects. Report prepared by the EKLIPSE expert working group on nature-based solutions to promote climate resilience in urban areas. Centre for Ecology & Hydrology, Wallingford, United Kingdom

Ribot JC, Agrawal A, Larson AM (2006) Recentralizing while decentralizing: how national governments reappropriate forest resources. World Dev 34(11):1864–1886. https://doi.org/10.1016/j.worlddev.2005.11.020

Rochette J, Billé R, Molenaar EJ, Drankier P, Chabason L (2015) Regional oceans governance mechanisms: a review. Mar Policy 60:9–19. https://doi.org/10.1016/j.marpol.2015.05.012

Rölfer L, Celliers L, Fernandes M, Rivers N, Snow B, Abson DJ (2024) Assessing collaboration, knowledge exchange, and stakeholder agency in coastal governance to enhance climate resilience. Reg Environ Change 24(1):6. https://doi.org/10.1007/s10113-023-02163-7

Romañach SS, DeAngelis DL, Koh HL, Li Y, Teh SY, Raja Barizan RS, Zhai L (2018) Conservation and restoration of mangroves: global status, perspectives, and prognosis. Ocean Coast Manag 154:72–82. https://doi.org/10.1016/j.ocecoaman.2018.01.009

Rosales RJC, Libranza LR, Roble JFZ, Jucdong FMB, Jumawan MC, Empeno S, Almagro JJN (2025) Economic valuation of a protected mangrove forest in Southern Cebu, Philippines. Biodiversitas J Biol Divers 26(5)

Rosenberg J (2020) Adaptation, official development assistance, and institution building: the case of the Caribbean community climate change centre. Sustainability 12(10):4269. https://doi.org/10.3390/su12104269

Rotich B, Mwangi E, Lawry S (2016) Where land meets the sea: a global review of the governance and tenure dimensions of coastal mangrove forests. CIFOR and USAID Tenure and Global Climate Change Program: Bogor, Indonesia and Washington, DC, USA

SAARC (2008) SAARC environment ministers Dhaka declaration on climate change. Dhaka, 3 July 2008

Sakthivel M, Khan N (2024) Protection of the Indian coastal ecosystem through coastal regulation zone (CRZ) notifications: an analysis. In: The Routledge companion to Indian ethics. Routledge India, pp 211–222

Sasmito SD, Basyuni M, Kridalaksana A, Saragi-Sasmito MF, Lovelock CE, Murdiyarso D (2023) Challenges and opportunities for achieving sustainable development goals through restoration of Indonesia's mangroves. Nat Ecol Evol 7(1):62–70. https://doi.org/10.1038/s41559-022-01926-5

Seddon N, Daniels E, Davis R, Chausson A, Harris R, Hou-Jones X, Huq S, Kapos V, Mace GM, Rizvi AR, Reid H, Roe D, Turner B, Wicander S (2020) Global recognition of the importance of nature-based solutions to the impacts of climate change. Glob Sustain 3:e15. https://doi.org/10.1017/sus.2020.8

Selvam V, Thamizoli P (2021) Science-based and community-centred approach to restore and sustain mangrove wetlands of India. Curr Sci 121(10):1288–1296

Sever SD, Tok E, Sellami AL (2025) Sustainable development goals in a transforming world: understanding the dynamics of localization. Sustainability 17(6):2763. https://doi.org/10.3390/su17062763

Shen X, Liu M, Hanson JO, Wang J, Locke H, Watson JE, Ma K (2023) Countries' differentiated responsibilities to fulfill area-based conservation targets of the Kunming-Montreal Global Biodiversity Framework. One Earth 6(5):548–559. https://doi.org/10.1016/j.oneear.2023.04.007

Sidik F, Lawrence A, Wagey T, Zamzani F, Lovelock CE (2022) Blue carbon: a new paradigm of mangrove conservation and management in Indonesia. Mar Policy 147:105388. https://doi.org/10.1016/j.marpol.2022.105388

Song AM, Dressler WH, Satizábal P, Fabinyi M (2021) From conversion to conservation to carbon: the changing policy discourse on mangrove governance and use in the Philippines. J Rural Stud 82:184–195. https://doi.org/10.1016/j.jrurstud.2021.01.008

Sudirman A, Idris Rabbani YP, Maulana M, Djuyandi Y (2023) The need for ASEAN environmental maritime security cooperations. PERSPEKTIF 12(3):1043–1053. https://doi.org/10.31289/perspektif.v12i3.9226

Thomas KV (2025) Coastal regulation zone: a tool for coastal zone management. In: Chattopadhyay S (ed) Marine and coastal resources of India. Sustainable development goals series. Springer, Cham. https://doi.org/10.1007/978-3-031-81276-7_30

Toledo I, Laino E, Iglesias G, Palazón A, Aragonés L (2025) Local authorities or national frameworks? A global review on coastal protection policies. Environ Develop 53:101119. https://doi.org/10.1016/j.envdev.2024.101119

Trouwborst A, Blackmore A, Boitani L, Bowman M, Caddell R, Chapron G, Cliquet A, Couzens E, Epstein Y, Fernández-Galiano E, Fleurke FM, Gardner R, Hunter L, Jacobsen K, Krofel M, Lewis M, López-Bao JV, Macdonald D, Redpath S, C Linnell JD (2017) International wildlife law: understanding and enhancing its role in conservation. Bioscience 67(9):784–790. https://doi.org/10.1093/biosci/bix086

Ty PH, Tan NQ, Dinh NC, Nhi TTP, Tien TTQ, Cuong LCH (2023) Uncovering gender disparities in payment for forest environmental services (PFES): a feminist political ecology view from Vietnam. Environ Sustain Indicators 19:100281. https://doi.org/10.1016/j.indic.2023.100281

UN (2011) Convention on biological diversity in the Nagoya protocol on access to genetic resources and the fair and equitable sharing of benefits arising from their utilization. Nagoya Protocol, Nagoya, Japan

UNDP (2019) Pathway for increasing nature-based solutions in NDCs: a seven-step approach for enhancing nationally determined contributions through nature-based solutions. UNDP, New York, USA, 20pp

UNDRR (2020) Ecosystem-based disaster risk reduction: implementing nature-based solutions for resilience, United Nations office for disaster risk reduction—regional office for Asia and the Pacific, Bangkok, Thailand

Vasseur L (2020) How ecosystem-based adaptation to climate change can help coastal communities through a participatory approach. Sustainability 13(4):2344. https://doi.org/10.3390/su13042344

Veettil BK, Pereira SFR, Quang NX (2018) Rapidly diminishing mangrove forests in Myanmar (Burma): a review. Hydrobiologia 822:19–35. https://doi.org/10.1007/s10750-018-3673-1

Veettil BK, Ward RD, Quang NX, Trang NTT, Giang TH (2019) Mangroves of Vietnam: historical development, current state of research and future threats. Estuar Coast Shelf Sci 218:212–236. https://doi.org/10.1016/j.ecss.2018.12.021

Veettil BK, Van DD, Quang NX, Hoai PN (2020) Spatiotemporal dynamics of mangrove forests in the Andaman and Nicobar Islands (India). Reg Stud Mar Sci 39:101455. https://doi.org/10.1016/j.rsma.2020.101455

Veettil BK, Ward RD, Dung NTK, Van DD, Quang NX, Hoai PN, Hoang N (2021) The use of bioshields for coastal protection in Vietnam: current status and potential. Reg Stud Mar Sci 47:101945. https://doi.org/10.1016/j.rsma.2021.101945

Veettil BK, Costi J, Quang NX, Van Nam T, Van DD, Hoai PN (2022) Anthropogenic influences on coastal environmental changes in the Mekong Delta: a study from Ben Tre Province, Southern Vietnam. Environ Monitor Assess 194(Suppl 2):773. https://doi.org/10.1007/s10661-022-10177-7

Velástegui-Montoya A, Brito-Matamoros R, Jaya-Montalvo M, Caiza-Morales L, Adami M, Morante-Carballo F, Carrión-Mero P (2025) Land use change and mangrove conservation strategies in the Gulf of Guayaquil Ecuador through spatial and Delphi based analysis. Discover Appl Sci 7(10):1121. https://doi.org/10.1007/s42452-025-07761-0

Voorberg W (2020) Co-management as a successful strategy for marine conservation. J Mar Sci Eng 8(7):491. https://doi.org/10.3390/jmse8070491

Vu TP, Pham, TT, Tran NMH, Nguyen TTA, Nguyen TVA, Tang TKH (2023) The blue carbon market in Vietnam: future development, potential and challenges. Occasional Paper 8. Center for International Forestry Research (CIFOR), Bogor, Indonesia; and World Agroforestry (ICRAF), Nairobi, Kenya

Walker JE, Ankersen T, Barchiesi S, Meyer CK, Altieri AH, Osborne TZ, Angelini C (2022) Governance and the mangrove commons: advancing the cross-scale, nested framework for the

global conservation and wise use of mangroves. J Environ Manage 312:114823. https://doi.org/10.1016/j.jenvman.2022.114823
Watts JD, Tacconi L, Irawan S, Wijaya AH (2019) Village transfers for the environment: lessons from community-based development programs and the village fund. Forest Policy Econ 108:101863. https://doi.org/10.1016/j.forpol.2019.01.008
Yi H, Huang C, Chen T, Xu X, Liu W (2018) Multilevel environmental governance: vertical and horizontal influences in local policy networks. Sustainability 11(8):2390. https://doi.org/10.3390/su11082390
Yuza AF, Manaf HA, Zainal Z (2023) Deforestation policy governance in Indonesia. CosmoGov: Jurnal Ilmu Pemerintahan 9(1):36–49. https://doi.org/10.24198/cosmogov.v9i1.44778
Zhang S, Butt MJ, Zulfiqar K (2023) Interconnected principles of ocean governance through science-policy integration under the judgements of the international dispute settlement bodies. Sci Prog 106(4). https://doi.org/10.1177/00368504231205395

Chapter 11
Community Participation and Traditional Knowledge in Coastal Bioshields

Abstract Community participation and traditional ecological knowledge are fundamental pillars for the successful establishment, management, and sustainability of coastal bioshields. This chapter explores the social and cultural dimensions of coastal protection by examining how local knowledge systems, customary practices, and participatory governance contribute to effective bioshield management. Drawing upon examples from Asia, Africa, and the Pacific, it highlights the role of indigenous and local communities in conserving mangroves, dune vegetation, and other coastal ecosystems through time-tested practices such as community forestry, rotational harvesting, and sacred site protection. The chapter analyzes participatory models that integrate traditional wisdom with modern ecological and engineering approaches, demonstrating how co-management frameworks enhance local ownership and long-term stewardship. It further addresses the challenges of balancing scientific knowledge with cultural values, particularly in contexts of rapid social and environmental change. Attention is given to gender inclusion, capacity building, and social equity as essential components of participatory bioshield governance. By emphasizing the complementarity between community-led initiatives and state or donor-supported programs, the chapter illustrates that sustainable coastal protection is most effective when rooted in local empowerment and social cohesion. Ultimately, it argues that recognizing and institutionalizing traditional knowledge within formal coastal management strategies not only strengthens ecological resilience but also reinforces the cultural integrity and adaptive capacity of coastal societies.

Keywords Community-based management · Indigenous practices · Local stewardship · Participatory governance · Social equity and inclusion · Traditional ecological knowledge (TEK)

B. Kozhikkodan Veettil, *The Potential of Bioshields for Coastal Protection*, Advances in Natural and Technological Hazards Research 55,
https://doi.org/10.1007/978-3-032-22888-8_11

11.1 Introduction

Community participation and traditional knowledge play a vital role in the sustainability and long-term success of bioshield projects for coastal protection (Loch and Riechers 2021; Stori et al. 2019). While scientific and engineering innovations provide technical frameworks for coastal bioshield design, it is the integration of local practices, indigenous wisdom, and community stewardship that ensures these systems remain resilient, adaptive, and contextually relevant (Nalau et al. 2018). In many coastal regions across the world, particularly in Asia, Africa, and the Pacific Islands, communities have long relied on traditional methods for managing coastal ecosystems such as mangroves, dunes, and salt marshes to mitigate erosion, enhance fisheries, and buffer storm impacts (Datta et al. 2012; Dorji et al. 2024; Siddique et al. 2024; Quevedo et al. 2025). These practices, grounded in centuries of experiential knowledge, are now recognized as essential to modern bioshield management and climate adaptation strategies (Giffin et al. 2020; Dorji et al. 2024).

The increasing frequency of extreme weather events and sea-level rise due to climate change has amplified the need for community-centered coastal protection approaches (Chen and Cheng 2020; Cooley et al. 2022). Global institutions, such as the Intergovernmental Panel on Climate Change (IPCC) and the United Nations Environment Programme (UNEP), emphasize the significance of locally led adaptation (LLA) strategies that prioritize community participation and incorporate traditional ecological knowledge (TEK) (Schipper et al. 2022). In many cases, government-driven or externally funded bioshield projects fail when they overlook local ecological knowledge, ownership, and socio-cultural dynamics (Arkhurst et al. 2023; Dorji et al. 2024; Adjaison et al. 2025; Lopes et al. 2025). Therefore, this chapter explores how communities and traditional knowledge systems contribute to the successful design, establishment, and maintenance of bioshields across different regions.

11.2 The Role of Communities in Coastal Protection

Community participation refers to the active engagement of local populations in planning, implementing, and monitoring bioshield initiatives. The rationale behind such involvement is multifaceted. First, communities possess a deep understanding of their local environment, including knowledge of tidal flows, sediment dynamics, plant species behavior, and historical shoreline changes (Amone-Mabuto et al. 2023; Risley et al. 2025). This information often complements scientific assessments and improves the contextual accuracy of bioshield design (Baustian et al. 2019). Second, communities act as custodians of these systems (Datta et al. 2012; Sathiyamoorthy and Sakurai 2024). When people are involved in bioshield projects, they develop a sense of ownership that encourages long-term maintenance and protection against encroachment or unsustainable exploitation (Sathiyamoorthy and Sakurai 2024; Siddiqui and Ishtiaque 2025).

In practice, community participation can occur at several levels, including consultative, collaborative, and empowered levels (Puskás et al. 2021; Hurlbert and Gupta 2024). At the consultative level, communities provide input into decisions made by external agencies or governments (Drake et al. 2021). At the collaborative level, they co-design and co-manage bioshield systems, sharing responsibilities and benefits (Sukuryadi et al. 2021; Asio et al. 2025). Finally, at the empowered level, communities take the lead in bioshield governance, supported by enabling policies and institutional frameworks (Damastuti et al. 2022; Dushkova and Ivlieva 2023; Sathiyamoorthy and Sakurai 2024). Empowered participation is often associated with the highest degree of sustainability, as seen in cases where local cooperatives, fisher associations, or indigenous councils have led mangrove restoration or dune stabilization efforts (Valenzuela et al. 2020; De Oliveira et al. 2024; Sathiyamoorthy and Sakurai 2024; Quevedo et al. 2025).

Community-driven bioshield initiatives also generate multiple co-benefits. These include livelihood diversification through eco-tourism and sustainable aquaculture, improved disaster preparedness, and strengthened social cohesion (Lin 2019; Moussa et al. 2024; Dulyakasem et al. 2026). For example, post-tsunami reconstruction programs in South and Southeast Asia revealed that communities with pre-existing local resource management groups were more capable of re-establishing mangrove barriers and restoring coastal vegetation compared to those reliant solely on top-down interventions (Kathiresan and Rajendran 2005).

Furthermore, integrating gender perspectives in community participation enhances project outcomes (Pearson et al. 2019; Jang et al. 2025). Women often play significant roles in mangrove seedling collection, nursery management, and environmental education, making their inclusion critical to holistic coastal protection (Pearson et al. 2019; Salampessy et al. 2025). Therefore, inclusive participation across gender, age, and occupational lines strengthens bioshield resilience and adaptability (Harper et al. 2024).

11.3 Traditional Ecological Knowledge in Coastal Bioshield Design and Management

Traditional Ecological Knowledge (TEK) refers to the cumulative body of knowledge, practices, and beliefs developed by communities through long-term interaction with their environment (Stori et al. 2019; Haq et al. 2023). In the context of coastal bioshields, TEK encompasses traditional planting methods, indigenous resource management systems, and cultural taboos that regulate coastal ecosystem use (Stori et al. 2019). Many of these traditional practices align remarkably well with modern ecological principles (Gómez-Baggethun et al. 2013; Stori et al. 2019).

For instance, indigenous communities in the Philippines, Vietnam, and Indonesia have long recognized the protective functions of mangroves and nipa palms against coastal erosion and storm surges (Veettil et al. 2019; Sofue et al. 2025). Local planting techniques often consider tidal regimes, soil salinity, and species zonation—factors that scientific projects sometimes overlook (Marquez and Olavides 2024; Sofue et al. 2025; Sunarto and Mamangkey 2025). In Vietnam's Mekong Delta, villagers traditionally planted *Avicennia marina* and *Rhizophora apiculata* in staggered patterns to dissipate wave energy and enhance sediment accretion (Veettil et al. 2019). These practices were later validated through remote sensing and hydrodynamic modeling, underscoring the empirical strength of TEK (Tinh et al. 2022).

Traditional knowledge also extends to social governance systems (Cinner and Aswani 2007). In many parts of Asia and Oceania, customary laws—such as the tabu in Fiji (Clarke and Jupiter 2010) or adat (Trialfhianty et al. 2025) in Indonesia—regulate the use of mangrove resources, ensuring regeneration and equitable access (Parsons et al. 2021). These community-imposed restrictions function similarly to modern conservation zoning or seasonal bans (Cinner and Aswani 2007; Parsons et al. 2021). The integration of TEK into bioshield projects not only strengthens ecological outcomes but also reinforces cultural identity and continuity (Gómez-Baggethun et al. 2013).

A critical challenge, however, lies in the documentation and transmission of traditional knowledge (Tran et al. 2025). Modernization, migration, and the erosion of cultural practices threaten the continuity of TEK systems (Sinthumule 2023; Tran et al. 2025). Collaborative programs that pair elders and local youth in bioshield restoration activities have proven effective in preserving and transferring this knowledge (Hosen et al. 2019; Haq et al. 2023). Moreover, partnerships between researchers and communities can formalize TEK into policy frameworks, elevating local practices to national adaptation strategies (Ciocco et al. 2024; Dorji et al. 2024).

11.4 Case Studies from Asia

Asia hosts some of the world's most successful examples of community-driven bioshield initiatives (Datta et al. 2012; Marquez and Olavides 2024). Owing to its vast coastlines, high population density, and exposure to tropical storms, countries such as Thailand (Kongkeaw et al. 2019), Vietnam (Veettil et al. 2021), and Indonesia (Damastuti et al. 2022) have developed innovative models that combine traditional knowledge with scientific restoration (Fig. 11.1). These Asian examples demonstrate that when community participation and traditional knowledge are harmonized with modern science, bioshields not only thrive ecologically but also deliver substantial socio-economic resilience (Gerona-Daga and Salmo 2022; Dulyakasem et al. 2026).

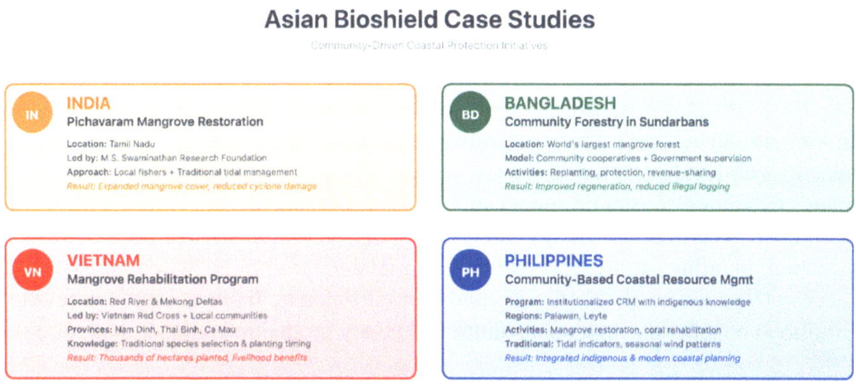

Fig. 11.1 Community driven bioshield case studies from Asia

11.4.1 India—The Pichavaram Mangrove Restoration Project

In Tamil Nadu, the Pichavaram mangrove ecosystem has been managed for decades through local-community involvement, primarily led by the M. S. Swaminathan Research Foundation (MSSRF) (Selvam et al. 2003; Ghosh et al. 2022). The project trained local fishers to plant and manage mangrove seedlings while integrating traditional practices of tidal channel management (Selvam and Thamizoli 2021). As a result, mangrove cover expanded significantly, reducing cyclone-related damages and improving fish yields (Selvam et al. 2003; Ghosh et al. 2022; Singha et al. 2024). The success of Pichavaram demonstrates how participatory restoration grounded in local practices can enhance both ecological and socio-economic resilience (Selvam et al. 2003; Ghosh et al. 2022).

11.4.2 Bangladesh—Community Forestry in the Sundarbans

The Sundarbans, the world's largest mangrove forest, spans across Bangladesh and India. In Bangladesh, community forestry programs empower local cooperatives to manage forest resources under government supervision (Mollick et al. 2022; Begum et al. 2023). Traditional fishing and honey-collecting communities are engaged in mangrove replanting and protection, with revenue-sharing mechanisms ensuring equitable benefits (Smith et al. 2021). This collaborative model has improved forest regeneration rates and reduced illegal logging (Begum et al. 2021, 2023).

11.4.3 Vietnam—Mangrove Rehabilitation in the Red River and Mekong Deltas

In Vietnam, mangrove restoration initiatives led by the Vietnam Red Cross, in partnership with local communities, have successfully planted thousands of hectares of mangroves along coastal provinces such as Nam Dinh, Thai Binh, and Ca Mau (Hai et al. 2020; Veettil et al. 2021; Pham et al. 2022). Traditional knowledge of species selection, planting timing, and community maintenance has been central to project success (Nguyen et al. 2016). Beyond coastal defense, these mangroves provide livelihood benefits through aquaculture and honey production (Ebeler et al. 2025).

11.4.4 The Philippines—Community-Based Coastal Resource Management (CBCRM)

The Philippines has institutionalized community-based management through its Coastal Resource Management (CRM) program, which integrates indigenous ecological knowledge with modern coastal zone planning (Pomeroy and Carlos 1997; Mudge 2018). Villagers participate in mangrove restoration, coral rehabilitation, and coastal clean-up activities (Balilo Jr et al. 2023). In regions such as Palawan and Leyte, local fisherfolk use traditional indicators—like tidal coloration and seasonal winds—to determine planting times for mangrove propagules (Marquez and Olavides 2024).

11.5 Case Studies from Other Regions

While Asia remains a global leader in community-based bioshield implementation, similar success stories exist in Africa, Latin America, and the Pacific. Collectively, these case studies highlight that community-based bioshield success is not confined to specific geographies. Across the globe, traditional ecological knowledge, when respected and integrated with modern conservation science, forms the foundation of effective, enduring coastal protection systems (Stori et al. 2019).

11.5.1 East Africa—Mangrove Management in Kenya and Tanzania

In Kenya, the Mikoko Pamoja project represents a pioneering community-led mangrove conservation initiative that combines traditional stewardship with carbon credit mechanisms (Huff and Tonui 2017; Kairo et al. 2018). Local communities plant and protect mangroves while generating income through the sale of carbon offsets (Huxham et al. 2018, 2023). Similarly, in Tanzania's Rufiji Delta, traditional fishing communities collaborate with NGOs to rehabilitate degraded mangrove stands using locally derived planting methods (Nyangoko et al. 2021, 2022; Ntibona et al. 2023). These projects demonstrate how indigenous practices can align with contemporary carbon finance mechanisms to promote sustainable bioshield management.

11.5.2 West Africa—Senegal's Mangrove Regeneration Programs

Senegal's Casamance region hosts one of the largest mangrove reforestation programs in Africa (Cormier-Salem and Panfili 2016). The project's success lies in community mobilization, traditional leadership involvement, and the revival of indigenous planting ceremonies (Diatta et al. 2020). Between 2006 and 2016, local communities restored over 10,000 hectares of mangroves (Fent et al. 2018), thereby significantly enhancing fish nursery grounds and protecting inland agricultural zones from saltwater intrusion (Gallup et al. 2020).

11.5.3 Latin America—Bioshields in Mexico and Ecuador

In Mexico, coastal communities in the Yucatán Peninsula have implemented mangrove rehabilitation using traditional "chinampa" farming techniques adapted to saline environments (Valderrama-Landeros et al. 2024; Marín-Muñiz 2025). These systems promote soil stabilization and nutrient cycling. In Ecuador, Afro-Ecuadorian communities in Esmeraldas Province protect mangrove forests through customary management rules that restrict overharvesting (Ocampo-Thomason 2006; Beitl et al. 2019; Jerez Columbié 2021). These community regulations function effectively as bioshield governance instruments (Ocampo-Thomason 2006).

11.5.4 Pacific Islands—Customary Marine Tenure Systems

In Fiji, Samoa, and Vanuatu, traditional systems of marine tenure—such as the "tabu areas"—regulate resource use and ensure mangrove regeneration (Clarke and Jupiter 2010; Love 2021). Communities employ indigenous planting techniques and use rituals to mark protected areas. These cultural frameworks maintain ecological balance while reinforcing local identity and governance (Clarke and Jupiter 2010; Love 2021; Kitolelei et al. 2022).

11.6 Integrating Traditional Knowledge with Modern Science

The integration of traditional ecological knowledge (TEK) with modern scientific approaches is central to the success and longevity of bioshield projects (Tengö et al. 2014; Maliao and Tóthmérész 2025). Both knowledge systems offer complementary strengths: while science provides standardized methodologies and predictive models, TEK offers site-specific insights derived from long-term interaction with the environment (Tengö et al. 2014; Haq et al. 2023). Effective integration requires mutual respect, co-learning, and participatory frameworks where communities and scientists work as equal partners rather than hierarchical actors (Stori et al. 2019).

Several models have emerged to facilitate such integration. The co-production of knowledge framework emphasizes joint problem definition, data collection, and interpretation by scientists and local knowledge holders (Tengö et al. 2017). For example, in Indonesia's North Java coast, researchers collaborated with local farmers and fishers to integrate traditional water-flow management techniques with GIS-based hydrodynamic modeling for mangrove rehabilitation (Damastuti et al. 2022; Van Bijsterveldt et al. 2022). This approach improved planting success rates and sediment retention compared to purely scientific designs (Damastuti et al. 2022).

Similarly, in the Philippines, community mapping exercises have merged traditional spatial knowledge—such as historical flood lines and sacred coastal groves—with satellite imagery to prioritize bioshield zones (Marquez and Olavides 2024). Such hybrid mapping approaches bridge the gap between lived experience and technological data, ensuring that bioshield designs reflect both ecological and cultural realities (James 2025).

In Vietnam, participatory monitoring programs for mangrove restoration in the Mekong and Red River deltas combine remote sensing data with local ecological indicators, including leaf coloration, crab population density, and tidal rhythm patterns observed by elders (Loc et al. 2021; Nguyen et al. 2021; Veettil et al. 2021). This dual monitoring approach not only improves ecological accuracy but also strengthens local ownership, ensuring the continuity of maintenance beyond project cycles (Veettil et al. 2021; Ebeler et al. 2025).

Beyond technical integration, blending TEK with science also enhances adaptive management—the ability to modify strategies based on real-time feedback. Traditional societies often rely on oral traditions and seasonal cues to adjust their practices; modern adaptive management borrows this principle by incorporating flexible, iterative learning (Haq et al. 2023; Dorji et al. 2024). When these systems converge, bioshield projects become more resilient to environmental variability and socio-economic change (Stori et al. 2019).

11.7 Policy Frameworks Supporting Community Participation

Effective bioshield implementation requires enabling policy environments that recognize and institutionalize community participation and traditional knowledge (Newing et al. 2024). Globally, several policy frameworks have begun to move in this direction.

At the international level, the Convention on Biological Diversity (CBD) and its Aichi Biodiversity Targets explicitly call for the inclusion of indigenous and local knowledge in ecosystem management (Fajardo et al. 2021). The UN Framework Convention on Climate Change (UNFCCC) further supports locally led adaptation strategies, highlighting that community-based actions can significantly reduce vulnerability to climate change impacts (Ravazzoli et al. 2025). The Sendai Framework for Disaster Risk Reduction (2015–2030) similarly acknowledges the role of indigenous knowledge in strengthening disaster resilience, particularly in coastal areas exposed to storms and tsunamis (Ali et al. 2021).

At the national level, several countries have embedded community participation in their coastal protection strategies. India's National Mangrove and Coral Reef Committee (NMCRC) mandates community participation in mangrove management through joint forest management models (Government of India 2021). Bangladesh's Coastal Greenbelt Project integrates local cooperatives into mangrove planting, using community contracts to ensure equitable benefit-sharing (Dey et al. 2021). Vietnam's Law on Environmental Protection in 2020 (Law No. 72/2020/QH14) emphasizes participatory environmental governance, encouraging local communities to take part in bioshield restoration and climate adaptation planning. The Philippines' Local Government Code (1991) empowers municipalities to engage in community-based coastal resource management, institutionalizing grassroots-level involvement (Ballad et al. 2016).

Beyond national frameworks, regional cooperation platforms—such as the ASEAN Working Group on Coastal and Marine Environment—promote regional knowledge exchange on community-based bioshields (Do et al. 2023). These networks facilitate cross-border learning, particularly among nations sharing similar coastal geomorphologies and climate challenges.

However, despite these positive developments, significant gaps remain in implementation. Many countries lack clear legal recognition of traditional tenure systems or indigenous governance structures (Ojong 2020; Lumosi et al. 2025). Bureaucratic complexities and top-down decision-making often marginalize local voices in coastal planning processes. To bridge these gaps, policy reform should prioritize:

a. Legal recognition of traditional ecological knowledge as a valid input into environmental management (Brondízio et al. 2021).
b. Devolution of authority to local and indigenous communities for bioshield management (Aleu et al. 2022).
c. Institutional frameworks that mandate free, prior, and informed consent (FPIC) in bioshield projects (Larsen and Chanthavisouk 2024).

d. Financial incentives for community stewardship, such as payments for ecosystem services (PES) or community-based carbon credits (Smith et al. 2019; Clifton and Mánez 2025).

When supported by policy, community-led bioshields can move from pilot projects to nationwide coastal resilience programs.

11.8 Challenges and Opportunities

While community participation and traditional knowledge have proven indispensable to bioshield success, several challenges persist that hinder their full potential.

11.8.1 Institutional and Governance Barriers

In many developing nations, overlapping jurisdictions between forestry, fisheries, and environmental agencies lead to fragmented management (Veettil et al. 2019, 2021). Communities often face bureaucratic hurdles in accessing coastal lands or receiving recognition for their stewardship (Botero et al. 2024). Without legal clarity, traditional resource users remain vulnerable to external encroachment, reducing their motivation to engage in long-term bioshield maintenance (Bhattacharjee et al. 2025).

11.8.2 Erosion of Traditional Knowledge Systems

Modernization, migration, and the loss of intergenerational knowledge transfer threaten the survival of TEK (Aswani et al. 2018). Younger generations may perceive traditional practices as outdated or economically unviable, leading to declining participation (Sinthumule 2023). Incorporating TEK into formal education and environmental curricula offers one solution, ensuring that local wisdom is passed down and validated within academic and policy frameworks (McCarter and Gavin 2011).

11.8.3 Economic Pressures and Opportunity Costs

Communities living along coastlines often depend on short-term income-generating activities such as shrimp farming, sand mining, or urban expansion—activities that may conflict with bioshield conservation. Creating economic incentives, such as ecotourism programs, sustainable aquaculture, or carbon credits, can align livelihood needs with environmental goals (Clifton and Mánez 2025; Da Silva et al. 2025).

11.8.4 Climate and Environmental Uncertainty

Climate change adds complexity to bioshield management. Rising sea levels, changing salinity regimes, and altered storm patterns can render traditional knowledge less predictive (Veettil et al. 2021). However, integrating TEK with scientific monitoring can help update traditional practices to align with new climatic realities (Tengö et al. 2017; Makondo and Thomas 2018).

11.8.5 Inclusivity and Power Dynamics

Even within communities, participation is not always equitable. Gender roles, caste, class, or ethnic hierarchies can marginalize certain groups from decision-making (Harper et al. 2024). Genuine participation requires deliberate inclusion mechanisms—such as women-led mangrove cooperatives or youth participation programs—to ensure diverse representation (Lima et al. 2025).

Despite these challenges, numerous opportunities exist for strengthening the interface between communities, knowledge, and bioshields:

a. Expanding citizen science initiatives, where community members collect environmental data using mobile applications or low-cost sensors (Cronemberger et al. 2023).
b. Promoting South–South cooperation through regional training and exchange programs between community leaders in Asia, Africa, and Latin America (Triulzi et al. 2024).
c. Encouraging transdisciplinary research that combines social sciences, ecology, and indigenous studies to co-design resilient bioshields (Spadon et al. 2025).
d. Mainstreaming community-driven bioshields within national adaptation plans and global climate finance mechanisms (e.g., Green Climate Fund) (Majlingova and Kádár 2025).

The growing recognition of community and traditional knowledge within climate discourse represents a major opportunity for future bioshield development.

11.9 Summary and Future Directions

Community participation and traditional knowledge are not peripheral components of bioshield development—they are foundational pillars that determine ecological effectiveness, socio-economic sustainability, and cultural legitimacy. The success stories from Asia, Africa, Latin America, and the Pacific reveal that bioshields rooted in community ownership and indigenous wisdom outperform externally imposed interventions in both durability and adaptability.

This chapter has shown that traditional ecological knowledge provides a detailed, site-specific understanding of coastal dynamics, species interactions, and management practices. When combined with modern scientific tools, such as remote sensing, GIS modeling, and hydrological analysis, these insights develop bioshields that are ecologically sound and socially just. Moreover, inclusive community participation fosters stewardship, accountability, and long-term maintenance, transforming bioshields from mere physical structures into living systems of resilience.

For the future, three directions stand out:

a. Institutionalizing TEK in National Policies: Governments should formally recognize TEK as a complementary knowledge system in environmental legislation and integrate it into national adaptation strategies.
b. Scaling Community-Based Models: Successful pilot projects should be upscaled through financial mechanisms, including carbon credits, ecosystem service payments, and international climate funds.
c. Enhancing Knowledge Exchange: Cross-regional learning platforms should be established to facilitate the exchange of community-based bioshield experiences across Asia, Africa, and the Pacific.

The next chapter, "Successful Bioshield Projects Around the World," will build upon these foundations by examining in detail how communities, policies, and technologies converge in practice to create globally recognized models of coastal protection. Together, these chapters underscore a central message that the future of coastal resilience lies not merely in engineering or finance, but in the power of people, culture, and collective ecological intelligence.

References

Adjaison D, Okwei R, Bandauko E, Nyantakyi-Frimpong H, Kutor SK, Arku G (2025) They ignored our ideas on the sea defence project: the socio-politics of coastal flood management and climate maladaptation in Keta and Glefe, Ghana. Int J Disaster Risk Reduct 130:105846. https://doi.org/10.1016/j.ijdrr.2025.105846

Aleu RB, Larsen RK, Methner N (2022) Participation and marginalization in water governance: probing the agency of powerholders. Ecol Soc 27(4):33. https://doi.org/10.5751/ES-13680-270433

Ali T, Buergelt PT, Paton D, Smith JA, Maypilama EL, Yuŋgirrŋa D, Dhamarrandji S, Gundjarranbuy R (2021) Facilitating sustainable disaster risk reduction in indigenous communities: reviving indigenous worldviews, knowledge and practices through two-way partnering. Int J Environ Res Public Health 18(3):855. https://doi.org/10.3390/ijerph18030855

Amone-Mabuto M, Mubai M, Bandeira S, Shalli M, Adams J, Lugendo B, Hollander J (2023) Coastal community's perceptions on the role of seagrass ecosystems for coastal protection and implications for management. Ocean Coast Manag 244:106811. https://doi.org/10.1016/j.ocecoaman.2023.106811

Arkhurst BM, Poku-Boansi M, Adarkwa KK (2023) Indigenous knowledge in climate change adaptation: choice of indigenous adaptation responses to coastal erosion in Ghanaian communities. Environ Sci Policy 147:326–335. https://doi.org/10.1016/j.envsci.2023.06.021

Asio MVS, Cvitanovic C, Lau J, Diedrich A (2025) Collaborative governance in marine and coastal management: a systematic review. Ocean Coast Manag 272:108020. https://doi.org/10.1016/j.ocecoaman.2025.108020

Aswani S, Lemahieu A, Sauer WHH (2018) Global trends of local ecological knowledge and future implications. PLoS ONE 13(4):e0195440. https://doi.org/10.1371/journal.pone.0195440

Balilo BB Jr, Candelaria AP, Dioneda RR Sr (2023) Assessment and perceived impact of Coastal Resource Management (CRM) programs in the southern part of Masbate Island, Philippines. J Geosci Environ Protect 11(3):87–103. https://doi.org/10.4236/gep.2023.113006

Ballad EL, Morooka Y, Shinbo T (2016) Factors inducing community participation in coastal resource management: case study of MPAs in Gonzaga, Cagayan, Philippines. J Rural Probl 52(4): 241–246. https://doi.org/10.7310/arfe.52.241

Baustian MM, Jung H, Bienn HC, Barra M, Hemmerling SA, Wang Y, White E, Meselhe E (2019) Engaging coastal community members about natural and nature-based solutions to assess their ecosystem function. Ecol Eng 143:100015. https://doi.org/10.1016/j.ecoena.2019.100015

Begum F, de Bruyn LL, Kristiansen P, Islam MA (2021) Institutionalising co-management activities for conservation of forest resources: evidence from the Sundarban mangrove forest management of Bangladesh. J Environ Manage 298:113504. https://doi.org/10.1016/j.jenvman.2021.113504

Begum F, De Bruyn LL, Kristiansen P, Islam MA (2023) Development pathways for co-management in the Sundarban mangrove forest: a multiple stakeholder perspective. Forest Policy Econ 148:102918. https://doi.org/10.1016/j.forpol.2023.102918

Beitl CM, Rahimzadeh-Bajgiran P, Bravo M, Ortega-Pacheco D, Bird K (2019) New valuation for defying degradation: Visualizing mangrove forest dynamics and local stewardship with remote sensing in coastal Ecuador. Geoforum 98:123–132. https://doi.org/10.1016/j.geoforum.2018.10.024

Bhattacharjee S, Sharma M, Tiwari A (2025) Change and continuity of coastal mangroves in Greater Mumbai, India: towards the sustainable governance of blue-green infrastructure. Land 14(9):1732. https://doi.org/10.3390/land14091732

Botero CM, Suman DO, Milanes CB (2024) The Multiple challenges faced by coastal and marine governance. Water 17(15):2322. https://doi.org/10.3390/w17152322

Brondízio ES, Aumeeruddy-Thomas Y, Bates P, Carino J, Fernández-Llamazares Á, Ferrari MF, Shrestha UB (2021) Locally based, regionally manifested, and globally relevant: indigenous and local knowledge, values, and practices for nature. Annu Rev Environ Resour 46(1):481–509. https://doi.org/10.1146/annurev-environ-012220-012127

Chen T, Cheng H (2020) Applying traditional knowledge to resilience in coastal rural villages. Int J Disaster Risk Reduct 47:101564. https://doi.org/10.1016/j.ijdrr.2020.101564

Cinner JE, Aswani S (2007) Integrating customary management into marine conservation. Biol Cons 140(3–4):201–216. https://doi.org/10.1016/j.biocon.2007.08.008

Ciocco TW, Miller BW, Tangen S, Crausbay SD, Oldfather MF (2024) Indigenous knowledge in climate adaptation planning: Reflections from initial effort. Front Climate 6:1393354. https://doi.org/10.3389/fclim.2024.1393354

Clarke P, Jupiter SD (2010) Law, custom and community-based natural resource management in Kubulau District (Fiji). Environ Conserv 37(1):98–106. https://doi.org/10.1017/S0376892910000354

Clifton J, Mánez KS (2025) A framework based on payments for ecosystem services to support the delivery of high integrity carbon and biodiversity credits. Ecosyst Serv 73:101724. https://doi.org/10.1016/j.ecoser.2025.101724

Cooley S, Schoeman D, Bopp L, Boyd P, Donner S, Ghebrehiwet DY, Ito SI, Kiessling W, Martinetto P, Ojea E, Racault MF, Rost B, Skern-Mauritzen M (2022) Oceans and Coastal Ecosystems and Their Services. In: Climate Change 2022: Impacts, Adaptation and Vulnerability. Contribution of Working Group II to the Sixth Assessment Report of the Intergovernmental Panel on Climate Change. Pörtner HO, Roberts DC, Tignor M, Poloczanska ES, Mintenbeck K, Alegría A, Craig M, Langsdorf S, Löschke S, Möller V, Okem A, Rama B (Eds.). Cambridge University Press,

Cambridge, UK and New York, NY, USA, pp. 379–550. https://doi.org/10.1017/9781009325844.005

Cormier-Salem M, Panfili J (2016) Mangrove reforestation: greening or grabbing coastal zones and deltas? Case studies in Senegal§. Afr J Aquat Sci 41(1):89–98. https://doi.org/10.2989/16085914.2016.1146122

Cronemberger C, Ribeiro KT, Acosta RK, de Andrade DFC, Marini-Filho OJ, Masuda LSM, de Tófoli CF (2023) Social participation in the Brazilian National Biodiversity monitoring program leads to multiple socioenvironmental outcomes. Citizen Science: Theory and Practice 8(1):32. https://doi.org/10.5334/cstp.582

Da Silva AE, Maracajá KF, Batalhão AC, Silva VF, Borges IM (2025) Ecotourism and co-management: strengthening socio-ecological resilience in local food systems. Sustainability 17(6):2443. https://doi.org/10.3390/su17062443

Damastuti E, De Groot R, Debrot AO, Silvius MJ (2022) Effectiveness of community-based mangrove management for biodiversity conservation: a case study from Central Java, Indonesia. Trees Forests People 7:100202. https://doi.org/10.1016/j.tfp.2022.100202

Datta D, Chattopadhyay R, Guha P (2012) Community based mangrove management: A review on status and sustainability. J Environ Manage 107:84–95. https://doi.org/10.1016/j.jenvman.2012.04.013

De Oliveira M, Morrison T, O'Brien KR, Lovelock CE (2024) Governance of coastal wetlands: beyond the community conservation paradigm. Ocean Coast Manag 255:107253. https://doi.org/10.1016/j.ocecoaman.2024.107253

Dey T, Rajib MGM, Islam MA (2021) People's Perceptions about the socio-economic and environmental impact of coastal green belt in Bangladesh. Indonesian Journal of Social and Environmental Issues (IJSEI) 2(2):155–162

Diatta CS, Diouf M, Sow AA, Karibuhoye C (2020) The Joola civilization and the management of mangroves in lower casamance, Senegal. Revue d'ethnoécologie (17). https://doi.org/10.4000/ethnoecologie.5855

Do HA, Pham TV, Van Le TT, Vu TM, Truong TL, Nguyen TH, Le TS (2023) ASEAN's soft law on marine environmental pollution. VMOST J Soc Sci Human 65(3):99–108. https://doi.org/10.31276/VMOSTJOSSH.65(3).99-108

Dorji T, Rinchen K, Morrison-Saunders A, Blake D, Banham V, Pelden S (2024) Understanding how Indigenous knowledge contributes to climate change adaptation and resilience: a systematic literature review. Environ Manage 74(6):1101–1123. https://doi.org/10.1007/s00267-024-02032-x

Drake AK, Perkovic A, Reeve C, Alexander SM, Nguyen VM, Dunmall KM (2022) Community participation in coastal and marine research and monitoring in Inuit Nunangat: a scoping literature review. Facets 7:891–911. https://doi.org/10.1139/facets-2021-0067

Dulyakasem R, Brooks C, Lehnert SL, Newton P (2026) Community-based natural resource management in coastal communities: the contribution of mangroves to household livelihoods in southern Thailand. Environmental Development 57:101340. https://doi.org/10.1016/j.envdev.2025.101340

Dushkova D, Ivlieva O (2023) Empowering communities to act for a change: a review of the community empowerment programs towards sustainability and resilience. Sustainability 16(19):8700. https://doi.org/10.3390/su16198700

Ebeler L, Balke T, Laurie E, Le H, Nguyen T, Renaud FG (2025) Local-scale impacts of mangrove restoration and conservation on coastal communities in two Vietnamese deltas: socio-economic and institutional dynamics. Ocean Coast Manage 269:107812. https://doi.org/10.1016/j.ocecoaman.2025.107812

Fajardo P, Beauchesne D, Carbajal-López A, Daigle RM, Fierro-Arcos LD, Goldsmit J, Zajderman S, Valdez-Hernández JI, Terán Maigua MY, Christofoletti RA (2021) Aichi target 18 beyond 2020: mainstreaming traditional biodiversity knowledge in the conservation and sustainable use of marine and coastal ecosystems. Peer J 9:e9616. https://doi.org/10.7717/peerj.9616

Fent A, Bardou R, Carney J, Cavanaugh K (2018) Transborder political ecology of mangroves in Senegal and the Gambia. Glob Environ Chang 54:214–226. https://doi.org/10.1016/j.gloenvcha.2019.01.003

Gallup L, Sonnenfeld DA, Dahdouh-Guebas F (2020) Mangrove use and management within the Sine-Saloum Delta. Ocean Coastal Manage 185:105001. https://doi.org/10.1016/j.ocecoaman.2019.105001

Gerona-Daga BME, Salmo SG (2022) A systematic review of mangrove restoration studies in Southeast Asia: challenges and opportunities for the United Nation's decade on ecosystem restoration. Front Mar Sci 9:987737. https://doi.org/10.3389/fmars.2022.987737

Ghosh S, Proisy C, Muthusankar G, Hassenrück C, Helfer V, Mathevet R, Andrieu J, Balachandran N, Narendran R (2022) Multiscale diagnosis of mangrove status in data-poor context using very high spatial resolution satellite images: a case study in Pichavaram mangrove forest, Tamil Nadu India. . Remote Sens 14(10):2317. https://doi.org/10.3390/rs14102317

Giffin AL, Brown CJ, Nalau J, Mackey BG, Connolly RM (2020) Marine and coastal ecosystem-based adaptation in Asia and Oceania: review of approaches and integration with marine spatial planning. Pac Conserv Biol 27(2):104–117. https://doi.org/10.1071/PC20025

Gómez-Baggethun E, Corbera E, Reyes-García V (2013) Traditional ecological knowledge and global environmental change: research findings and policy implications. Ecol Soc J Integr Sci Resil Sustain 18(4):72. https://doi.org/10.5751/ES-06288-180472

Government of India (2021) Induction Material – Internal Work Study Unit. Ministry of Environment, Forest and Climate Change. https://moef.gov.in/uploads/2021/12/Induction-Material-2021.pdf. Accessed on 09.01.2026

Hai NT, Dell B, Phuong VT, Harper RJ (2020) Towards a more robust approach for the restoration of mangroves in Vietnam. Ann for Sci 77(1):18. https://doi.org/10.1007/s13595-020-0921-0

Haq SM, Pieroni A, Bussmann RW, Abd-ElGawad AM, El-Ansary HO (2023) Integrating traditional ecological knowledge into habitat restoration: implications for meeting forest restoration challenges. J Ethnobiol Ethnomed 19(1):33. https://doi.org/10.1186/s13002-023-00606-3

Harper SJ, Gurney GG, Darling E, Mangubhai S, Jupiter S, Lestari WP, Holmes KE, Sumaryati S, Sulisyati R, Fox M, Ban NC (2024) Gender differences in the perceived impacts of coastal management and conservation. NPJ Ocean Sustain 3(1):34. https://doi.org/10.1038/s44183-024-00070-w

Hosen N, Nakamura H, Hamzah A (2019) Adaptation to climate change: does traditional ecological knowledge hold the key? Sustainability 12(2):676. https://doi.org/10.3390/su12020676

Huff A, Tonui C (2017) Making 'Mangroves Together': carbon, conservation and co-management in Gazi Bay, Kenya, STEPS Working Paper 95, Brighton: STEPS Centre, 33pp.

Hurlbert M, Gupta J (2024) The split ladder of participation: a literature review and dynamic path forward. Environ Sci Policy 157:103773. https://doi.org/10.1016/j.envsci.2024.103773

Huxham M, Kairu A, Kivugo R, Mwafrica M, Huff A, Shilland R (2023) Rawls in the mangrove: perceptions of justice in nature-based solutions projects. People Nat 5(5):1497–1511. https://doi.org/10.1002/pan3.10498

Huxham M, Whitlock D, Githaiga M, Dencer-Brown A (2018) Carbon in the coastal seascape: how interactions between mangrove forests, seagrass meadows and tidal marshes influence carbon storage. Current Forestry Reports 4(2):101–110. https://doi.org/10.1007/s40725-018-0077-4

James I (2025) Participatory seascape mapping: a community-based approach to ocean governance and marine conservation. Ocean Coast Manag 261:107531. https://doi.org/10.1016/j.ocecoaman.2024.107531

Jang N, Caswell C, McIvor S, Ball T (2025) Advancing gender-responsive and socially inclusive practices in nature-based solutions for adaptation. International Institute for Sustainable Development, 32 pp

Jerez Columbié Y (2021) People of the mangrove: a lens into socioecological interactions in the Ecuadorian Black Pacific. Etropic Electron J Stud Tropics 20(2): 74–94. https://doi.org/10.25120/etropic.20.2.2021.3808

Kairo JG, Hamza AJ, Wanjiru C (2018) Mikoko Pamoja: a demonstrably effective community-based blue carbon project in Kenya. In: A Blue Carbon Primer, pp 341–350. CRC Press, 550 pp
Kathiresan K, Rajendran N (2005) Coastal mangrove forests mitigated tsunami. Estuar Coast Shelf Sci 65(3):601–606. https://doi.org/10.1016/j.ecss.2005.06.022
Kitolelei S, Lowry JH, Qaqara N, Ryle J, Veitayaki J, Piovano S (2022) Spatial use of marine resources in a rural village: A case study from Qoma. . Front Marine Sci 9:993103. https://doi.org/10.3389/fmars.2022.993103
Kongkeaw C, Kittitornkool J, Vandergeest P, Kittiwatanawong K (2019) Explaining success in community based mangrove management: four coastal communities along the Andaman Sea. . Ocean Coastal Manage 178:104822. https://doi.org/10.1016/j.ocecoaman.2019.104822
Larsen PB, Chanthavisouk C (2024) Free, prior, and informed consent, local officials, and changing biodiversity governance in Hin Nam No. Conservation Biology 38(6):e14388. https://doi.org/10.1111/cobi.14388
Lima MH, Salehin M, Chowdhury MA, Hasan MH, Hossain MJ, Bala SK (2025) From participation to empowerment: the case of women in community-based water management in hydrologically diverse southwest coastal Bangladesh. Environ Dev Sustain 1–28. https://doi.org/10.1007/s10668-025-05994-y
Lin PS (2019) Building resilience through ecosystem restoration and community participation: Post-disaster recovery in coastal island communities. International Journal of Disaster Risk Reduction 39:101249. https://doi.org/10.1016/j.ijdrr.2019.101249
Loc HH, Park E, Thu TN, Diep NTH, Can NT (2021) An enhanced analytical framework of participatory GIS for ecosystem services assessment applied to a Ramsar wetland site in the Vietnam Mekong Delta. Ecosyst Serv 48:101245. https://doi.org/10.1016/j.ecoser.2021.101245
Loch TK, Riechers M (2021) Integrating indigenous and local knowledge in management and research on coastal ecosystems in the Global South: A literature review. Ocean Coast Manag 212:105821. https://doi.org/10.1016/j.ocecoaman.2021.105821
Lopes PFM, Constantinescu Ş, Silvano RAM, Tatui F, Teampău P, Vaidianu N (2025) Unintended and overlooked consequences of exclusionary marine conservation. *ICES Journal of Marine Science* 82: fsae190, https://doi.org/10.1093/icesjms
Love MW (2021) Tabu, MPA and community: nomenclature and the political ecology of marine conservation in Vanuatu. Journal of Political Ecology 28(1):91–110. https://doi.org/10.2458/jpe.2301
Lumosi C, Hazin C, Fitzsimons J, Qin S (2025) Understanding the Role and Challenges for Indigenous and Community-Governed Lands in Contributing to Target 3 of the Global Biodiversity Framework. Land 14(7):1493. https://doi.org/10.3390/land14071493
Majlingova A, Kádár TS (2025) From Risk to Resilience: Integrating Climate Adaptation and Disaster Reduction in the Pursuit of Sustainable Development. Sustainability 17(12):5447. https://doi.org/10.3390/su17125447
Makondo CC, Thomas DS (2018) Climate change adaptation: Linking indigenous knowledge with western science for effective adaptation. Environ Sci Policy 88:83–91. https://doi.org/10.1016/j.envsci.2018.06.014
Maliao RJ, Tóthmérész B (2025) Braiding traditional ecological knowledge and Western science in the management of freshwater social-ecological systems: a systematic map protocol. Environ Evid 14:17. https://doi.org/10.1186/s13750-025-00371-8
Marín-Muñiz JL (2025) Wetlands of Mexico. In: *Wetlands of Tropical and Subtropical South and Central America: Biodiversity, Livelihoods and Conservation*, p155–172.
Marquez GPB, Olavides RD (2024) Integrating science-based and local ecological knowledge: A case study of mangrove restoration and rehabilitation projects in the Philippines. Disasters 48:e12630. https://doi.org/10.1111/disa.12630
McCarter J, Gavin MC (2011) Perceptions of the value of traditional ecological knowledge to formal school curricula: opportunities and challenges from Malekula Island. Vanuatu. J Ethnobiology Ethnomedicine 7:38. https://doi.org/10.1186/1746-4269-7-38

Mollick AS, Roy M, Khan NI, Islam W, Sadath N, Nath TK (2022) Assessing Good Governance in Protected Areas (PA) Co-management: A Case Study of the Sundarbans Mangrove Forests of Bangladesh. J Sustain for 41(3–5):277–301. https://doi.org/10.1080/10549811.2021.1923531

Moussa LG, Mohan M, Burmeister N, King SA, Burt JA, Rog SM, Watt MS, Udagedara S, Sujud L, Montenegro JF, Heng JE, Carvalho SA, Ali T, Veettil BK, Arachchige PS, Albanai JA, Sidik F, Shaban A, Peñaranda ML, Ewane EB (2024) Mangrove Ecotourism along the Coasts of the Gulf Cooperation Council Countries: A Systematic Review. Land 13(9):1351. https://doi.org/10.3390/land13091351

Mudge L (2018) Use of community perceptions to evaluate and adapt coastal resource management practices in the Philippines. Ocean Coast Manag 163:304–322. https://doi.org/10.1016/j.ocecoaman.2018.07.008

Nalau J, Becken S, Schliephack J, Parsons M, Brown C, Mackey B (2018) The role of indigenous and traditional knowledge in ecosystem-based adaptation: A review of the literature and case studies from the Pacific Islands. Weather, Climate, and Society 10(4):851–865. https://doi.org/10.1175/WCAS-D-18-0032.1

Newing H, Brittain S, Buchadas A, Del Giorgio O, Grasham CF, Ferritto R, Garcia Marquez JR, Khanyari M, König B, Kulkarni A, Murali R, Qin S, Rakowski J, Winn F, Ghoddousi A (2024) 'Participatory' conservation research involving indigenous peoples and local communities: Fourteen principles for good practice. Biol Cons 296:110708. https://doi.org/10.1016/j.biocon.2024.110708

Nguyen T, Van Tam N, Quoi LP, Parnell KE (2016) Community perspectives on an internationally funded mangrove restoration project: Kien Giang province, Vietnam. Ocean Coast Manag 119:146–154. https://doi.org/10.1016/j.ocecoaman.2015.10.008

Nguyen HT, Hardy GE, Le TV, Nguyen HQ, Nguyen HH, Nguyen TV, Dell B (2021) Mangrove Forest Landcover Changes in Coastal Vietnam: A Case Study from 1973 to 2020 in Thanh Hoa and Nghe An Provinces. Forests 12(5):637. https://doi.org/10.3390/f12050637

Nguyen T, Van Tam N, Quoi LP, Parnell KE (2015) Community perspectives on an internationally funded mangrove restoration project: Kien Giang province. Vietnam. Ocean & Coastal Management 119:146–154. https://doi.org/10.1016/j.ocecoaman.2015.10.008

Ntibona LN, Shalli MS, Mangora MM (2023) Willingness and drivers of community participation in mangrove conservation in the Rufiji Delta, Tanzania. Western Indian Ocean Journal of Marine Science 22(1):31–45. https://doi.org/10.4314/wiojms.v22i1

Nyangoko BP, Berg H, Mangora MM, Gullström M, Shalli MS (2021) Community Perceptions of Mangrove Ecosystem Services and Their Determinants in the Rufiji Delta. Tanzania. Sustainability 13(1):63. https://doi.org/10.3390/su13010063

Nyangoko BP, Berg H, Mangora MM, Shalli MS, Gullström M (2022) Community perceptions of climate change and ecosystem-based adaptation in the mangrove ecosystem of the Rufiji Delta. Tanzania. Climate and Development 14(10):896–908. https://doi.org/10.1080/17565529.2021.2022449

Ocampo-Thomason P (2006) Mangroves, people and cockles: impacts of the shrimp farming industry on mangrove communities in Esmeraldas Province, Ecuador. In: International Conference on Environment and Livelihoods in Coastal Zones: Managing Agriculture-Fishery-Aquaculture Conflicts, Bac Lieu, Vietnam (pp. 140–153).

Ojong N (2020) Indigenous land rights: where are we today and where should the research go in the future? Settler Colonial Studies 10(2):193–215. https://doi.org/10.1080/2201473X.2020.1726149

Parsons M, Taylor L, Crease R (2021) Indigenous Environmental Justice within Marine Ecosystems: A Systematic Review of the Literature on Indigenous Peoples' Involvement in Marine Governance and Management. Sustainability 13:4217. https://doi.org/10.3390/su13084217

Pearson J, McNamara KE, Nunn PD (2019) Gender-specific perspectives of mangrove ecosystem services: Case study from Bua Province. Fiji Islands. Ecosystem Services 38:100970. https://doi.org/10.1016/j.ecoser.2019.100970

Pham TT, Vu TP, Hoang TL, Dao TL, Nguyen DT, Pham DC, Dao LH, Nguyen VT, Hoang NV (2022) The Effectiveness of Financial Incentives for Addressing Mangrove Loss in Northern Vietnam. Frontiers in Forests and Global Change 4:709073. https://doi.org/10.3389/ffgc.2021.709073

Pomeroy RS, Carlos MB (1997) Community-based coastal resource management in the Philippines: A review and evaluation of programs and projects, 1984–1994. Mar Policy 21(5):445–464. https://doi.org/10.1016/S0308-597X(97)00016-X

Puskás N, Abunnasr Y, Naalbandian S (2021) Assessing deeper levels of participation in nature-based solutions in urban landscapes – A literature review of real-world cases. Landscape and Urban Planning 210: 104065. https://doi.org/10.1016/j.landurbplan.2021.104065

Quevedo JMD, Miller MA, Gevaña DT, Marks D, Friess DA, Tonoto P, Taylor D (2025) Perceptions of coastal vegetated ecosystems: A systematic review across geographical and sectoral dimensions. Ambio 1–18. https://doi.org/10.1007/s13280-025-02193-x

Ravazzoli E, Lavarello-Schettini R, Oberti B, Maino F (2025) Community-based approaches in disaster risk reduction and climate change adaptation: an analysis of applied participatory processes. Clim Change 178(12):225. https://doi.org/10.1007/s10584-025-04078-0

Risley SC, Britsch ML, Stoll JS, Leslie HM (2025) Mapping local knowledge supports science and stewardship. Ambio 54:1648–1665. https://doi.org/10.1007/s13280-025-02170-4

Salampessy ML, Lidiawati I, Metak A, Fitrianti D (2025) Women's Participation in Mangrove Conservation in Pantai Bahagia Village, Muara Gembong Regency. Bekasi. Forest and Nature 1(2):56–66

Sathiyamoorthy S (2024) Sakurai T (2024) Effectiveness of community participation in mangrove restoration: the evidence from northern Sri Lanka. Environ Econ Policy Stud 26:759–779. https://doi.org/10.1007/s10018-024-00397-1

Schipper ELF, Revi A, Preston BL, Carr ER, Eriksen SH, Fernandez-Carril LR, Glavovic BC, Hilmi NJM, Ley D, Mukerji R, Muylaert de Araujo MS, Perez R, Rose SK, Singh PK (2022) Climate Resilient Development Pathways. In: Climate Change 2022: Impacts, Adaptation and Vulnerability. Contribution of Working Group II to the Sixth Assessment Report of the Intergovernmental Panel on Climate Change. Pörtner HO, Roberts DC, Tignor M, Poloczanska ES, Mintenbeck K, Alegría A, Craig M, Langsdorf S, Löschke S, Möller V, Okem A, Rama B (Eds.). Cambridge University Press, Cambridge, UK and New York, NY, USA, pp 2655–2807. https://doi.org/10.1017/9781009325844.027

Selvam V, Ravichandran KK, Gnanappazham L, Navamuniyammal M (2003) Assessment of community-based restoration of Pichavaram mangrove wetland using remote sensing data. Curr Sci 85:794–798

Selvam V, Thamizoli P (2021) Science-based and community-centred approach to restore and sustain mangrove wetlands of India. Curr Sci 121(10):1288–1296

Siddique MR, Hossain M, Rashid AM, Khan NA, Shuvo SN, Hassan MZ (2024) Evaluating co-management in the Sundarbans mangrove forest of Bangladesh: success and limitations from local forest users' perspectives. Ecol Soc 29(2):8. https://doi.org/10.5751/ES-14905-290208

Siddiqui MR, Ishtiaque A (2025) Roles and nature of public participation in climate change adaptation in coastal South Asia. Discov Environ 3:166. https://doi.org/10.1007/s44274-025-00360-1

Singha P, Dasgupta R, Hashimoto S, Mitra BK (2024) An integrated coastal exposure modelling approach to assist mangrove ecosystem based disaster risk reduction (Eco-DRR) in Tamil Nadu. India. Progress in Disaster Science 24:100388. https://doi.org/10.1016/j.pdisas.2024.100388

Sinthumule NI (2023) Traditional ecological knowledge and its role in biodiversity conservation: A systematic review. Front Environ Sci 11:1164900. https://doi.org/10.3389/fenvs.2023.1164900

Smith AC, Tasnim T, Irfanullah HM, Turner B, Chausson A, Seddon N (2021) Nature-based Solutions in Bangladesh: Evidence of Effectiveness for Addressing Climate Change and Other Sustainable Development Goals. Front Environ Sci 9:737659. https://doi.org/10.3389/fenvs.2021.737659

Smith T, Bulkan J, Zerriffi H, Tansey J (2019) Indigenous peoples, local communities, and Payments for Ecosystem Services. The Canadian Geographer / Le Géographe Canadien 63(4):616–630. https://doi.org/10.1111/cag.12553

Spadon G, Oyebode O, Botero C, Sharma T, Goerlandt F, Pelot R (2025) Community-Centered Spatial Intelligence for Climate Adaptation at Nova Scotia's Eastern Shore. In: Proceedings of the 1st ACM SIGSPATIAL International Workshop on Spatial Intelligence for Smart and Connected Communities (pp. 72–75). https://doi.org/10.1145/3764924.3770892

Sofue Y, Quevedo JMD, Lukman KM, Kohsaka R (2025) Identifying changes in mangrove landscapes in the Philippines and Indonesia using remote sensing and community perceptions: towards ecosystem services management. Regional Studies in Marine Science 82:104023. https://doi.org/10.1016/j.rsma.2025.104023

Stori FT, Peres CM, Turra A, Pressey RL (2019) Traditional ecological knowledge supports ecosystem-based management in disturbed coastal marine social-ecological systems. Front Mar Sci 6:452035. https://doi.org/10.3389/fmars.2019.00571

Sukuryadi, Harahab N, Primyastanto M, Semedi B (2021) Collaborative-based mangrove ecosystem management model for the development of marine ecotourism in Lembar Bay, Lombok, Indonesia. Environ Dev Sustain 23: 6838–6868. https://doi.org/10.1007/s10668-020-00895-8

Sunarto S, Mamangkey J (2025) Mapping the knowledge Landscape of Local and Traditional Ecological Knowledge (LEK/TEK) in mangrove ecosystems: a bibliometric analysis in the past decade. Int J Sci Technol Manage 6(5):1006–1011. https://doi.org/10.46729/ijstm.v6i5.1360

Tengö M, Brondizio ES, Elmqvist T, Malmer P, Spierenburg M (2014) Connecting diverse knowledge systems for enhanced ecosystem governance: the multiple evidence base approach. Ambio 43(5):579–591. https://doi.org/10.1007/s13280-014-0501-3

Tengö M, Hill R, Malmer P, Raymond CM, Spierenburg M, Danielsen F, Elmqvist T, Folke C (2017) Weaving knowledge systems in IPBES, CBD and beyond—lessons learned for sustainability. Curr Opinion Environ Sustain 26–27:17–25. https://doi.org/10.1016/j.cosust.2016.12.005

Tinh PH, MacKenzie RA, Hung TD, Vinh TV, Ha HT, Lam MH, Hanh NT, Tung NX, Hai PM, Huyen BT (2022) Mangrove restoration in Vietnamese Mekong delta during 2015–2020: achievements and challenges. Front Mar Sci 9:1043943. https://doi.org/10.3389/fmars.2022.1043943

Tran TA, Rigg J, Pham VHT, Yong ML, Van Vo O, Pham PX, Vo TD (2025) Erosion of traditional ecological knowledge under conditions of hydrosocial rupture: insights from the Mekong floodplains communities. Ambio 54:1666–1682. https://doi.org/10.1007/s13280-025-02172-2

Trialfhianty TI, Quinn CH, Beger M (2025) Engaging customary law to improve the effectiveness of marine protected areas in Indonesia. Ocean Coast Manag 261:107543. https://doi.org/10.1016/j.ocecoaman.2025.107543

Triulzi I, Kabra R, Allagh KP, Kiarie J (2024) Strengths and weaknesses of the south-south learning exchange: a qualitative analysis of experts' perspectives. Gates Open Res 7:116. https://doi.org/10.12688/gatesopenres.14699.2

Valderrama-Landeros L, Troche-Souza C, Alcántara-Maya JA, Velázquez-Salazar S, Vázquez-Balderas B, Villeda-Chávez E et al (2024) An assessment of mangrove forest in northwestern Mexico using the Google earth engine cloud computing platform. PLoS ONE 19(12):e0315181. https://doi.org/10.1371/journal.pone.0315181

Valenzuela BR, Yeo-Chang Y, Park MS, Chun J-N (2020) Local people's participation in mangrove restoration projects and impacts on social capital and livelihood: a case study in the Philippines. Forests 11(5):580. https://doi.org/10.3390/f11050580

Van Bijsterveldt CE, Debrot AO, Bouma TJ, Maulana MB, Pribadi R, Schop J, Tonneijck FH, Van Wesenbeeck BK (2022) To plant or not to plant: when can planting facilitate mangrove restoration? Front Environ Sci 9:690011. https://doi.org/10.3389/fenvs.2021.690011

Veettil BK, Ward RD, Dung NTK, Van DD, Quang NX, Hoai PN, Hoang N (2021) The use of bioshields for coastal protection in Vietnam: current status and potential. Reg Stud Marine Sci 47:101945. https://doi.org/10.1016/j.rsma.2021.101945

Veettil BK, Ward RD, Quang NX, Trang NTT, Giang TH (2019) Mangroves of Vietnam: historical development, current state of research and future threats. Estuar Coast Shelf Sci 218:212–236. https://doi.org/10.1016/j.ecss.2018.12.021

Part V
Coastal Bioshields: Global and Regional Case Studies

Chapter 12
Successful Coastal Bioshield Projects Around the World

Abstract Across the globe, bioshield initiatives have demonstrated the potential of nature-based solutions to protect coastlines while enhancing biodiversity and supporting livelihoods. This chapter presents a comprehensive synthesis of successful bioshield projects implemented in diverse geographic and socioeconomic contexts. Drawing from case studies in Asia, Africa, Latin America, and the Pacific, it identifies the ecological, technical, and governance factors that contribute to long-term project success. Key examples include mangrove restoration in Vietnam and Bangladesh, dune stabilization in India, coral-mangrove synergy projects in the Philippines, and integrated coastal vegetation management in Kenya and Mexico. The chapter examines success through multiple lenses—ecological effectiveness, economic efficiency, and social inclusiveness—to evaluate how these projects achieve measurable reductions in coastal vulnerability while generating co-benefits such as carbon sequestration, fisheries enhancement, and ecotourism opportunities. It emphasizes adaptive planning, local participation, policy integration, and science-based monitoring as critical determinants of sustainability. Furthermore, it explores the role of international cooperation, donor funding, and multi-sectoral partnerships in scaling up bioshield initiatives. By synthesizing lessons from global experiences, this chapter highlights replicable models and best practices that can guide future bioshield development. It concludes that successful bioshield projects share a common foundation: the fusion of ecological restoration with community empowerment and institutional support, offering a blueprint for sustainable coastal resilience worldwide.

Keywords Best management practices · Coastal restoration success · Community resilience · Ecosystem-based adaptation · Global case studies · Nature-based solutions

B. Kozhikkodan Veettil, *The Potential of Bioshields for Coastal Protection*, Advances in Natural and Technological Hazards Research 55,
https://doi.org/10.1007/978-3-032-22888-8_12

12.1 Introduction

Accelerating climate change, coupled with rapid coastal development and ecosystem degradation, has significantly increased the exposure of coastal regions to hazards such as storm surges, coastal erosion, flooding, and saline intrusion (Veettil et al. 2021; Perricone et al. 2023). In this context, coastal bioshields, which are vegetated coastal systems including mangrove forests, salt marshes, seagrass meadows, and coastal dunes, have emerged as critical components of sustainable coastal defense strategies. Functioning as nature-based solutions (NbS), bioshields attenuate wave energy, stabilize sediments, and reduce shoreline retreat, while simultaneously delivering a wide range of co-benefits (Perricone et al. 2023; Gijón Mancheño et al. 2024; Daro Justine and Seenath 2025). These include biodiversity conservation, carbon sequestration, water quality regulation, and the sustenance of coastal livelihoods, thereby aligning risk reduction objectives with broader goals of climate adaptation and sustainable development (Paxton et al. 2024).

Over the past two decades, a growing body of empirical evidence and practical experience from around the world has demonstrated that well-designed and properly managed bioshields can effectively complement, and in some contexts, partially substitute for conventional hard-engineering structures, such as seawalls, breakwaters, and groynes (Narayan et al. 2016; Huynh et al. 2024). Unlike rigid infrastructure, bioshields are inherently dynamic systems capable of self-repair and adaptation to changing environmental conditions, including sea-level rise and shifting sediment regimes (Huynh et al. 2024; Paxton et al. 2024). As a result, they are increasingly recognized within international policy frameworks and national adaptation strategies as cost-effective, resilient, and socially inclusive approaches to coastal protection (Paxton et al. 2024).

Despite substantial variations in environmental conditions and cultural contexts, successful bioshield initiatives tend to exhibit several common characteristics. These include meaningful community participation in planning, implementation, and stewardship (Damastuti et al. 2023); strong institutional backing and integration within coastal management and climate adaptation policies (Rahman et al. 2018); the use of native species adapted to local hydrodynamic, geomorphological, and soil conditions (Veettil et al. 2021; Sloey et al. 2025); and long-term monitoring coupled with adaptive management (Chee et al. 2021). By refining these shared elements and lessons learned, this chapter aims to identify transferable principles and best practices that can inform the design, scaling, and replication of bioshield projects worldwide, thereby strengthening global efforts toward resilient and nature-positive coastal protection (Sloey et al. 2025).

This chapter synthesizes and critically examines a selection of successful coastal bioshield projects from diverse geographic and socio-political contexts worldwide. By highlighting cases from different continents, the chapter illustrates how ecological processes, socio-economic incentives, and governance arrangements interact to determine bioshield performance and long-term sustainability. Focus is given to the institutional policies that have enabled these initiatives to move beyond short-term restoration projects toward durable, landscape-scale interventions.

12.2 Successful Bioshield Projects in Asia

Asia, home to some of the most densely populated and hazard-prone coastal zones on the planet, has been at the forefront of bioshield implementation (Monioudi et al. 2025). Countries such as Vietnam, India, Indonesia, and Bangladesh have made significant advancements in restoring mangroves and coastal vegetation for disaster risk reduction (Hai et al. 2020; Islam et al. 2020; Rudianto et al. 2022; Ray et al. 2024). Across Asia, the success of bioshield projects is tied to long-term institutional commitment, participatory governance, and adaptive management strategies that align with local ecological and socio-economic contexts (Damastuti et al. 2023; Siddiqui and Ishtiaque 2025).

12.2.1 Vietnam: Integrating Mangrove Restoration and Coastal Defense

Vietnam's long coastline, stretching over 3,260 km, is highly vulnerable to tropical storms and sea-level rise (Takagi et al. 2015; Veettil et al. 2021). The Mangrove Plantation and Disaster Risk Reduction Project, initiated in the 1990s and supported by the Vietnam Red Cross and the Danish Red Cross, is often cited as a landmark example of bioshield success (IFRC 2022). The program planted tens of thousands of hectares of mangroves across coastal provinces such as Nam Dinh, Thai Binh, and Thanh Hoa (Hai et al. 2020).

The restored mangroves provided a cost-effective defense against storm surges, reducing maintenance costs of sea dikes by up to 50% and protecting over 350,000 people in vulnerable coastal areas (GDPC 2022; Andreasson and Bechauf 2025). Beyond physical protection, the project also generated socio-economic co-benefits such as enhanced fisheries, carbon storage, and ecotourism opportunities (GDPC 2022; Andreasson and Bechauf 2025). The integration of mangrove management into provincial and national climate adaptation strategies ensured long-term sustainability (Nguyen et al. 2017).

12.2.2 Bangladesh: Mangrove Bioshields in the Sundarbans

The Sundarbans Reserve Forest, spanning the Bangladesh-India border, represents one of the world's largest natural bioshields (Sen 2019). In Bangladesh, government-led programs have focused on rehabilitating degraded mangroves to enhance cyclone resilience (Khanal et al. 2024). After the devastation of Cyclone Sidr in 2007, Bangladesh launched a large-scale Coastal Greenbelt Project, which restored mangroves along embankments and chars (riverine islands) (Government of Bangladesh 2008).

These efforts reduced cyclone damage in restored areas by more than 30% compared to unprotected zones (Iftekhar and Islam 2004). The integration of bioshield restoration with community-based co-management systems allowed local populations to share resource benefits, fostering stewardship and reducing illegal logging (Begum et al. 2021).

12.2.3 India: The Tamil Nadu Coastal Vegetation Project

In India, the 2004 Indian Ocean tsunami served as a wake-up call, exposing the vulnerability of coastal settlements. The Tamil Nadu government and the M.S. Swaminathan Research Foundation initiated coastal bioshield projects that restored mangroves, casuarina belts, and dune vegetation (Selvam et al. 2005; Sridhar 2006). Villages that maintained intact bioshields reported lower casualties and property losses than those without vegetation buffers (Mascarenhas and Jayakumar 2008). These findings prompted the inclusion of bioshields into India's national coastal zone management framework (Selvam et al. 2012). The program emphasized local participation, especially women's self-help groups in seedling nurseries and maintenance, creating both ecological and livelihood benefits (Selvam et al. 2012).

12.2.4 The Philippines and Indonesia: Community-Led Bioshields

In the Philippines, the Department of Environment and Natural Resources (DENR) launched the National Greening Program, restoring mangroves along storm-prone coasts such as Leyte and Samar after Typhoon Haiyan. Community-based organizations played a central role in planning, planting, and monitoring, resulting in improved survival rates of mangroves and increased local income from eco-fisheries (Primavera and Esteban 2008).

Similarly, Indonesia's Mangrove for the Future (MFF) initiative, coordinated by IUCN and Wetlands International, has successfully rehabilitated over 25,000 ha of degraded mangrove forests across Java and Sumatra (Arifanti et al. 2022; Indrajaya et al. 2022; Sasmito et al. 2022). The use of mixed-species plantations and zoning approaches, where protective mangroves buffer aquaculture ponds, proved more resilient than monocultures (Lukman et al. 2021; Widayanti et al. 2023).

12.3 Successful Bioshield Projects in Africa

African coastal nations are increasingly recognizing the value of nature-based defenses to complement or replace costly engineering solutions (Cortes 2020; Hagedoorn et al. 2021; Knight 2024). Mangroves, dunes, and coastal forests are now central components of national adaptation plans across the African continent (March et al. 2024; Santos et al. 2024). African bioshield projects have demonstrated that integrating ecosystem restoration with community incentives and sustainable finance mechanisms can yield tangible benefits, both for resilience and livelihoods (Awazi 2025).

12.3.1 Senegal and the Gambia: Mangrove Regeneration in the Saloum Delta

In West Africa, the Saloum Delta Biosphere Reserve in Senegal and Tanbi Wetlands in The Gambia have been sites of extensive mangrove restoration since the early 2000s (Carney et al. 2014; Fent et al. 2019; Diouf et al. 2025). These efforts, led by NGOs such as Oceanium and supported by the Global Environment Facility (GEF), have resulted in the planting of over 80 million mangrove propagules covering more than 10,000 ha (Livelihoods 2019).

The regenerated mangroves stabilized shorelines, reduced salinity intrusion into rice fields, and increased fish and oyster stocks (Cormier-Salem and Panfili 2016; Livelihoods 2019; World Bank 2024). Importantly, these initiatives combined ecological restoration with carbon offset programs through the blue carbon mechanism, allowing communities to benefit financially from carbon credits under voluntary markets (Livelihoods 2019).

12.3.2 Kenya and Tanzania: Coastal Reforestation for Community Resilience

In East Africa, Kenya's Mikoko Pamoja project, recognized by the UN as the first community-led mangrove carbon offset project, has linked mangrove conservation to sustainable financing (Wylie et al. 2016; Nyamasege 2019; Rasowo et al. 2024). Revenue from carbon credit sales supports education, water, and health projects in local villages (Wylie et al. 2016).

Similarly, in Tanzania's Rufiji Delta, the National Mangrove Management Plan integrates bioshield restoration with sustainable wood harvesting zones. By granting communities regulated access to mangrove products, the program has improved compliance and reduced illegal cutting (Wood et al. 2013; Nyangoko et al. 2022).

12.3.3 Egypt: Coastal Dune Stabilization in the Nile Delta

The Nile Delta, one of the most densely populated coastal regions in Africa, is highly susceptible to sea-level rise (Torresan et al. 2020; Hassan et al. 2025). In northern Egypt, dune stabilization projects along the Mediterranean coast have been implemented using native halophytic plants such as *Ammophila arenaria* and *Zygophyllum album* (Ahmed et al. 2014; El-Khalafy et al. 2025). These vegetation belts act as natural buffers, protecting agricultural lands and coastal settlements from erosion and saltwater intrusion (Masria et al. 2015).

12.4 Successful Bioshield Projects in Europe

Europe, while less frequently associated with tropical mangrove systems, has a long tradition of using vegetated dunes, salt marshes, and wetlands for coastal defense (van Loon-Steensma 2015; Galluccio et al. 2024). The region has pioneered the concept of "managed realignment," allowing the sea to reclaim selected low-lying areas to restore natural buffers (Pender and Faulkner 2011; Van den Hoven et al. 2022).

12.4.1 The Netherlands: Room for the River and Coastal Resilience

The Netherlands' Room for the River program exemplifies the European approach to integrating bioshields within engineered systems (Rijke et al. 2012). Implemented since 2006, the program involves setting back dikes and creating floodplains that allow controlled inundation (Van Alphen 2020). Re-vegetated floodplains with reeds and saltmarsh grasses dissipate storm surges, reduce peak flood levels, and enhance biodiversity (Rijke et al. 2012). Although not tropical, these vegetated buffers function similarly to mangrove bioshields by providing flexibility and ecological value to flood management systems (Rijke et al. 2012; Van Alphen 2020).

12.4.2 United Kingdom: Managed Realignment at Medmerry, Sussex

The Medmerry Managed Realignment Scheme on England's south coast, completed in 2013, transformed a flood-prone agricultural area into 183 ha of intertidal habitat with salt marshes and lagoons (Burgess et al. 2016; Dale et al. 2016, 2018). The project, managed by the Environment Agency, has become one of the UK's most successful nature-based coastal defense initiatives (Thomas 2014; Oliver

2024). The restored salt marsh reduced flood risk to over 300 homes and provided habitat for migratory birds (UK Environment Agency 2014; Krawiec 2017). The Medmerry project illustrates how combining hard engineering (breach construction and controlled flooding) with bioshield restoration can create cost-effective and ecologically rich coastal defenses (Waryszak et al. 2021).

12.4.3 Spain and France: Dune and Marsh Rehabilitation

In Spain's Doñana National Park and France's Camargue region, dune stabilization and saltmarsh rehabilitation have been central to coastal management (Rodríguez-Vidal et al. 2014; Muñoz-Reinoso et al. 2020; Davranche et al. 2024). Using native dune grasses and halophytes, authorities have stabilized shifting sands and promoted sediment accretion. These projects not only protect against storm surges but also sustain tourism and local biodiversity (Silva et al. 2016).

12.5 Successful Bioshield Projects in the Americas

The Americas offer diverse examples of coastal bioshield initiatives, ranging from mangrove rehabilitation in the Caribbean and Latin America to marsh restoration in North America (Moritsch et al. 2022; Rowland et al. 2024). These projects demonstrate the adaptability of bioshields across climatic zones and socio-economic settings (Friess et al. 2024).

12.5.1 The United States: Living Shorelines and Marsh Restoration

In the United States, the concept of "living shorelines" has gained traction as an alternative to conventional seawalls and bulkheads (Isdell et al. 2021; Mason et al. 2026). Living shoreline projects integrate vegetation, shellfish reefs, and coir logs to stabilize coasts while maintaining natural processes (Davenport et al. 2018; Bilkovic et al. 2017; Smith et al. 2020).

One notable example is the Chesapeake Bay Living Shoreline Initiative, which has restored several kilometers of tidal marsh and oyster reef habitat in Maryland and Virginia (Davenport et al. 2018; Isdell et al. 2021). The restored wetlands absorb wave energy, enhance water quality through nutrient filtration, and provide nursery habitats for fish and shellfish (O'Donnell 2017; Isdell et al. 2021; Roney et al. 2025).

Similarly, in Louisiana's Coastal Master Plan, the state government has prioritized marsh creation and barrier island restoration using dredged sediments (Louisiana

State 2012; Peyronnin et al. 2013). These vegetated wetlands act as bioshields that attenuate storm surges, reducing hurricane damage to inland communities (Rezaie et al. 2020; Gupta et al. 2025).

12.5.2 Mexico and the Caribbean: Mangrove Protection and Ecotourism Synergies

In Mexico's Yucatán Peninsula, mangrove restoration has been integrated into ecotourism and fisheries management (Cinti et al. 2025). The Sian Ka'an Biosphere Reserve and Celestún Lagoon projects restored thousands of hectares of mangroves degraded by shrimp farming and coastal development (Claudino-Sales 2019; Gayosso-Soto et al. 2024). These efforts not only improved storm protection but also enhanced the tourism appeal of the region, creating a win–win for conservation and local livelihoods (Arceo-Carranza et al. 2024; Cinti et al. 2025).

Across the Caribbean, bioshield restoration has been promoted by the Caribbean Community Climate Change Centre (CCCCC) as part of regional climate adaptation (Mohan 2025). Jamaica, Belize, and Dominica have launched mangrove planting and coral reef restoration programs that reinforce coastal resilience (Krauss et al. 2023). The combination of coral and mangrove bioshields has proven particularly effective in reducing coastal erosion and wave run-up during hurricanes (Krauss et al. 2023).

12.5.3 South America: Coastal Forests and Wetlands in Brazil and Colombia

Brazil's Restinga and other coastal forests, dominated by salt-tolerant vegetation, serve as natural bioshields against storm surges and dune erosion (Zamith and Scarano 2006; Garbin et al. 2018; Santos et al. 2023; Silva et al. 2024). Restoration efforts in Rio de Janeiro and São Paulo states have rehabilitated several hectares of degraded dunes and wetlands (Zamith and Scarano 2006; Loureiro and Nascimento 2022). The projects, supported by Brazil's Chico Mendes Institute for Biodiversity Conservation (ICMBio), emphasize biodiversity conservation and carbon sequestration in addition to protection (Schmidt et al. 2023).

In Colombia, mangrove restoration in the Ciénaga Grande de Santa Marta lagoon complex has reversed decades of degradation caused by hydrological alterations (Elster et al. 1999; Perdomo-Trujillo et al. 2021). The reconnection of natural water flows and reforestation with *Rhizophora mangle* and *Avicennia germinans* have significantly improved coastal resilience and fish productivity (Elster and Polanía 2017; Samaniego et al. 2021).

12.6 Successful Bioshield Projects in Oceania

Oceania's small island nations are among the most vulnerable to sea-level rise and extreme weather events (Forbes et al. 2013; Weir et al. 2023). Consequently, they have pioneered innovative, community-based bioshield initiatives that combine traditional knowledge with modern ecological science (Weir et al. 2023; Nunn et al. 2024).

12.6.1 Fiji: Community-Led Mangrove and Coral Bioshields

Fiji's Integrated Coastal Management (ICM) program has become a model for the Pacific (Hastings et al. 2012; Weeks and Jupiter 2013). Through the Locally Managed Marine Area (LMMA) network, villages have restored mangroves and adjacent coral reefs to buffer storm impacts (Hastings et al. 2012; UNDP 2012). These bioshields not only reduce wave energy but also support subsistence fisheries and tourism (Nurse et al. 2014; Giffin 2024).

By blending traditional no-take areas with scientific monitoring, Fijian communities have achieved long-term ecosystem recovery (Weeks and Jupiter 2013; Jupiter et al. 2014). The integration of mangrove and coral bioshields represents a holistic approach to coastal defense in island ecosystems, including those in Oceania (Jones et al. 2012; Carlson et al. 2021).

12.6.2 Vanuatu and the Solomon Islands: Ecosystem-Based Adaptation

In Vanuatu, the Pacific Ecosystem-based Adaptation to Climate Change (PEBACC) project, funded by SPREP and GIZ, has implemented mangrove restoration, coastal forest rehabilitation, and community training (Mackey et al. 2017; Weir et al. 2023). The project demonstrated that ecosystem-based measures can reduce storm surge heights considerably compared to unprotected coasts (Mackey et al. 2017; Sahin et al. 2021).

In the Solomon Islands, local governments and NGOs have restored mangroves around Honiara and Choiseul Province, reducing vulnerability to cyclones and sea-level rise (Warren-Rhodes et al. 2011; Ahn et al. 2025). These programs have also addressed gender inclusion, empowering women's groups to lead nursery and planting activities (Albert and Schwarz 2013; Bruckner and Paia 2025).

12.6.3 Australia: Saltmarsh and Mangrove Rehabilitation

Australia has a well-established record of coastal wetland conservation. In New South Wales and Queensland, the Coastal Wetlands Restoration Program has restored several hectares of salt marsh and mangrove habitats since 2015 (Rogers et al. 2023; Rowland et al. 2023). The restoration of tidal exchange and sediment supply has enhanced coastal resilience and biodiversity through vertical accretion (Saintilan et al. 2024).

In northern Australia, Indigenous ranger programs integrate traditional ecological knowledge with bioshield maintenance (Maclean et al. 2022; Robinson et al. 2022). These co-managed projects have strengthened both ecological outcomes and cultural continuity, demonstrating how inclusive governance supports sustainability (Jarvis et al. 2021; Wright et al. 2021).

12.7 Cross-cutting Factors Behind Success

A review of successful bioshield projects worldwide reveals several recurring factors (Fig. 12.1) that contribute to long-term success. These factors span ecological, socio-economic, institutional, and technological dimensions (Liu et al. 2021; Louarn et al. 2025; Nienaber and Imrie-Kuzu 2025; Paxton et al. 2024; Sloey et al. 2025).

12.7.1 Ecological Design and Site Suitability

Matching species composition and planting techniques to local hydrological, sediment, and salinity conditions is crucial (Silva and Amarasinghe 2021; Su et al. 2021). Projects that used native species adapted to site-specific conditions, such as *Rhizophora apiculata* in Vietnam, *Avicennia marina* in Kenya, and dune grasses in Europe, achieved higher survival rates and ecosystem stability (Lewis III 2005; Zhao et al. 2015).

Successful bioshields also incorporate natural geomorphological processes. Allowing tidal exchange, sediment accretion, and lateral migration ensures the long-term adaptability of vegetated buffers under changing climate and sea-level conditions (Wang et al. 2019; Breda et al. 2021; Liu et al. 2021; Sadat-Noori et al. 2021; Zhi et al. 2025).

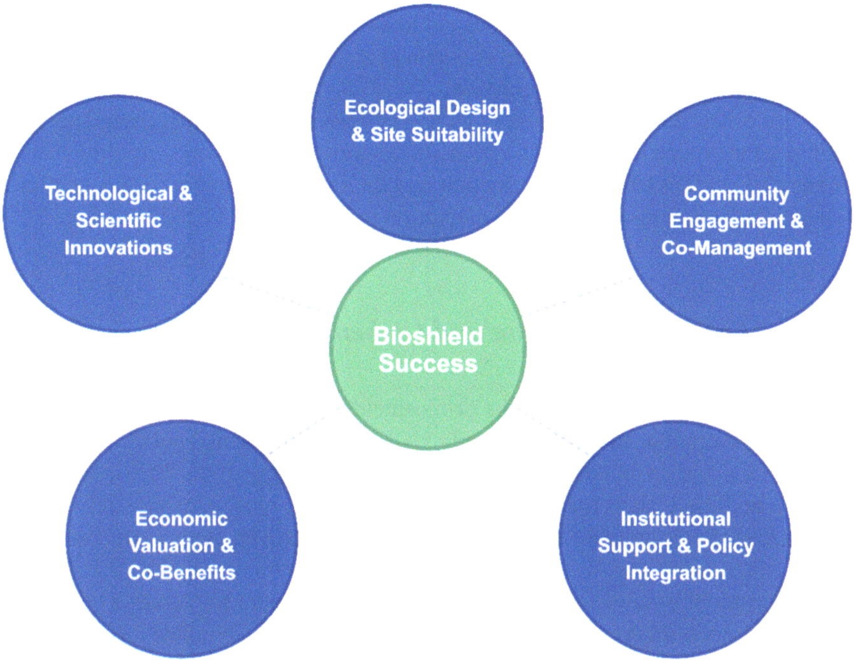

Fig. 12.1 Factors affecting the success of coastal bioshields

12.7.2 Community Engagement and Co-management

Virtually all enduring bioshield projects involve strong community participation (Parsons et al. 2025). Local ownership enhances protection, reduces vandalism, and supports maintenance (Parsons et al. 2025). For instance, Vietnam's Red Cross-led mangrove projects (Hoa 2012) and Kenya's Mikoko Pamoja initiative (Huff and Tonui 2017) empowered communities with both ecological and financial incentives.

Participatory governance models—where communities, NGOs, and local governments share responsibilities—promote transparency and adaptive management (Parsons et al. 2025; Ravazzoli et al. 2025). Projects that neglected social inclusion often faced land-use conflicts or low community support (Ebeler et al. 2025; Parsons et al. 2025).

12.7.3 Institutional Support and Policy Integration

Embedding bioshields within broader coastal zone management and climate adaptation policies ensures sustainability (Islam et al. 2021; MEFCC 2022). National strategies, such as Bangladesh's Coastal Greenbelt (MEFCC 2022) and the Philippines' National Greening Program (Lozano et al. 2025), provide legal and financial backing for large-scale implementation.

Policy coherence across sectors (forestry, fisheries, disaster management) prevents contradictory mandates (OECD 2020). International frameworks like the Sendai Framework for Disaster Risk Reduction and the Paris Agreement further reinforce bioshield adoption as a recognized adaptation measure (OECD 2020).

12.7.4 Economic Valuation and Co-benefits

Quantifying the ecosystem services of bioshields, such as avoided flood damages, carbon sequestration, and fisheries productivity, has strengthened policy support and funding (Narayan et al. 2016; Jakovac et al. 2020; Zhang et al. 2025). Studies in Vietnam and the United States have demonstrated that bioshields can deliver high benefit–cost ratios compared to hard infrastructure (Narayan et al. 2016; Pang et al. 2023).

Integrating bioshields into blue carbon markets, payment for ecosystem services (PES), and ecotourism provides financial sustainability (Choudhary et al. 2024). Community-based enterprises linked to mangrove products or eco-lodges further reinforce conservation incentives (Choudhary et al. 2024).

12.7.5 Technological and Scientific Innovations

Remote sensing, GIS, and drone technologies now play a critical role in monitoring bioshield health and spatial dynamics (McCarthy et al. 2017; Christofi et al. 2025). Countries like Australia have used satellite-based change detection to guide adaptive management (Lymburner et al. 2020).

Advances in hydrodynamic modeling have also improved bioshield design by quantifying wave attenuation, sediment transport, and flood risk reduction (Kamil et al. 2021). These technologies bridge the gap between scientific understanding and practical implementation.

12.8 Summary and Key Lessons

The global experience of bioshield implementation demonstrates that nature-based solutions can deliver effective, adaptive, and multifunctional coastal protection. From the mangrove belts of Southeast Asia to the dune systems of Europe and the salt marshes of North America, bioshields have proven capable of reducing wave energy, stabilizing shorelines, and safeguarding lives and livelihoods. Several key lessons emerge from this global review:

a. ***Integration over isolation***: Bioshields are most effective when integrated with engineered structures and spatial planning frameworks, creating hybrid systems that combine natural and artificial defenses.
b. ***Local engagement***: Long-term success depends on empowering local communities through co-management, benefit-sharing, and capacity building.
c. ***Ecological authenticity***: Using native species and respecting natural geomorphic processes enhances resilience and ecosystem integrity.
d. ***Policy and finance alignment***: Mainstreaming bioshields into climate adaptation and coastal management policies ensures institutional support, while innovative finance (carbon credits, PES, green bonds) secures long-term funding.
e. ***Monitoring and adaptive management***: Scientific monitoring using modern tools—satellite imagery, hydrodynamic models, and citizen science—supports data-driven adaptation.
f. ***Global cooperation***: Sharing best practices across regions through South-South and North–South collaborations accelerates learning and innovation.

As climate change intensifies, the global coastal community faces a critical choice between reactive hard engineering and proactive ecological resilience. Bioshields embody a pragmatic, scientifically grounded, and socially inclusive path forward.

This synthesis of successful projects around the world provides a robust foundation for understanding what works, why it works, and how bioshield strategies can be scaled up to meet the urgent challenges of coastal protection in the twenty-first century. The next chapter, “Challenges, Failures, and Lessons Learned”, will explore the other side of this narrative—examining the limitations, shortcomings, and hard-won lessons that complement the successes outlined here, offering a balanced perspective for future bioshield implementation.

References

Ahmed DA, Shaltout KH, Kamal SA (2014) Mediterranean sand dunes in Egypt: threatened habitat and endangered flora. Life Sci J 11(10):946–956

Ahn HK, Kwon S, Song C, Lim C (2025) Enhanced rapid mangrove habitat mapping approach to setting protected areas using satellite indices and deep learning: a case study of the Solomon Islands. Remote Sensing 17(14):2512. https://doi.org/10.3390/rs17142512

Albert JA, Schwarz AJ (2013) Mangrove management in Solomon Islands: case studies from Malaita Province. CGIAR research program on aquatic agricultural systems. Penang, Malaysia. Policy brief: AAS-2013-14

Andreasson E, Bechauf R (2025) Sustainable Asset Valuation (Savi) of Mangroves in Vietnam: an economic valuation of mangroves for flood protection and erosion control in the DEEP C Industrial Zone. NBI Report. https://nbi.iisd.org/wp-content/uploads/2025/06/mangroves-protecting-coastal-areas-vietnam.pdf. Accessed 27 Nov 2025

Arceo-Carranza D, Hernández Mendoza L, Teutli-Hernández C, Herrera-Silveira J, Caballero Vázquez J, Chiappa-Carrara X (2024) Mangrove ecosystems as fundamental habitats for fish from the Mexican Caribbean: an evaluation between a conserved and restoration zone. Reg Stud Mar Sci 77:103650. https://doi.org/10.1016/j.rsma.2024.103650

Arifanti VB, Kauffman JB, Ilman M, Tosiani A, Novita N (2022) Contributions of mangrove conservation and restoration to climate change mitigation in Indonesia. Glob Change Biol 28(15):4523–4538. https://doi.org/10.1111/gcb.16216

Awazi NP (2025) Building climate resilient communities along Africa's Coasts—the role of mangroves. Palgrave Macmillan Cham, 165 pp

Begum F, Lobry de Bruyn L, Kristiansen P, Islam MA (2021) Institutionalising co-management activities for conservation of forest resources: evidence from the Sundarban mangrove forest management of Bangladesh. J Environ Manag 298:113504. https://doi.org/10.1016/j.jenvman.2021.113504

Bilkovic DM, Mitchell MM, La Peyre MK, Toft JD (2017) Living shorelines: the science and management of nature-based coastal protection. CRC Press, 519 pp. https://doi.org/10.1201/9781315151465

Breda A, Saco PM, Sandi SG, Saintilan N, Riccardi G, Rodríguez JF (2021) Accretion, retreat and transgression of coastal wetlands experiencing sea-level rise. Hydrol Earth Syst Sci 25:769–786. https://doi.org/10.5194/hess-25-769-2021

Bruckner HK, Paia MT (2025) From mangroves to womangroves to feminist foodscapes: (en)gendering research on indigenous food livelihoods in the Solomon Islands. Agric Hum Values 42:507–525. https://doi.org/10.1007/s10460-024-10634-8

Burgess H, Kilkie P, Callaway T (2016) Understanding the physical processes occurring within a new coastal managed realignment site, Medmerry, Sussex, UK. In: Coastal management: changing coast, changing climate, changing minds. ICE Publishing, pp 263–272

Carlson RR, Evans LJ, Foo SA, Grady BW, Li J, Seeley M, Xu Y, Asner GP (2021) Synergistic benefits of conserving land-sea ecosystems. Glob Ecol Conserv 28:e01684. https://doi.org/10.1016/j.gecco.2021.e01684

Carney J, Gillespie TW, Rosomoff R (2014) Assessing forest change in a priority West African mangrove ecosystem: 1986–2010. Geoforum 53:126–135. https://doi.org/10.1016/j.geoforum.2014.02.013

Chee SY, Firth LB, Then AY, Yee JC, Mujahid A, Affendi YA, Amir AA, Lau CM, Ooi JL, Quek YA, Tan CE, Yap TK, Yeap CA (2021) Enhancing uptake of nature-based solutions for informing coastal sustainable development policy and planning: a Malaysia case study. Front Ecol Evol 9:708507. https://doi.org/10.3389/fevo.2021.708507

Choudhary B, Dhar V, Pawase AS (2024) Blue carbon and the role of mangroves in carbon sequestration: Its mechanisms, estimation, human impacts and conservation strategies for economic incentives. J Sea Res 199:102504. https://doi.org/10.1016/j.seares.2024.102504

Christofi D, Mettas C, Evagorou E, Stylianou N, Eliades M, Theocharidis C, Chatzipavlis A, Hasiotis T, Hadjimitsis D (2025) A review of open remote sensing data with GIS, AI, and UAV support for shoreline detection and coastal erosion monitoring. Appl Sci 15(9):4771. https://doi.org/10.3390/app15094771

Cinti A, Ramirez L, Castrejón M, Aburto JA, Loto L, Fulton S, Parma AM et al (2025) Small-scale fisheries in ecologically sensitive areas in Latin America and the Caribbean: do marine protected areas benefit fisheries governance? Ambio 54(1):20–42. https://doi.org/10.1007/s13280-024-02062-z

Claudino-Sales V (2019) Sian Ka'an, Mexico. In: Coastal world heritage sites. Coastal research library, vol 28. Springer, Dordrecht. https://doi.org/10.1007/978-94-024-1528-5_20

Cormier-Salem M, Panfili J (2016) Mangrove reforestation: greening or grabbing coastal zones and deltas? Case studies in Senegal. Afr J Aquat Sci 41(1):89–98. https://doi.org/10.2989/16085914.2016.1146122

Cortes J (2020) Back to mother nature: the potential of using nature-based solutions to mitigate climate change vulnerabilities along the coast of Dar es Salaam, Tanzania. Master thesis in Environmental studies and sustainability science. LUCSUS (Lund University Centre for Sustainability Studies), Sweden

Dale J, Burgess HM, Cundy AB (2016) Sedimentation rhythms and hydrodynamics in two engineered environments in an open coast managed realignment site. Mar Geol 383:120–131. https://doi.org/10.1016/j.margeo.2016.12.001

Dale J, Burgess HM, Nash DJ, Cundy AB (2018) Hydrodynamics and sedimentary processes in the main drainage channel of a large open coast managed realignment site. Estuar Coast Shelf Sci 215:100–111. https://doi.org/10.1016/j.ecss.2018.10.007

Damastuti E, Van Wesenbeeck BK, Leemans R, De Groot RS, Silvius MJ (2023) Effectiveness of community-based mangrove management for coastal protection: a case study from Central Java, Indonesia. Ocean Coast Manag 238:106498. https://doi.org/10.1016/j.ocecoaman.2023.106498

Daro Justine YE, Seenath A (2025) Vegetative nature-based solutions for coastal flood risk management: benefits, challenges, and uncertainties. Ocean Coast Manag 261:107520. https://doi.org/10.1016/j.ocecoaman.2024.107520

Davenport TM, Seitz RD, Knick KE, Jackson N (2018) Living shorelines support nearshore benthic communities in upper and lower Chesapeake Bay. Estuaries Coasts 41(Suppl 1):197–206. https://doi.org/10.1007/s12237-017-0361-8

Davranche A, Arzel C, Pouzet P, Carrasco AR, Lefebvre G, Lague D, Thibault M, Newton A, Fleurant C, Maanan M, Poulin B (2024) A multi-sensor approach to monitor the ongoing restoration of edaphic conditions for salt marsh species facing sea level rise: an adaptive management case study in Camargue, France. Sci Total Environ 908:168289. https://doi.org/10.1016/j.scitotenv.2023.168289

Diouf AA, Bernard Mingou PA, Ba T, Gueye M (2025) Wetlands of Senegal. In: Pullaiah T (ed) Wetlands of tropical and subtropical Asia and Africa: biodiversity, livelihoods and conservation, pp 341–369. https://doi.org/10.1002/9781394235278.ch15

Ebeler L, Balke T, Laurie E, Le H, Nguyen T, Renaud FG (2025) Local-scale impacts of mangrove restoration and conservation on coastal communities in two Vietnamese deltas: socio-economic and institutional dynamics. Ocean Coast Manag 269:107812. https://doi.org/10.1016/j.ocecoaman.2025.107812

El-Khalafy MM, Mostafa W, Al-Sodany YM, Shaltout KH, Toto SM (2025) Investigating sand dunes flora conservation based on remote sensing and predictive modeling in the Mediterranean coastal region, Egypt. Sci Rep 15:43583. https://doi.org/10.1038/s41598-025-29710-y

Elster C, Polanía J (2017) Posibilidades de recuperación del manglar en la Ciénaga Grande de Santa Marta (Colombia). Actualidades Biológicas 22(72):29–36. https://doi.org/10.17533/udea.acbi.329697

Elster C, Perdomo L, Schnetter ML (1999) Impact of ecological factors on the regeneration of mangroves in the Ciénaga Grande de Santa Marta, Colombia. Hydrobiologia 413:35–46. https://doi.org/10.1023/A:1003838809903

Fent A, Bardou R, Carney J, Cavanaugh K (2019) Transborder political ecology of mangroves in Senegal and The Gambia. Glob Environ Change 54:214–226. https://doi.org/10.1016/j.gloenvcha.2019.01.003

Forbes DL, James TS, Sutherland M, Nichols SE (2013) Physical basis of coastal adaptation on tropical small islands. Sustain Sci 8(3):327–344. https://doi.org/10.1007/s11625-013-0218-4

Friess DA, Shribman ZI, Stankovic M, Iram N, Baustian MM, Ewers Lewis CJ (2024) Restoring blue carbon ecosystems. Camb Prisms: Coast Futures 2:e9. https://doi.org/10.1017/cft.2024.9

Galluccio G, Hinkel J, Fiorini Beckhauser E, Bisaro A, Biancardi Aleu R, Campostrini P, Casas MF, Espin O, Vafeidis AT (2024) Sea level rise in Europe: adaptation measures and decision-making principles. In: van den Hurk B, Pinardi N, Kiefer T, Larkin K, Manderscheid P, Richter K (eds) Sea level rise in Europe: 1st assessment report of the knowledge hub on sea level rise (SLRE1). Copernicus Publications, State Planet, 3-slre1, 6. https://doi.org/10.5194/sp-3-slre1-6-2024

Garbin ML, Misaki F, Ferreira PF, Soares RB, Mariotte P, Sansevero JB, Rocha PG, Silva AG (2018) Long-term regeneration of a tropical plant community after sand mining. Ecol Evol 8(11):5712. https://doi.org/10.1002/ece3.4111

Gayosso-Soto E, Cohuo S, Sánchez-Sánchez JA, Macario-González L, Villegas-Sánchez CA, Medina-Quej A, Castro-Pérez JM et al (2024) The environmental legal framework of Mexican

Caribbean dunes: a retrospective case study of vegetation and coastal dune loss in the Sian Ka'an biosphere reserve. Land 13(9):1533. https://doi.org/10.3390/land13091533
Giffin AL (2024) Marine and coastal ecosystem-based adaptation. PhD thesis, Griffith University, Australia. https://doi.org/10.25904/1912/5193
Gijón Mancheño A, Jafino BA, Hofland B, Van Wesenbeeck BK, Kazi S, Urrutia I (2024) Nature meets infrastructure: the role of Mangroves in strengthening Bangladesh's coastal flood defenses. Sustainability 17(4):1567. https://doi.org/10.3390/su17041567
Global Disaster Preparedness Center (GDPC) (2022) Mangrove restoration and protection in North-Eastern Viet Nam. https://preparecenter.org/story/mangrove-restoration-and-protection-in-north-eastern-viet-nam/. Accessed 27 Nov 2025
Government of Bangladesh (2008) Cyclone Sidr in Bangladesh: damage, loss and needs assessment for disaster recovery and reconstruction. Prepared by the Government of Bangladesh, Assisted by the International Development Community with Financial Support from the European Commission
Gupta A, Bilskie MV, Woodson CB (2025) The potential of wetlands and barrier islands as a coastal defense in mitigating the storm surge. Front Ecol Evol 13:1524570. https://doi.org/10.3389/fevo.2025.1524570
Hagedoorn LC, Appeaning Addo K, Koetse MJ, Kinney K, Van Beukering PJ (2021) Angry waves that eat the coast: an economic analysis of nature-based and engineering solutions to coastal erosion. Ocean Coast Manag 214:105945. https://doi.org/10.1016/j.ocecoaman.2021.105945
Hai N, Dell B, Phuong V, Haroer RJ (2021) Towards a more robust approach for the restoration of mangroves in Vietnam. Ann For Sci 77:18 (2020). https://doi.org/10.1007/s13595-020-0921-0
Hassan S, Saleh M, Mohamed B, Elhebiry MS, Abdeldayem A, Issawy E, Zahran K, Kamh S (2025) Environmental risk assessment of the Nile Delta, Egypt, based on radar interferometry, altimetry, and geodetic measurements. Sci Rep 15(1):19209. https://doi.org/10.1038/s41598-025-03831-w
Hastings JG, Gruby RL, Sievanen LS (2012) Science-based coastal management in Fiji: two case studies from the NGO sector. Mar Policy 36(4):907–914. https://doi.org/10.1016/j.marpol.2012.01.002
Hoa SLT (2012) Mangrove community forestry in Vietnam. RECOFTC publication summarizing Vietnam mangrove community forestry, including Red Cross involvement. https://www.recoftc.org/sites/default/files/publications/resources/recoftc-0000121-0005-en.pdf. Accessed 18 Dec 2025
Huff A, Tonui C (2017) Making 'mangroves together': carbon, conservation and co-management in Gazi Bay, Kenya, STEPS Working Paper 95. STEPS Centre, Brighton
Huynh LT, Su J, Wang Q, Stringer LC, Switzer AD, Gasparatos A (2024) Meta-analysis indicates better climate adaptation and mitigation performance of hybrid engineering-natural coastal defence measures. Nat Commun 15(1):2870. https://doi.org/10.1038/s41467-024-46970-w
IFRC (2022) The multiple benefits of mangrove restoration: lessons from Viet Nam. The International Federation of Red Cross and Red Crescent Societies. https://disasterlaw.ifrc.org/media/3886. Accessed 27 Nov 2025
Iftekhar MS, Islam MR (2004) Managing mangroves in Bangladesh: a strategy analysis. J Coast Conserv 10:139–146. https://doi.org/10.1652/1400-0350(2004)010[0139:MMIBAS]2.0.CO;2
Indrajaya Y, Yuwati TW, Lestari S, Winarno B, Narendra BH, Nugroho HY, Rachmanadi D, Turjaman M, Adi RN, Savitri E, Putra PB, Santosa PB, Nugroho NP, Cahyono SA, Wahyuningtyas RS, Prayudyaningsih R, Halwany W, Siarudin M, Widiyanto A, Mendham D et al (2022) Tropical forest landscape restoration in Indonesia: a review. Land 11(3):328. https://doi.org/10.3390/land11030328
Isdell RE, Bilkovic DM, Guthrie AG, Mitchell MM, Chambers RM, Leu M, Hershner C (2021) Living shorelines achieve functional equivalence to natural fringe marshes across multiple ecological metrics. PeerJ 9:e11815. https://doi.org/10.7717/peerj.11815
Islam MA, Paull DJ, Griffin AL, Murshed S (2020) Assessing ecosystem resilience to a tropical cyclone based on ecosystem service supply proficiency using geospatial techniques and social

responses in coastal Bangladesh. Int J Disaster Risk Reduct 49:101667. https://doi.org/10.1016/j.ijdrr.2020.101667
Islam S, Anzum F, Kamal AB, Akter M, Rezaie AM, Khan MR (2021) The integration of nature-based solutions into climate adaptation policy and planning in Bangladesh. J Sci Policy Govern 18(2):1–17. https://doi.org/10.38126/JSPG180204
Jakovac CC, Latawiec AE, Lacerda E, Leite Lucas I, Korys KA, Iribarrem A, Malaguti GA, Turner RK, Luisetti T, Baeta Neves Strassburg B (2020) Costs and carbon benefits of mangrove conservation and restoration: a global analysis. Ecol Econ 176:106758. https://doi.org/10.1016/j.ecolecon.2020.106758
Jarvis D, Stoeckl N, Larson S, Grainger D, Addison J, Larson A (2021) The learning generated through indigenous natural resources management programs increases quality of life for indigenous people—improving numerous contributors to wellbeing. Ecol Econ 180:106899. https://doi.org/10.1016/j.ecolecon.2020.106899
Jones HP, Hole DG, Zavaleta ES (2012) Harnessing nature to help people adapt to climate change. Nat Clim Change 2(7):504–509. https://doi.org/10.1038/nclimate1463
Jupiter SD, Cohen PJ, Weeks R, Tawake A, Govan H (2014) Locally-managed marine areas: multiple objectives and diverse strategies. Pac Conserv Biol 20(2):165–179. https://doi.org/10.1071/PC140165
Kamil EA, Takaijudin H, Hashim AM (2021) Mangroves as coastal bio-shield: a review of mangroves performance in wave attenuation. Civ Eng J 7(11):1964–1981. https://doi.org/10.28991/cej-2021-03091772
Khanal S, Koirala M, Shang Z (2024) Ecological conservation and restoration in Bangladesh. In: Sustainable ecological restoration and conservation in the Hindu Kush Himalayan Region: a comprehensive review. CABI, GB, pp 73–83. https://doi.org/10.1079/9781800622579.0006
Knight J (2024) Nature-based solutions for coastal resilience in South Africa. South Afr Geogr J 106(1):21–50. https://doi.org/10520/ejc-sageo_v106_n1_a21
Krauss KW, T Whelan KR, Kennedy JP, Friess DA, Rogers CS, Stewart HA, Grimes KW, Trench CA, Ogurcak DE, Toline CA, Ball LC, From AS (2023) Framework for facilitating mangrove recovery after hurricanes on Caribbean islands. Restor Ecol 31(7):e13885. https://doi.org/10.1111/rec.13885
Krawiec K (2017) Medmerry, West Sussex, UK: coastal evolution from the Neolithic to the Medieval period and community resilience to environmental change. Hist Environ Policy Pract 8(2):101–112. https://doi.org/10.1080/17567505.2017.1317081
Lewis RR III (2005) Ecological engineering for successful management and restoration of mangrove forests. Ecol Eng 24(4):403–418. https://doi.org/10.1016/j.ecoleng.2004.10.003
Liu Z, Fagherazzi S, Cui B (2021) Success of coastal wetlands restoration is driven by sediment availability. Commun Earth Environ 2(1):44. https://doi.org/10.1038/s43247-021-00117-7
Livelihoods (2019) The proof by 10: results of the study on the social impacts of the largest mangrove restoration project of the Livelihoods Carbon Fund in Senegal. https://livelihoods.eu/the-proof-by-10-results-of-the-study-on-the-social-impacts-of-the-largest-mangrove-restoration-project-of-the-carbon-livelihoods-fund-in-senegal/. Accessed 27 Nov 2025
Louarn A, Meur-Ferec C, Hervé-Fournereau N (2025) The concept of 'nature-based solutions' applied to urban coastal risks: a bibliometric and content analysis review. Ocean Coast Manag 261:107530. https://doi.org/10.1016/j.ocecoaman.2024.107530
Louisiana State (2012) Louisiana comprehensive master plan for a sustainable coast. Coastal protection and restoration authority of Louisiana. Executive summary. https://coastal.la.gov/wp-content/uploads/2023/05/230418_CPRA_Executive-Summary_final.pdf. Accessed 27 Nov 2025
Loureiro N, Nascimento MT (2022) Survival, seedlings growth and natural regeneration in areas under ecological restoration in a sandy coastal plain (Restinga) of southeastern Brazil. Austral Ecol 47(2):326–340. https://doi.org/10.1111/aec.13114
Lozano AL, Cortado JMG, Cosep PA, Javier JAL (2025) National Greening Program in the Philippines: a thematic review of policy frameworks, implementation practices and outcomes. Int J Environ Clim Change 15(7):125–139. https://doi.org/10.9734/ijecc/2025/v15i74918

Lukman KM, Uchiyama Y, Kohsaka R (2021) Sustainable aquaculture to ensure coexistence: perceptions of aquaculture farmers in East Kalimantan, Indonesia. Ocean Coast Manag 213:105839. https://doi.org/10.1016/j.ocecoaman.2021.105839

Lymburner L, Bunting P, Lucas R, Scarth P, Alam I, Phillips C, Ticehurst C, Held A (2020) Mapping the multi-decadal mangrove dynamics of the Australian coastline. Remote Sens Environ 238:111185. https://doi.org/10.1016/j.rse.2019.05.004

Mackey B, Ware D, Nalau J, Buckwell A, Sahin O, Fleming C, Smart J, Connolly R, Hallgren W (2017)Vanuatu Ecosystem and Socio-economic Resilience Analysis and Mapping (ESRAM). http://hdl.handle.net/10072/395716. Accessed 18 Dec 2025

Maclean K, Robinson C, Bock E, Rist P (2022) Reconciling risk and responsibility on Indigenous country: bridging the boundaries to guide knowledge sharing for cross-cultural biosecurity risk management in northern Australia. J Cult Geogr 39(1):32–54. https://doi.org/10.1080/08873631.2021.1911078

March A, Woolley M, Failler P (2024) Integration of climate change mitigation and adaptation in Blue Economy planning in Africa. Mitig Adapt Strateg Glob Change 29:38. https://doi.org/10.1007/s11027-024-10133-5

Mascarenhas A, Jayakumar S (2008) An environmental perspective of the post-tsunami scenario along the coast of Tamil Nadu, India: role of sand dunes and forests. J Environ Manag 89(1):24–34. https://doi.org/10.1016/j.jenvman.2007.01.053

Mason PA, Bilkovic DM, Guthrie AG, Nunez K, Jones S, Boyd C (2026) Encouraging living shorelines in the United States—a review of state definitions and policies. Mar Policy 185:106973. https://doi.org/10.1016/j.marpol.2025.106973

Masria A, Iskander M, Negm A (2015) Coastal protection measures, case study (Mediterranean zone, Egypt). J Coast Conserv 19:281–294. https://doi.org/10.1007/s11852-015-0389-5

McCarthy MJ, Colna KE, El-Mezayen MM, Laureano-Rosario AE, Méndez-Lázaro P, Otis DB, Muller-Karger FE et al (2017) Satellite remote sensing for coastal management: a review of successful applications. Environ Manag 60(2):323–339. https://doi.org/10.1007/S00267-017-0880-X

Ministry of Environment, Forest and Climate Change (MEFCC) (2022) National Adaptation Plan of Bangladesh (2023–2050) Ministry of Environment, Forest and Climate Change, Government of the People's Republic of Bangladesh

Mohan PS (2025) Climate change adaptation in small island developing states: evidence from the nationally determined contributions of Caribbean States. Reg Environ Change 25(1):32. https://doi.org/10.1007/s10113-025-02369-x

Monioudi IN, Vousdoukas MI, Giardino A, Stocchino A, Mentaschi L, Feyen L (2025) Impacts of sea level rise and adaptation across Asia and the Pacific. Sci Rep 15(1):35742. https://doi.org/10.1038/s41598-025-11517-6

Moritsch M, Manrow M, Raes L (2022) Coastal ecosystem services modeling in Latin America to guide conservation and restoration strategies: the case of mangroves in Guatemala and El Salvador. Front Ecol Evol 10:843145. https://doi.org/10.3389/fevo.2022.843145

Muñoz-Reinoso JC, Jordán RV, Tejada-Tejada M (2020) Analysis of spatio-temporal changes in the vegetation cover of the coastal landscape of Doñana. J Coast Res 95(SI):113–117

Narayan S, Beck MW, Reguero BG, Losada IJ, van Wesenbeeck B, Pontee N et al (2016) The effectiveness, costs and coastal protection benefits of natural and nature-based defences. PLoS ONE 11(5):e0154735. https://doi.org/10.1371/journal.pone.0154735

Nguyen TT, Pittock J, Nguyen BH (2017) Integration of ecosystem-based adaptation to climate change policies in Viet Nam. Climat Change 142:97–111. https://doi.org/10.1007/s10584-017-1936-x

Nienaber A-M, Imrie-Kuzu D (2025) Transformative change in coastal biodiversity conservation: a systematic literature review of governance, social-ecological, and cultural pathways. Sustainability 17(24):11186. https://doi.org/10.3390/su172411186

Nunn PD, Kumar R, Barrowman HM, Chambers L, Fifita L, Gegeo D, Gomese C, McGree S, Rarai A, Cheer K, Esau D, Fong T, Fong-Lomavatu M, Geraghty P, Heorake T, Kekeubata E,

Korovulavula I, Kubunavanua E, Lui S, Waiwai M (2024) Traditional knowledge for climate resilience in the Pacific Islands. Wiley Interdiscip Rev Clim Change 15(4):e882. https://doi.org/10.1002/wcc.882

Nurse LA, McLean RF, Agard J, Briguglio LP, Duvat-Magnan V, Pelesikoti N, Webb A et al (2014) Small islands. Climate change 2014: impacts, adaptation, and vulnerability. Part B: Regional aspects. Contribution of working group II to the fifth assessment report of the intergovernmental panel on climate change, p 1613

Nyamasege RM (2019) Influence of conservation interventions by Mikoko Pamoja project on the socioeconomic wellbeing of local communities living in Gazi Bay, Kwale County, Kenya. Doctoral dissertation, Africa Nazarene University, Kenya

Nyangoko BP, Berg H, Mangora MM, Shalli MS, Gullström M (2022) Local perceptions of changes in mangrove ecosystem services and their implications for livelihoods and management in the Rufiji Delta, Tanzania. Ocean Coast Manag 219:106065. https://doi.org/10.1016/j.ocecoaman.2022.106065

O'Donnell JE (2017) Living shorelines: a review of literature relevant to New England coasts. J Coast Res 33(2):435–451

OECD (2020) Common ground between the Paris Agreement and the Sendai Framework: climate change adaptation and disaster risk reduction. OECD Publishing, Paris. https://doi.org/10.1787/3edc8d09-en

Oliver S (2024) Inventing the managed realignment of the coast: trying 'to live with nature not defeat her.' Geogr J 190(1):e12529. https://doi.org/10.1111/geoj.12529

Pang S, Abdul Majid M, Perera HA, Sarkar MS, Ning J, Zhai W, Guo R, Deng Y, Zhang H (2023) A systematic review and global trends on blue carbon and sustainable development: a bibliometric study from 2012 to 2023. Sustainability 16(6):2473. https://doi.org/10.3390/su16062473

Parsons M, Godden NJ, Henrique KP, Tschakert P, Gonda N, Atkins E, Crease RP et al (2025) Participatory approaches to climate adaptation, resilience, and mitigation: a systematic review. Ambio 1–16. https://doi.org/10.1007/s13280-025-02202-z

Paxton AB, Riley TN, Steenrod CL, Puckett BJ, I BA, Paliotti ST, Adler AM, Exar L, T McLean JE, Kelley J, Zhang YS, Smith CS, Gittman RK, Silliman BR (2024) Evidence on the performance of nature-based solutions interventions for coastal protection in biogenic, shallow ecosystems: a systematic map. Environ Evid 13:28. https://doi.org/10.1186/s13750-024-00350-5

Pender G, Faulkner H (eds) (2011) Flood risk science and management. Wiley-Blackwell, Chichester, West Sussex, 528 pp

Perdomo-Trujillo LV, Mancera-Pineda JE, Medina-Calderón JH, Sánchez-Núñez DA, Schnetter ML (2021) Effect of restoration actions on organic carbon pools in the Lagoon—Delta Ciénaga Grande de Santa Marta, Colombian Caribbean. Water 13(9):1297. https://doi.org/10.3390/w13091297

Perricone V, Mutalipassi M, Mele A, Buono M, Vicinanza D, Contestabile P (2023) Nature-based and bioinspired solutions for coastal protection: an overview among key ecosystems and a promising pathway for new functional and sustainable designs. ICES J Mar Sci 80(5):1218–1239. https://doi.org/10.1093/icesjms/fsad080

Peyronnin N, Green M, Richards CP, Owens A, Reed D, Chamberlain J, Belhadjali K et al (2013) Louisiana's 2012 coastal master plan: overview of a science-based and publicly informed decision-making process. J Coast Res 67:1–15. https://doi.org/10.2112/SI_67_1.1

Primavera JH, Esteban JMA (2008) A review of mangrove rehabilitation in the Philippines: successes, failures and future prospects. Wetl Ecol Manag 16:345–358. https://doi.org/10.1007/s11273-008-9101-y

Rahman HM, Sherren K, Van Proosdij D (2018) Institutional innovation for nature-based coastal adaptation: lessons from salt marsh restoration in Nova Scotia, Canada. Sustainability 11(23):6735. https://doi.org/10.3390/su11236735

Rasowo JO, Nyonje B, Olendi R, Orina P, Odongo S (2024) Towards environmental sustainability: further evidences from decarbonization projects in Kenya's Blue Economy. Front Mar Sci 11:1239862. https://doi.org/10.3389/fmars.2024.1239862

Ravazzoli E, Lavarello-Schettini R, Oberti B, Maino F (2025) Community-based approaches in disaster risk reduction and climate change adaptation: an analysis of applied participatory processes. Clim Change 178(12):225. https://doi.org/10.1007/s10584-025-04078-0

Ray K, Basak SK, Giri CK, Kotal HN, Mandal A, Chatterjee K, Saha S, Biswas B, Mondal S, Das I, Ghosh A, Bhadury P, Joshi R (2024) Ecological restoration at pilot-scale employing site-specific rationales for small-patch degraded mangroves in Indian Sundarbans. Sci Rep 14(1):12952. https://doi.org/10.1038/s41598-024-63281-8

Rezaie AM, Loerzel J, Ferreira CM (2020) Valuing natural habitats for enhancing coastal resilience: wetlands reduce property damage from storm surge and sea level rise. PLoS ONE 15(1):e0226275. https://doi.org/10.1371/journal.pone.0226275

Rijke J, van Herk S, Zevenbergen C, Ashley R (2012) Room for the River: delivering integrated river basin management in the Netherlands. Int J River Basin Manag 10(4):369–382. https://doi.org/10.1080/15715124.2012.739173

Robinson CJ, Macdonald JM, Douglas M, Perry J, Setterfield S, Cooper D, Bangalang NG et al (2022) Using knowledge to care for country: indigenous-led evaluations of research to adaptively co-manage Kakadu National Park, Australia. Sustain Sci 17(2):377–390. https://doi.org/10.1007/s11625-021-01015-9

Rodríguez-Vidal J, Bardají T, Zazo C, Goy JL, Borja F, Dabrio CJ, Abad M et al (2014) Coastal dunes and marshes in Doñana National Park. In: Landscapes and landforms of Spain. Springer Netherlands, Dordrecht, pp. 229–238. https://doi.org/10.1007/978-94-017-8628-7_19

Rogers K, Lal KK, Asbridge EF, Dwyer PG (2023) Coastal wetland rehabilitation first-pass prioritisation for blue carbon and associated co-benefits. Mar Freshw Res 74(3):177–199. https://doi.org/10.1071/MF22014

Roney SH, Haas K, Bliss T, Weissburg MJ (2025) Restored oyster reefs function as living shorelines to reduce wave energy in intertidal marshes. Ecol Eng 221:107774. https://doi.org/10.1016/j.ecoleng.2025.107774

Rowland PI, Hagger V, Lovelock CE (2023) Opportunities for blue carbon restoration projects in degraded agricultural land of the coastal zone in Queensland, Australia. Reg Environ Change 23:42. https://doi.org/10.1007/s10113-022-02013-y

Rowland PI, Wartman M, Bursic J, Carnell P (2024) Restored and created tidal marshes recover ecosystem services over time. Environ Sustain Indic 24:100539. https://doi.org/10.1016/j.indic.2024.100539

Rudianto R, Darmawan V, Isdianto A, Bintoro J (2022) Restoration of coastal ecosystems as an approach to the integrated mangrove ecosystem management and mitigation and adaptation to climate changes in north coast of East Java. J Coast Conserv 26:37. https://doi.org/10.1007/s11852-022-00865-4

Sadat-Noori M, Rankin C, Rayner D, Heimhuber V, Gaston T, Drummond C, Glamore W et al (2021) Coastal wetlands can be saved from sea level rise by recreating past tidal regimes. Sci Rep 11(1):1196. https://doi.org/10.1038/s41598-021-80977-3

Sahin O, Hadwen WL, Buckwell A, Fleming C, Ware D, Smart JC, Mackey B et al (2021) Assessing how ecosystem-based adaptations to climate change influence community wellbeing: a Vanuatu case study. Reg Environ Change 21(4):90. https://doi.org/10.1007/s10113-021-01809-8

Saintilan N, Sun Y, Lovelock CE, Rogers K, Goddard M, Hutley LB, Jones A et al (2024) Vertical accretion trends in Australian tidal wetlands. Estuaries Coasts 47(7):2057–2070. https://doi.org/10.1007/s12237-023-01267-x

Samaniego J, Schmidt KU, Carlino H et al (2021) Current understanding of the potential impact of carbon dioxide removal approaches on the SDGs in selected countries in Latin America and the Caribbean. Final Report, Carnegie Climate Governance Initiative (C2G)/Economic Commission for Latin America and the Caribbean (ECLAC), March 2021

Santos VJ, Santos CR, Pires CS, Rocha HA, Zickel CS, Almeida EB Jr (2023) Restingas no litoral sul da Bahia: diversidade, estrutura e distribuição da vegetação lenhosa. J Environ Anal Prog 8(3):172–183. https://doi.org/10.24221/jeap.8.3.2023.5523.172-183

Santos MM, Ferreira AV, Lanzinha JCG (2024) Climate change, coastal vulnerability, and mangrove protection in Africa. In: Proceedings of the international conference on changing cities VI: spatial, design, landscape, heritage and socio-economic dimensions, Rhodes Island, Greece
Sasmito SD, Basyuni M, Kridalaksana A, Lovelock CE, Murdiyarso D (2022) Challenges and opportunities for achieving sustainable development goals through restoration of Indonesia's mangroves. Nat Ecol Evol 7(1):62–70. https://doi.org/10.1038/s41559-022-01926-5
Schmidt AJ, Delgado-Mendez JM, Cunha-Lignon M (2023) Mangrove and salt marsh protected areas in Brazil. In: Brazilian mangroves and salt marshes. Springer International Publishing, Cham, pp 335–353. https://doi.org/10.1007/978-3-031-13486-9_16
Selvam V, Ravishankar T, Karunagaran VM, Ramasubramanian R, Eganathan P, Parida AK (2005) Toolkit for establishing coastal bioshield. M. S. Swaminathan Research Foundation, Centre for Research on Sustainable Agriculture and Rural Development, 120 pp
Selvam V, Ramasubramanian R, Sivakumar A, Sophia JD (2012) Role of coastal bioshields in strengthening the coastal resilience. In: Gonsalves JF, Mohan P (eds) Strengthening resilience in post-disaster situations: stories, experience and lessons learned. International Development Research Centre, Ottawa, pp 538–541
Sen H (2019) The Sundarbans: a disaster-prone eco-region. Springer, Berlin, p 654
Siddiqui MR, Ishtiaque A (2025) Roles and nature of public participation in climate change adaptation in coastal South Asia. Discov Environ 3:166. https://doi.org/10.1007/s44274-025-00360-1
Silva WD, Amarasinghe M (2021) Response of mangrove plant species to a saline gradient: implications for ecological restoration. Acta Botanica Brasilica 35(1):151–160. https://doi.org/10.1590/0102-33062020abb0170
Silva R, Martínez M, Odériz I, Mendoza E, Feagin R (2016) Response of vegetated dune–beach systems to storm conditions. Coast Eng 109:53–62. https://doi.org/10.1016/j.coastaleng.2015.12.007
Silva DFD, Silva TSD, Esteves LS, Sfreddo GA (2024) The dunes contribution to the ecosystem service of coastal protection on the northern coast of RS. Soc Nat 36:e70074. https://doi.org/10.14393/SN-v36-2024-70074x
Sloey TM, Hildebrandt S, Morris RL, Bilskie MV, Bland A, Bushek D, Bilkovic DM et al (2025) A blueprint to greener shorelines: advancing the effectiveness, sustainability, and widespread adoption of coastal nature-based solutions through transdisciplinary research. Estuaries Coasts 48(5):133. https://doi.org/10.1007/s12237-025-01571-8
Smith CS, Rudd ME, Gittman RK, Melvin EC, Patterson VS, Renzi JJ, Wellman EH, Silliman BR (2020) Coming to terms with living shorelines: a scoping review of novel restoration strategies for shoreline protection. Front Mar Sci 7:541502. https://doi.org/10.3389/fmars.2020.00434
Sridhar A (2006) Pre-and post-tsunami coastal planning and land-use policies and issues in India. In: Proceedings of the workshop on coastal area planning and management in Asian tsunami-affected countries, Bangkok, Thailand, 25 pp. Organized by FAO
Su J, Friess DA, Gasparatos A (2021) A meta-analysis of the ecological and economic outcomes of mangrove restoration. Nat Commun 12(1):5050. https://doi.org/10.1038/s41467-021-25349-1
Takagi H, Thao ND, Esteban M, Mikami T, Van Cong L, Ca VT (2015) Coastal disasters in Vietnam. In: Handbook of coastal disaster mitigation for engineers and planners. Butterworth-Heinemann, pp 235–255. https://doi.org/10.1016/B978-0-12-801060-0.00012-5
Thomas K (2014) Managed realignment in the UK: the role of the Environment Agency. In: Managed realignment: A viable long-term coastal management strategy? Springer Netherlands, Dordrecht, pp. 83–94. https://doi.org/10.1007/978-94-017-9029-1_8
Torresan S, Furlan E, Critto A, Michetti M, Marcomini A (2020) Egypt's coastal vulnerability to sea-level rise and storm surge: present and future conditions. Integr Environ Assess Manag 16(5):761–772. https://doi.org/10.1002/ieam.4280
UK Environment Agency (2014) Environment Agency flood risk schemes win 3 construction awards. https://www.gov.uk/government/news/environment-agency-flood-risk-schemes-win-3-construction-awards. Accessed 27 Dec 2025

United Nations Development Programme (UNDP) (2012) Fiji locally-managed marine area network, Fiji. Equator initiative case study series. New York, USA
Van Alphen S (2020) Room for the river: innovation, or tradition? The case of the Noordwaard. In: Hein C (ed) Adaptive strategies for water heritage, pp 309–323. https://doi.org/10.1007/978-3-030-00268-8_15
Van den Hoven K, Kroeze C, Van Loon-Steensma JM (2022) Characteristics of realigned dikes in coastal Europe: overview and opportunities for nature-based flood protection. Ocean Coast Manag 222:106116. https://doi.org/10.1016/j.ocecoaman.2022.106116
van Loon-Steensma JM (2015) Salt marshes to adapt the flood defences along the Dutch Wadden Sea coast. Mitig Adapt Strateg Glob Change 20:929–948. https://doi.org/10.1007/s11027-015-9640-5
Veettil BK, Ward RD, Dung NTK, Van DD, Quang NX, Hoai PN, Hoang N (2021) The use of bioshields for coastal protection in Vietnam: current status and potential. Reg Stud Mar Sci 47:101945. https://doi.org/10.1016/j.rsma.2021.101945
Wang F, Lu X, Sanders CJ, Tang J (2019) Tidal wetland resilience to sea level rise increases their carbon sequestration capacity in United States. Nat Commun 10:5434. https://doi.org/10.1038/s41467-019-13294-z
Warren-Rhodes K, Schwarz AM, Boyle LN, Albert J, Agalo SS, Warren R, Duke N et al (2011) Mangrove ecosystem services and the potential for carbon revenue programmes in Solomon Islands. Environ Conserv 38(4):485–496. https://doi.org/10.1017/S0376892911000373
Waryszak P, Gavoille A, Whitt AA, Kelvin J, Macreadie PI (2021) Combining gray and green infrastructure to improve coastal resilience: lessons learnt from hybrid flood defenses. Coast Eng J 63(3):335–350. https://doi.org/10.1080/21664250.2021.1920278
Weeks R, Jupiter SD (2013) Adaptive comanagement of a marine protected area network in Fiji. Conserv Biol 27(6):1234–1244. https://doi.org/10.1111/cobi.12153
Weir T, Dovey L, Okrupa M (2023) Ecosystem-based adaptation in the Pacific Islands. In: Leal Filho W, Nagy GJ, Ayal D (eds) Handbook of nature-based solutions to mitigation and adaptation to climate change. Springer, Cham. https://doi.org/10.1007/978-3-030-98067-2_26-1
Widayanti TF, Irfan AM, Djafar EM, Hakim MZ, Muin AM, Riza M, Aswan M (2023) The role of the biological diversity convention in mangroves rehabilitation in Indonesia. IOP Conf Ser Earth Environ Sci 1181:012006. https://doi.org/10.1088/1755-1315/1181/1/012006
Wood A, Stedman-Edwards P, Mang J (2013) Tanzania: Rufiji, Ruvu and Wami Mangroves. In: The root causes of biodiversity loss. Routledge, pp 309–336
World Bank (2024) Country climate and development report Senegal. The World Bank Group, Washington, DC, USA, 136 pp
Wright A, Yap M, Jones R, Richardson A, Davis V, Lovett R (2021) Examining the associations between indigenous rangers, culture and wellbeing in Australia, 2018–2020. Int J Environ Res Public Health 18(6):3053. https://doi.org/10.3390/ijerph18063053
Wylie L, Sutton-Grier AE, Moore A (2016) Keys to successful blue carbon projects: lessons learned from global case studies. Mar Policy 65:76–84. https://doi.org/10.1016/j.marpol.2015.12.020
Zamith LR, Scarano FR (2006) Restoration of a Restinga sandy coastal plain in Brazil: survival and growth of planted woody species. Restor Ecol 14(1):87–94. https://doi.org/10.1111/j.1526-100X.2006.00108.x
Zhang J, Lu Z, Zhou J, Qin G, Bai Y, Sanders CJ, Macreadie PI, Yuan J, Huang X, Wang F (2025) Getting the best of carbon bang for mangrove restoration buck. Nat Commun 16(1):1297. https://doi.org/10.1038/s41467-025-56587-2
Zhao Q, Bai J, Huang L, Gu B, Lu Q, Gao Z (2015) A review of methodologies and success indicators for coastal wetland restoration. Ecol Ind 60:442–452. https://doi.org/10.1016/j.ecolind.2015.07.003
Zhi L, Shao D, Xie X, Li X, Cui B, Bai J (2025) Natural process-based coastal wetland restoration by relocating river flow paths in a river-dominated delta. Commun Earth Environ 6(1):608. https://doi.org/10.1038/s43247-025-02612-7

Chapter 13
Coastal Bioshields: Challenges, Failures, and Lessons Learned

Abstract Despite growing global recognition of bioshields as effective nature-based solutions, numerous initiatives across Asia have encountered significant challenges and mixed outcomes. This chapter critically examines the failures, limitations, and lessons learned from bioshield projects in the region, drawing on case examples from Southeast Asia. It identifies key ecological, technical, and socio-political factors that have constrained success, including inappropriate site selection, monoculture planting, insufficient community involvement, and weak institutional coordination. The chapter also explores how overreliance on donor-driven short-term projects and inadequate post-implementation monitoring often lead to ecosystem degradation and loss of local trust. Ecologically, failed projects reveal the consequences of ignoring hydrological regimes, soil salinity, and species–site compatibility. From a governance perspective, fragmented policies, overlapping mandates, and inconsistent enforcement hinder sustainable outcomes. Socially, neglecting traditional knowledge and local participation has frequently undermined legitimacy and maintenance. Yet, these failures offer valuable insights. The chapter synthesizes key lessons learned, emphasizing adaptive management, stakeholder inclusion, and integrated coastal planning as essential pathways for future bioshield success. By comparing failures and recoveries across Asian nations, it highlights the evolution of bioshield strategies from experimental restoration to mature ecosystem-based management. Ultimately, this chapter underscores that recognizing and learning from past shortcomings is critical to advancing resilient, inclusive, and scientifically grounded coastal protection frameworks across the region.

Keywords Adaptive management · Bioshield failures · Ecosystem restoration challenges · Lessons learned in Asia · Project evaluation · Stakeholder engagement

B. Kozhikkodan Veettil, *The Potential of Bioshields for Coastal Protection*, Advances in Natural and Technological Hazards Research 55,
https://doi.org/10.1007/978-3-032-22888-8_13

13.1 Introduction

Bioshields, such as mangrove belts, *Casuarina* plantations, and dune vegetation, have gained prominence as nature-based solutions for mitigating coastal hazards and enhancing climate resilience (Veettil et al. 2021; Jayson-Quashigah et al. 2025; Marino et al. 2025). Across Asia, however, the practical implementation of these measures has frequently fallen short of expectations (Feagin et al. 2010; Veettil et al. 2021). Numerous coastal bioshield projects have been initiated with strong political or donor support but have underperformed or failed due to inappropriate species selection, insufficient consideration of local hydrodynamic and sedimentary conditions, limited community engagement, and fragmented institutional arrangements (Feagin et al. 2010; Wodehouse 2020; Veettil et al. 2021). Although such shortcomings are rarely documented in detail, they provide critical evidence for refining bioshield design and governance.

This chapter critically examines the challenges, failures, and constraints encountered by coastal bioshield initiatives in Vietnam, Southeast Asia, and adjacent regions. It adopts an interdisciplinary perspective, assessing ecological and technical limitations alongside socio-economic and governance factors that shape project outcomes. By synthesizing insights from scientific studies, field evaluations, and policy analyses, the chapter gathers key lessons learned from unsuccessful or underperforming interventions. These lessons can inform strategic recommendations aimed at enhancing the effectiveness, sustainability, and social equity of future bioshield programs, thereby strengthening their contribution to long-term coastal resilience under increasing climate and development pressures.

13.2 Ecological and Biophysical Challenges

Several ecological and biophysical factors influence the success of establishing a coastal bioshield (Fig. 13.1), including species, site, hydrology, and regional climate.

13.2.1 Species Selection and Site Suitability

One of the most persistent ecological failures in bioshield projects stems from the inappropriate selection of species. Across Vietnam and the Philippines, the large-scale planting of *Rhizophora apiculata* in high-energy zones has frequently resulted in high mortality, as this species thrives in sheltered mudflats rather than on exposed coasts (Ellison 2000; Marchand 2008; Primavera and Esteban 2008). Similarly, *Avicennia alba* plantations in Central Vietnam have failed in sandy substrates rather than silty ones (Nguyen et al. 2023; Chatterjee and Bhandari 2025).

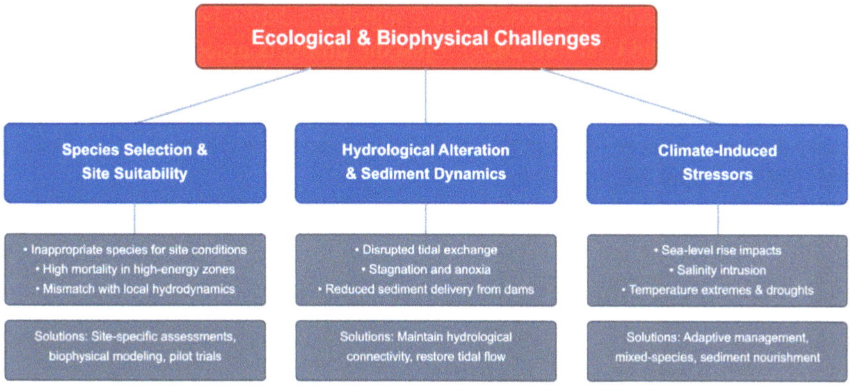

Fig. 13.1 Ecological and biophysical challenges faced by coastal bioshields

Ecological mismatch often arises from top-down planning that disregards local hydrodynamics, salinity, and sediment regimes (Sidik et al. 2025). Lessons learned emphasize the importance of site-specific ecological assessments before planting, using biophysical modeling, and pilot trials to determine suitable species assemblages (Lewis and Brown 2014; Beselly et al. 2023; Twomey et al. 2024).

13.2.2 Hydrological Alteration and Sediment Dynamics

The success of coastal bioshields depends on natural tidal exchange and sediment supply (Tamura et al. 2020). Yet, many restoration sites or naturally occurring mangrove areas have been compromised by dike construction, aquaculture ponds, or canalization, disrupting hydrology (Besset et al. 2019; Newton et al. 2020). For example, in Vietnam's Mekong Delta, restricted tidal flushing has led to stagnation, anoxia, and seedling die-off (Newton et al. 2020; Tamura et al. 2020; Thu Thuy et al. 2022; Tinh et al. 2022).

Moreover, upstream damming on the Mekong and Red Rivers has drastically reduced sediment delivery, undermining mangrove accretion and leading to erosion even in replanted areas (Anthony et al. 2015; Piman and Manish 2017; Quang and Viet 2023; Laonamsai et al. 2025). Maintaining hydrological connectivity is thus essential for the long-term stability of bioshields (Besset et al. 2019; Li et al. 2023).

13.2.3 Climate-Induced Stressors

Sea-level rise, salinity intrusion, and temperature extremes increasingly threaten young plantations (Ward et al. 2016; Ferreira et al. 2023a, b). Studies show high mortality rates in seedlings exposed to prolonged inundation beyond 10 h per tidal cycle (Henderson and Glamore 2024). Episodic droughts associated with El Niño events further reduce growth (Hickey et al. 2021). Adaptive management—such as staggered planting schedules, mixed-species configurations, and sediment nourishment—has proven more resilient than monoculture schemes (He et al. 2021; Hsiung et al. 2024; Li et al. 2025).

13.3 Technical and Design-Related Failures

13.3.1 Monoculture Plantations

Many early bioshield projects emphasized rapid coverage, leading to monospecific stands of *Rhizophora apiculata* or *Casuarina equisetifolia* (Hai et al. 2020). While these projects initially provided structural protection, they lacked ecological diversity and resilience (Su et al. 2021; Li et al. 2025). Monocultures are vulnerable to pests (e.g., *Brontispa longissima* beetle in casuarina belts) and storms that cause extensive die-offs (Edwards and Cerullo 2024). Current best practice favors polyculture and zonation, mimicking natural succession from pioneer species (*Avicennia*, *Sonneratia*) seaward to more robust taxa landward (Van Loon et al. 2016; Zimmer et al. 2022; Ferreira et al. 2023a, b).

13.3.2 Improper Spacing and Planting Techniques

Over-dense planting, often one seedling per 0.5 m^2, has resulted in competition, stunted growth, and high mortality (Ferreira et al. 2023a, b). Conversely, too sparse spacing reduces the efficiency of wave attenuation (Sreeranga and Lei 2026). Technical guidelines now recommend variable densities depending on species and site energy (Stubbs and Saenger 2002).

Failure to anchor propagules properly or to protect seedlings from wave scouring has also contributed to early losses (Kinya et al. 2024). Simple innovations such as bamboo fences or brushwood barriers can greatly enhance survival in high-energy areas (Mai Van et al. 2021; Shu et al. 2023). However, it is seen that several mangrove restoration projects have failed in Vietnam (e.g., Bac Lieu Province, Mekong Delta) due to high-energy waves or where the fences were eaten by wood-boring molluscs (Fig. 13.2).

Fig. 13.2 Attempt to plant mangroves using bamboo fences and failure due to the loss of fences by high-energy waves and wood-boring molluscs in Bac Lieu Province, Mekong Delta. Very few mangrove seedlings survived after 3 years of planting. *Photographs* Bijeesh Kozhikkodan Veettil

13.3.3 Maintenance Neglect

Post-planting care, including replanting, weeding, and monitoring, is often overlooked once donor funding ends (Hai et al. 2020). For example, in Vietnam, several mangrove rehabilitation projects did not survive beyond 3 months due to a lack of follow-up maintenance (Hai et al. 2020). Sustained financial and institutional commitment is vital for the maturation of bioshields (Gorman et al. 2022).

13.4 Socio-economic and Governance Challenges

13.4.1 Land Tenure and Conflicts of Use

Unclear land tenure remains a major obstacle (Orchard et al. 2015). In coastal Vietnam, overlapping claims among forestry departments, aquaculture cooperatives, and private leaseholders create tension (Orchard et al. 2016; Hai et al. 2020; Luom et al. 2021). Farmers are often reluctant to allocate productive shrimp ponds for mangrove restoration without compensation (Nguyen et al. 2017).

Community-based co-management models, which grant user rights in exchange for maintenance obligations, have achieved better outcomes but require transparent legal frameworks (Schmitt et al. 2013; Jhaveri et al. 2018; Phong et al. 2023).

13.4.2 Community Participation Deficits

Many bioshield initiatives were technocrat-driven, implemented without adequate local engagement (Hai et al. 2020). Projects designed externally tend to neglect indigenous knowledge and livelihood needs, resulting in poor ownership and eventual degradation (Hai et al. 2020). Conversely, participatory schemes, such as those in Nam Dinh and Soc Trang in Vietnam, have shown higher survival rates and socio-economic benefits (Spelchan et al. 2011; Hai et al. 2020; Nguyen 2022). In other words, bioshield success hinges not merely on ecological suitability but on social legitimacy (Rodríguez-Rodríguez et al. 2024).

13.4.3 Economic Pressures and Opportunity Costs

Shrimp farming and tourism offer immediate profits that outcompete long-term ecological gains from bioshields (Anh et al. 2024). The conversion of mangrove belts into ponds continues despite regulatory bans because enforcement is weak and alternative income sources are scarce (Hai et al. 2020; Ebeler et al. 2025).

Economic valuation studies indicate that the total ecosystem service value of mangroves may exceed USD 10 000 ha^{-1} yr^{-1}, yet these benefits are diffuse and long-term (Getzner and Islam 2020; Van Zanten et al. 2025). Effective payment for ecosystem services (PES) mechanisms could internalize these values and incentivize conservation (Zhang et al. 2024).

13.5 Institutional and Policy-Related Failures

13.5.1 Fragmented Governance

Institutional fragmentation remains a major barrier to effective bioshield implementation. For example, in Vietnam, the responsibilities for coastal management are distributed across multiple agencies: the Ministry of Agriculture and Rural Development (MARD) oversees mangrove and protection forests, the Ministry of Natural Resources and Environment (MONRE) is responsible for coastal zoning and environmental management, while provincial People's Committees retain authority over land-use planning (Orchard et al. 2015; Veettil et al. 2019, 2021). This division of mandates frequently results in overlapping responsibilities, policy inconsistencies, and weak coordination across sectors (Orchard et al. 2015). Consequently, bioshield initiatives are often planned and implemented in isolation from broader coastal development strategies (Pham et al. 2022). Establishing an integrated National Coastal Coordination Council could help harmonize policies, clarify institutional roles, and strengthen cross-sectoral monitoring and accountability (Pham et al. 2022).

13.5.2 Project-Based and Donor-Dependent Approaches

Many bioshield initiatives have been implemented as short-term, donor-funded projects, typically spanning three to five years (Hai et al. 2020; Stewart-Sinclair et al. 2020). While such projects can catalyze rapid action, their impacts are often difficult to sustain once external funding ends. Several early mangrove restoration projects experienced degradation following project completion due to the absence of dedicated local budget allocations and long-term management arrangements (Hai et al. 2020). Ensuring sustainability, therefore, requires embedding bioshields within national and subnational development plans and budgetary frameworks, rather than treating them as stand-alone interventions (Friess et al. 2022; Gerona-Daga and Salmo 2022).

13.5.3 Weak Monitoring and Evaluation

Monitoring and evaluation systems for coastal bioshield projects remain limited in scope and rigor. Indicators commonly focus on short-term outputs, such as seedling survival rates, while neglecting broader measures of ecological functionality, including carbon sequestration, biodiversity support, and coastal protection performance (Cadier et al. 2020). The lack of standardized metrics constrains comparative assessment and adaptive learning across projects (Cadier et al. 2020). Although national mangrove monitoring protocols incorporating remote sensing and field-based validation have begun to emerge, their application remains uneven across regions (Rondon et al. 2023; Sunkur et al. 2024). Strengthening and standardizing monitoring frameworks is essential for evidence-based policy learning and long-term effectiveness (Cadier et al. 2020).

13.6 Case Examples of Coastal Bioshield Failures and Lessons Learned: Vietnam

a. ***Mangroves in Tam Giang–Cau Hai Lagoon (Central Vietnam)***: A 2006–2012 restoration program planted *Rhizophora apiculata* in sandy substrates along lagoon margins. More than 80% of the seedlings died within two years. Various studies revealed that site hydrodynamics favored salt marsh grasses rather than mangroves in Central Vietnam (Veettil et al. 2019). The lesson: conducting baseline geomorphological and hydrodynamic assessments before species selection is important.

b. ***Casuarina Belts in Binh Thuan (South-Central Vietnam)***: Introduced in the 1990s to stabilize coastal dunes, *Casuarina* plantations suffered heavy storm damage and pest outbreaks. Monoculture and the absence of understory vegetation reduced resilience. Later projects incorporated mixed native species (*Pandanus*, *Acacia*) and achieved better results. The lesson: promoting ecological diversity and structural layering is important.
c. ***Mangrove-Shrimp Systems in Ca Mau (Southern Vietnam)***: While hailed as sustainable, some integrated mangrove-shrimp models failed due to market volatility and weak certification schemes. Farmers prioritized pond expansion, leading to canopy loss below the mandated 50% threshold (Ha et al. 2012). The lesson: aligning economic incentives with ecological compliance through verified eco-labels and PES is important.

13.7 Knowledge Gaps and Research Limitations

a. ***Long-Term Data Deficiency*** (World Bank 2016; Van Hespen et al. 2023): A persistent limitation in bioshield research is the shortage of long-term monitoring data. Most studies evaluate project performance within five years of implementation, providing limited insight into ecological succession, structural development, and long-term carbon dynamics. This constrains the understanding of bioshield durability and effectiveness under changing climatic conditions.
b. ***Socio-economic Impact Metrics*** (Paxton et al. 2023; Liu et al. 2025): Socio-economic outcomes of bioshield projects remain inadequately quantified. There is a clear need for rigorous cost–benefit analyses and standardized indicators to assess livelihood enhancement, equity, and poverty-alleviation impacts alongside ecological performance.
c. ***Hydrodynamic Interactions*** (Kamil et al. 2021; Best et al. 2022): Empirical evidence on hydrodynamic processes is limited, particularly regarding wave attenuation, current modification, and sediment trapping across varying vegetation densities and species compositions in Southeast Asia. This knowledge gap restricts the optimization of bioshield design for hazard reduction.
d. ***Restoration Ecology*** (Granado et al. 2018): Research on genetic diversity, provenance, and adaptive traits of planted species is insufficient. The absence of robust genetic and ecological guidance for seed sourcing may undermine long-term resilience and restoration success.
e. ***Cross-country Comparisons*** (Fortes 2018; Van Hespen et al. 2023): Lessons from bioshield initiatives are often fragmented and context-specific. The lack of systematic cross-country comparisons limits regional learning, underscoring the need for coordinated knowledge-sharing networks and harmonized research frameworks.

13.8 Emerging Lessons Learned

13.8.1 Adopt a Landscape and Systems Approach

Bioshields function best when integrated into wider coastal landscape mosaics, including wetlands, polders, and agricultural hinterlands (Ferreira et al. 2023a, b; Paxton et al. 2024). Restoration must consider catchment-scale processes rather than isolated plots (Yang et al. 2022; de Leeuw et al. 2024).

13.8.2 Strengthen Participatory Governance

Co-management arrangements involving local communities, NGOs, and authorities foster stewardship (Bennett et al. 2018; Khan et al. 2021). Capacity-building and gender-inclusive participation enhance social equity and durability (Armitage et al. 2020).

13.8.3 Incorporate Adaptive Management and Monitoring

Coastal bioshield projects should employ iterative learning—i.e., plan, implement, monitor, and adjust (Thom 2000; Garmestani et al. 2023). Integrating remote sensing, drones, and AI-based change detection supports evidence-based decisions (Veettil et al. 2021; Christofi et al. 2025; Wang et al. 2025).

13.8.4 Blend Engineering with Ecology

Hybrid solutions—such as mangrove-friendly breakwaters or porous fences—buffer young plantations and accelerate sediment deposition (Shu et al. 2023; Huynh et al. 2024). Experience in Soc Trang and Bac Lieu (Vietnamese Mekong Delta) shows that combining soft and hard measures achieves faster shoreline stabilization (Schmitt et al. 2013; Mai Van et al. 2021).

13.8.5 Secure Long-Term Financing

Financial sustainability can be achieved through blue-carbon credits, PES schemes, and integration into climate-adaptation budgets (Clifton and Mánez 2025; Farahmand et al. 2025; Li and Liu 2025; Purushotham and Thompson 2026). Ensuring recurrent funding prevents project decay after donor exit (Muluh et al. 2018).

13.9 Regional Cooperation and Policy Learning

Regional cooperation plays a critical role in advancing effective bioshield planning and implementation by facilitating knowledge exchange, capacity building, and policy harmonization (Nagabhatla et al. 2018). Regional organizations, such as the Association of Southeast Asian Nations (ASEAN), the Partnerships in Environmental Management for the Seas of East Asia (PEMSEA), and the United Nations Environment Programme's Mangroves for the Future (MFF) initiative, can provide important platforms for cross-country learning and collaboration. Through joint training programs, technical workshops, and pilot projects on ecosystem-based adaptation, countries including Vietnam, Thailand, and Indonesia have enhanced institutional capacity and shared practical experience in bioshield design, monitoring, and governance (Miller and Taylor 2023).

Despite these advances, regional learning remains fragmented and uneven (Morita and Matsumoto 2021). A key policy lesson emerging from these experiences is the need to institutionalize cooperation through a dedicated regional coastal bioshield knowledge hub (Karcher et al. 2024). Such a mechanism could support the dissemination of standardized technical guidelines, performance indicators, and monitoring protocols, while also promoting the exchange of best practices and lessons from both successes and failures. In addition, a regional platform could facilitate access to innovative and blended financing mechanisms, strengthening the scalability and long-term sustainability of bioshield initiatives across Asia (Atmaja et al. 2025).

13.10 Future Outlook

The future effectiveness of coastal bioshields in Asia will depend largely on the ability of institutions and practitioners to transform lessons from past failures into systematic learning and adaptive governance (Twomey et al. 2025). Advances in digital monitoring tools, remote sensing, and decision-support systems, together with strengthening legal and policy frameworks, provide new opportunities to improve planning, implementation, and evaluation (Palutikof et al. 2019; Ruckelshaus et al. 2020; Huynh et al. 2024; Sunkur et al. 2024; Terton et al. 2024). The growing recognition of nature-based solutions within national adaptation plans and climate strategies further signals a favorable policy environment for scaling bioshield approaches (Paxton et al. 2024; Terton et al. 2024).

Nevertheless, these opportunities will remain limited unless fundamental governance and socio-economic constraints are addressed (Rahman et al. 2023). Persistent challenges related to land tenure insecurity, inadequate and uneven community participation, and the absence of sustained and predictable financing continue to undermine long-term success (Sánchez-Arcilla et al. 2022; Rahman et al. 2023; Elrick-Barr et al. 2024; Favero and Hinkel 2024). Without resolving these structural issues, bioshields risk remaining isolated pilot interventions rather than evolving into robust, scalable components of coastal risk management (Sánchez-Arcilla et al. 2022; Nienaber 2024). Bridging scientific knowledge, policy processes, and community-based action is therefore essential to achieving transformative change and ensuring that bioshields contribute meaningfully to long-term coastal resilience under accelerating climate and development pressures (Corgo et al. 2024).

13.11 Conclusions

The trajectory of bioshield initiatives in Vietnam and across the broader Asian region reflects a complex interplay of notable achievements and recurring shortcomings. Many project failures have not stemmed from a lack of commitment or investment, but rather from limited ecological understanding, weak alignment with local socioeconomic realities, and fragmented institutional arrangements that undermine long-term effectiveness.

Several key lessons emerge from this experience. Successful coastal bioshields require site-specific ecological design that accounts for local hydrodynamics, sediment regimes, and species suitability, coupled with genuine community ownership and participation. Equally important is the need for integrated coastal governance frameworks that align sectoral policies, clarify responsibilities, and support adaptive management over time. Bioshields are most effective when conceptualized as living infrastructure—dynamic systems that evolve through natural processes and sustained human stewardship—rather than as static or purely engineering-based interventions.

By internalizing these lessons, future bioshield programs can convert past failures into opportunities for learning and improvement. In doing so, they can strengthen coastal resilience while contributing to the restoration of ecosystems and livelihoods, ultimately realizing the full potential of bioshields as both natural defenses and catalysts for long-term socio-ecological renewal.

References

Anh HH, Beaulieu A, Da Hanh TM, Tru LC (2024) Economic effects of a controlled mangrove-to-pond coverage ratio policy on mangrove conservation and shrimp farming: a case study in Vietnam using instrumental regression analysis. Trees For People 16:100579. https://doi.org/10.1016/j.tfp.2024.100579

Anthony EJ, Brunier G, Besset M, Goichot M, Dussouillez P, Nguyen VL (2015) Linking rapid erosion of the Mekong River delta to human activities. Sci Rep 5(1):14745. https://doi.org/10.1038/srep14745

Armitage D, Mbatha P, Muhl K, Rice W, Sowman M (2020) Governance principles for community-centered conservation in the post-2020 global biodiversity framework. Conserv Scie Pract 2(2):e160. https://doi.org/10.1111/csp2.160

Atmaja T, Yulianti W, Arini EY, Santoso H, Setiawati MD, Akita N (2025) Scaling nature-based solutions (NbS): lessons from global progress and Indonesia's path to sustainability. Earth Syst Environ 1–23. https://doi.org/10.1007/s41748-025-00842-5

Bennett NJ, Whitty TS, Finkbeiner E, Pittman J, Bassett H, Gelcich S, Allison EH (2018) Environmental stewardship: a conceptual review and analytical framework. Environ Manag 61(4):597. https://doi.org/10.1007/s00267-017-0993-2

Beselly S, Grueters U, Van Der Wegen M, Reyns J, Dijkstra J, Roelvink D (2023) Modelling mangrove-mudflat dynamics with a coupled individual-based-hydro-morphodynamic model. Environ Model Softw 169:105814. https://doi.org/10.1016/j.envsoft.2023.105814

Besset M, Gratiot N, Anthony EJ, Bouchette F, Goichot M, Marchesiello P (2019) Mangroves and shoreline erosion in the Mekong River delta, Viet Nam. Estuar Coast Shelf Sci 226:106263. https://doi.org/10.1016/j.ecss.2019.106263

Best ÜSN, van der Wegen M, Dijkstra J, Reyns J, van Prooijen BC, Roelvink D (2022) Wave attenuation potential, sediment properties and mangrove growth dynamics data over Guyana's intertidal mudflats: assessing the potential of mangrove restoration works. Earth Syst Sci Data 14:2445–2462. https://doi.org/10.5194/essd-14-2445-2022

Cadier C, Bayraktarov E, Piccolo R, Adame MF (2020) Indicators of coastal wetlands restoration success: a systematic review. Front Mar Sci 7:600220. https://doi.org/10.3389/fmars.2020.600220

Chatterjee S, Bhandari G (2025) A prospective evaluation on mangrove regeneration: case specific assessment from Indian part of Sundarbans. Discov Appl Sci 7:1304. https://doi.org/10.1007/s42452-025-07722-7

Christofi D, Mettas C, Evagorou E, Stylianou N, Eliades M, Theocharidis C, Chatzipavlis A, Hasiotis T, Hadjimitsis D (2025) A review of open remote sensing data with GIS, AI, and UAV support for shoreline detection and coastal erosion monitoring. Appl Sci 15(9):4771. https://doi.org/10.3390/app15094771

Clifton J, Mánez KS (2025) A framework based on payments for ecosystem services to support the delivery of high integrity carbon and biodiversity credits. Ecosyst Serv 73:101724. https://doi.org/10.1016/j.ecoser.2025.101724

Corgo J, Cruz SS, Conceição P (2024) Nature-based solutions in spatial planning and policies for climate change adaptation: a literature review. Ambio 53:1599–1617. https://doi.org/10.1007/s13280-024-02052-1

de Leeuw A, Tromp E, de Boer F (2024) Flood resilient landscapes: area-based solutions combine added value for society with flood risk management. Proc IAHS 386:173–179. https://doi.org/10.5194/piahs-386-173-2024

Ebeler L, Balke T, Laurie E, Le H, Nguyen T, Renaud FG (2025) Local-scale impacts of mangrove restoration and conservation on coastal communities in two Vietnamese deltas: socio-economic and institutional dynamics. Ocean Coast Manag 269:107812. https://doi.org/10.1016/j.ocecoaman.2025.107812

Edwards DP, Cerullo GR (2024) Biodiversity is central for restoration. Curr Biol 34(9):R371–R379. https://doi.org/10.1016/j.cub.2024.02.032

Ellison AM (2000) Mangrove restoration: do we know enough? Restor Ecol 8(3):219–229. https://doi.org/10.1046/j.1526-100x.2000.80033.x

Elrick-Barr CE, Thomsen DC, Smith TF (2024) Governance innovations in the coastal zone: towards social-ecological resilience. Environ Sci Policy 153:103687. https://doi.org/10.1016/j.envsci.2024.103687

Farahmand S, Hilmi N, Duarte CM (2025) The rise and flows of blue carbon credits advance global climate and biodiversity goals. NPJ Ocean Sustain 4(1):39. https://doi.org/10.1038/s44183-025-00141-6

Favero F, Hinkel J (2024) Key innovations in financing nature-based solutions for coastal adaptation. Climate 12(4):53. https://doi.org/10.3390/cli12040053

Feagin RA, Mukherjee N, Shanker K, Baird AH, Cinner J, Kerr AM, Dahdouh-Guebas F et al (2010) Shelter from the storm? Use and misuse of coastal vegetation bioshields for managing natural disasters. Conserv Lett 3(1):1–11. https://doi.org/10.1111/j.1755-263X.2009.00087.x

Ferreira AC, De Lacerda LD, Rodrigues JVM, Bezerra LEA (2023a) New contributions to mangrove rehabilitation/restoration protocols and practices. Wetl Ecol Manag 31(1):89–114. https://doi.org/10.1007/s11273-022-09903-2

Ferreira CS, Kašanin-Grubin M, Solomun MK, Sushkova S, Minkina T, Zhao W, Kalantari Z (2023b) Wetlands as nature-based solutions for water management in different environments. Curr Opin Environ Sci Health 33:100476. https://doi.org/10.1016/j.coesh.2023.100476

Fortes MD (2018) Seagrass ecosystem conservation in Southeast Asia needs to link science to policy and practice. Ocean Coast Manag 159:51–56. https://doi.org/10.1016/j.ocecoaman.2018.01.028

Friess DA, Gatt YM, Ahmad R, Brown BM, Sidik F, Wodehouse D (2022) Achieving ambitious mangrove restoration targets will need a transdisciplinary and evidence-informed approach. One Earth 5(5):456–460. https://doi.org/10.1016/j.oneear.2022.04.013

Garmestani A, Allen CR, Angeler DG, Gunderson L, Ruhl JB (2023) Multiscale adaptive management of social–ecological systems. Bioscience 73(11):800–807. https://doi.org/10.1093/biosci/biad096

Gerona-Daga MEB, Salmo SG (2022) A systematic review of mangrove restoration studies in Southeast Asia: challenges and opportunities for the United Nation's decade on ecosystem restoration. Front Mar Sci 9:987737. https://doi.org/10.3389/fmars.2022.987737

Getzner M, Islam MS (2020) Ecosystem services of mangrove forests: results of a meta-analysis of economic values. Int J Environ Res Public Health 17(16):5830. https://doi.org/10.3390/ijerph17165830

Gorman D, Vanderklift MA, Lafratta A (2022) Quantitative analysis of methodological and environmental influences on survival of planted mangroves in restoration and afforestation. Forests 13(3):404. https://doi.org/10.3390/f13030404

Granado R, Pinto Neta LC, F, A, Voloch CM, Lira CF (2018) Assessing genetic diversity after mangrove restoration in Brazil: why is it so important? Diversity 10(2):27. https://doi.org/10.3390/d10020027

Ha TTT, Bush SR, Mol AP, Van Dijk H (2012) Organic coasts? Regulatory challenges of certifying integrated shrimp–mangrove production systems in Vietnam. J Rural Stud 28(4):631–639. https://doi.org/10.1016/j.jrurstud.2012.07.001

Hai NT, Dell B, Phuong VT, Harper RJ (2020) Towards a more robust approach for the restoration of mangroves in Vietnam. Ann For Sci 77(1):18. https://doi.org/10.1007/s13595-020-0921-0

He Z, Sun H, Yu X, Yin Z, Wu M, Zhao L, Hu Z, Peng Y, Lee SY (2021) Monoculture or mixed culture? Relevance of fine root dynamics to carbon sequestration oriented mangrove afforestation and restoration. Front Mar Sci 8:763922. https://doi.org/10.3389/fmars.2021.763922

Henderson B, Glamore W (2024) A lifecycle model approach for predicting mangrove extent. Sci Total Environ 952:175962. https://doi.org/10.1016/j.scitotenv.2024.175962

Hickey SM, Radford B, Callow JN, Phinn SR, Duarte CM, Lovelock CE (2021) ENSO feedback drives variations in dieback at a marginal mangrove site. Sci Rep 11(1):8130. https://doi.org/10.1038/s41598-021-87341-5

Hsiung AR, Ong OX, Teo XS, Friess DA, Todd PA, Swearer SE, Morris RL (2024) Determinants of mangrove seedling survival incorporated within hybrid living shorelines. Ecol Eng 202:107235. https://doi.org/10.1016/j.ecoleng.2024.107235

Huynh LT, Su J, Wang Q, Stringer LC, Switzer AD, Gasparatos A (2024) Meta-analysis indicates better climate adaptation and mitigation performance of hybrid engineering-natural coastal defence measures. Nat Commun 15(1):2870. https://doi.org/10.1038/s41467-024-46970-w

Jayson-Quashigah PN, Staneva J, Chen W, Djath B, Mahu E, Appeaning Addo K (2025) Evaluating mangroves as nature-based solutions for coastal protection under current and future sea level rise scenarios. Front Mar Sci 12:1526082. https://doi.org/10.3389/fmars.2025.1526082

Jhaveri N, Dzung NT, Dung NK (2018) Mangrove collaborative management in Vietnam and Asia. Tenure and Global Climate Change (TGCC) report. USAID, Washington, DC

Kamil EA, Takaijudin H, Hashim AM (2021) Mangroves as coastal bio-shield: a review of mangroves performance in wave attenuation. Civ Eng J 7(11):1964–1981. https://doi.org/10.28991/cej-2021-03091772

Karcher DB, Cvitanovic C, Colvin R, van Putten I (2024) Enabling successful science-policy knowledge exchange between marine biodiversity research and management: an Australian case study. Environ Policy Gov 34(3):291–306. https://doi.org/10.1002/eet.2078

Khan NA, Choudhury JK, Rashid AZ, Siddique MR, Sinha K (2021) Co-management practices by Non-Government Organizations (NGOs) in selected coastal forest zones of Bangladesh: a focus on sustainability. Sustainability 14(22):14885. https://doi.org/10.3390/su142214885

Kinya G, Kairo JG, Nyoike RN, Nguu JG, Githinji BK, Githaiga MN (2024) Eco-engineering mangrove restoration at Gazi Bay, Kenya. Diversity 16(3):135. https://doi.org/10.3390/d16030135

Laonamsai J, Innanchai K, Xu M, Sriariyawat A, Pakoksung K, Inseeyong N, Polsomboon P, Chuenchum P (2025) Long-term sediment decline in the Mekong River along Thailand's border under hydropower development. Sci Total Environ 1007:180950. https://doi.org/10.1016/j.scitotenv.2025.180950

Lewis RR, Brown B (2014) Ecological mangrove rehabilitation—a field manual for practitioners. Mangrove action project, Canadian International Development Agency, and OXFAM, 151 pp

Li P, Liu D (2025) How blue carbon financing can sustain blue carbon ecosystems protection and restoration: a proposed conceptual framework for the blue carbon financing mechanism. Ocean Coast Manag 265:107644. https://doi.org/10.1016/j.ocecoaman.2025.107644

Li B, Liu JP, Jia Y (2023) Comparison of the causes of erosion-deposition between Yellow River, Yangtze River and Mekong River Subaqueous Deltas II: comparative analysis. Water 15(1):38. https://doi.org/10.3390/w15010038

Li Z, Li Y, Fan B (2025) Diverse restored mangrove stands enhance carbon storage compared to monospecific plantation: a meta-analysis. For Ecol Manag 578:122446. https://doi.org/10.1016/j.foreco.2024.122446

Liu T, Constantz B, Hale G, Beck MW (2025) Financial value of nature: coastal housing markets, mangroves, and climate resilience. Rev Finance. https://doi.org/10.1093/rof/rfaf061

Luom TT, Phong NT, Smithers S, Van Tai T (2021) Protected mangrove forests and aquaculture development for livelihoods. Ocean Coast Manag 205:105553. https://doi.org/10.1016/j.ocecoaman.2021.105553

Mai Van C, Ngo A, Mai T, Dao HT (2021) Bamboo fences as a nature-based measure for coastal wetland protection in Vietnam. Front Mar Sci 8:756597. https://doi.org/10.3389/fmars.2021.756597

Marchand M (2008) Mangrove restoration in Vietnam: key considerations and a practical guide. https://resolver.tudelft.nl/uuid:98b5ba43-1452-4631-81dc-ad043ef3992c. Accessed 9 Dec 2025

Marino M, Baptist MJ, Alkharoubi AI, Nasca S, Cavallaro L, Foti E, Musumeci RE (2025) Nature-based solutions as building blocks for coastal flood risk reduction: a model-based ecosystem service assessment. Sci Rep 15(1):12070. https://doi.org/10.1038/s41598-025-95230-4

Miller MA, Taylor D (2023) A transboundary agenda for nature-based solutions across sectors, scales and disciplines: insights from carbon projects in Southeast Asia. Ambio 53(4):534. https://doi.org/10.1007/s13280-023-01961-x

Morita K, Matsumoto KI (2021) Governance challenges for implementing nature-based solutions in the Asian region. Polit Gov 9(4):102–113. https://doi.org/10.17645/pag.v9i4.4420

Muluh GN, Kimengsi JN, Azibo NK (2018) Challenges and prospects of sustaining donor-funded projects in rural Cameroon. Sustainability 11(24):6990. https://doi.org/10.3390/su11246990

Nagabhatla N, Hung NT, Tuyen LTT, Cam VTN, Dhanraj J, Thien NT, Swierczek F (2018) Ecosystem-based approach for planning research and capacity development for integrated coastal zone management in Southeast Asia. APN Sci Bull 9(1). https://doi.org/10.30852/sb.2019.537

Newton A, Icely J, Cristina S, Perillo GM, Turner RE, Ashan D, Cragg S, Luo Y, Tu C, Li Y, Zhang H, Ramesh R, Forbes DL, Solidoro C, Béjaoui B, Gao S, Pastres R, Kelsey H, Taillie D, Kuenzer C (2020) Anthropogenic, direct pressures on coastal wetlands. Front Ecol Evol 8:512636. https://doi.org/10.3389/fevo.2020.00144

Nguyen TH (2022) Climate change adaptation in fisheries livelihoods associated with mangrove forests in Xuan Thuy National Park, Vietnam: a case study in Giao An Commune, Giao Thuy District, Nam Dinh Province. In: Ito T, Tamura M, Kotera A, Ishikawa-Ishiwata Y (eds) Interlocal adaptations to climate change in East and Southeast Asia. SpringerBriefs in climate studies. Springer, Cham. https://doi.org/10.1007/978-3-030-81207-2_7

Nguyen TP, Luom TT, Parnell KE (2017) Mangrove allocation for coastal protection and livelihood improvement in Kien Giang province, Vietnam: constraints and recommendations. Land Use Policy 63:401–407. https://doi.org/10.1016/j.landusepol.2017.01.048

Nguyen LTM, Hoang HT, Choi E, Park PS (2023) Distribution of mangroves with different aerial root morphologies at accretion and erosion sites in Ca Mau Province, Vietnam. Estuar Coast Shelf Sci 287:108324. https://doi.org/10.1016/j.ecss.2023.108324

Nienaber A (2024) Transformative change in coastal biodiversity conservation: a systematic literature review of governance, social-ecological, and cultural pathways. Sustainability 17(24):11186. https://doi.org/10.3390/su172411186

Orchard SE, Stringer LC, Quinn CH (2015) Environmental entitlements: institutional influence on mangrove social-ecological systems in Northern Vietnam. Resources 4(4):903–938. https://doi.org/10.3390/resources4040903

Orchard SE, Stringer LC, Quinn CH (2016) Mangrove system dynamics in Southeast Asia: linking livelihoods and ecosystem services in Vietnam. Reg Environ Change 16:865–879. https://doi.org/10.1007/s10113-015-0802-5

Palutikof JP, Rissik D, Webb S, Tonmoy FN, Boulter SL, Leitch AM, Campbell MJ et al (2019) CoastAdapt: an adaptation decision support framework for Australia's coastal managers. Clim Change 153(4):491–507. https://doi.org/10.1007/s10584-018-2200-8

Paxton AB, Riley TN, Steenrod CL, Puckett BJ, Paliotti ST, Adler AM, Exar L, McLean JE, Kelley J, Zhang YS, Smith CS, Gittman RK, Silliman BR (2024) Evidence on the performance of nature-based solutions interventions for coastal protection in biogenic, shallow ecosystems: a systematic map. Environ Evid 13:28. https://doi.org/10.1186/s13750-024-00350-5

Paxton AB, Riley TN, Steenrod CL, Smith CS, Zhang YS, Gittman RK, Silliman BR, Buckel CA, Viehman TS, Puckett BJ, Davis J (2023) What evidence exists on the performance of nature-based solutions interventions for coastal protection in biogenic, shallow ecosystems? A systematic map protocol. Environ Evid 12:11. https://doi.org/10.1186/s13750-023-00303-4

Pham TT, Vu TP, Hoang TL, Dao TL, Nguyen DT, Pham DC, Dao LH, Nguyen VT, Hoang NV (2022) The effectiveness of financial incentives for addressing mangrove loss in Northern Vietnam. Front For Glob Change 4:709073. https://doi.org/10.3389/ffgc.2021.709073

Phong NT, Nuong CT, Quang NH (2023) Local perceptions of mangrove protection and livelihood improvement in co-management: lessons learnt and recommendations. Ocean Coast Manag 237:106530. https://doi.org/10.1016/j.ocecoaman.2023.106530

Piman T, Manish S (2017) Case study on sediment in the Mekong River basin: current state and future trends. Project Report 2017-03. Stockholm Environment Institute. https://www.sei.org/publications/sediment-mekong-river/

Primavera JH, Esteban JMA (2008) A review of mangrove rehabilitation in the Philippines: successes, failures and future prospects. Wetl Ecol Manag 16:345–358. https://doi.org/10.1007/s11273-008-9101-y

Purushotham MK, Thompson BS (2026) Incorporating blue carbon into climate change mitigation policies: multi-level governance challenges for carbon credits and NDCs. Ecol Econ 240:108800. https://doi.org/10.1016/j.ecolecon.2025.108800

Quang NH, Viet TQ (2023) Long-term analysis of sediment load changes in the Red River system (Vietnam) due to dam-reservoirs. J Hydro-Environ Res 51:48–66. https://doi.org/10.1016/j.jher.2023.10.002

Rahman THM, Manuel P, Sherren K, Rapaport E, Van Proosdij D (2023) Characterizing social barriers to nature-based coastal adaptation approaches. Nat Based Solut 4:100099. https://doi.org/10.1016/j.nbsj.2023.100099

Rodríguez-Rodríguez JA, de Paula D, Costa M, Wartman M, Rasheed AR, Palacios M, Macreadie P (2024) Global trends in applying decision science in mangrove restoration: are we missing some dimensions? Ocean Coast Manag 254:107172. https://doi.org/10.1016/j.ocecoaman.2024.107172

Rondon M, Ewane EB, Abdullah MM, Watt MS, Blanton A, Abulibdeh A, Burt JA, Rogers K, Ali T, Reef R, Mohtar R, Sidik F, Fahrenberg M, Galgamuwa GA, Charabi YA, Arachchige PS, King S, Srinivasan S, Mohan M et al (2023) Remote sensing-based assessment of mangrove ecosystems in the Gulf Cooperation Council countries: a systematic review. Front Mar Sci 10:1241928. https://doi.org/10.3389/fmars.2023.1241928

Ruckelshaus M, Reguero BG, Arkema K, Compeán RG, Weekes K, Bailey A, Silver J (2020) Harnessing new data technologies for nature-based solutions in assessing and managing risk in coastal zones. Int J Disaster Risk Reduct 51:101795. https://doi.org/10.1016/j.ijdrr.2020.101795

Sánchez-Arcilla A, Cáceres I, Roux XL, Hinkel J, Schuerch M, Nicholls RJ, Otero DM, Staneva J, De Vries M, Pernice U, Briere C, Caiola N, Gracia V, Ibáñez C, Torresan S (2022) Barriers and enablers for upscaling coastal restoration. Nat Based Solut 2:100032. https://doi.org/10.1016/j.nbsj.2022.100032

Schmitt K, Albers T, Pham TT, Dinh SC (2013) Site-specific and integrated adaptation to climate change in the coastal mangrove zone of Soc Trang Province, Viet Nam. J Coast Conserv 17:545–558. https://doi.org/10.1007/s11852-013-0253-4

Shu A, Zhu J, Cui B, Wang L, Zhang Z, Pi C (2023) Coastal wave-energy attenuation by artificial wooden fences deployed for mangrove restoration: an experimental study. Front Mar Sci 10:1165048. https://doi.org/10.3389/fmars.2023.1165048

Sidik F, Basyuni M, Pradisty NA, Hanintyo R, Widagti N, Priyono B, Amelia R, Macklin P, Surana N, Slamet B (2025) Hydrological suitability for mangrove species in restored mangroves in abandoned ponds. Aquat Bot 199:103873. https://doi.org/10.1016/j.aquabot.2025.103873

Spelchan DG, Nicoll IA, Nguyen TPH (2011) Co-management/shared governance of natural resources and protected areas in Viet Nam. Deutsche Gesellschaft für Internationale Zusammenarbeit (GIZ), Vietnam, 169 pp

Sreeranga S, Lei J (2026) Evaluating portable reef for mangrove sapling protection: an experimental study using live mangroves. Coast Eng 204:104903. https://doi.org/10.1016/j.coastaleng.2025.104903

Stewart-Sinclair PJ, Purandare J, Bayraktarov E, Waltham N, Reeves S, Statton J, Sinclair EA, Brown BM, Shribman ZI, Lovelock CE (2020) Blue restoration—building confidence and overcoming barriers. Front Mar Sci 7:541700. https://doi.org/10.3389/fmars.2020.541700

Stubbs BJ, Saenger P (2002) The application of forestry principles to the design, execution and evaluation of mangrove restoration projects. Bois Et Forêts des Tropiques 273:5–21

Su J, Friess DA, Gasparatos A (2021) A meta-analysis of the ecological and economic outcomes of mangrove restoration. Nat Commun 12(1):5050. https://doi.org/10.1038/s41467-021-25349-1

Sunkur R, Kantamaneni K, Bokhoree C, Rathnayake U, Fernando M (2024) Mangrove mapping and monitoring using remote sensing techniques towards climate change resilience. Sci Rep 14(1):6949. https://doi.org/10.1038/s41598-024-57563-4

Tamura T, Nguyen VL, Ta TK, Bateman MD, Gugliotta M, Anthony EJ, Nakashima R, Saito Y (2020) Long-term sediment decline causes ongoing shrinkage of the Mekong megadelta, Vietnam. Sci Rep 10(1):8085. https://doi.org/10.1038/s41598-020-64630-z

Terton A, Qi J, Jang N (2024) Tracking progress on the integration of nature-based solutions and ecosystem-based adaptation in national adaptation plan processes. NAP Global Network synthesis report. International Institute for Sustainable Development

Thom RM (2000) Adaptive management of coastal ecosystem restoration projects. Ecol Eng 15(3–4):365–372. https://doi.org/10.1016/S0925-8574(00)00086-0

Thu Thuy P, Ngoc Nam V, Tuan VQ, Hong TTK, Tan Loi N, Hoa TNM, Thi Thuy Anh N, Van Anh NT, Quang NN (2022) Opportunities and challenges for mangrove restoration in the Mekong Delta: status, policies and stakeholder outlook. CIFOR-ICRAF. https://doi.org/10.17528/cifor/008610

Tinh PH, MacKenzie RA, Hung TD, Vinh TV, Ha HT, Lam MH, Hanh NT, Tung NX, Hai PM, Huyen BT (2022) Mangrove restoration in Vietnamese Mekong Delta during 2015–2020: achievements and challenges. Front Mar Sci 9:1043943. https://doi.org/10.3389/fmars.2022.1043943

Twomey AJ, Nunez K, Carr JA, Crooks S, Friess DA, Glamore W, Orr M, Reef R, Rogers K, Waltham NJ, Lovelock CE (2024) Planning hydrological restoration of coastal wetlands: key model considerations and solutions. Sci Total Environ 915:169881. https://doi.org/10.1016/j.scitotenv.2024.169881

Twomey AJ, Erickson K, Bishop MJ, Boody K, Callaghan DP, Cannard T, Lovelock CE, Mayer-Pinto M, Morris RL, Pomeroy AW, Saunders MI, Steven A, Waltham NJ, Bugnot A (2025) Interdisciplinary solutions to enable nature-based solutions for coastal protection achieve ecological and engineering outcomes. Environ Sci Policy 171:104157. https://doi.org/10.1016/j.envsci.2025.104157

Van Hespen R, Hu Z, Borsje B, De Dominicis M, Friess DA, Jevrejeva S, Kleinhans MG, Maza M, Van Bijsterveldt CE, Van der Stocken T, Van Wesenbeeck B, Xie D, Bouma TJ (2023) Mangrove forests as a nature-based solution for coastal flood protection: biophysical and ecological considerations. Water Sci Eng 16(1):1–13. https://doi.org/10.1016/j.wse.2022.10.004

Van Loon AF, Te Brake B, Van Huijgevoort MHJ, Dijksma R (2016) Hydrological classification, a practical tool for mangrove restoration. PLoS ONE 11(3):e0150302. https://doi.org/10.1371/journal.pone.0150302

Van Zanten B, Brander LM, Castaneda J, Herrera D, Kaczan D, Brown BM, Smith IA, Jongman B (2025) Integrated spatial cost-benefit analysis of large-scale mangrove conservation and restoration in Indonesia. Front Environ Sci 13:1459034. https://doi.org/10.3389/fenvs.2025.1459034

Veettil BK, Ward RD, Quang NX, Trang NTT, Giang TH (2019) Mangroves of Vietnam: historical development, current state of research and future threats. Estuar Coast Shelf Sci 218:212–236. https://doi.org/10.1016/j.ecss.2018.12.021

Veettil BK, Ward RD, Dung NTK, Van DD, Quang NX, Hoai PN, Hoang N (2021) The use of bioshields for coastal protection in Vietnam: current status and potential. Reg Stud Mar Sci 47:101945. https://doi.org/10.1016/j.rsma.2021.101945

Wang R, Calderon CV, Cohen J, Smyth E (2025) An integrative framework for AI-supported coastal hydrodynamics monitoring and analysis. Sci Rep 15(1):12403. https://doi.org/10.1038/s41598-025-94791-8

Ward RD, Friess DA, Day RH, MacKenzie RA (2016) Impacts of climate change on mangrove ecosystems: a region by region overview. Ecosyst Health Sustain 2(4):e01211. https://doi.org/10.1002/ehs2.1211

Wodehouse DC (2020) Towards successful community mangrove management and rehabilitation. PhD thesis, Bangor University, United Kingdom

World Bank (2016) Managing coasts with natural solutions: guidelines for measuring and valuing the coastal protection services of mangroves and coral reefs. In: Beck MW, Lange G-M (eds) Wealth accounting and the valuation of ecosystem services partnership (WAVES). World Bank, Washington, DC

Yang Q, He C, Xu Y, Ye Y, Guo H, Zhang S, Li F, Liu K (2022) Coastal wetlands play an important role in the ecological security pattern of the coastal zone. Land 11(12):2317. https://doi.org/10.3390/land11122317

Zhang Y, Pan L, Liu Y, Tao Y, Su Z, Huang Q, Wei J, Wu E, Fan H, Qiu G (2024) Determining the payment for ecosystem services of mangrove forests: the approach combining ecosystem services and social-economic level and application in Shankou National Nature Reserve, Guangxi, China. Glob Ecol Conserv 54:e03146. https://doi.org/10.1016/j.gecco.2024.e03146

Zimmer M, Ajonina GN, Amir AA, Cragg SM, Crooks S, Duke NC, Fratini S, Friess DA, Helfer V, Huxham M, Kathiresan K, Kodikara KA, Koedam N, Lee SY, Mangora MM, Primavera J, Satyanarayana B, Yong JW, Wodehouse D (2022) When nature needs a helping hand: different levels of human intervention for mangrove (re-)establishment. Front For Glob Change 5:784322. https://doi.org/10.3389/ffgc.2022.784322

Chapter 14
Coastal Bioshields in Vietnam: A National Perspective

Abstract Vietnam, with its extensive 3,260 km coastline, is among the most climate-vulnerable countries in Southeast Asia, making coastal bioshields a critical component of its national coastal protection strategy. This chapter provides a comprehensive national perspective on the development, implementation, and management of bioshields in Vietnam. It traces the historical evolution of mangrove rehabilitation, coastal afforestation, and community-based protection programs since the 1980s, highlighting their integration into broader frameworks such as the National Target Program on Climate Change, the National Strategy for Natural Disaster Prevention, and coastal zone management policies. The chapter examines key ecological and socio-economic outcomes of bioshield projects, emphasizing their dual role in shoreline stabilization and livelihood enhancement through aquaculture, fisheries, and ecotourism in Vietnam. It also discusses regional differences in bioshield performance—particularly among the Mekong Delta, Central Coast, and Northern estuaries—driven by hydrological variability and land-use pressures. Institutional challenges, including fragmented management, limited financing, and land tenure conflicts, are analyzed alongside emerging policy innovations such as payment for ecosystem services (PES) and blue carbon initiatives. Drawing on national reports and scientific studies, the chapter concludes that while Vietnam's bioshield programs have achieved measurable ecological and social benefits, their long-term sustainability depends on strengthening governance coherence, scientific monitoring, and community stewardship. Vietnam's experience thus offers a valuable model for other developing coastal nations seeking to integrate bioshields into climate adaptation and sustainable development frameworks.

Keywords Blue carbon initiatives · Climate adaptation policy · Coastal afforestation · Ecosystem-based coastal protection · Mangrove rehabilitation · Mekong Delta management

B. Kozhikkodan Veettil, *The Potential of Bioshields for Coastal Protection*, Advances in Natural and Technological Hazards Research 55, https://doi.org/10.1007/978-3-032-22888-8_14

14.1 Introduction

Vietnam is a coastal nation of pronounced geographical diversity and socio-economic importance, yet it remains among the countries most vulnerable to natural hazards and the impacts of climate change (Veettil et al. 2020, 2024; Duijndam et al. 2023). With a coastline extending approximately 3,260 km, Vietnam encompasses a complex mosaic of deltas, estuaries, lagoons, sandy shores, and offshore archipelagos that support dense populations and critical economic activities (Veettil et al. 2021; Lappe et al. 2022). These coastal zones are routinely exposed to typhoons, storm surges, coastal erosion, saltwater intrusion, and flooding, while long-term processes, such as sea-level rise, increasingly exacerbate these risks (Veettil et al. 2021). Rapid urbanization, population growth, and intensive land-use conversion, particularly for aquaculture, industry, and tourism, have further heightened exposure and sensitivity, placing substantial pressure on both natural ecosystems and engineered coastal infrastructure (Nguyen and Parnell 2019; Veettil et al. 2022a, 2025).

In response to these growing challenges, Vietnam has progressively shifted from a predominantly hard-engineering approach toward more integrated and sustainable coastal protection strategies (Cuong and Van Cu 2014; Cong et al. 2020). Over the past two decades, nature-based solutions (NbS) have gained prominence within national adaptation and disaster risk reduction agendas (Nguyen et al. 2017; Dang et al. 2021). Among these, coastal bioshields, defined as strategically managed belts of natural or semi-natural vegetation such as mangroves, seagrass beds, casuarina plantations, coastal forests, and dune vegetation, have emerged as a cornerstone of ecosystem-based coastal defense (Veettil et al. 2021). Empirical evidence from Vietnam and comparable tropical regions demonstrates that bioshields can significantly attenuate wave energy, reduce storm surge impacts, stabilize sediments, and mitigate shoreline erosion, often at lower long-term costs than conventional grey infrastructure (Veettil et al. 2021; Tien et al. 2025).

Beyond their protective function, coastal bioshields offer a wide array of co-benefits that are particularly relevant in the Vietnamese context (Veettil et al. 2021; Veettil and Puri 2025a). Mangrove forests and other coastal vegetation in Vietnam support high levels of biodiversity, provide nursery habitats for fisheries, enhance carbon sequestration, and contribute to water quality regulation (Veettil et al. 2019; Hai et al. 2020). For coastal communities, these ecosystems underpin livelihoods through fisheries, aquaculture support, non-timber forest products, and emerging opportunities linked to eco-tourism and carbon markets (Tran et al. 2023; Vu et al. 2023; Veettil and Puri 2025a). As such, coastal bioshields in Vietnam represent multifunctional systems that align environmental protection with socio-economic development objectives, reinforcing their appeal within national policy frameworks (Veettil et al. 2021).

This chapter provides a national-level assessment of coastal bioshields in Vietnam, examining their ecological basis, spatial distribution, and historical evolution from post-war rehabilitation initiatives to contemporary climate adaptation efforts. It analyses the institutional and policy frameworks governing bioshield planning, management, and financing, with particular emphasis on their integration into national strategies for climate change adaptation, disaster risk reduction, and integrated coastal zone management. The chapter further explores socio-economic dimensions, including community participation, co-management mechanisms, and trade-offs between conservation objectives and competing land uses. Recent advances in scientific research, monitoring technologies, and restoration practices are reviewed, alongside persistent challenges such as fragmented governance, financial sustainability, and climate-induced ecological uncertainty. Collectively, these insights position Vietnam's experience as an instructive case for mainstreaming ecosystem-based adaptation in vulnerable coastal regions.

14.2 Vietnam's Coastal Environment and the Need for Bioshields

14.2.1 Geographic and Climatic Context

Vietnam's coastline spans a diverse range of geomorphological and ecological settings, reflecting the country's pronounced latitudinal and climatic gradients. From the expansive alluvial plains of the Red River Delta in the north and the Mekong Delta in the south to the narrow sandy beaches, lagoons, and rocky headlands of Central Vietnam, the coastal environment is both highly productive and inherently dynamic (Lappe et al. 2022; Veettil et al. 2024). Approximately 28 provinces and several centrally governed cities are located along the coast, accommodating nearly half of the national population and serving as engines of economic growth. Coastal and marine sectors, including capture fisheries, aquaculture, tourism, port services, and coastal industries, collectively contribute more than 50% of the national gross domestic product, underscoring the strategic importance of coastal zones to Vietnam's development trajectory (PEMSEA 2020).

Climatically, Vietnam lies within the tropical monsoon region and is among the countries most frequently affected by extreme weather events in the Western Pacific (Minh et al. 2024). Coastal areas are exposed to an average of six to ten typhoons and tropical storms annually, often accompanied by intense rainfall, storm surges, flooding, shoreline erosion, and saline intrusion (Trinh et al. 2020; Minh et al. 2024). These impacts are particularly acute in low-lying deltaic regions, where even moderate storm events can disrupt livelihoods and critical infrastructure (Trinh et al. 2020). The Mekong Delta is widely recognized as one of the most climate-vulnerable regions globally due to its low elevation, dense population, and dependence on climate-sensitive livelihoods (Minderhoud et al. 2019; Tran et al. 2021).

Within this geographic and climatic context, coastal bioshields play an increasingly important role as natural buffers against hydro-meteorological hazards in

Vietnam (Veettil et al. 2021). Vegetated systems such as mangrove forests, coastal forests, and dune vegetation can attenuate wave energy, trap sediments, and reduce erosion, thereby complementing conventional engineered structures, including sea dikes, revetments, and breakwaters (Veettil et al. 2021; Daro Justine and Seenath 2025). Their capacity to function across multiple spatial scales makes bioshields particularly well-suited to Vietnam's heterogeneous coastal landscape (Veettil et al. 2021).

14.2.2 Historical Context of Coastal Protection

Historically, coastal protection in Vietnam has relied predominantly on hard-engineering solutions (Fig. 14.1), pervasive dike systems constructed to protect agricultural land and settlements from flooding and saline intrusion (van Ledden et al. 2020). Many of these structures date back centuries and reflect a long tradition of hydraulic management (Hung et al. 2020). While dikes and seawalls have provided essential protection, their effectiveness has increasingly been challenged by rising sea levels, land subsidence, and the growing intensity of storms, as well as by high construction and maintenance costs (Veettil et al. 2021).

Since the early 1990s, there has been a gradual shift in policy and practice toward more ecosystem-based approaches to coastal defense in Vietnam (Veettil et al. 2021). Mangrove-based bioshields have gained prominence as sustainable and cost-effective alternatives or complements to grey infrastructure (Schmitt et al. 2013). This transition was accelerated by a series of extreme events that exposed the limitations of relying solely on hard structures.

Fig. 14.1 Hard engineering structures and geotubes used for coastal protection in Vietnam. *Photographs* Bijeesh Kozhikkodan Veettil

In response, large-scale mangrove rehabilitation and coastal afforestation programs were initiated nationwide, with substantial support from international and national organizations, including the Vietnam Red Cross, the United Nations Development Programme, and the Japan International Cooperation Agency. Evaluations of these initiatives demonstrated that well-established mangrove belts could significantly reduce wave impacts, lower dike repair and maintenance costs, and simultaneously enhance fisheries productivity and household incomes (e.g., UNDP 2024). These outcomes strengthened the evidence base for bioshields and laid the foundation for their integration into national strategies for disaster risk reduction, climate change adaptation, and sustainable coastal zone management.

14.3 Types and Distribution of Bioshields in Vietnam

14.3.1 Mangrove Bioshields

Mangrove forests constitute the most prominent and widely studied form of coastal bioshields in Vietnam, owing to their extensive distribution (Fig. 14.2), structural complexity, and proven capacity to mitigate coastal hazards. These ecosystems are primarily concentrated along the northern coast associated with the Red River Delta, the estuarine and lagoon systems of Central Vietnam, and the low-lying southern coast of the Mekong Delta (Veettil et al. 2019). Historically, Vietnam supported more than 400,000 ha of mangroves before 1945; however, widespread deforestation, wartime impacts, rapid expansion of aquaculture, and accelerating urbanization led to a dramatic decline in mangrove extent (Veettil et al. 2019; Nguyen et al. 2021). According to a 2025 national inventory brief on blue carbon accounting, the total mangrove forest area in Vietnam in 2024 was estimated at 168,741.5 ha, representing approximately 3% of the country's total forest area (Thuy et al. 2025).

In northern Vietnam, mangrove bioshields are primarily distributed in Quang Ninh, Hai Phong, Thai Binh, and Nam Dinh provinces (Veettil et al. 2019; Hong Tinh et al. 2020; Vu et al. 2022). These systems are typically characterized by relatively simple species assemblages dominated by *Kandelia obovata* and *Sonneratia caseolaris*, which are well adapted to cooler winter temperatures and high sediment loads (Hong Tinh et al. 2020; Le et al. 2025a, b). Mangrove belts in this region play a crucial role in protecting sea dikes from wave action and erosion while simultaneously providing nursery habitats for commercially important fish, shrimp, and crab species (Veettil et al. 2019, 2021).

Central Vietnam supports comparatively narrow and fragmented mangrove belts, reflecting the region's steep coastal topography, high wave energy, and limited estuarine areas (Veettil et al. 2019, 2020). Mangroves are mainly confined to sheltered lagoons and estuaries, such as the Tam Giang–Cau Hai lagoon system in Thua Thien Hue Province and Thi Nai Lagoon in Binh Dinh Province (Tin et al. 2010; Veettil et al.

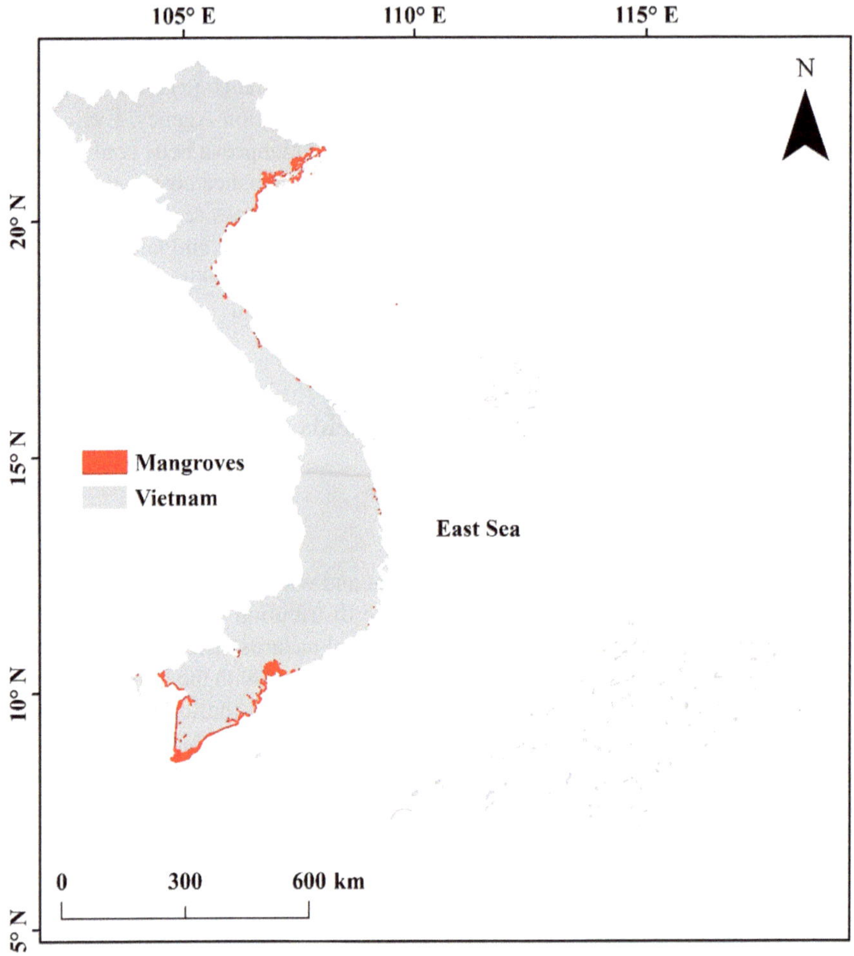

Fig. 14.2 Spatiotemporal distribution of mangrove vegetation in Vietnam (*Source* Bunting et al. 2022)

2019, 2021). Although limited in spatial extent, these mangroves offer locally significant protection against storm surges and erosion and provide important ecosystem services to adjacent fishing communities (Veettil et al. 2020). Several mangrove plantation areas, such as the Bau Ca Cai mangrove forest in Quang Ngai Province, have developed successfully in Central Vietnam (Veettil et al. 2024).

The southern region, particularly the Mekong Delta, hosts the largest and most continuous mangrove ecosystems in the country (Veettil et al. 2019; Tinh et al. 2022). Extensive mangrove belts occur in Ca Mau, Bac Lieu, Soc Trang, and Tra Vinh provinces, where low wave gradients and high sediment inputs favor mangrove establishment and expansion (Tinh et al. 2022). Dominant species include *Rhizophora apiculata*, *Avicennia alba*, and *Bruguiera gymnorhiza* (Tran et al. 2020). These mangrove forests are widely recognized for their effectiveness in attenuating wave energy, promoting sediment accretion, and stabilizing shorelines, thereby forming a critical first line of defense against coastal hazards (Veettil et al. 2021).

14.3.2 Casuarina and Dune Vegetation Bioshields

In areas where mangroves are ecologically unsuitable, particularly along exposed sandy coasts, alternative forms of vegetative bioshields have been developed (Fig. 14.3). *Casuarina equisetifolia* plantations are widely distributed along the south-central, central, and northern coasts of Vietnam, where they function as windbreaks and erosion control measures (Hoang Xuan 1996; Tong and Nguyen 2021). Their tolerance to saline soils, strong winds, and nutrient-poor substrates makes them well-suited for stabilizing coastal dunes and protecting agricultural land, settlements, and transport infrastructure (Hoang Xuan 1996; Zimmermann et al. 2016).

Complementing *Casuarina* plantations, natural dune vegetation also plays a significant bioshield role. Pioneer species such as *Ipomoea pes-caprae*, *Spinifex littoreus*, and *Pandanus tectorius* are common along sandy shorelines in provinces

Fig. 14.3 Dune vegetation, including *Casuarina equisetifolia*, in Long Hai Beach, Ba Ria-Vung Tau Province, Vietnam. *Photographs* Bijeesh Kozhikkodan Veettil

including Ba Ria-Vung Tau, Binh Thuan, Ninh Thuan, and Quang Nam (Veettil et al. 2021). These species help bind sand, reduce wind erosion, and enhance dune resilience, particularly under conditions of increasing storm intensity (Lee et al. 2020).

14.3.3 Coral Reefs and Seagrass Meadows

Although located primarily in subtidal zones, coral reefs and seagrass meadows in Vietnam (Fig. 14.4) contribute indirectly to coastal bioshield functions by dissipating wave energy and stabilizing sediments before waves reach the shoreline (Veettil et al. 2021, 2022b; Veettil and Puri 2025b). Fringing and barrier reefs around Ly Son Island, Nha Trang Bay, Phu Quoc, and the Spratly Archipelago reduce wave height and protect adjacent coastlines and nearshore ecosystems (Veettil and Puri 2025b). Similarly, seagrass meadows attenuate currents and trap sediments, enhancing coastal stability (Potouroglou et al. 2017; Contti Neto et al. 2022; Veettil et al. 2022b). Despite their importance, these ecosystems face growing pressures from climate change, coastal development, and pollution, highlighting the need for integrated management that recognizes their role within broader bioshield systems (Veettil et al. 2022b, 2025).

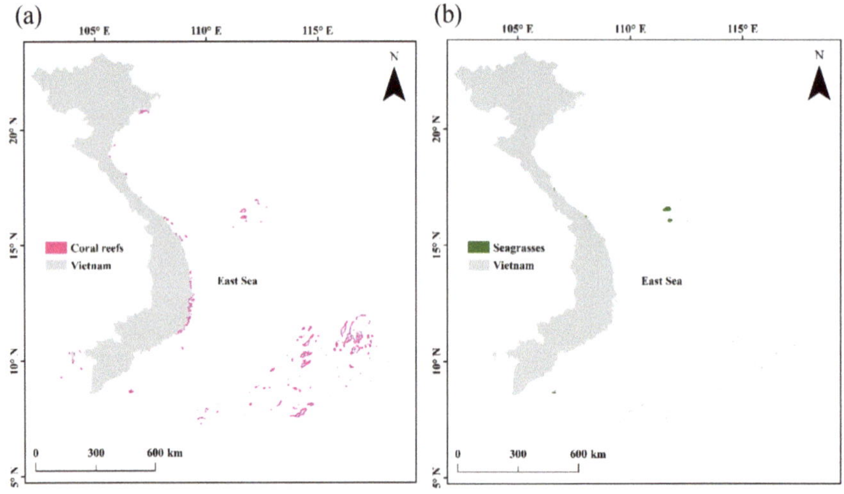

Fig. 14.4 Spatial distribution of **a** coral reefs in Vietnam (*Source* UNEP-WCMC et al. 2021) and **b** seagrasses (*Source* UNEP-WCMC and Short 2021)

14.4 Ecological Functions and Protective Effectiveness

14.4.1 Wave Attenuation and Sediment Stabilization

One of the most critical ecological functions of coastal bioshields in Vietnam (Fig. 14.5) is their capacity to attenuate wave energy and stabilize sediments, thereby reducing shoreline erosion and damage to coastal infrastructure (Quartel et al. 2007; Veettil et al. 2021; Tien et al. 2025). Numerous field measurements and numerical modeling studies have demonstrated that mangrove forests are highly effective in dissipating wave energy through their complex above- and below-ground structures, including trunks, branches, pneumatophores, and prop roots (Vinh 2015; Tien et al. 2025). Depending on forest width, tree density, and species composition, mangroves can reduce incoming wave energy by approximately 50–90% over distances ranging from 100 to 500 m (Mazda et al. 2006; Van Wesenbeeck et al. 2025). Empirical observations indicate that mangrove belts approximately 100–200 m wide can considerably reduce wave heights during moderate storm conditions, thereby significantly lowering stress on adjacent sea dikes and shorelines (Quartel et al. 2007; Bao 2011).

In addition to wave attenuation, mangroves enhance sediment trapping and accretion by slowing water flow and promoting the deposition of suspended sediments (Fagherazzi et al. 2017). This process contributes to vertical land building and shoreline stabilization, which is particularly important in deltaic regions experiencing subsidence and relative sea-level rise (Phan et al. 2015; Besset et al. 2019). Over time, these functions can improve the resilience of coastal landscapes and reduce the frequency and cost of maintenance required for engineered defenses.

Vegetative bioshields along sandy coasts, such as *Casuarina equisetifolia* plantations and natural dune vegetation, also provide important protective functions. In Central Vietnam, casuarina belts have been shown to reduce near-surface wind speeds in adjacent agricultural and residential areas, thereby limiting wind erosion, sand encroachment, and damage during storm events (Chaudhari et al. 2009). When

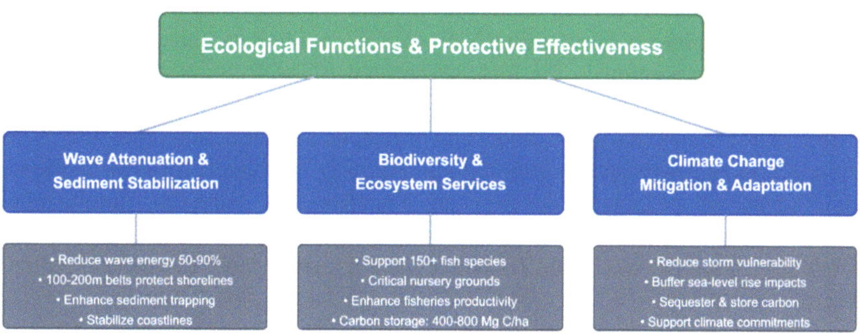

Fig. 14.5 Ecological functions and protective effectiveness of coastal bioshields in Vietnam

combined with well-developed dune systems, these vegetative barriers contribute to the overall stability of sandy shorelines.

14.4.2 Biodiversity and Ecosystem Services

Beyond their physical protective role, bioshields support high levels of biodiversity and deliver a broad suite of ecosystem services that underpin coastal livelihoods and food security (Veettil et al. 2021; Bimrah et al. 2022). Mangrove ecosystems in southern Vietnam alone are estimated to support more than 150 fish species and over 200 species of invertebrates, many of which are of significant commercial and subsistence value (RIMF 2025). These habitats function as critical nursery and feeding grounds, enhancing nearshore fisheries productivity and contributing to the sustainability of coastal economies (Veettil et al. 2019).

Bioshields also play a substantial role in regulating ecosystem services, including nutrient cycling, water quality improvement, and carbon sequestration. Mangrove forests are among the most carbon-dense ecosystems globally, with soil carbon stocks in Vietnam commonly ranging from 400 to 800 Mg C ha^{-1} (Hong Tinh et al. 2020). This blue carbon storage not only enhances ecosystem resilience but also represents an important contribution to national (NDCs) and global climate mitigation efforts (Vu et al. 2023; Veettil and Puri 2025a).

14.4.3 Climate Change Mitigation and Adaptation Synergies

Vietnam's coastal bioshields exemplify the strong synergies between climate change adaptation and mitigation (UNDP 2024). From an adaptation perspective, bioshields reduce exposure and vulnerability to storms, erosion, and sea-level rise by buffering physical impacts and stabilizing coastal landforms (Veettil et al. 2021). From a mitigation perspective, they sequester and store significant amounts of carbon in both biomass and sediments, supporting Vietnam's commitments under its Nationally Determined Contribution to the Paris Agreement (UNDP 2024; Veettil and Puri 2025a).

Integrating bioshields into national and provincial climate action plans, therefore, offers a cost-effective and multifunctional pathway to enhance coastal resilience while advancing sustainable development objectives. By delivering simultaneous ecological, protective, and socio-economic benefits, bioshields represent a strategic investment in long-term climate resilience for Vietnam's vulnerable coastal regions (Veettil et al. 2021).

14.5 National Policies and Institutional Frameworks

14.5.1 Policy Evolution

Over the past two decades, Vietnam has progressively integrated coastal bioshields into its national policy architecture, reflecting a growing recognition of ecosystem-based approaches as integral to climate resilience and disaster risk reduction (MAE 2024). This evolution marks a shift from sectorally fragmented and infrastructure-centric coastal protection toward more integrated and sustainable frameworks that acknowledge the multifunctional value of coastal ecosystems (Veettil et al. 2019, 2021).

The 2011 National Strategy on Climate Change (Decision No. 2139/QĐ-TTg) represents one of the earliest high-level policy instruments to explicitly emphasize mangrove restoration and coastal afforestation as key measures for reducing vulnerability to climate-related hazards, particularly storms, flooding, and coastal erosion (Thuy et al. 2014). This strategic orientation was further reinforced through legal instruments governing forest management. The 2004 Law on Forest Protection and Development (Law No. 29/2004/QH11), revised in 2017 (Law No. 16/2017/QH14), formally classifies mangrove forests as special-use and protection forests, thereby prioritizing their conservation and protective functions over extractive uses. This legal classification has provided an important basis for restricting the conversion of mangrove areas and for allocating state resources toward restoration and protection.

Building on these foundations, the National Action Plan on Mangrove Management for the period 2017–2030 articulated more concrete targets and implementation mechanisms. The plan aims to restore approximately 20,000 ha of degraded mangrove forests by 2030, while also improving management effectiveness, strengthening community participation, and enhancing the contribution of mangroves to disaster risk reduction and livelihood support (MAE 2024). More recently, the National Adaptation Plan (2021) has consolidated bioshields within Vietnam's broader climate adaptation agenda, explicitly identifying mangroves and other coastal ecosystems as core nature-based solutions. Within the NAP, bioshields are positioned as cost-effective and scalable measures that complement engineered defenses and contribute to long-term resilience under conditions of climate uncertainty (Veettil et al. 2021).

Collectively, these policy instruments demonstrate an increasingly coherent national commitment to mainstreaming bioshields into climate change adaptation, forestry, and coastal management strategies, while aligning domestic objectives with international frameworks such as the Paris Agreement and the Sendai Framework for Disaster Risk Reduction.

14.5.2 *Institutional Arrangements*

From an institutional perspective, responsibility for the planning, implementation, and management of coastal bioshields in Vietnam is distributed across multiple government agencies (Orchard et al. 2015). The Ministry of Agriculture and Rural Development (MARD) plays a central role, particularly through its mandates related to forestry, mangrove restoration, sea dike protection, and disaster risk management (Decree No. 119/2016/ND-CP). In parallel, the Ministry of Natural Resources and Environment (MONRE) is responsible for coastal and marine spatial planning, land-use management, climate change policy, and the coordination of Integrated Coastal Zone Management (ICZM) initiatives (Decision No. 2295/QD-TTg dated 2014; UN/DOALOS-NPA 2023).

At the subnational level, provincial People's Committees are tasked with translating national policies into local action, including land allocation, enforcement, and the integration of bioshields into provincial development and coastal protection plans (Decree 119/2016 on coastal forests, Vietnamese Government). In principle, coordination among MARD, MONRE, and provincial authorities enables bioshields to be incorporated into ICZM programs that balance environmental protection with socio-economic development objectives (Law on Marine and Island Resources and Environment, Law No. 82/2015/QH13; Decision No. 1117/QD-TTg on the Master Plan for Sustainable Use of Coastal Zone Resources in 2024).

In practice, however, implementation is often constrained by overlapping mandates, fragmented planning processes, and limited or short-term funding (Veettil et al. 2019, 2021). Ambiguities in institutional responsibilities—particularly at the interface between forestry management and coastal zone governance—can lead to inefficiencies and gaps in enforcement (Orchard et al. 2015; Beresnev et al. 2016). Addressing these challenges will require stronger inter-agency coordination mechanisms, clearer delineation of roles, and the development of a more unified national framework for bioshield management. Enhanced financial mechanisms and long-term investment strategies will also be essential to ensure the sustained effectiveness of bioshields as a cornerstone of Vietnam's coastal resilience strategy (Veettil et al. 2021; Pham et al. 2022).

14.6 Socio-economic Dimensions of Bioshield Programs in Vietnam

14.6.1 *Community Participation and Livelihoods*

The long-term effectiveness and sustainability of bioshield programs in Vietnam are closely linked to the degree of community participation in their planning, implementation, and management (Veettil et al. 2021). Many of the country's most successful bioshield initiatives have adopted community-based or co-management approaches that recognize local residents as key stewards of coastal ecosystems rather than passive beneficiaries (Hai et al. 2020). Such approaches are particularly important in densely populated coastal areas, where livelihoods are directly dependent on mangroves and nearshore resources (Hai et al. 2020).

Notable examples include mangrove rehabilitation programs supported by the Vietnam Red Cross in Thai Binh and Nam Dinh provinces, where local communities (especially women's unions) were actively engaged in seedling production, planting, and maintenance. These initiatives are believed to restore protective mangrove belts and generate short-term employment, thereby strengthening social cohesion and disaster preparedness at the community level (IFRC 2022). By linking ecosystem restoration with livelihood support, these programs may enhance local ownership and compliance, reducing pressures on rehabilitated areas.

In the Mekong Delta, integrated mangrove–shrimp farming systems have emerged as a particularly effective model for reconciling conservation and economic objectives (Truong and Do 2018; Nguyen et al. 2022). Under these systems, farmers maintain a specified proportion of mangrove cover within aquaculture ponds, allowing shrimp production to coexist with ecosystem protection (Lai et al. 2022; Nguyen et al. 2022). Empirical studies indicate that such integrated systems can stabilize incomes, reduce disease risks, and improve environmental sustainability compared to intensive monoculture practices (To et al. 2017; Truong and Do 2018; Lai et al. 2022; Nguyen et al. 2022). These livelihood-compatible models demonstrate how bioshields can be embedded within broader coastal production systems rather than treated as land-use constraints.

14.6.2 Economic Valuation and Cost–Benefit Analysis

Economic valuation studies provide strong evidence in support of bioshields as cost-effective investments in coastal protection. Analyses conducted in Vietnam and comparable coastal settings indicate that mangrove restoration can generate benefit–cost ratios ranging from approximately 3:1 to as high as 10:1, depending on local conditions and the range of benefits considered (Su et al. 2021; Zhang et al. 2025). These benefits include avoided storm and erosion damage, enhanced fisheries productivity, livelihood diversification, and the emerging value of carbon sequestration under blue carbon frameworks (Veettil et al. 2021; Veettil and Puri 2025a).

From a public finance perspective, coastal bioshields in Vietnam are also attractive due to their relatively low maintenance costs (Aerts 2018; Veettil et al. 2021). Annual expenditures for maintaining mangrove belts are commonly estimated at USD 327–800 per hectare, which is substantially lower than the costs associated with repairing, reinforcing, or upgrading sea dikes and other hard-engineering structures per hectare per year (Gijón Mancheño et al. 2025). When evaluated over long time horizons, these cost differentials further strengthen the economic rationale for integrating bioshields into national and provincial coastal protection strategies (Aerts 2018).

14.6.3 Gender and Social Equity

Gender and social equity considerations are increasingly recognized as critical dimensions of effective bioshield governance in Vietnam. Women play central roles in many restoration activities, including nursery management, seedling propagation, planting, monitoring, and community outreach (Nguyen and Dang 2018). Despite their substantial contributions, women's roles are often informal and underrepresented in decision-making processes and benefit-sharing mechanisms (Nguyen and Dang 2018).

Research suggests that gender-inclusive governance arrangements can enhance both ecological outcomes and social justice by improving knowledge exchange, accountability, and local stewardship (Le et al. 2025a, b). In several coastal communes (e.g., Quang Tri Province), community cooperatives and microfinance initiatives have successfully empowered women-led restoration groups, enabling them to access credit, formalize livelihoods linked to mangrove management, and strengthen their influence in local resource governance (Nguyen and Dang 2018). Integrating gender equity into bioshield programs is, therefore, not only a matter of social inclusion but also a strategic approach to improve the effectiveness and sustainability of coastal bioshield initiatives in Vietnam.

14.7 Technological and Research Innovations

14.7.1 Remote Sensing and GIS Applications

Advances in remote sensing and geographic information systems (GIS) technologies have significantly enhanced the capacity to monitor, assess, and manage coastal bioshields in Vietnam (Veettil et al. 2019, 2021). Over the past decade, researchers and management agencies from Vietnam have increasingly relied on satellite-based datasets, particularly from Landsat and Sentinel-2 platforms, to analyze changes in mangrove extent, canopy cover, and vegetation health at multiple spatial and temporal scales (e.g., Veettil et al. 2020, 2022a, 2024, 2025; Vu et al. 2022). These data have been complemented by high-resolution unmanned aerial vehicle (UAV) imagery, which enables detailed mapping of species composition, seedling survival, biomass, and fine-scale shoreline dynamics (e.g., Ngo et al. 2023; Tong et al. 2023).

Applications of remote sensing and GIS have proven especially valuable in large and remote mangrove systems, where field-based monitoring alone is often logistically challenging and costly (Kuenzer et al. 2011; Veettil et al. 2021). Case studies from Ca Mau and Hai Phong provinces demonstrate that integrating satellite imagery with systematic field surveys improves the accuracy and reliability of assessments of restoration success, degradation hotspots, and rates of shoreline change (Tran Thi et al. 2014; Hoa et al. 2020; Thi Huyen et al. 2024). Such integrated approaches support evidence-based decision-making by providing timely and spatially explicit information to managers and policymakers (Kuenzer et al. 2011).

14.7.2 Modelling and Forecasting

Numerical modelling has become an increasingly important tool for evaluating the protective effectiveness of bioshields and optimizing their design under present and future climate conditions (Magdalena et al. 2021). Hydrodynamic and wave models, such as Delft3D and MIKE 21, are widely applied in Vietnam to simulate wave propagation, current patterns, and sediment transport across mangrove belts of varying widths, densities, and species compositions (Vinh 2015). These models allow researchers to quantify the extent to which mangroves attenuate wave energy and reduce nearshore hydrodynamic forces (Vinh 2015).

Modelling outputs are increasingly used to inform coastal planning and engineering decisions at the provincial level (Duy et al. 2025). In erosion-prone regions such as Soc Trang and Bac Lieu, scenario-based simulations have been employed to identify optimal mangrove belt widths and seedling planting configurations that maximize protective benefits while minimizing land-use conflicts (Vinh 2015; Thinh et al. 2025). By linking biophysical processes with spatial planning, modelling tools contribute to more efficient and resilient bioshield implementation.

14.7.3 Integration of AI and IoT

Emerging research in Vietnam is exploring the integration of artificial intelligence (AI) and Internet of Things (IoT) technologies to support more adaptive and data-driven bioshield management. Research groups at institutions such as the Vietnam Academy of Science and Technology are developing AI-based predictive models that analyze multi-source datasets to forecast mangrove growth, mortality risks, and responses to environmental stressors (VAST 2025).

At the same time, IoT-based sensor networks are being piloted globally to monitor key environmental parameters in mangrove zones, including salinity, soil moisture, water levels, and microclimatic conditions (Castillo-Atoche et al. 2025). These real-time data streams enable early detection of stress events and improve the capacity to evaluate restoration performance at fine temporal scales (Castillo-Atoche et al. 2025), and such methods are yet to be applied in Vietnam. Together, AI and IoT innovations offer promising pathways to enhance monitoring efficiency, strengthen early warning systems, and support adaptive management of bioshields in the face of increasing climate variability and uncertainty (Ayadi et al. 2025).

14.8 Case Examples of Bioshields in Vietnam

14.8.1 Can Gio Mangrove Biosphere Reserve (Ho Chi Minh City)

The Can Gio Mangrove Biosphere Reserve, located in Ho Chi Minh City, is widely regarded as one of Vietnam's most successful and internationally recognized examples of large-scale mangrove rehabilitation (Veettil et al. 2019, 2021). Once extensively degraded by wartime (1955–1975) defoliation and subsequent conversion to aquaculture and other land uses, the area has undergone sustained restoration efforts since the early 1980s (Veettil et al. 2019). Through coordinated replanting, protection, and management programs, Can Gio mangrove forests (Fig. 14.6) have been transforming the landscape into a resilient coastal ecosystem.

In recognition of its ecological significance and management achievements, Can Gio was designated as a UNESCO Biosphere Reserve in 2000 (Van Cuong et al. 2019). Today, the mangrove system functions as a critical bioshield, attenuating wave energy, reducing storm surge impacts, and protecting the rapidly urbanizing coastline of southern Vietnam (Tuan and Kuenzer 2012). Beyond its protective role, the reserve delivers substantial co-benefits, including carbon sequestration, biodiversity conservation, and livelihood diversification (Tuan and Kuenzer 2012). Ecotourism has become an increasingly important economic activity, generating revenues and providing income opportunities for local communities (Son et al. 2022). Can Gio thus exemplify how long-term investment in bioshields can yield sustained ecological and socio-economic returns.

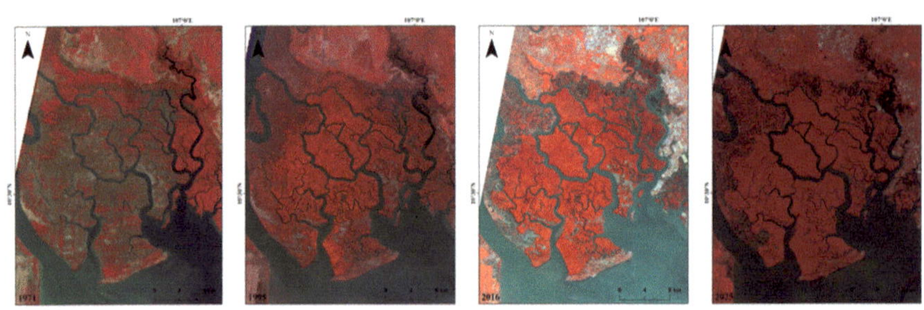

Fig. 14.6 Landsat series images (1971, 1995, 2016, and 2025) showing the deforestation of mangroves in Can Gio due to the application of defoliating agents during the Vietnam War and subsequent regeneration and reforestation during the post-war period

14.8.2 Ca Mau Peninsula: Integrating Bioshields with Aquaculture

The Ca Mau Peninsula in the Mekong Delta provides a compelling example of how bioshields can be integrated with productive land-use systems. In this region, extensive mangrove–shrimp integrated farming systems have been developed (Lai et al. 2022; Nguyen et al. 2022). These systems maintain a minimum proportion of mangrove cover within aquaculture areas, thereby preserving the protective and ecological functions of mangroves while allowing for commercial shrimp production (Nguyen et al. 2022).

Supported by international eco-certification schemes such as Naturland, mangrove–shrimp systems in Ca Mau have demonstrated multiple benefits (Ha et al. 2012; Baumgartner and Nguyen 2017). Mangrove belts enhance sediment deposition and shoreline stability, reduce wave impacts on dikes, and lower maintenance costs for coastal defenses (Jayson-Quashigah et al. 2025). At the same time, eco-certification provides access to premium export markets and price incentives for sustainably produced seafood, improving farmer incomes and reinforcing compliance with conservation requirements (Xuan 2021). The Ca Mau case highlights the potential for bioshields to be embedded within market-based mechanisms that align environmental protection with economic development (McEwin and McNally 2014).

14.9 Challenges and Constraints

Despite growing recognition of their ecological and socio-economic value, the planning, implementation, and long-term sustainability of coastal bioshields in Vietnam continue to face a range of interrelated challenges and constraints (Hai et al. 2020; Tinh et al. 2022; Veettil et al. 2025). These issues operate across ecological, institutional, and socio-economic dimensions, often limiting the effectiveness and durability of bioshield interventions (Le 2021; Pham et al. 2022; Ebeler et al. 2025).

One of the most persistent challenges is land-use conflict in coastal zones (Le 2021; Tran et al. 2024). Rapid expansion of shrimp aquaculture, coastal tourism, industrial zones, and urban infrastructure has increasingly encroached upon mangrove forests and other coastal vegetation (Veettil et al. 2019, 2021). In many cases, short-term economic incentives associated with intensive aquaculture or real estate development outweigh the perceived long-term benefits of maintaining bioshields, leading to mangrove clearance or fragmentation (Nguyen and Parnell 2019; Ebeler et al. 2025). Such conflicts are particularly acute in the Mekong Delta and peri-urban coastal areas, where competition for land is intense and regulatory enforcement remains uneven (Nguyen and Parnell 2019).

Financial constraints also pose a significant barrier to coastal bioshield sustainability in Vietnam. While substantial progress in mangrove restoration has been achieved through donor-funded projects and international cooperation, long-term financing for maintenance, monitoring, and adaptive management is often inadequate (Pham and Nguyen 2021; Pham et al. 2022). Local government budgets are frequently insufficient to support post-restoration activities, resulting in degraded outcomes once external funding cycles end (Pham and Nguyen 2021). The absence of stable and diversified financing mechanisms undermines the continuity of bioshield management and limits the scaling up of successful models (Friess et al. 2022).

Climate change itself introduces additional ecological uncertainty (Hülsen et al. 2025). Accelerating sea-level rise, increased storm intensity, altered sediment dynamics, and saline intrusion threaten the survival of young plantations, particularly in exposed or subsiding coastal areas of Vietnam (Veettil et al. 2020, 2022a, 2024). Inappropriate species selection, inadequate site preparation, and limited consideration of future climate scenarios can further exacerbate plantation failure, reducing the protective effectiveness of bioshields over time (Hai et al. 2020; Hülsen et al. 2025).

Institutional fragmentation remains a structural constraint (Pham et al. 2022). Overlapping mandates and unclear divisions of responsibility among the Ministry of Agriculture and Rural Development, the Ministry of Natural Resources and Environment, and local authorities may lead to inconsistencies in planning, implementation, and enforcement (Veettil et al. 2019). Weak coordination across sectors and administrative levels often hampers the integration of bioshields into broader coastal development and climate adaptation strategies (Zinzani 2018).

Finally, monitoring and knowledge gaps constrain adaptive management in Vietnam. Long-term, systematic data on ecological performance, socio-economic benefits, and governance outcomes are limited, particularly beyond the lifespan of individual projects (Hai et al. 2020). The lack of standardized indicators and sustained monitoring programs reduces the ability of managers to evaluate effectiveness, learn from experience, and adjust interventions in response to changing environmental and socio-economic conditions (Tran et al. 2025). Addressing these challenges will be critical for strengthening the resilience and long-term viability of bioshields in Vietnam as a national coastal protection strategy.

14.10 Conclusions

Vietnam's experience with coastal bioshields demonstrates that these systems extend well beyond their role as ecological interventions, functioning instead as integrated socio-economic infrastructures embedded within dynamic coastal landscapes. Across the country, areas that were once heavily degraded by deforestation, conflict, and unsustainable land-use practices are increasingly being re-envisioned as living infrastructures that provide multiple, mutually reinforcing benefits. By attenuating waves and storm surges, stabilizing shorelines, supporting biodiversity, sustaining livelihoods, and sequestering carbon, coastal bioshields contribute simultaneously to disaster risk reduction, climate change adaptation and mitigation, and sustainable development.

The evidence reviewed in this chapter highlights the effectiveness of bioshields when they are designed and managed in ways that are ecologically appropriate, institutionally supported, and socially inclusive. Vietnam's diverse coastal settings—from deltaic mangrove systems in the Mekong and Red River deltas to sandy and reef-associated coasts in central and island regions—illustrate the adaptability of bioshields to different geomorphological and socio-economic contexts. Case studies from Can Gio mangrove forests and Ca Mau Peninsula underscore the importance of long-term commitment, community participation, and integration with livelihood systems in achieving durable outcomes.

At the same time, significant challenges remain. Persistent land-use competition, particularly from aquaculture, tourism, and urban expansion, continues to threaten bioshield integrity in many coastal areas. Institutional complexity and overlapping mandates among agencies complicate planning and implementation, while reliance on short-term or externally funded projects constrains long-term maintenance and monitoring. Climate change introduces additional uncertainty, with sea-level rise, salinity intrusion, and increasing storm intensity posing risks to both existing and newly established bioshields.

Despite these constraints, Vietnam's combination of growing scientific capacity, policy recognition of nature-based solutions, and accumulated experience in community-based management provides a strong foundation for the future expansion and consolidation of bioshield programs. Continued investment in research, innovative financing mechanisms such as blue carbon, and stronger integration of bioshields into national and provincial planning frameworks will be essential to sustain and scale these efforts.

As Vietnam advances its transition towards a green, inclusive, and climate-resilient economy, coastal bioshields are poised to remain a cornerstone of sustainable coastal protection. More broadly, they symbolize a development pathway that seeks harmony between nature, people, and progress, offering valuable lessons for other coastal nations confronting the shared challenges of climate change and coastal vulnerability.

References

Aerts JC (2018) A review of cost estimates for flood adaptation. Water 10(11):1646. https://doi.org/10.3390/w10111646

Ayadi R, Forouheshfar Y, Moghadas O (2025) Enhancing system resilience to climate change through artificial intelligence: a systematic literature review. Front Clim 7:1585331. https://doi.org/10.3389/fclim.2025.1585331

Bao QT (2011) Effect of mangrove forest structures on wave attenuation in coastal Vietnam. Oceanologia 53(3):807–818. https://doi.org/10.5697/oc.53-3.807

Baumgartner U, Nguyen TH (2017) Organic certification for shrimp value chains in Ca Mau, Vietnam: a means for improvement or an end in itself? Environ Dev Sustain 19:987–1002. https://doi.org/10.1007/s10668-016-9781-z

Beresnev N, Phung T, Broadhead J (2016) Mangrove-related policy and institutional frameworks in Pakistan. Thailand and Viet Nam. Food and Agriculture Organization of the United Nations Regional Office for Asia and the Pacific International Union for Conservation of Nature (IUCN) Bangkok

Besset M, Gratiot N, Anthony EJ, Bouchette F, Goichot M, Marchesiello P (2019) Mangroves and shoreline erosion in the Mekong River delta, Viet Nam. Estuar Coast Shelf Sci 226:106263. https://doi.org/10.1016/j.ecss.2019.106263

Bimrah K, Dasgupta R, Hashimoto S, Saizen I, Dhyani S (2022) Ecosystem services of mangroves: a systematic review and synthesis of contemporary scientific literature. Sustainability 14(19):12051. https://doi.org/10.3390/su141912051

Bunting P, Rosenqvist A, Hilarides L, Lucas RM, Thomas N, Tadono T, Worthington TA, Spalding M, Murray NJ, Rebelo LM (2022) Global mangrove extent change 1996–2020: global mangrove watch version 3.0. Remote Sens 14(15):3657. https://doi.org/10.3390/rs14153657

Castillo-Atoche A, García NC, Atoche-Enseñat R, Estrada-López JJ, Quijano-Cetina R, Chávez L, Vázquez-Castillo J, Castillo-Atoche A (2025) Sustainable GIoT-based mangrove monitoring system for smart coastal cities with energy harvesting from SMFCs. Technologies 13(12):549. https://doi.org/10.3390/technologies13120549

Chaudhari S, Devi Prasad KV, Shanker K (2009) Impact of Casuarina plantations on Olive Ridley Turtle Nesting along the Northern Tamil Nadu Coast, India. ATREE, Bangalore and MCBT, Mamallapuram, India, p 44

Cong NV, Thạnh TD, Nhon DH (2020) Integrated coastal zone management in Vietnam—status and emerging problems. Vietnam J Mar Sci Technol 20(4):369–381. https://doi.org/10.15625/1859-3097/15316

Contti Neto N, Pomeroy A, Lowe R, Ghisalberti M (2022) Seagrass meadows reduce wind-wave driven sediment resuspension in a sheltered environment. Front Mar Sci 8:733542. https://doi.org/10.3389/fmars.2021.733542

Cuong NQ, Van Cu N (2014) Integrated coastal management in Vietnam: current situation and orientation. Vietnam J Mar Sci Technol 14(1):89–96. https://doi.org/10.15625/1859-3097/14/1/4034

Dang NA, Benavidez R, Tomscha SA, Nguyen H, Tran DD, Nguyen DT, Loc HH, Jackson BM (2021) Ecosystem service modelling to support nature-based flood water management in the Vietnamese Mekong River Delta. Sustainability 13(24):13549. https://doi.org/10.3390/su132413549

Daro Justine YE, Seenath A (2025) Vegetative nature-based solutions for coastal flood risk management: benefits, challenges, and uncertainties. Ocean Coast Manag 261:107520. https://doi.org/10.1016/j.ocecoaman.2024.107520

Duijndam SJ, Botzen WW, Hagedoorn LC, Bubeck P, Haer T, Pham M, Aerts JC (2023) Drivers of migration intentions in coastal Vietnam under increased flood risk from sea level rise. Clim Change 176(2):12. https://doi.org/10.1007/s10584-022-03479-9

Duy DV, Ty TV, Phat LT, Minh HV, Nam ND, Downes NK, Avtar R, Tanaka H (2025) Simulating the coastal protection performance of breakwaters in the Mekong Delta: insights from the Western Coast of Ca Mau Province, Vietnam. J Mar Sci Eng 13(8):1559. https://doi.org/10.3390/jmse13081559

Ebeler L, Balke T, Laurie E, Le H, Nguyen T, Renaud FG (2025) Local-scale impacts of mangrove restoration and conservation on coastal communities in two Vietnamese deltas: socio-economic and institutional dynamics. Ocean Coast Manag 269:107812. https://doi.org/10.1016/j.ocecoaman.2025.107812

Fagherazzi S, Bryan KR, Nardin W (2017) Buried alive or washed away: the challenging life of mangroves in the Mekong Delta. Oceanography 30(3):48–59. https://doi.org/10.5670/oceanog.2017.313

Friess DA, Gatt YM, Ahmad R, Brown BM, Sidik F, Wodehouse D (2022) Achieving ambitious mangrove restoration targets will need a transdisciplinary and evidence-informed approach. One Earth 5(5):456–460. https://doi.org/10.1016/j.oneear.2022.04.013

Gijón Mancheño A, Jafino BA, Hofland B, Van Wesenbeeck BK, Kazi S, Urrutia I (2025) Nature meets infrastructure: the role of mangroves in strengthening Bangladesh's coastal flood defenses. Sustainability 17(4):1567. https://doi.org/10.3390/su17041567

Ha TTT, Bush SR, Mol AP, Van Dijk H (2012) Organic coasts? Regulatory challenges of certifying integrated shrimp–mangrove production systems in Vietnam. J Rural Stud 28(4):631–639. https://doi.org/10.1016/j.jrurstud.2012.07.001

Hai NT, Dell B, Phuong VT, Harper RJ (2020) Towards a more robust approach for the restoration of mangroves in Vietnam. Ann for Sci 77(1):18. https://doi.org/10.1007/s13595-020-0921-0

Hoa NH, Ngoc TLT, Le An T, Nghia NH, Khanh DLV, Thu NHT, Premnath CFS et al (2020) Monitoring changes in coastal mangrove extents using multi-temporal satellite data in selected communes, Hai Phong City, Vietnam. For Soc 4:256–270. https://doi.org/10.24259/fs.v4i1.8486

Hoang Xuan T (1996) Relation between the properties of coastal sandy soil and planting potential of Casuarina equisetifolia in Vietnam. Recent Casuarina research and development. Proceedings of the third international Casuarina workshop, Da Nang, Vietnam, 4–7 March 1996, pp 214–217, ref. 4

Hong Tinh P, Van Thanh V, Sy Tuan M, Van Quang P, Sharma S, MacKenzie RA (2020) A comparison of soil carbon stocks of intact and restored mangrove forests in Northern Vietnam. Forests 11(6):660. https://doi.org/10.3390/f11060660

Hülsen S, Dee LE, Kropf CM, Meiler S, Bresch DN (2025) Mangroves and their services are at risk from tropical cyclones and sea level rise under climate change. Commun Earth Environ 6(1):262. https://doi.org/10.1038/s43247-025-02242-z

Hung NT, Duc DM, Quynh DT, Cuong VD (2020) Nearshore topographical changes and coastal stability in Nam Dinh Province, Vietnam. J Mar Sci Eng 8(10):755. https://doi.org/10.3390/jmse8100755

IFRC (2022) The multiple benefits of mangrove restoration: lessons from Viet Nam. The International Federation of Red Cross and Red Crescent Societies. https://disasterlaw.ifrc.org/media/3886. Accessed 27 Nov 2025

Jayson-Quashigah PN, Staneva J, Chen W, Djath B (2025) Assessing the role of mangroves in shoreline stabilisation through a What-If modelling framework. Discov Appl Sci 7(9):982. https://doi.org/10.1007/s42452-025-07623-9

Kuenzer C, Bluemel A, Gebhardt S, Quoc TV, Dech S (2011) Remote sensing of mangrove ecosystems: a review. Remote Sens 3(5):878–928. https://doi.org/10.3390/rs3050878

Lai QT, Tuan VA, Thuy NTB, Huynh LD, Duc NM (2022) A closer look into shrimp yields and mangrove coverage ratio in integrated mangrove-shrimp farming systems in Ca Mau, Vietnam. Aquacult Int 30(2):863–882. https://doi.org/10.1007/s10499-021-00831-1

Lappe R, Ullmann T, Bachofer F (2022) State of the Vietnamese coast—assessing three decades (1986 to 2021) of coastline dynamics using the Landsat archive. Remote Sens 14(10):2476. https://doi.org/10.3390/rs14102476

Le H (2021) Competing for land, mangroves and marine resources in coastal Vietnam, vol 24. Springer, Netherlands, p 171

Le NN, Nguyen KT, Dinh TD, Pham TMH, Le THV, Nguyen XT, Pham TD, Pham TD (2025a) Distribution of heavy metals in mangrove soils of Red River Delta, Northern of Vietnam. Mar Environ Res 213:107631. https://doi.org/10.1016/j.marenvres.2025.107631

Le TL, Ngo HS, Nguyen HM (2025b) Integrating gender equality considerations into the community engagement framework in renewable energy infrastructure projects: a suggestion for Vietnam. In: Green transformation in the context of global change. CRC Press, pp 488–497

Lee J, Yen L, Chu M, Lin Y, Chang C, Lin R, Chao K, Lee M (2020) Growth characteristics and anti-wind erosion ability of three tropical foredune pioneer species for sand dune stabilization. Sustainability 12(8):3353. https://doi.org/10.3390/su12083353

Magdalena I, Kusnowo V, Azis MI (2021) 1D–2D numerical model for wave attenuation by mangroves as a porous structure. Computation 9(6):66. https://doi.org/10.3390/computation9060066

Mazda Y, Magi M, Ikeda Y, Kurokawa T, Asano T (2006) Wave reduction in a mangrove forest dominated by *Sonneratia* sp. Wetl Ecol Manag 14:365–378. https://doi.org/10.1007/s11273-005-5388-0

McEwin A, McNally R (2014) Organic Shrimp certification and carbon financing: an assessment for the mangroves and markets project in Ca Mau Province, Vietnam. SNV Netherlands Development Organisation, Hague

Minderhoud PS, Coumou L, Erkens G, Middelkoop H, Stouthamer E (2019) Mekong delta much lower than previously assumed in sea-level rise impact assessments. Nat Commun 10(1):3847. https://doi.org/10.1038/s41467-019-11602-1

Minh HV, Lien BT, Hong Ngoc DT, Ty TV, Ngan NV, Cong NP, Downes NK, Meraj G, Kumar P (2024) Understanding rainfall distribution characteristics over the Vietnamese Mekong Delta: a comparison between coastal and inland localities. Atmosphere 15(2):217. https://doi.org/10.3390/atmos15020217

Ministry of Agriculture and Environment (MAE) (2024) Dual solutions for climate change response. https://en.mae.gov.vn/Pages/chi-tiet-tin-Eng.aspx?ItemID=8340

Ngo D, Nguyen H, Nguyen K, Dang C, Nguyen H, Dang N, Pham T (2023) Application of multispectral UAV to estimate mangrove biomass in Vietnam: a case study in Dong Rui commune, Quang Ninh Province. One Ecosyst 8:e103760. https://doi.org/10.3897/oneeco.8.e103760

Nguyen THM, Dang TH (2018) Gender role in mangrove resource management: case study in Trieu Phong district of Quang Tri province, Vietnam. J Vietnam Environ 9:92–98. https://doi.org/10.13141/jve.vol9.no2.pp92-98

Nguyen TP, Parnell KE (2019) Coastal land use planning in Ben Tre, Vietnam: constraints and recommendations. Heliyon 5(4):e01487. https://doi.org/10.1016/j.heliyon.2019.e01487

Nguyen TT, Pittock J, Nguyen BH (2017) Integration of ecosystem-based adaptation to climate change policies in Viet Nam. Clim Change 142:97–111. https://doi.org/10.1007/s10584-017-1936-x

Nguyen HT, Hardy GE, Le TV, Nguyen HQ, Nguyen HH, Nguyen TV, Dell B (2021) Mangrove forest landcover changes in coastal Vietnam: a case study from 1973 to 2020 in Thanh Hoa and Nghe An Provinces. Forests 12(5):637. https://doi.org/10.3390/f12050637

Nguyen H, Chu L, Harper RJ, Dell B, Hoang H (2022) Mangrove-shrimp farming: a triple-win approach for communities in the Mekong River Delta. Ocean Coast Manag 221:106082. https://doi.org/10.1016/j.ocecoaman.2022.106082

Orchard SE, Stringer LC, Quinn CH (2015) Environmental entitlements: institutional influence on mangrove social-ecological systems in Northern Vietnam. Resources 4(4):903–938. https://doi.org/10.3390/resources4040903

PEMSEA, Viet Nam Administration for Seas and Islands (VASI) and Ministry of Environment and Natural Resources (MONRE) (2020) National state of oceans and coasts 2018: blue economy growth of Viet Nam. Partnerships in Environmental Management for the Seas of East Asia (PEMSEA), Quezon City, Philippines, 232 pp

Pham TT, Nguyen VD (2021) A decade of mangrove conservation achievements and challenges in Vietnam. Center for International Forestry Research (CIFOR), Bogor, Indonesia

Pham TT, Vu TP, Hoang TL, Dao TL, Nguyen DT, Pham DC, Dao LH, Nguyen VT, Hoang NV (2022) The effectiveness of financial incentives for addressing mangrove loss in Northern Vietnam. Front For Glob Change 4:709073. https://doi.org/10.3389/ffgc.2021.709073

Phan LK, de Vries JSVT, Stive MJ (2015) Coastal mangrove squeeze in the Mekong Delta. J Coast Res 31(2):233–243. https://doi.org/10.2112/JCOASTRES-D-14-00049.1

Potouroglou M, Bull JC, Krauss KW, Kennedy HA, Fusi M, Daffonchio D, Mangora MM, Githaiga MN, Diele K, Huxham M (2017) Measuring the role of seagrasses in regulating sediment surface elevation. Sci Rep 7(1):11917. https://doi.org/10.1038/s41598-017-12354-y

Quartel S, Kroon A, Augustinus P, Van Santen P, Tri N (2007) Wave attenuation in coastal mangroves in the Red River Delta, Vietnam. J Asian Earth Sci 29(4):576–584. https://doi.org/10.1016/j.jseaes.2006.05.008

Research Institute for Marine Fisheries, Vietnam (RIMF) (2025) Biodiversity in marine ecosystems in Vietnam. https://www.rimf.org.vn/bantin/chitiet/biodiversity-in-marine-ecosystems-in-vietnam. Accessed 02 Feb 2025

Schmitt K, Albers T, Pham TT, Dinh SC (2013) Site-specific and integrated adaptation to climate change in the coastal mangrove zone of Soc Trang Province, Viet Nam. J Coast Conserv 17(3):545–558. https://doi.org/10.1007/s11852-013-0253-4

Son LV, Phong NV, Kien TT, Hoa LT, Le Phu V (2022) Impacts of livelihood activities on the environment and natural resources of Can Gio Mangrove Forest, Ho Chi Minh City. IOP Conf Ser Earth Environ Sci 964:012014. https://doi.org/10.1088/1755-1315/964/1/012014

Su J, Friess DA, Gasparatos A (2021) A meta-analysis of the ecological and economic outcomes of mangrove restoration. Nat Commun 12(1):5050. https://doi.org/10.1038/s41467-021-25349-1

Thi Huyen N, Hoang TuL, Thi Ha P, Kim Duyet N, Kim Loi N (2024) Application of remote sensing in analyzing mangrove forest changes and drivers in Mui Ca Mau National Park, Vietnam. Appl Environ Res 46(4):052. https://doi.org/10.35762/AER.2024052

Thinh NV, Dung NT, Hiep NT, Luu DP, An DT (2025) Assessment of mangrove cover change based on combining remote sensing technique and hydrodynamic model simulation. J Environ Earth Sci 7(2):151–160. https://doi.org/10.30564/jees.v7i2.7730

Thuy PT, Moeliono M, Locatelli B, Brockhaus M, Gregorio MD, Mardiah S (2014) Integration of adaptation and mitigation in climate change and forest policies in Indonesia and Vietnam. Forests 5(8):2016–2036. https://doi.org/10.3390/f5082016

Thuy PT, Swails E, Dat PT (2025) Towards comprehensive blue carbon accounting in Vietnam. The Center for International Forestry Research and World Agroforestry (CIFOR-ICRAF). https://doi.org/10.17528/cifor-icraf/009382

Tien LNH, Dung TX, Phuoc VLH (2025) Coastal protection in Cu Lao Dung Mangroves (Soc Trang Province, Vietnam): quantifying wave energy dissipation. J Environ Earth Sci 7(1):514–526. https://doi.org/10.30564/jees.v7i1.7399

Tin HC, Pho MV, Phap TT (2010) Current status of mangrove vegetation in Tam Giang-Cau Hai wetland area, Thua Thien Hue province (in Vietnamese). J Res Dev 1(78):88–94

Tinh PH, MacKenzie RA, Hung TD, Vinh TV, Ha HT, Lam MH, Hanh NT, Tung NX, Hai PM, Huyen BT (2022) Mangrove restoration in Vietnamese Mekong Delta during 2015–2020: achievements and challenges. Front Mar Sci 9:1043943. https://doi.org/10.3389/fmars.2022.1043943

To LP, Nguyen HT, Roel B, Lam TS, Dang KN (2017) Sustainability and livelihood opportunities for mangrove-shrimp farmers in the coastal areas of SocTrang province. CTU J Innov Sustain Dev 5:18–29. https://doi.org/10.22144/ctu.jen.2017.003

Tong TMT, Nguyen NH (2021) Ecosystem-based approaches and policy perspectives: towards an integrated blue–green solutions in Vietnam. In: Mukherjee M, Shaw R (eds) Ecosystem-based disaster and climate resilience. Disaster and risk research: GADRI book series. Springer, Singapore. https://doi.org/10.1007/978-981-16-4815-1_6

Tong SS, Pham-Duc B, Phan TH, Bui VT, Le VC, Pham TL, Tong THA (2023) Investigation of estuarine mangrove ecosystem changes using unmanned aerial vehicle images: case study in Xuan Thuy National Park (Vietnam). Reg Stud Mar Sci 62:102910. https://doi.org/10.1016/j.rsma.2023.102910

Tran TT, Yen NTM, Son HN, Quang NX (2020) Multi-correlation between nematode communities and environmental variables in mangrove-shrimp ponds, Ca Mau Province, Southern Vietnam. J Biol/tạp Chí Sinh Học 42(3):15–29. https://doi.org/10.15625/2615-9023/v42n3.14546

Tran DD, Dang MM, Du Duong B, Sea W, Vo TT (2021) Livelihood vulnerability and adaptability of coastal communities to extreme drought and salinity intrusion in the Vietnamese Mekong Delta. Int J Disaster Risk Reduct 57:102183. https://doi.org/10.1016/j.ijdrr.2021.102183

Tran TT, Takahashi K, Nishikawa H, Nguyen VH, Hoang TT, Nguyen TH, Dau TK, Vo VT, Tran DD (2023) Evaluation of mangrove ecosystem importance for local livelihoods in different landscapes: a case study of the Hau and Hoang Mai River Estuaries in Nghe An, North-Central Vietnam. Sustainability 15(4):3843. https://doi.org/10.3390/su15043843

Tran TV, Reef R, Zhu X (2024) Long-term changes of mangrove distribution and its response to anthropogenic impacts in the Vietnamese Southern Coastal Region. J Environ Manag 370:122658. https://doi.org/10.1016/j.jenvman.2024.122658

Tran TV, Reef R, Zhu X, Mohan M (2025) Spectral analysis for monitoring mangrove restoration: a case study in the Vietnamese Southern Coastline. Camb Prisms Coast Futures 3:e8. https://doi.org/10.1017/cft.2025.5

Tran Thi V, Tien Thi Xuan A, Phan Nguyen H, Dahdouh-Guebas F, Koedam N (2014) Application of remote sensing and GIS for detection of long-term mangrove shoreline changes in Mui Ca Mau, Vietnam. Biogeosciences 11:3781–3795. https://doi.org/10.5194/bg-11-3781-2014

Trinh TT, Pattiaratchi C, Bui T (2020) The contribution of forerunner to storm surges along the Vietnam coast. J Mar Sci Eng 8(7):508. https://doi.org/10.3390/jmse8070508

Truong TD, Do LH (2018) Mangrove forests and aquaculture in the Mekong River Delta. Land Use Policy 73:20–28. https://doi.org/10.1016/j.landusepol.2018.01.029

Tuan VQ, Kuenzer C (2012) Can Gio mangrove biosphere reserve evaluation 2012: current status, dynamics, and ecosystem services. IUCN, Hanoi, Viet Nam, p 102

UN/DOALOS-Norad Programmes of Assistance, Hai Dang V, Binh NQ (2023) Ocean governance study of Viet Nam. www.un.org/oceancapacity/norway/NOGS. Accessed 18 Nov 2025

UNDP (2024) Mangroves, the lifeline of Viet Nam's coasts: stewards of climate, biodiversity, and communities. https://www.undp.org/vietnam/undp-helps-vietnams-coastal-communities-adapt-climate-changen. Accessed 27 Dec 2025

UNEP-WCMC, Short FT (2021) Global distribution of seagrasses (version 7.1). Seventh update to the data layer used in Green and Short (2003). https://doi.org/10.34892/x6r3-d211

UNEP-WCMC, WorldFish Centre, WRI, TNC (2021) Global distribution of coral reefs, compiled from multiple sources including the Millennium Coral Reef Mapping Project. Version 4.1, updated by UNEP-WCMC. Includes contributions from IMaRS-USF and IRD (2005), IMaRS-USF (2005) and Spalding et al. (2001). UN Environment Programme World Conservation Monitoring Centre, Cambridge (UK). Data https://doi.org/10.34892/t2wk-5t34

Van Cuong C, Dart P, Ha NM, Le VTM, Hockings M (2019) Biosphere reserves in Vietnam: management challenges. In: UNESCO biosphere reserves. Routledge, pp 201–212

van Ledden M, Tung TT, Nguyen DH, Nguyen LT (2020) Coastal development between opportunity and disaster risk. An assessment of the coastal protection system in Vietnam. Policy Research Working Paper 9351. World Bank

Van Wesenbeeck BK, Van Zelst VT, Antolinez JA, De Boer WP (2025) Quantifying uncertainty in wave attenuation by mangroves to inform coastal green belt policies. Commun Earth Environ 6(1):258. https://doi.org/10.1038/s43247-025-02178-4

VAST (2025) Remote sensing and artificial intelligence applied to monitoring functional traits of mangrove vegetation in Vietnam. https://vast.gov.vn/web/vietnam-academy-of-science-and-technology/tin-chi-tiet/-/chi-tiet/remote-sensing-and-artificial-intelligence-applied-to-monitoring-functional-traits-of-mangrove-vegetation-in-vietnam-139953-871.html? Accessed 18 Dec 2025

Veettil BK, Puri V (2025a) Blue carbon ecosystems in Vietnam: a review. Ocean Coast Manag 267:107711. https://doi.org/10.1016/j.ocecoaman.2025.107711

Veettil BK, Puri V (2025b) Coral reefs in Vietnam: current state of research and future perspectives. Estuar Coast Shelf Sci 326:109554. https://doi.org/10.1016/j.ecss.2025.109554

Veettil BK, Ward RD, Quang NX, Trang NTT, Giang TH (2019) Mangroves of Vietnam: historical development, current state of research and future threats. Estuar Coast Shelf Sci 218:212–236. https://doi.org/10.1016/j.ecss.2018.12.021

Veettil BK, Costi J, Marques WC, Tran X, Quang NX, Van DD, Hoai PN (2020) Coastal environmental changes in Southeast Asia: a study from Quang Nam Province, Central Vietnam. Reg Stud Mar Sci 39:101420. https://doi.org/10.1016/j.rsma.2020.101420

Veettil BK, Ward RD, Dung NTK, Van DD, Quang NX, Hoai PN, Hoang N (2021) The use of bioshields for coastal protection in Vietnam: current status and potential. Reg Stud Mar Sci 47:101945. https://doi.org/10.1016/j.rsma.2021.101945

Veettil BK, Quang NX, Dung NTK, Bao DQ, Xuan-Linh T, Nhat-Duc H (2022a) Mangrove forest degradation and shoreline erosion in the Mekong delta: a case study from Ben Tre province. Vietnam J Sci Technol 60(5):825–836. https://doi.org/10.15625/2525-2518/16555

Veettil BK, Ward RD, Van DD, Quang NX, Hoai PN (2022b) Seagrass ecosystems along the Vietnamese coastline: current state of research and future perspectives. Estuar Coast Shelf Sci 277:108085. https://doi.org/10.1016/j.ecss.2022.108085

Veettil BK, Tran X, Lopes MS (2024) Mangrove vegetation changes and shoreline erosion along the central coastline of Vietnam: a study from Quang Ngai province. Environ Dev 52:101065. https://doi.org/10.1016/j.envdev.2024.101065

Veettil BK, Acharki S, Puri V, Teck V (2025) Coastal environmental changes in Ninh Thuan Province, South-Central Vietnam. Plos ONE 20(2):e0313382. https://doi.org/10.1371/journal.pone.0313382

Vinh VD (2015) Applying numerical model to assess role of mangrove forest in wave attenuation in Hai Phong coastal area. Vietnam J Mar Sci Technol 15(1):67–76. https://doi.org/10.15625/1859-3097/15/1/6082

Vu TT, Pham TD, Saintilan N, Skidmore A, Luu HV, Vu QH, Le NN, Nguyen HQ, Matsushita B (2022) Mapping multi-decadal mangrove extent in the northern coast of Vietnam using Landsat time-series data on Google Earth Engine platform. Remote Sens 14(18):4664. https://doi.org/10.3390/rs14184664

Vu TP, Pham TT, Tran NMH, Nguyen TTA, Nguyen TVA, Tang Thi KH (2023) Blue carbon market in Viet Nam: potential and challenges for future development. Occasional Paper 8. Center for International Forestry Research (CIFOR), Bogor, Indonesia; and World Agroforestry (ICRAF), Nairobi, Kenya

Xuan BB (2021) Consumer preference for eco-labelled aquaculture products in Vietnam. Aquaculture 532:736111. https://doi.org/10.1016/j.aquaculture.2020.736111

Zhang J, Lu Z, Zhou J, Qin G, Bai Y, Sanders CJ, Macreadie PI, Yuan J, Huang X, Wang F (2025) Getting the best of carbon bang for mangrove restoration buck. Nat Commun 16(1):1297. https://doi.org/10.1038/s41467-025-56587-2

Zimmermann TG, Andrade AC, Richardson DM (2016) Experimental assessment of factors mediating the naturalization of a globally invasive tree on sandy coastal plains: a case study from Brazil. AoB PLANTS 8:plw042. https://doi.org/10.1093/aobpla/plw042

Zinzani A (2018) International development policies and coastalscape metabolism: the case of the Mekong Delta, Vietnam. Soc Sci 7(2):19. https://doi.org/10.3390/socsci7020019

Part VI
Coastal Bioshields: Future Directions

Chapter 15
Technological Innovations and Research Frontiers

Abstract Technological advancement has become a critical driver in strengthening the design, monitoring, and evaluation of bioshields as nature-based coastal protection systems. This chapter examines emerging technologies and research frontiers that are reshaping both the understanding and practical application of bioshields within coastal resilience planning. It reviews recent innovations in remote sensing, geographic information systems (GIS), unmanned aerial vehicles (UAVs), and artificial intelligence (AI) that enable more accurate mapping of coastal vegetation dynamics, enhanced assessment of ecosystem condition, and improved prediction of shoreline change. Advances in hydrodynamic and ecological modeling are also highlighted as essential tools for optimizing bioshield configuration, placement, and performance across diverse climatic and geomorphological contexts. The chapter also explores the expanding role of biotechnology and ecological engineering in enhancing plant tolerance to salinity, sedimentation, and sea-level rise. Integrated monitoring networks—incorporating Internet of Things (IoT) sensors, automated data platforms, and real-time analytics—are examined as mechanisms for strengthening early-warning systems and supporting adaptive management. The interface between technological innovation and community science, including participatory data collection and the use of digital platforms, is highlighted as an emerging opportunity to advance inclusive and evidence-based decision-making. The chapter further identifies key research gaps, such as the need for standardized assessment frameworks, interdisciplinary modeling approaches, and improved socio-technical integration to scale bioshield innovations globally. By bridging ecological science with technological advancements, the chapter envisions a new research frontier in which digital tools and ecosystem-based principles converge to support the development of smarter, more resilient coastal protection systems for the future.

Keywords Artificial intelligence (AI) · Biotechnology in restoration · Ecological engineering · Hydrodynamic modeling · Remote sensing and GIS · Smart coastal monitoring

B. Kozhikkodan Veettil, *The Potential of Bioshields for Coastal Protection*,
Advances in Natural and Technological Hazards Research 55,
https://doi.org/10.1007/978-3-032-22888-8_15

15.1 Introduction

The increasing vulnerability of coastal zones to erosion, storm surges, and sea-level rise has amplified the demand for innovative, sustainable, and scientifically grounded protection measures. While traditional hard engineering solutions, such as seawalls and groynes, have been widely implemented, their ecological and socio-economic drawbacks have prompted a shift toward nature-based solutions (Veettil et al. 2021). Among these, bioshields, which are vegetation-based coastal protection systems comprising mangroves, salt marshes, casuarinas, and dune vegetation, represent an adaptive, multifunctional, and ecologically harmonious approach (Athikalam and Vaideeswaran 2022).

Parallel to the global rise of nature-based approaches, rapid advancements in environmental technology are transforming how bioshields are conceptualized, designed, and managed (Fig. 15.1). High-resolution remote sensing, varying from multispectral satellites to UAV-based Light Detection and Ranging (LiDAR), offers unprecedented capabilities for mapping vegetation structure, sediment distribution, and shoreline change (Veettil et al. 2020a, b, 2021; Tian et al. 2023). Machine learning and artificial intelligence (AI) now allow automated classification of bioshield health, prediction of erosion hotspots, and early detection of degradation patterns (Tsiakos and Chalkias 2023; Wei et al. 2023). Hydrodynamic and morphodynamic models, increasingly coupled with ecological data, simulate how bioshields interact with waves, tides, and storms under present and future climate scenarios (e.g., Kumbier et al. 2022; Xie et al. 2022). These models help practitioners optimize planting configurations, species mixes, and landscape designs to maximize the protective functions of coastal bioshields globally.

Advances in genetics and biotechnology are opening additional frontiers. Genomic tools enhance understanding of species adaptation to salinity, inundation, and heat stress, which are traits as crucial as climate extremes intensify (Natarajan et al. 2021; Nizam et al. 2022). Research on phenotypic plasticity and assisted gene flow is improving restoration strategies by identifying resilient genotypes suited to local hydrological and sedimentary conditions (Miryeganeh et al. 2022). Biotechnological interventions, such as microbial inoculants that promote root growth or nutrient uptake, are being explored to increase early survival rates of planted mangroves and dune vegetation (Farrer et al. 2022).

Digital innovations are also reshaping governance and community participation. Mobile-based monitoring platforms enable citizen scientists and local stakeholders to document growth performance, report disturbances, and participate in adaptive management (Okarda et al. 2025). Cloud-based decision-support systems integrate ecological, socio-economic, and engineering datasets to inform planners and policy-makers (Xie et al. 2016; Abdelmajeed et al. 2023; Ciliberti and Coro 2025). These tools are particularly valuable in Asia, where densely populated deltas, estuaries, and island systems depend heavily on bioshields for both livelihood security and disaster risk reduction.

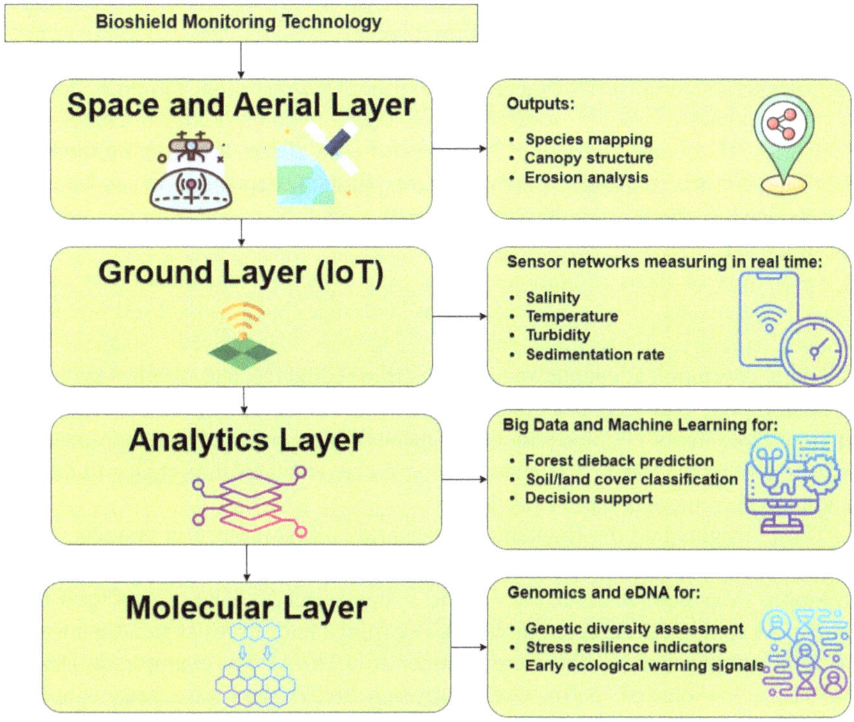

Fig. 15.1 Various technological innovations in coastal bioshield monitoring and assessment

This chapter synthesizes emerging technological innovations and research trends shaping the future of bioshields. By highlighting state-of-the-art approaches in mapping, modelling, monitoring, and management, it outlines opportunities to embed scientific advances into policy frameworks and implementation strategies. Ultimately, strengthening the integration of ecological knowledge and technological tools will be critical for scaling bioshields as robust, sustainable, and climate-resilient coastal protection systems.

15.2 Evolution of Technological Approaches in Coastal Protection

15.2.1 From Hard Engineering to Hybrid Systems

For much of the twentieth century, coastal protection strategies were dominated by hard engineering interventions, including concrete seawalls, bulkheads, groynes, and rock revetments (Williams et al. 2016). These structures were designed to offer rapid and reliable defense against erosion and storm-induced flooding. While often effective in the short term, they imposed substantial environmental costs. Rigid barriers interrupt natural littoral drift, exacerbate scouring at their base, and diminish sediment supply to adjacent beaches, leading to unintended geomorphological changes and habitat loss (Dean and Dalrymple 2004; Veettil et al. 2020a, b, 2021). Furthermore, their high construction and maintenance costs, limited adaptability to sea-level rise, and reduced aesthetic value have increasingly called their long-term sustainability into question.

Awareness of these limitations catalyzed a scientific re-evaluation of natural coastal systems, which are not as passive as landscapes but as active, self-organizing defense structures. Ecosystems such as mangroves, salt marshes, seagrass beds, and dune vegetation attenuate wave energy, trap sediments, and stabilize shorelines through biophysical feedbacks (Jordan and Fröhle 2022). The protective functions of these ecosystems, coupled with their capacity for regeneration and their contributions to biodiversity and carbon storage, have elevated their importance in modern coastal management (Duarte et al. 2013).

These insights laid the foundation for hybrid coastal protection systems, which integrate engineering principles with ecosystem processes to achieve multifunctional resilience (Van Zelst et al. 2021). Hybrid systems seek to balance structural reliability with ecological integrity, moving away from a built-to-resist paradigm toward one emphasizing build-with-nature (Slinger et al. 2021). For example, in Demak, Indonesia, low-crested, permeable brushwood breakwaters have been combined with mangrove rehabilitation efforts (van Wesenbeeck et al. 2015). This approach slows currents, promotes sediment deposition, and creates favorable conditions for natural mangrove recolonization (van Wesenbeeck et al. 2015). Similar pilot projects in Vietnam have demonstrated that hybrid interventions can outperform traditional infrastructure by enhancing long-term stability while reducing environmental footprints (Duy et al. 2025).

The development of hybrid systems also reflects advances in coastal engineering theory. Concepts such as dynamic equilibrium, sediment-budget analysis, and system-based resilience have replaced static design norms (Georgiou et al. 2024). In this coastal engineering design, species selection must be based on root architecture, salinity tolerance, growth rates, and sediment-trapping efficiency (Stachew et al. 2021). This multidisciplinary design philosophy acknowledges that natural systems change over time, and effective solutions must accommodate such variability.

15.2.2 Technological Integration in Bioshield Research

The past two decades have witnessed an unprecedented rise in the application of advanced technologies to coastal ecosystem research, particularly in the study and management of bioshields. This transformation is driven by the need for spatially detailed, temporally continuous, and scientifically rigorous data to understand how vegetated systems respond to hydrodynamic forces, climate stressors, and anthropogenic disturbances.

High-resolution remote sensing has become a cornerstone of bioshield assessment. Satellite missions such as Landsat series and Sentinel-2, together with commercial platforms (e.g., PlanetScope, WorldView series), provide multispectral and hyperspectral data with increasing spatial resolution, enabling accurate mapping of species composition, canopy structure, leaf area index, and shoreline change (Kuenzer et al. 2011; Veettil et al. 2020a, 2024). Unmanned Aerial Vehicles (UAVs) equipped with LiDAR, optical, and thermal sensors provide sub-centimeter imagery for analyzing seedling survival, erosion patterns, and storm impacts (Morgan et al. 2022; Zhou et al. 2023). LiDAR-derived elevation models are particularly valuable for quantifying micro-topographic features that influence tidal inundation patterns and sediment deposition, which are key determinants of bioshield resilience (Taddia et al. 2021).

In parallel, the Internet of Things (IoT) has revolutionized in situ monitoring. Networks of sensor nodes can measure salinity, temperature, turbidity, pore-water pressure, and sedimentation rates in real time (Xu et al. 2014). These continuous datasets can improve our understanding of fine-scale ecological processes, such as pneumatophore respiration, root biomechanics, and nutrient fluxes. When integrated with hydrodynamic models, IoT data can enhance predictions of wave attenuation, flood mitigation capacity, and vegetation–flow interactions under varying climate scenarios (Wang et al. 2025).

Big data analytics and machine learning further extend the analytical capacity of bioshield research. Algorithms can classify land cover, detect degradation, and forecast mangrove dieback or dune destabilization using historical and real-time datasets (Farzanmanesh et al. 2025; Lin et al. 2025; Peng et al. 2025). Predictive models trained on climate indices, sea-level anomalies, and vegetation metrics help identify future vulnerability hotspots and guide targeted restoration efforts (Payus et al. 2022). Machine learning-supported decision-support platforms are now being incorporated into coastal planning systems, enabling data-driven scenario testing and adaptive management (Pourzangbar et al. 2023).

Interdisciplinary research has amplified these technological advancements by linking fields previously isolated from one another. Genomic technologies, such as environmental DNA (eDNA) sequencing and population genomics, provide insights into the genetic diversity, stress tolerance, and adaptive potential of key bioshield species (e.g., Ma et al. 2022). These findings inform species selection for restoration, ensuring the use of resilient genotypes suited to site-specific salinity gradients, inundation regimes, and temperature variability.

Climate and morphodynamic modelling, once the domain of engineering, are now integrated with ecological data to simulate bioshield performance under future climate pathways. Coupled biophysical models evaluate how changes in storm intensity, sea-level rise, and sediment supply affect vegetation establishment and long-term stability. Such models are essential for designing bioshield configurations that remain effective in rapidly evolving coastal environments (e.g., Gourgue et al. 2022; Kumbier et al. 2022).

Collectively, these technological innovations mark a major transition in the science and practice of coastal protection. They enable more accurate characterization of bioshield functions, support evidence-based planning, and facilitate dynamic management approaches in the face of climate uncertainty. As hybrid and nature-based systems expand globally, the integration of engineering, ecology, and digital technologies will continue to shape the next generation of sustainable coastal protection strategies.

15.3 Remote Sensing and Geospatial Technologies

Remote sensing and geospatial technologies have rapidly evolved into indispensable tools for understanding, monitoring, and managing bioshields (Veettil et al. 2021). Their ability to generate spatially explicit, temporally consistent, and cost-effective data makes them central to mapping vegetation dynamics, assessing coastal vulnerability, and informing adaptive management strategies (Pham et al. 2019a). Recent technological advances (Fig. 15.2), ranging from high-resolution satellites to drone-based platforms, LiDAR, and synthetic aperture radar, enable researchers to capture complex biophysical parameters that govern bioshield performance (Pham et al. 2019a, b; Windle et al. 2023).

15.3.1 Satellite-Based Mapping and Monitoring

Satellite remote sensing represents one of the most powerful and widely adopted approaches for large-scale bioshield assessment (Veettil et al. 2021). Sensors such as Landsat, Sentinel-2, and MODIS provide multi-temporal, multispectral datasets that allow researchers to quantify vegetation cover, detect health anomalies, evaluate biomass trends, and track shoreline changes across decadal timescales (Pham et al. 2019a; Elmahdy and Ali 2022; Cavalli 2023). Landsat's 30-m spatial resolution and 40-year archive enable long-term vegetation change analysis, while Sentinel-2's higher spatial (10–20 m) and spectral resolution enhance the discrimination of different coastal vegetation types (Veettil 2022; Veettil et al. 2020a, 2020b, 2024).

In Vietnam, long-term analyses using Landsat imagery have been instrumental in tracking mangrove cover changes in the Mekong Delta, Ca Mau Peninsula, and Red River Delta (e.g., Hauser et al. 2017). Pham et al. (2019b) discussed how time-series image classification can quantify both natural regeneration and anthropogenic degradation, exposing hotspots of loss linked to aquaculture expansion and coastal erosion.

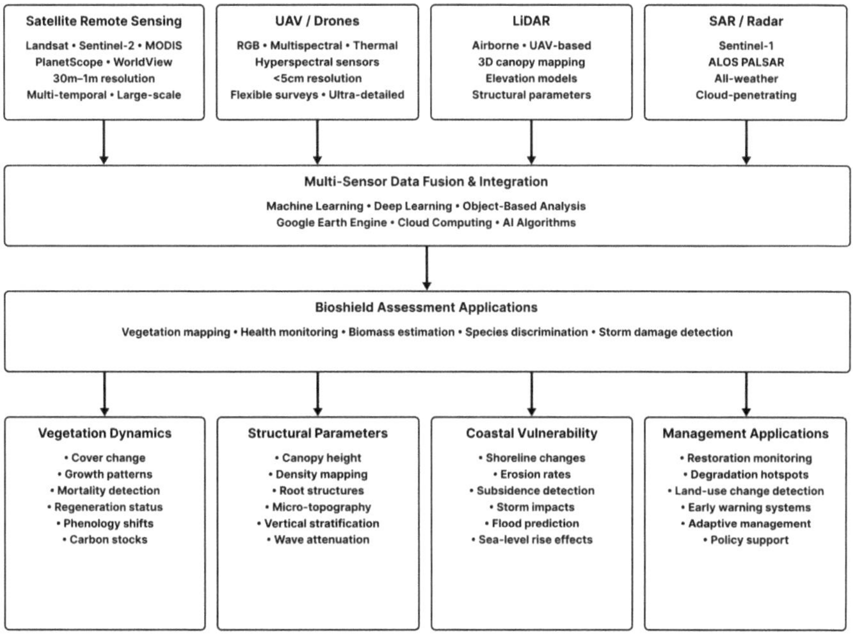

Fig. 15.2 Remote sensing applications in coastal bioshield establishment and management

Similar approaches in Bangladesh and India have mapped rapid bioshield degradation due to shrimp aquaculture, embankment construction, and urban encroachment, demonstrating how remote sensing supports evidence-based coastal zoning and habitat restoration prioritization (Karim et al. 2019; Paul and Mahmood 2025).

Commercial satellite constellations, such as PlanetScope, RapidEye, and WorldView, now offer sub-10 m (sometimes sub-1 m) spatial resolutions with daily revisit cycles. These ultrahigh-frequency observations allow near-real-time monitoring of vegetation mortality, sediment plume dynamics, storm impacts, and illegal land conversion. When combined with digital elevation models (DEMs), satellite imagery enables quantitative assessments of topographic controls on vegetation zonation, tidal inundation regimes, and shoreline recession rates (Moffett et al. 2015). Integration with DEMs derived from LiDAR or TanDEM-X missions also supports hydrodynamic modeling and flood prediction (Muhadi et al. 2020).

Machine-learning algorithms, particularly random forests, convolutional neural networks (CNNs), and support vector machines (SVMs), are increasingly applied to satellite data to automate classification, improve species-level discrimination, and detect subtle stress indicators such as chlorophyll decline or canopy thinning (e.g., Jiang et al. 2021; Maurya et al. 2021; Wei et al. 2023; Seydi et al. 2024). These analytical advances significantly enhance the ability of resource managers to identify early-warning signals of ecosystem degradation and plan timely interventions.

15.3.2 Unmanned Aerial Vehicles (UAVs) and Drone Applications

Unmanned Aerial Vehicles (UAVs), commonly known as drones, have revolutionized coastal ecosystem monitoring by offering centimeter-level spatial resolution and unparalleled survey flexibility. UAVs equipped with RGB, multispectral, hyperspectral, or thermal sensors provide ultra-high-resolution (<5 cm) imagery, enabling highly detailed mapping of individual seedlings, canopy health, root structures, and geomorphic microfeatures (Ruwaimana et al. 2018; Zhu et al. 2024).

In Southeast Asia, UAV applications have expanded rapidly as part of large-scale mangrove rehabilitation and monitoring initiatives (Basyuni et al. 2025). In Vietnam, drones have been used to quantify mangrove gap dynamics, assess seedling survival rates, and identify areas requiring enrichment planting (Ngo et al. 2023, 2024). In the Ayeyarwady Delta of Myanmar, UAV-based monitoring has proven effective for detecting storm-induced damage, mapping encroachment by agriculture, and guiding community-based replanting programs (Vogel et al. 2022).

The ability to collect data at flexible intervals (weekly, monthly, or after major storms) makes UAVs particularly valuable for detecting short-term disturbances and verifying restoration success. The integration of UAV imagery with GIS platforms and AI-based classification algorithms (e.g., deep learning-based semantic segmentation) increases mapping accuracy, enabling automated identification of species, growth stages, and stress patterns (e.g., Pádua et al. 2017; Conti et al. 2025). Combined with ground-based ecological surveys, UAV data support high-resolution digital terrain model (DTM) generation, structural mapping of dune vegetation, and fine-scale sediment-budget assessments—critical components for bioshield design and long-term management (Tysiąc et al. 2025).

15.3.3 LiDAR and Radar Applications

LiDAR has significantly advanced the characterization of bioshield structure by enabling 3D mapping of canopy height, density, and spatial heterogeneity. Airborne and UAV-based LiDAR systems generate high-resolution elevation models that capture root platforms, micro-topography, and the vertical stratification of vegetation, which are parameters essential for assessing wave attenuation capacity and hydrological flow patterns (Lin et al. 2019; Niwa et al. 2023). LiDAR-derived elevations also support hydrodynamic modeling by improving predictions of tidal inundation, storm-surge pathways, and sediment deposition patterns in mangrove and salt marsh ecosystems (Hladi and Alber 2012; Smolders et al. 2020).

Radar sensors, particularly Synthetic Aperture Radar (SAR), provide another powerful tool for bioshield monitoring, especially in tropical regions where persistent cloud cover limits optical remote sensing (Bunting et al. 2018). Instruments such as Sentinel-1 (C-band SAR), ALOS PALSAR (L-band SAR), and upcoming NISAR missions offer cloud-penetrating capabilities and all-weather imaging, allowing continuous monitoring during monsoon seasons and extreme weather (Torres et al. 2012; Bunting et al. 2018; Rosen et al. 2025). SAR backscatter signatures enable the discrimination of mangroves from upland vegetation, detection of inundation extent, and mapping of biomass variations based on dielectric properties (Canisius et al. 2019; Singh and Mahajan 2022). L-band SAR, with its deeper canopy penetration capacity, is particularly effective for estimating mangrove structural parameters, soil moisture, and regeneration status (Hamdan et al. 2014; Pradisty et al. 2025).

Advanced techniques such as Polarimetric SAR (PolSAR) and Interferometric SAR (InSAR) further enhance the ability to detect soil subsidence, track changes in canopy roughness, and quantify vertical land motion—key factors influencing bioshield resilience under sea-level rise (Kim et al. 2014; Wdowinski 2016; Fatnassi et al. 2023).

15.3.4 Integration of Multi-sensor Data and Data Fusion Approaches

The increasing diversity of remote sensing platforms has prompted the development of advanced data fusion techniques that integrate optical, radar, and LiDAR datasets to overcome the limitations of individual sensors. Multi-sensor integration enhances the spatial, spectral, and temporal resolution of bioshield monitoring systems and enables a more comprehensive understanding of vegetation dynamics, sediment processes, and hydrodynamic interactions (Pham et al. 2019b).

Optical sensors such as Landsat or Sentinel-2 provide detailed spectral information useful for vegetation health assessment, species identification, and chlorophyll quantification (Verrelst et al. 2015; Pham et al. 2019a, 2019b). However, cloud cover and atmospheric interference often restrict data availability, especially in monsoon-dominated regions (Veettil et al. 2020a, b). SAR systems, such as Sentinel-1, ALOS PALSAR, or TerraSAR-X, address this limitation by providing cloud-penetrating, all-weather imaging. Combining optical and SAR data through object-based image analysis (OBIA), Bayesian fusion, or machine-learning algorithms enhances classification accuracy, particularly in complex estuarine and deltaic landscapes where mixed pixels and tidal variability present significant challenges (Kuenzer et al. 2011).

Integration of LiDAR data introduces a third dimension to bioshield assessment. LiDAR-derived canopy height models (CHMs) and digital elevation models (DEMs) can be fused with multispectral imagery to improve biomass estimation and distinguish between vegetation structural types (Zhu et al. 2024). For example, CHM data combined with Sentinel-2 vegetation indices (e.g., NDVI, EVI, red-edge indices) allow researchers to disentangle spectral confusion among mangrove species that exhibit similar reflectance patterns but differ in canopy height or structural density (Wang et al. 2022). This fusion also supports accurate quantification of carbon stocks, facilitating inclusion of mangrove and coastal bioshield ecosystems in national greenhouse gas (GHG) inventories and blue carbon accounting frameworks (Alongi 2012; Verrelst et al. 2015).

Temporal fusion approaches, such as the STARFM (Spatial and Temporal Adaptive Reflectance Fusion Model) and its derivatives, allow the merging of high-temporal, moderate-resolution imagery (e.g., MODIS) with episodic high-resolution imagery (e.g., Landsat, Sentinel-2) (Gao et al. 2006; Zhu et al. 2010). This enables near-continuous monitoring of bioshield conditions at fine spatial scales, which is essential for detecting seasonal leaf loss, storm damage, algal blooms, or sediment burial events (Gao et al. 2006). Such time-series analyses also help in identifying long-term phenological trends driven by climate change, such as shifts in mangrove flowering cycles, changes in productivity under rising temperatures, or increases in canopy stress under prolonged salinity exposure (Pirasteh et al. 2021; Long et al. 2024; Jin et al. 2025).

SAR–LiDAR fusion has proven especially valuable for assessing subsidence and vertical land movement in low-lying coastal areas (Zhang et al. 2025). Interferometric SAR (InSAR) can detect millimeter-scale ground deformation, while LiDAR provides a stable reference for validating elevation changes. In rapidly subsiding deltas like the Mekong, Mississippi, and Ayeyarwady (Syvitski et al. 2009), such integrated analyses reveal how land subsidence interacts with sea-level rise to shape bioshield vulnerability and regeneration potential (Kuenzer et al. 2011). This information is critical for determining where sediment augmentation, embankment removal, or managed realignment may be requireed to maintain ecosystem functionality.

Machine learning and deep learning algorithms significantly enhance multi-sensor data integration by automating complex classification tasks. Convolutional neural networks (CNNs), recurrent neural networks (RNNs), and hybrid models can ingest raw imagery, topographic data, spectral indices, and textural metrics to generate highly accurate land-cover maps (Maggiori et al. 2016; Zhu et al. 2017). These models can identify subtle changes in canopy structure, detect invasive species encroachment, and map early-stage dieback events that may not be visually apparent in single-sensor imagery (Kerr and Ostrovsky 2003; Zhu et al. 2017).

Geospatial platforms such as Google Earth Engine (GEE), NASA SERVIR, and cloud-based GIS systems have democratized access to multi-sensor data and high-performance computing (Gorelick et al. 2017). Researchers can now process decades of imagery, run time-series analyses, and develop predictive models without the need for local high-end hardware (Kaur et al. 2023). These platforms can also facilitate collaboration among scientists, policymakers, and coastal managers, providing shared interfaces for visualizing bioshield conditions, forecasting degradation risks, and evaluating restoration outcomes.

Multi-sensor integration not only enhances analytical precision but also supports a holistic, process-based understanding of how bioshields evolve under natural and anthropogenic pressures. The combination of spectral, structural, hydrological, and geomorphological data offers unprecedented insight into feedbacks among vegetation, sediment transport, storm surges, and sea-level rise. As coastal hazards intensify under climate change, such comprehensive data fusion approaches will be essential for designing resilient bioshields, optimizing species selection, and implementing adaptive management strategies that account for dynamic environmental conditions.

15.4 Modelling Innovations

Modelling approaches have become central to understanding and optimizing the performance of bioshields under current and future climate conditions (Mendez and Losada 2004; Bao et al. 2025). Advances in hydrodynamic, ecological, biogeochemical, and AI-driven models now allow researchers to evaluate complex feedbacks between vegetation, sediment processes, and coastal hazards at multiple spatial and temporal scales (Van Wesenbeeck et al. 2025). These innovations provide the quantitative evidence needed to design effective bioshield configurations, assess resilience under climate change, and support policy decision-making.

15.4.1 Hydrodynamic and Sediment Transport Modelling

Hydrodynamic modelling represents one of the most critical tools for evaluating the protective efficacy of vegetated bioshields. Numerical models, such as SWAN (Lashley et al. 2019), MIKE 21 (Thai et al. 2022), Delft3D (Geng et al. 2025), and XBeach (James et al. 2021), simulate wave transformation, storm surge propagation, and sediment dynamics under various hydrological conditions. By incorporating vegetation parameters, including stem density, diameter, canopy height, flexibility, and drag coefficients, these models quantify the impact of vegetative resistance on flow modification and wave attenuation (e.g., Mendez and Losada 2004).

Studies have demonstrated that vegetation can reduce wave height by 20–60% depending on species characteristics, water depth, and incoming wave energy (Mazda et al. 1997; Narayan et al. 2016). Delft3D and XBeach enable researchers to simulate wave–vegetation interactions at fine scales, capturing feedback between hydrodynamics and morphodynamics (James et al. 2021; Geng et al. 2025). These models reveal how root structures stabilize sediments, how wave dissipation reduces erosion, and how vegetation influences sediment deposition patterns over time (Dissanayake et al. 2022).

In Vietnam, hydrodynamic models have been extensively used to guide bioshield planning along vulnerable coasts. For example, San et al. (2022) integrated MIKE 21 simulations with field surveys in the Mekong Delta to determine the optimal width of protective structures (e.g., breakwaters) required to reduce storm-surge impacts. Dang et al. (2022) used the Sea Level Affects Marshes Model (SLAMM) to investigate potential changes in mangrove distribution in response to future sea level rise scenarios in the coastal area in the south of the Mekong Delta. These studies have informed government guidelines for mangrove restoration and coastal defense planning.

Hydrodynamic models are increasingly being linked with climate projections to evaluate bioshield performance under scenarios of accelerated sea-level rise, subsidence, and intensifying storm regimes. Such coupling provides critical insight into when and where bioshields may remain effective, where sediment supply constraints may limit resilience, and where hybrid or engineering-assisted solutions may be necessary.

15.4.2 Ecological and Carbon Modelling

Coastal bioshields, particularly mangrove forests, salt marshes, and seagrass meadows, play a dual role in climate adaptation and climate change mitigation. To quantify their carbon storage potential, researchers increasingly rely on carbon accounting models such as InVEST (Chen et al. 2025), CO2FIX (Masera et al. 2003), COMET-Planner (Swan et al. 2015), and specialized blue carbon assessment frameworks.

These tools are widely used to estimate carbon sequestration in aboveground biomass, belowground root systems, and long-term soil organic carbon pools, which together represent the major carbon reservoirs in mangrove forests, salt marshes, and seagrass ecosystems (Chen et al. 2025). Mangrove ecosystems, for instance, store carbon at rates up to 4–5 times higher than upland tropical forests due to high deposition of organic-rich sediments and slow decomposition under anoxic conditions (Chatting et al. 2022; Hilmi et al. 2023). Carbon models help identify areas with high mitigation potential, evaluate carbon losses from land conversion, and quantify greenhouse gas (GHG) reduction benefits under restoration scenarios.

On the ecological side, dynamic vegetation models, such as FOREST-GEO (Condit 1998; Anderson-Teixeira et al. 2015), CENTURY (Parton 1996), ED (Moorcroft et al. 2001), and MANGRO-GRO (Berger et al. 2008), simulate growth, mortality, recruitment, and competitive interactions under varying environmental conditions. These tools are valuable for predicting how bioshields will respond to stressors such as salinity intrusion, temperature extremes, nutrient limitation, and prolonged inundation.

For example, CENTURY-based simulations reveal that increasing salinity and reduced sediment supply may slow mangrove growth in many deltaic regions, potentially diminishing wave attenuation capacity over time (Chen and Twilley 1999). FOREST-GEO models can project shifts in species composition (Anderson-Teixeira et al. 2015), such as transitions from *Rhizophora*-dominated stands to *Avicennia*-dominated assemblages, under scenarios of altered hydrology or increased storm frequency.

Increasingly, ecological models are coupled with socio-economic and land-use change models to evaluate trade-offs between coastal protection, fisheries productivity, livelihood dependence, and agricultural expansion. This integrated approach helps planners determine where bioshields offer the highest multi-benefit outcomes, and where competing land uses may constrain restoration potential. Multi-criteria decision frameworks combining ecological, economic, and engineering indicators are becoming standard tools for coastal zone planning.

15.4.3 AI-Driven Predictive Modelling

Artificial intelligence (AI) has emerged as a transformative force in coastal bioshield research (Fig. 15.3), offering unprecedented capability to analyze high-volume datasets and predict coastal environmental changes. Machine learning (ML) and deep learning (DL) algorithms can utilize variables derived from remote sensing, sensor networks, climate data, and field observations to model vegetation dynamics, degradation risk, and erosion susceptibility.

Supervised ML algorithms, such as random forests (RF), support vector machines (SVMs), and gradient boosting models (GBM), are widely used to classify coastal land cover, estimate blue carbon biomass, and identify coastal degradation hotspots (Wang et al. 2019; Pham et al. 2020; He et al. 2023). RF models, for example, have been applied in Thailand and Indonesia to map species distribution, distinguish between healthy and stressed mangrove stands, and predict zones of future

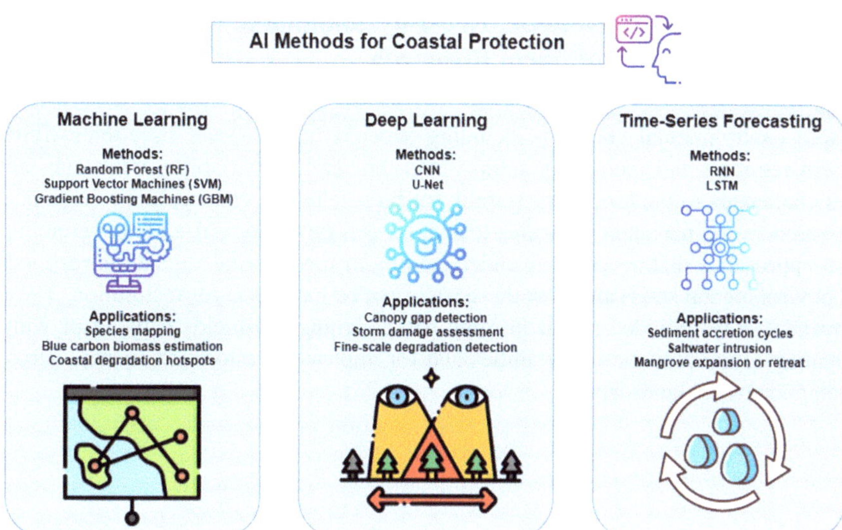

Fig. 15.3 Applications of AI in coastal protection

dieback (Purwanto et al. 2023; Pinkeaw et al. 2024). These models often outperform traditional regression-based approaches due to their ability to handle nonlinear relationships and high-dimensional datasets (Maxwell et al. 2018).

Deep learning approaches, especially convolutional neural networks (CNNs) and U-Net–based semantic segmentation models, allow automated detection of canopy gaps, storm damage, sediment scouring, and early degradation patterns that may be invisible to human interpreters (Weinstein et al. 2020; Tsiakos and Chalkias 2023). CNNs are particularly effective at processing multispectral and hyperspectral imagery from UAVs and satellites, enabling fine-scale species discrimination and canopy condition assessment.

Emerging AI applications also integrate time-series forecasting algorithms, such as recurrent neural networks (RNNs) and long short-term memory (LSTM) models. These models can be used to predict several parameters, such as future mangrove expansion or retreat, sediment accretion and erosion cycles, salinity intrusion patterns, climate-driven changes in productivity and mortality, and storm or flood vulnerability at landscape scales (e.g., Ramadhan et al. 2021; Farzanmanesh et al. 2025). These predictive capabilities enhance adaptive management by allowing local authorities to identify high-risk areas, prioritize restoration sites, and test management interventions through virtual simulations before implementation. AI-driven modelling is increasingly being incorporated into early-warning systems for coastal degradation, combining real-time sensor data with predictive algorithms to detect anomalies in hydrology, vegetation growth, or water quality (e.g., Chen et al. 2022; Xu et al. 2023).

15.5 Advances in Genetic, Eco-Physiological, and Biotechnological Research

Rapid environmental changes, including accelerating sea-level rise, intensifying storm regimes, and increasing salinity and temperature variability, pose substantial challenges to the long-term stability of bioshields. To enhance their resilience, researchers are increasingly turning to genetic, eco-physiological, and biotechnological approaches that deepen our understanding of how coastal vegetation responds to environmental stress and how these traits can be harnessed for restoration. These advances may provide critical insights for designing bioshields capable of withstanding future climatic uncertainties and for improving restoration success across dynamic coastal landscapes.

15.5.1 Genetic Diversity and Adaptive Traits

Genetic diversity underpins ecological resilience, shaping species' ability to adapt to environmental stressors and recover from disturbances (Reusch et al. 2005). In bioshields, which are often composed of mangroves, salt marsh plants, and coastal dune vegetation, genetic variability determines tolerance to salinity, inundation frequency, sedimentation, and temperature extremes. Modern molecular tools, including microsatellites, amplified fragment length polymorphisms (AFLPs), and single-nucleotide polymorphisms (SNPs), have been widely applied to assess genetic structure and adaptive variation in dominant mangrove species such as *Rhizophora apiculata*, *Bruguiera gymnorrhiza*, *Avicennia marina*, and *Sonneratia alba* (e.g., Yan et al. 2016; Amade et al. 2021).

Studies conducted in Vietnam, Malaysia, Indonesia, and northern Australia indicate that mangrove populations exhibit considerable genetic differentiation between estuarine, deltaic, and open-coast environments (e.g., Wee et al. 2020; Mantiquilla et al. 2021). In the Mekong Delta, for example, downstream populations of mangroves have shown specific genetic signatures (e.g., associated with tolerance to prolonged inundation and soft mud substrates), while upstream populations of similar species exhibit alleles linked to salinity resilience (Triest et al. 2025). These patterns reflect not only geographic isolation but also strong local adaptation driven by micro-environmental gradients.

Understanding such genetic structure is vital for the restoration of coastal vegetation bioshields. Historically, restoration programs often relied on propagules sourced from limited or non-local populations, occasionally leading to maladaptation, poor survival, or reduced ecosystem function (Veettil et al. 2019). Genetic insights may allow practitioners to select propagules with traits suited to specific hydrological and edaphic conditions. Furthermore, maintaining genetic diversity across restored stands increases long-term adaptability, enhancing resistance to pests, diseases, and climate variability. As bioshield planning increasingly incorporates climate projections, genetic data will play an essential role in identifying populations most likely to persist under future conditions.

15.5.2 Ecophysiological Research on Stress Tolerance

Ecophysiology bridges plant physiology and environmental science by examining how plants regulate water use, photosynthesis, nutrient uptake, and stress tolerance under real-world environmental conditions. In coastal ecosystems, where vegetation (e.g., mangroves, salt marshes) is exposed to salinity fluctuations, tidal inundation, hypoxia, high winds, sediment instability, and temperature extremes, ecophysiological approaches provide critical insights into how species acclimate or fail to acclimate to these dynamic stressors (Ball and Sobrado 2002). Characterizing their physiological responses helps identify species and genotypes best suited for bioshield design in a specific coastal area.

Recent studies across Thailand, India, Vietnam, and the Philippines have assessed traits such as gas exchange parameters, stomatal conductance, water-use efficiency, photosynthetic capacity, and chlorophyll fluorescence (e.g., Nandy et al. 2009; Paliyavuth et al. 2009; Nguyen et al. 2017; Barnuevo and Asaeda 2018). Measurements of $\delta^{13}C$ and $\delta^{15}N$ isotopic signatures have been used to infer long-term water-use strategies and nutrient acquisition patterns under saline stress (Reef et al. 2010).

Other species exhibit morphological or anatomical adaptations that enhance their suitability for bioshields. *Rhizophora* species, with their extensive prop roots, stabilize sediments and attenuate wave energy, while *Avicennia* species rely on salt glands to excrete excess ions (Parida and Jha 2010). Integrating ecophysiological insights into restoration planning may help match species to local stress regimes, improving establishment success and minimizing long-term maintenance requirements. These findings also provide essential data for predicting how coastal plant communities will respond to stressors such as warming temperatures, episodic heat waves, and shifts in tidal dynamics.

15.5.3 Biotechnology and Assisted Restoration

Biotechnological innovations are accelerating the development and implementation of restoration strategies by enabling efficient propagation, enhancing stress tolerance, and improving seedling performance in degraded environments (Al-Bahrany and Al-Khayri 2003; Stella et al. 2025). Tissue culture and micropropagation techniques have been particularly transformative, allowing for the large-scale production of genetically uniform, high-quality seedlings of mangrove species that are otherwise difficult to cultivate (Al-Bahrany and Al-Khayri 2003). These techniques are especially valuable in regions where natural propagule supply is limited due to habitat loss, overharvesting, or poor fruiting seasons. They also allow the propagation of rare or threatened species, contributing to biodiversity conservation while supporting bioshield establishment.

In parallel, bioengineering approaches are gaining traction. Researchers are exploring the use of beneficial microbial consortia, including nitrogen-fixing bacteria, mycorrhizal fungi, and halotolerant plant growth-promoting rhizobacteria (PGPR), to enhance seedling vigor, nutrient uptake, and stress tolerance (Sagar et al. 2022; Mishra et al. 2023). Studies have shown that inoculating seedlings with PGPR improves survival in saline and nutrient-poor substrates by enhancing root growth and improving osmotic balance (e.g., Kumar Arora et al. 2020).

Biodegradable planting structures, such as coconut-coir and jute logs and mats, are increasingly used in coastal and riparian restoration (Kalibová et al. 2016; Villanueva et al. 2022). These materials stabilize sediments and reduce erosion while providing a moist, soil-holding substrate that facilitates root establishment and seedling survival in erosion-prone areas (e.g., Villanueva et al. 2022). These structures gradually decompose, reducing environmental impact while enhancing early growth conditions.

Collectively, advances in biotechnology and assisted restoration hold immense potential for scaling up bioshield establishment, particularly in regions facing acute land loss and limited natural regeneration.

15.6 Technological Innovations for Monitoring and Evaluation

Effective monitoring and evaluation are essential for ensuring that bioshields function as intended, remain resilient under changing environmental conditions, and deliver long-term protective and ecological benefits. Traditional monitoring, which is often based on occasional field surveys, has been constrained by limited spatial coverage, high labor and logistics costs, and vulnerability to extreme weather conditions. In recent years, however, rapid developments in digital technology, remote sensing techniques, and participatory platforms have transformed how researchers and practitioners observe coastal vegetation dynamics, hydrodynamic processes, and early indicators of stress (Veettil et al. 2019, 2020b; 2021). These innovations not only improve the precision and frequency of data collection but also support adaptive coastal management, enabling bioshields to remain robust in the face of climate-driven disturbances.

15.6.1 Internet of Things (IoT) and Sensor Networks

IoT-based environmental monitoring systems have emerged as a cornerstone of contemporary bioshield assessment (Boonrat et al. 2024; Subbiah et al. 2024). These systems consist of distributed wireless sensors that collect real-time measurements of soil salinity, porewater conductivity, tidal inundation frequency, sediment deposition, and microclimatic variables (Chavanne et al. 2024; Garcia et al. 2025). Such fine-scale environmental data are crucial for understanding ecosystem health, growth rates, stress thresholds, and hydrological performance of coastal vegetation.

IoT systems are particularly valuable for detecting early warning indicators of stress, such as sudden saltwater incursions during storm surges or prolonged drought-induced salinization (Ariza-Colpas et al. 2021; Alruwais et al. 2025). Integrating these sensors with cloud-based dashboards and automated data analytics may facilitate continuous, remote access for researchers, local authorities, and coastal managers. In coastal Vietnam, pilot deployments in the Mekong Delta's mangrove rehabilitation zones have demonstrated that IoT sensors can identify salinity spikes up to 72 h before visible leaf damage occurs, enabling timely adaptive interventions (Phan et al. 2023).

Low-power wide-area networks (LPWAN), such as Long Range Wide Area Network (LoRaWAN), extend sensor communication over several kilometers with minimal energy consumption, making them ideal for remote coastal landscapes where conventional telecommunications infrastructure is limited (Parri et al. 2019). Coupled with solar-powered charging systems, these networks may provide a cost-effective, scalable, and climate-resilient monitoring approach (Malek et al. 2023).

15.6.2 Data Analytics and Cloud Computing

The rise of cloud-based geospatial platforms has revolutionized environmental monitoring by enabling the processing of large-scale datasets that were previously inaccessible or computationally prohibitive (Gorelick et al. 2017). Platforms such as Google Earth Engine (GEE) and Microsoft Planetary Computer allow researchers to access petabytes of satellite imagery, climate reanalysis data, digital elevation models, and time-series products without needing local high-performance computing facilities (Gorelick et al. 2017; Augspurger 2021).

These tools play a central role in bioshield research and management. Analysts can rapidly compute long-term trends in shoreline change, vegetation indices (e.g., NDVI, EVI), land subsidence, and sediment plume movement. For example, GEE has been widely used to track mangrove loss and gain in Indonesia, Myanmar, and Vietnam, revealing spatial patterns driven by aquaculture expansion, erosion hotspots, and successful restoration (e.g., Richards and Friess 2016).

Cloud platforms also support near-real-time change detection using Sentinel-1 and Sentinel-2 imagery, enabling managers to detect storm-induced damage or illegal encroachment within days rather than months (Huang and Jin 2021; Pacheco-Pascagaza et al. 2022). When cloud computing is coupled with AI-based classifiers, such as random forest (RF), support vector machines (SVM), or convolutional neural networks (CNN), monitoring workflows become faster, more accurate, and increasingly automated (Gorelick et al. 2017; Lomeo and Singh 2022). These improvements democratize access to advanced analytics, allowing government agencies, NGOs, and small research institutions to engage in rigorous, data-driven coastal management without substantial computational infrastructure. Furthermore, cloud-based platforms facilitate integration across multiple data sources, such as field sensors, drones, satellites, and hydrodynamic models, supporting multi-scale analyses that are vital for evaluating bioshield performance under complex and changing coastal dynamics.

15.6.3 Citizen Science and Participatory Technology

Alongside advances in remote sensing and computing, technological innovation has also enabled broader community participation in bioshield monitoring. Coastal communities are often the primary beneficiaries of bioshield protection and possess intimate knowledge of local environmental conditions (Danielsen et al. 2005). Participatory technologies help mobilize this knowledge while enhancing the inclusiveness and social legitimacy of monitoring processes.

Mobile applications, low-cost sensors, and participatory GIS (PGIS) platforms allow community members to collect information on tree survival, pest outbreaks, shoreline erosion, illegal cutting, or pollution events (Danielsen et al. 2005; Dema et al. 2019; Bullen and Miles 2024). Such initiatives can significantly improve data coverage in remote areas where formal monitoring agencies have limited access (Damastuti and de Groot 2019). Mobile-based reporting tools, which are often linked with cloud databases, enable crowdsourced observations that complement scientific datasets (Gorelick et al. 2017). For example, community observations of seedling mortality or erosion hotspots can be cross-referenced with drone imagery or IoT data, providing a multi-layered understanding of bioshield performance. The integration of local knowledge into digital platforms also supports adaptive governance, ensuring that management responses incorporate both scientific evidence and community priorities.

Participatory technology ultimately strengthens social resilience by building local capacity, fostering stewardship, and reducing dependency on external monitoring teams (Danielsen et al. 2005; Dema et al. 2019). It bridges gaps between science, policy, and society, ensuring that technological innovation supports equitable and context-sensitive coastal governance.

15.7 Emerging Frontiers in Research and Innovation

As climate change accelerates the frequency and intensity of coastal hazards, the need for innovative, multi-scalar approaches to bioshield management has never been more urgent. Emerging scientific and technological frontiers are pushing beyond traditional ecological monitoring, offering transformative tools for understanding, modeling, and financing nature-based coastal protection. Three revolutionary areas—AI-enabled digital ecosystems, blue carbon accounting, and interdisciplinary transboundary collaborations—are believed to shape the next decade of bioshield research and implementation across Asia and beyond.

15.7.1 Integration of AI, Big Data, and Digital Twins

The convergence of artificial intelligence (AI), big data analytics, and digital twin technology is redefining how coastal systems are studied and managed. Digital twins, which are dynamic virtual models that mirror real-world ecosystems, integrate real-time sensor data, remote sensing inputs, hydrodynamic models, and ecological growth functions to simulate how bioshields react to environmental changes and human interventions (Chen et al. 2023; Skákala et al. 2023; Masria and Abouelsaad 2025). Unlike conventional static models, digital twins continually update their parameters, enabling near-instantaneous forecasting of shoreline erosion, sediment transport, carbon fluxes, or vegetation stress under different climatic or anthropogenic scenarios (Haupt et al. 2025; Reinert et al. 2025). This capability significantly enhances predictive risk assessment. Managers can now test virtual scenarios such as mangrove belt widening, sediment nourishment, setback lines, or land-use zoning changes before implementing them (e.g., Pillai et al. 2022; Scala et al. 2025). AI-driven analytics improve model precision by detecting nonlinear relationships, such as the observation that salinity peaks after successive tidal cycles (e.g., Saccotelli et al. 2024; Duc et al. 2025). The integration of machine-learning algorithms further automates anomaly detection, a capability increasingly central to early-warning systems for identifying vegetation dieback (Paul et al. 2025), shoreline-erosion hotspots (Vos et al. 2019), or the early emergence of invasive species in coastal landscapes (Zaka and Samat 2023).

Big data plays a foundational role in this transformation. The proliferation of IoT sensors, drone-generated imagery, citizen science inputs, and satellite time-series datasets produces unprecedented data volumes. Cloud-based platforms allow processing and visualization at regional or national scales, enabling researchers to monitor bioshield dynamics with high frequency and spatial resolution.

15.7.2 Blue Carbon Accounting and Climate Finance

Growing recognition of the climate mitigation potential of mangroves, seagrass meadows, and coastal wetlands has placed blue carbon at the center of global climate policy (Perera et al. 2022). Advances in Earth observation, geospatial modeling, and AI-assisted carbon estimation are enabling more accurate and scalable quantification of bioshields' carbon stocks and sequestration rates (Byrd et al. 2018). High-resolution satellite imagery combined with LiDAR and hyperspectral datasets allows estimation of aboveground biomass, while machine learning models trained on soil core data are improving predictions of belowground carbon pools (Fatoyinbo and Simard 2013; Maxwell et al. 2023).

Improved carbon accounting is particularly significant for the Asian region, which is home to the world's largest mangrove extent, because it facilitates participation in climate finance mechanisms such as carbon markets, REDD+, and blue carbon crediting schemes (Maxwell et al. 2023; Li and Liu 2025). Incorporating bioshields into national greenhouse gas inventories and carbon pricing policies can unlock new revenue streams to support large-scale restoration (Li and Liu 2025).

Financial innovations are also emerging. Blended finance models, blue bonds, and climate resilience funds increasingly recognize the protective and mitigation value of bioshields (Dasgupta 2021; Châles et al. 2023). Projects in Indonesia and the Philippines have begun integrating carbon payments with community co-benefits, enabling local stewardship while scaling restoration efforts (Gevaña et al. 2018). Moreover, the integration of carbon modeling with socio-economic valuation frameworks allows policymakers to compare the benefits of bioshields with hard engineering alternatives, often revealing higher long-term economic returns due to carbon sequestration, biodiversity conservation, and reduced disaster recovery costs (Mcleod et al. 2011; Narayan et al. 2017).

Future research will focus on consolidating methodological standards for soil carbon sampling, improving uncertainty quantification, and integrating carbon models into digital twin systems to simulate future carbon budgets under shifting climatic regimes.

15.7.3 Interdisciplinary Collaboration and Transboundary Research

Coastal ecosystems, sediment pathways, and climatic drivers do not respect political boundaries. Thus, advancing bioshield science requires robust transboundary and interdisciplinary collaboration. Initiatives such as the Global Mangrove Watch offer platforms for data harmonization, remote sensing standardization, and capacity building across countries (Bunting et al. 2018). These networks provide open-access datasets, training modules, and monitoring tools that facilitate coordinated restoration planning at landscape and seascape scales.

Regional partnerships across ASEAN nations are increasingly common. Universities, research institutes, and governments collaborate on biodiversity assessments, hydrodynamic modeling, and community-based adaptation studies (Friess et al. 2016; Gerona-Daga and Salmo 2022). These knowledge exchanges ensure that innovations in one country, whether drone-based seed dispersal, sediment-trapping structures, or AI-enabled restoration planning, can be tested and adapted elsewhere. Such collaborations also help develop standardized methodologies for monitoring bioshields, essential for climate reporting, financing mechanisms, and cross-country comparisons.

Interdisciplinary research is expanding beyond natural sciences to include social sciences, economics, governance studies, and participatory planning (Friess et al. 2016). Understanding local perceptions, land tenure complexities, inequities in resource distribution, and gendered impacts of coastal hazards is vital for designing inclusive bioshield strategies (Prakash et al. 2022). Future directions point toward a new generation of integrated coastal resilience science, where digital technologies, carbon finance, community knowledge, and cross-border governance converge to enhance adaptive capacity in rapidly changing coastal landscapes (Narayan et al. 2017; Bunting et al. 2018).

15.8 Future Directions

Technological progress has opened new horizons for bioshield research, monitoring, and management, offering unprecedented capabilities for understanding coastal ecosystem dynamics. However, the rapid diversification of methods and tools also brings challenges related to data harmonization, governance, inclusivity, and long-term accessibility. To maximize the contribution of bioshields to sustainable coastal protection, future research and practice must strategically address these emerging complexities.

A key priority is the standardization of methodologies for remote sensing, ecological modelling, and performance evaluation across regions. At present, bioshield assessments often rely on disparate indicators, processing workflows, and classification schemes, which complicate cross-country comparisons and undermine the scalability of regional conservation strategies (Kuenzer et al. 2011; Hamilton and Casey 2016; Bunting et al. 2018). Standardized protocols for vegetation mapping, biomass estimation, hydrodynamic modelling, sediment dynamics assessment, and hazard mitigation metrics would allow more consistent evaluation of bioshield functionality. Establishing such standards also supports regional collaboration, enabling countries to pool expertise, validate methods across ecological gradients, and jointly monitor transboundary coastal systems.

A second major direction involves the integration of multi-source datasets, including satellite imagery, UAV photogrammetry, IoT-based environmental sensors, and field-based ecological surveys, into unified analytical platforms. Coastal bioshields operate across multiple spatial and temporal scales, and no single data source can capture their full complexity. Developing integrative platforms and interoperable data architectures will allow researchers to merge long-term historical datasets with high-frequency sensor data, improving the resolution of forecasts, trend analyses, and ecological risk assessments (Wang et al. 2022; Dong et al. 2025). Data fusion techniques, supported by AI and machine learning, will be essential for producing coherent, decision-ready insights from increasingly large and heterogeneous datasets.

Ensuring open-access databases and transparent data governance will be fundamental for advancing scientific collaboration and policy development. Open data platforms enable researchers, government agencies, and community groups to share information, replicate analyses, and track progress in restoration and adaptation initiatives. Transparent data governance also enhances accountability in climate reporting, supports evidence-based policymaking, and facilitates the uptake of new methods in countries with limited technical resources. As global initiatives such as

Global Mangrove Watch have demonstrated (Bunting et al. 2018), publicly accessible datasets catalyze rapid innovation and broaden participation.

Equally important is the ethical and equitable use of technology. While digital tools and advanced analytics can significantly improve coastal management, they risk reinforcing power imbalances if not deployed responsibly. Future approaches must ensure that local communities, especially those in vulnerable coastal regions, benefit directly from data-driven decision-making. This includes providing access to user-friendly monitoring tools, ensuring participatory interpretation of results, and integrating community priorities into adaptation planning. Respect for data privacy, cultural sensitivities, and traditional ecological knowledge should guide the design of monitoring systems and digital platforms.

In Asia, where dense coastal populations, high exposure to climate-related hazards, and rapid socio-economic transformation intersect, the stakes are especially high (Neumann et al. 2015). The region's diverse coastal landscapes, varying from deltaic mangroves to coral-fringed islands, require context-specific tools that align technological innovation with local realities. Future research must therefore embrace inclusive, co-produced frameworks that weave together indigenous knowledge systems, community-based monitoring, and advanced scientific methods (Tengö et al. 2014; Zamboni et al. 2022). Such hybrid approaches not only strengthen ecological outcomes but also enhance social legitimacy, ensuring that bioshields serve as both ecological infrastructure and community assets (Wyborn 2015).

Ultimately, the future of bioshields lies in integrating cutting-edge technology with collaborative governance and grassroots engagement (Masria 2024). By advancing methodological consistency, strengthening data integration, promoting transparency, and centering equity, researchers and policymakers can harness innovation to build resilient, adaptive, and inclusive coastal protection systems for the decades ahead (Wyborn 2015; Seddon et al. 2021; Wu et al. 2024).

15.9 Conclusions

Technological innovation is fundamentally transforming contemporary approaches to coastal protection, reshaping the ways in which bioshields are conceptualized, designed, monitored, and assessed. Recent advancements in high-resolution remote sensing, ecological and hydrodynamic modelling, biotechnology, and AI-driven analytics have markedly improved the accuracy, efficiency, and scalability of ecosystem-based protection strategies. Collectively, these tools not only advance scientific understanding of coastal processes and ecosystem functioning but also facilitate more adaptive, anticipatory, and evidence-based decision-making in the context of accelerating climate change.

Asia, which is characterized by some of the world's most dynamic, densely populated, and hazard-exposed coastlines, stands at the forefront of this technological transition. Countries such as Vietnam, Indonesia, the Philippines, India, and Bangladesh are leading efforts to integrate digital environmental monitoring networks, drone-based coastal mapping, species-selection research, and hybrid grey–green infrastructure into large-scale restoration and bioshield engineering programs. These initiatives illustrate how technological innovation, when aligned with local ecological conditions and socio-economic contexts, can substantially enhance the resilience, effectiveness, and multifunctionality of coastal ecosystems.

Nevertheless, technological progress on its own is insufficient to ensure the long-term sustainability of bioshield interventions. The effectiveness of such innovations is fundamentally contingent upon coherent governance arrangements, transparent data-sharing mechanisms, and equitable access to information. Equally vital is the meaningful participation of local communities, whose situated knowledge, stewardship practices, and sustained engagement are central to the success of restoration and protection efforts. Addressing ethical considerations related to data governance, benefit-sharing, and the accessibility of emerging technologies is essential to prevent innovation from reinforcing existing socio-economic inequalities.

Looking ahead, the future of bioshields will depend on the effective harmonization of technological precision with socio-ecological inclusiveness. By coupling advanced scientific methodologies with community-driven stewardship and locally grounded knowledge systems, bioshields can evolve from isolated pilot initiatives into integral components of sustainable coastal management. This holistic vision is anchored in innovation, resilience, and the principles of coexistence with nature, which offers a compelling pathway for safeguarding coastal ecosystems and the livelihoods that rely upon them.

References

Abdelmajeed AY, Rastogi A, Juszczak R (2023) Cloud-based remote sensing for wetland monitoring—a review. Remote Sens 15(6):1660. https://doi.org/10.3390/rs15061660

Al-Bahrany AM (2003) Al-Khayri JM (2003) Micropropagation of grey mangrove *Avicennia marina*. Plant Cell,Tissue Organ Cult 72:87–93. https://doi.org/10.1023/A:1021205731719

Alongi DM (2012) Carbon sequestration in mangrove forests. Carbon Manag 3:313–322

Alruwais N, Hasanin T, Burle T, Marzouk R, Vijayalakshmi J, Srinivasan S (2025) Integrating remote sensing, IoT, and machine learning for tracking saltwater intrusion in coastal aquifers due to over-extraction: a geospatial approach. Desalin Water Treat 323:101403. https://doi.org/10.1016/j.dwt.2025.101403

Amade FM, Oosthuizen CJ, Chirwa PW (2021) Genetic diversity and contemporary population genetic structure of Avicennia marina from Mozambique. Aquat Bot 171:103374. https://doi.org/10.1016/j.aquabot.2021.103374

Anderson-Teixeira KJ, Davies SJ, Bennett AC, Gonzalez-Akre EB, Muller-Landau HC, Wright SJ, Salim KA, Almeyda Zambrano AM, Alonso A, Baltzer JL, Basset Y, Bourg NA, Broadbent EN, Brockelman WY, Bunyavejchewin SP. Burslem FR, Butt N, Cao M, Cardenas D, Zimmerman J (2015). CTFS-ForestGEO: a worldwide network monitoring forests in an era of global change. Global Change Biol 21(2): 528–549. https://doi.org/10.1111/gcb.12712

Ariza-Colpas PP, Ayala-Mantilla CE, Shaheen Q, Piñeres-Melo MA, Villate-Daza DA, Morales-Ortega RC, De-la-Hoz-Franco E, Sanchez-Moreno H, Aziz BS, Afzal M (2021) SISME, Estuarine monitoring system based on IOT and machine learning for the detection of salt wedge in aquifers: case study of the Magdalena River Estuary. Sensors 21(7):2374. https://doi.org/10.3390/s21072374

Athikalam PT, Vaideeswaran KA (2022) Vegetation bioshield for coastal protection in South Asia: Status and way forward. J Coast Conserv 26(1):3. https://doi.org/10.1007/s11852-022-00850-x

Augspurger T (2021) Scalable sustainability with the planetary computer. In: AGU fall meeting abstracts, vol 2021, pp U51B-14

Ball MC, Sobrado MA (2002) Ecophysiology of mangroves: challenges in linking physiological processes with patterns. In: Physiological plant ecology: 39th symposium of the British ecological society, p 331. Cambridge University Press

Bao D, George Xue Z, Hiatt M, Xu K, Harris CK, Trepanier JC (2025) A machine learning-based prediction-to-map framework for rapid and accurate spatial flood prediction. NPJ Nat Hazards 2(1):71. https://doi.org/10.1038/s44304-025-00122-2

Barnuevo A, Asaeda T (2018) Integrating the ecophysiology and biochemical stress indicators into the paradigm of mangrove ecology and a rehabilitation blueprint. PLoS ONE 13(8):e0202227. https://doi.org/10.1371/journal.pone.0202227

Basyuni M, Mubaraq A, Amelia R, Wirasatriya A, Iryanthony SB, Slamet B, Al Mustaniroh SS, Pradisty NA, Sidik F, Hanintyo R, Sumarga E, Larekeng SH, Salmo SG, Kajita T, Ali HM, Sakti AD, Arifanti VB (2025) Mangrove aboveground biomass estimation using UAV imagery and a constructed height model in Budeng-Perancak, Bali. Indonesia. Ecol Informat 86:103037. https://doi.org/10.1016/j.ecoinf.2025.103037

Berger U, Rivera-Monroy VH, Doyle TW, Dahdouh-Guebas F, Duke NC, Fontalvo-Herazo ML, Hildenbrandt H, Koedam N, Mehlig U, Piou C, Twilley RR (2008) Advances and limitations of individual-based models to analyze and predict dynamics of mangrove forests: a review. Aquat Bot 89(2):260–274. https://doi.org/10.1016/j.aquabot.2007.12.015

Boonrat P, Boonrat P, Aharari A (2024) Precision rehabilitation: IoT-Based monitoring in mangrove ecosystems. In: 2024 IEEE 4th international conference on electronic communications, Internet of Things and Big Data (ICEIB), pp 145–149. IEEE. https://doi.org/10.1109/ICEIB61477.2024.10602576

Bullen J, Miles A (2024) Exploring local perspectives on flood risk: a participatory GIS approach for bridging the gap between modelled and perceived flood risk zones. Appl Geogr 163:103176. https://doi.org/10.1016/j.apgeog.2023.103176

Bunting P, Rosenqvist A, Lucas RM, Rebelo L, Hilarides L, Thomas N, Hardy A, Itoh T, Shimada M, Finlayson CM (2018) The global Mangrove watch—a new 2010 global baseline of Mangrove extent. Remote Sens 10(10):1669. https://doi.org/10.3390/rs10101669

Byrd KB, Ballanti L, Thomas N, Nguyen D, Holmquist JR, Simard M, Windham-Myers L (2018) A remote sensing-based model of tidal marsh aboveground carbon stocks for the conterminous United States. ISPRS J Photogramm Remote Sens 139:255–271. https://doi.org/10.1016/j.isprsjprs.2018.03.019

Canisius F, Brisco B, Murnaghan K, Keizer E (2019) SAR backscatter and InSAR coherence for monitoring wetland extent, flood pulse and vegetation: a study of the amazon lowland. Remote Sens 11(6):720. https://doi.org/10.3390/rs11060720

Cavalli RM (2023) Remote data for mapping and monitoring coastal phenomena and parameters: a systematic review. Remote Sens 16(3):446. https://doi.org/10.3390/rs16030446

Châles F, Bellanger M, Bailly D, Dutra LX, Pendleton L (2023) Using standards for coastal nature-based solutions in climate commitments: applying the IUCN Global standard to the case of pacific small island developing states. Nature Based Solutions 3:100034. https://doi.org/10.1016/j.nbsj.2022.100034

Chatting M, Walton M, Skov MW, Kennedy H, Husrevoglu YS, Le Vay L (2022) Future mangrove carbon storage under climate change and deforestation. Front Mar Sci 9:781876. https://doi.org/10.3389/fmars.2022.781876

Chavanne X, Frangi J (2024) A sensor to monitor soil moisture, salinity, and temperature profiles for wireless networks. J Sens Actuator Netw 13(3):32. https://doi.org/10.3390/jsan13030032

Chen G, Yang J, Huang B, Ma C, Tian F, Ge L, Li J (2023) Toward digital twin of the ocean: from digitalization to cloning. Intell Marine Technol Syst 1(1):3. https://doi.org/10.1007/s44295-023-00003-2

Chen J, Lu Y, Liu F, Gao G, Xie M (2025) Remote sensing evidence of blue carbon stock increase and attribution of its drivers in coastal China. Remote Sens 17(15):2559. https://doi.org/10.3390/rs17152559

Chen K, Kuang C, Wang L, Chen K, Han X, Fan J (2022) Storm Surge Prediction Based on Long Short-Term Memory Neural Network in the East China Sea. Appl Sci 12(1):181. https://doi.org/10.3390/app12010181

Chen R, Twilley RR (1999) A simulation model of organic matter and nutrient accumulation in mangrove wetland soils. Biogeochemistry 44(1):93–118. https://doi.org/10.1007/BF00993000

Ciliberti S, Coro G (2025) Distributed environments for ocean forecasting: the role of cloud computing. State Planet Discussions, 1–10. https://doi.org/10.5194/sp-5-opsr-24-2025

Condit R (1998) Tropical forest census plots: methods and results from Barro Colorado Island, Panama and a comparison with other plots. Springer Science & Business Media, pp 211

Conti LA, Barcellos RL, Oliveira P, Nascimento Neto FC (2025) Geographic object-oriented analysis of UAV Multispectral images for tree distribution mapping in mangroves. Remote Sens 17(9):1500. https://doi.org/10.3390/rs17091500

Damastuti E (2019) De Groot R (2019) Participatory ecosystem service mapping to enhance community-based mangrove rehabilitation and management in Demak. Indonesia. Reg Environ Change 19:65–78. https://doi.org/10.1007/s10113-018-1378-7

Dang ATN, Reid M, Kumar L (2022) Assessing potential impacts of sea level rise on mangrove ecosystems in the Mekong Delta. Vietnam. Reg Environ Change 22:70. https://doi.org/10.1007/s10113-022-01925-z

Danielsen F, Burgess ND, Balmford A (2005) Monitoring Matters: Examining the potential of locally-based approaches. Biodivers Conserv 14:2507–2542. https://doi.org/10.1007/s10531-005-8375-0

Dasgupta P (2021) The economics of biodiversity: The Dasgupta Review. HM Treasury, London

Dean RG, Dalrymple RA (2004) Coastal processes with engineering applications. Cambridge University Press, pp. 487

Dema T, Brereton M, Roe P (2019). Designing participatory sensing with remote communities to conserve endangered species. In: Proceedings of the 2019 CHI conference on human factors in computing systems, pp 1–16. https://doi.org/10.1145/3290605.3300894

Dissanayake P, Brown J, Yates M (2022) Morphodynamic Evolution and Sustainable Development of Coastal Systems. Journal of Marine Science and Engineering 10(5):647. https://doi.org/10.3390/jmse10050647

Dong D, Huang H, Gao Q, Li K, Zhang S, Yan R (2025) Integrating Remote Sensing and Field Data to Quantify Mangrove Biomass Carbon Hotspots and Protection Gaps in Guangdong. China. Forests 16(10):1612. https://doi.org/10.3390/f16101612

Duarte CM, Losada IJ, Hendriks IE, Mazarrasa I, Marbà N (2013) The role of coastal plant communities for climate change mitigation and adaptation. Nat Clim Chang 3(11):961–968. https://doi.org/10.1038/nclimate1970

Duc PN, Duc TT, Van GP, Van HN, Minh TT (2025) Predicting salinity levels in the Mekong delta (Viet Nam): analysis of machine learning and deep learning models. Discover Artificial Intelligence 5(1):79. https://doi.org/10.1007/s44163-025-00336-3

Duy DV, Ty TV, Phat LT, Minh HV, Nam ND, Downes NK, Avtar R, Tanaka H (2025) Simulating the Coastal Protection Performance of Breakwaters in the Mekong Delta: Insights from the Western Coast of Ca Mau Province. Vietnam. Journal of Marine Science and Engineering 13(8):1559. https://doi.org/10.3390/jmse13081559

Elmahdy SI, Ali TA (2022) Monitoring Changes and Soil Characterization in Mangrove Forests of the United Arab Emirates Using the Canonical Correlation Forest Model by Multitemporal of Landsat Data. Frontiers in Remote Sensing 3:782869. https://doi.org/10.3389/frsen.2022.782869

Farrer EC, Van Bael SA, Clay K, Smith MK (2022) Plant-microbial symbioses in coastal systems: their ecological importance and role in coastal restoration. Estuaries Coasts 45(7):1805–1822. https://doi.org/10.1007/s12237-022-01052-2

Farzanmanesh R, Khoshelham K, Volkova L et al (2025) Using a long short-term memory neural network model to forecast mangrove change in two blue forests conservation projects. Discov. for. 1:10. https://doi.org/10.1007/s44415-025-00009-0

Fatnassi S, Yahia M, Ali T, Mortula M (2023) Polarimetric SAR characterization of mangrove forest environment in the United Arab Emirates (UAE). *International Journal of Advanced Computer Science and Applications* 14(3): https://doi.org/10.14569/IJACSA.2023.0140380

Fatoyinbo TE, Simard M (2013) Height and biomass of mangroves in Africa from ICESat/GLAS and SRTM. Int J Remote Sens 34(2):668–681. https://doi.org/10.1080/01431161.2012.712224

Friess DA, Thompson BS, Brown B, Amir AA, Cameron C, Koldewey HJ, Sasmito SD, Sidik F (2016) Policy challenges and approaches for the conservation of mangrove forests in Southeast Asia. Conserv Biol 30(5):933–949. https://doi.org/10.1111/cobi.12784

Fu B, Wei Y, Jiang L, Yao H, Li X, Yang Y, Jia M, Sun W (2025) Estimation of mangrove heights and aboveground biomass using UAV-LiDAR, Sentinel-1 and ZY-3 stereo images. Eco Inform 88:103160. https://doi.org/10.1016/j.ecoinf.2025.103160

Gao F, Masek J, Schwaller M, Hall F (2006) On the blending of the Landsat and MODIS surface reflectance: Predicting daily Landsat surface reflectance. IEEE Trans Geosci Remote Sens 44(8):2207–2218. https://doi.org/10.1109/TGRS.2006.872081

García NC, J J, Chávez L (2025) Sustainable GIoT-Based Mangrove Monitoring System for Smart Coastal Cities with Energy Harvesting from SMFCs. *Technologies* 13(12): 549. https://doi.org/10.3390/technologies13120549

Geng L, Dou Z, Gu Y, Cao H, Yao P, Zhou Z, Shao J (2025) The synergistic effect between engineering measures and ecological measures in coastal erosion prevention. Front Mar Sci 12:1671752. https://doi.org/10.3389/fmars.2025.1671752

Georgiou IY, FitzGerald DM, Hanegan KC (2024) Storm and tidal interactions control sediment exchange in mixed-energy coastal systems. *PNAS Nexus* 3(2). https://doi.org/10.1093/pnasnexus/pgae042

Gerona-Daga MEB, Salmo SG III (2022) A systematic review of mangrove restoration studies in Southeast Asia: Challenges and opportunities for the United Nation's Decade on Ecosystem Restoration. Front Mar Sci 9:987737. https://doi.org/10.3389/fmars.2022.987737

Gevaña DT, Camacho LD, Pulhin JM (2018) Conserving mangroves for their blue carbon: Insights and prospects for community-based mangrove management in Southeast Asia. In: *Threats to Mangrove Forests: Hazards, Vulnerability, and Management* (pp. 579–588). Cham: Springer International Publishing. https://doi.org/10.1007/978-3-319-73016-5_26

Gorelick N, Hancher M, Dixon M, Ilyushchenko S, Thau D, Moore R (2017) Google Earth Engine: Planetary-scale geospatial analysis for everyone. Remote Sens Environ 202:18–27. https://doi.org/10.1016/j.rse.2017.06.031

Gourgue O, van Belzen J, Schwarz C, Vandenbruwaene W, Vanlede J, Belliard J-P, Fagherazzi S, Bouma TJ, van de Koppel J, Temmerman S (2022) Biogeomorphic modeling to assess the resilience of tidal-marsh restoration to sea level rise and sediment supply. Earth Surf. Dynam. 10:531–553. https://doi.org/10.5194/esurf-10-531-2022

Hamdan O, Khali Aziz H, Mohd Hasmadi I (2014) L-band ALOS PALSAR for biomass estimation of Matang Mangroves, Malaysia. Remote Sens Environ 155:69–78. https://doi.org/10.1016/j.rse.2014.04.029

Hamilton SE, Casey D (2016) Creation of a high spatio-temporal resolution global database of continuous mangrove forest cover for the 21st century (CGMFC-21). Glob Ecol Biogeogr 25(6):729–738. https://doi.org/10.1111/geb.12449

Haupt S, Sibolla B, Molapo R, Mdakane L, Fourie N (2025) Exploring the Use of Data in a Digital Twin for the Marine and Coastal Environment. ISPRS Int J Geo Inf 14(4):140. https://doi.org/10.3390/ijgi14040140

Hauser LT, Nguyen VuG, Nguyen BA, Dade E, Nguyen HM, Nguyen TTQ, Le TQ, Vu LH, Tong ATH, Pham HV (2017) Uncovering the spatio-temporal dynamics of land cover change and fragmentation of mangroves in the Ca Mau peninsula, Vietnam using multi-temporal SPOT satellite imagery (2004–2013). Appl Geogr 86:197–207. https://doi.org/10.1016/j.apgeog.2017.06.019

He K, Li W, Zhang Y, Sun G, McNulty SG, Flanagan NE, Richardson CJ (2023) Identifying driving hydrogeomorphic factors of coastal wetland downgrading using random forest classification models. Sci Total Environ 894:164995. https://doi.org/10.1016/j.scitotenv.2023.164995

Hilmi N, Benitez Carranco MB, Broussard D, Mathew M, Djoundourian S, Cassotta S, Safa A, Maliki S, Allemand D, Berthomieu C, MJ (2023) Tropical blue carbon: Solutions and perspectives for valuations of carbon sequestration. *Frontiers in Climate* 5: 1169663. https://doi.org/10.3389/fclim.2023.1169663

Hladik C, Alber M (2012) Accuracy assessment and correction of a LIDAR-derived salt marsh digital elevation model. Remote Sens Environ 121:224–235. https://doi.org/10.1016/j.rse.2012.01.018

Huang M, Jin S (2021) Backscatter Characteristics Analysis for Flood Mapping Using Multi-Temporal Sentinel-1 Images. Remote Sensing 14(15):3838. https://doi.org/10.3390/rs14153838

James RK, Lynch A, Herman PM, Van Katwijk MM, Van Tussenbroek BI, Dijkstra HA, Bouma TJ (2021) Tropical biogeomorphic seagrass landscapes for coastal protection: Persistence and wave attenuation during major storms events. Ecosystems 24(2):301–318. https://doi.org/10.1007/s10021-020-00519-2

Jiang Y, Zhang L, Yan M, Qi J, Fu T, Fan S, Chen B (2021) High-Resolution Mangrove Forests Classification with Machine Learning Using Worldview and UAV Hyperspectral Data. Remote Sensing 13(8):1529. https://doi.org/10.3390/rs13081529

Jin Y, Zhu J, Yin Z, Zhu W, Lee D (2025) A phenology-informed framework for detecting deforestation in North Korea Using Fused Satellite Time-Series. Remote Sensing 17(16):2789. https://doi.org/10.3390/rs17162789

Jordan P, Fröhle P (2022) Bridging the gap between coastal engineering and nature conservation? A review of coastal ecosystems as nature-based solutions for coastal protection. J Coast Conserv 26(2):4. https://doi.org/10.1007/s11852-021-00848-x

Kalibová J, Jačka L, Petrů J (2016) The effectiveness of jute and coir blankets for erosion control in different field and laboratory conditions. Solid Earth 7:469–479. https://doi.org/10.5194/se-7-469-2016

Karim MF, Zhang X, Li R (2019) Dynamics of shrimp farming in the southwestern coastal districts of Bangladesh using a Shrimp Yield Dataset (SYD) and landsat satellite archives. Sustainability 11(17):4635. https://doi.org/10.3390/su11174635

Kaur H, Tyagi S, Mehta M (2023) Singh D (2023) Time series (2001/2002–2021) analysis of Earth observation data using Google Earth Engine (GEE) for detecting changes in land use land cover (LULC) with specific reference to forest cover in East Godavari Region, Andhra Pradesh. India. J Earth Syst Sci 132:86. https://doi.org/10.1007/s12040-023-02099-w

Kerr JT, Ostrovsky M (2003) From space to species: Ecological applications for remote sensing. Trends Ecol Evol 18(6):299–305. https://doi.org/10.1016/S0169-5347(03)00071-5

Kim J, Lu Z, Jones JW, Shum C, Lee H, Jia Y (2014) Monitoring everglades freshwater marsh water level using L-band synthetic aperture radar backscatter. Remote Sens Environ 150:66–81. https://doi.org/10.1016/j.rse.2014.03.031

Kumar Arora N, Fatima T, Mishra J, Mishra I, Verma S, Verma R, Verma M, Bhattacharya A, Verma P, Mishra P, Bharti C (2020) Halo-tolerant plant growth promoting rhizobacteria for improving productivity and remediation of saline soils. J Adv Res 26:69–82. https://doi.org/10.1016/j.jare.2020.07.003

Kumbier K, Rogers K, Hughes MG, Lal KK, Mogensen LA, Woodroffe CD (2022) An eco-morphodynamic modelling approach to estuarine hydrodynamics & wetlands in response to sea-level rise. Front Mar Sci 9:860910. https://doi.org/10.3389/fmars.2022.860910

Kuenzer C, Bluemel A, Gebhardt S, Quoc TV, Dech S (2011) Remote sensing of mangrove ecosystems: a review. Remote Sens 3(5):878–928. https://doi.org/10.3390/rs3050878

Lashley CH, Garzon J, Miesse T, Ferreira C, Bricker JD (2019) Comparison of implicit and explicit vegetation representations in SWAN hindcasting wave dissipation by coastal Wetlands in Chesapeake Bay. Geosciences 9(1):8. https://doi.org/10.3390/geosciences9010008

Li P, Liu D (2025) How blue carbon financing can sustain blue carbon ecosystems protection and restoration: A proposed conceptual framework for the blue carbon financing mechanism. Ocean Coast Manag 265:107644. https://doi.org/10.1016/j.ocecoaman.2025.107644

Lin X, Hei J, Wang Y, Zhang A (2025) Research on intelligent classification of coastal land cover by integrating remote sensing images and deep learning. Front Environ Sci 13:1612446. https://doi.org/10.3389/fenvs.2025.1612446

Lin Y, Cheng Y, Zhou T, Ravi R, Hasheminasab SM, Flatt JE, Troy C, Habib A (2019) Evaluation of UAV LiDAR for mapping coastal environments. Remote Sens 11(24):2893. https://doi.org/10.3390/rs11242893

Lomeo D, Singh M (2022) Cloud-based monitoring and evaluation of the spatial-temporal distribution of Southeast Asia's mangroves using deep learning. Remote Sens 14(10):2291. https://doi.org/10.3390/rs14102291

Long K, Chen Z, Zhang H, Zhang M (2024) Spatiotemporal disturbances and attribution analysis of mangrove in southern China from 1986 to 2020 based on time-series Landsat imagery. Sci Total Environ 912:169157. https://doi.org/10.1016/j.scitotenv.2023.169157

Ma D, Ding Q, Guo Z, Xu C, Liang P, Zhao Z, Zheng HL (2022) The genome of a mangrove plant, *Avicennia marina*, provides insights into adaptation to coastal intertidal habitats. Planta 256(1):6. https://doi.org/10.1007/s00425-022-03916-0

Maggiori E, Tarabalka Y, Charpiat G, Alliez P (2016) Convolutional neural networks for large-scale remote-sensing image classification. IEEE Trans Geosci Remote Sens 55(2):645–657. https://doi.org/10.1109/TGRS.2016.2612821

Malek K, Ortíz Rodríguez E, Lee Y, Murillo J, Mohammadkhorasani A, Vigil L, Zhang S, Moreu F (2023) Design and implementation of sustainable solar energy harvesting for low-cost remote sensors equipped with real-time monitoring systems. J Infrastruct Intell Resil 2(3):100051. https://doi.org/10.1016/j.iintel.2023.100051

Mantiquilla JA, Shiao M, Shih H, Chen W, Chiang Y (2021) A review on the genetic structure of ecologically and economically important mangrove species in the Indo-West Pacific. Ecol Genet Genom 18:100078. https://doi.org/10.1016/j.egg.2020.100078

Masera OR, Garza-Caligaris J, Kanninen M, Karjalainen T, Liski J, Nabuurs G, Pussinen A, De Jong B, Mohren G (2003) Modeling carbon sequestration in afforestation, agroforestry and forest management projects: The CO2FIX vol 2 approach. Ecol Model 164(2–3):177–199. https://doi.org/10.1016/S0304-3800(02)00419-2

Masria A (2024) Bridging coastal challenges: the role of remote sensing and future research. Regional studies in marine science 73:103502. https://doi.org/10.1016/j.rsma.2024.103502

Masria A, Abouelsaad O (2025) Artificial intelligence applications in coastal engineering and its challenges – a review. Cont Shelf Res 286:105425. https://doi.org/10.1016/j.csr.2025.105425

Maurya K, Mahajan S, Chaube N (2021) Remote sensing techniques: mapping and monitoring of mangrove ecosystem—a review. Complex Intell. Syst. 7:2797–2818. https://doi.org/10.1007/s40747-021-00457-z

Maxwell AE, Warner TA, Fang F (2018) Implementation of machine-learning classification in remote sensing: an applied review. Int J Remote Sens 39(9):2784–2817. https://doi.org/10.1080/01431161.2018.1433343

Maxwell TL, Hengl T, Parente LL, Minarik R, Worthington TA, Bunting P, Smart LS, Spalding MD, Landis E (2023) Global mangrove soil organic carbon stocks dataset at 30 m resolution for the year 2020 based on spatiotemporal predictive machine learning. Data Brief 50:109621. https://doi.org/10.1016/j.dib.2023.109621

Mazda Y, Magi M, Kogo M, Hong PN (1997) Mangroves as a coastal protection from waves in the Tong King delta. Vietnam. Mangroves Salt Marshes 1(2):127–135. https://doi.org/10.1023/A:1009928003700

Mcleod E, Chmura GL, Bouillon S, Salm R, Björk M, Duarte CM, Lovelock CE, Schlesinger WH, Silliman BR (2011) A blueprint for blue carbon: toward an improved understanding of the role of vegetated coastal habitats in sequestering CO_2. Front Ecol Environ 9(10):552–560. https://doi.org/10.1890/110004

Mendez FJ, Losada IJ (2004) An empirical model to estimate the propagation of random breaking and nonbreaking waves over vegetation fields. Coast Eng 51(2):103–118. https://doi.org/10.1016/j.coastaleng.2003.11.003

Miryeganeh M, Marlétaz F, Gavriouchkina D, Saze H (2022) De novo genome assembly and in natura epigenomics reveal salinity-induced DNA methylation in the mangrove tree Bruguiera gymnorhiza. New Phytol 233(5):2094–2110. https://doi.org/10.1111/nph.17738

Mishra RK, Sahu PK, Mishra V, Jamal H, Varma A, Tripathi S (2023) Isolation and characterization of halotolerant plant growth promoting rhizobacteria from mangrove region of Sundarbans, India for enhanced crop productivity. Front Plant Sci 14:1122347. https://doi.org/10.3389/fpls.2023.1122347

Moffett KB, Nardin W, Silvestri S, Wang C, Temmerman S (2015) Multiple stable states and catastrophic shifts in coastal wetlands: progress, challenges, and opportunities in validating theory using remote sensing and other methods. Remote Sens 7(8):10184–10226. https://doi.org/10.3390/rs70810184

Moorcroft PR, Hurtt GC, Pacala SW (2001) A method for scaling vegetation dynamics: the ecosystem demography model (ED). Ecol Monogr 71(4):557–586. https://doi.org/10.1890/0012-9615(2001)071[0557:AMFSVD]2.0.CO;2

Morgan RG, Hodgson ME, Wang C, Schill RS (2022) Unmanned aerial remote sensing of coastal vegetation: a review. Ann GIS 28(3):385–399. https://doi.org/10.1080/19475683.2022.2026476

Muhadi NA, Abdullah AF, Bejo SK, Mahadi MR, Mijic A (2020) The use of LiDAR-Derived DEM in flood applications: a review. Remote Sens 12(14):2308. https://doi.org/10.3390/rs12142308

Nandy P, Dasgupta N, Das S (2009) Differential expression of physiological and biochemical characters of some Indian mangroves towards salt tolerance. Physiol Mol Biol Plants 15(2):151–160. https://doi.org/10.1007/s12298-009-0017-7

Narayan S, Beck MW, Reguero BG, Losada IJ, van Wesenbeeck B, Pontee N et al (2016) The effectiveness, costs and coastal protection benefits of natural and nature-based defences. PLoS ONE 11(5):e0154735. https://doi.org/10.1371/journal.pone.0154735

Natarajan P, Murugesan AK, Govindan G, Gopalakrishnan A, Kumar R, Duraisamy P, Balaji R, Shyamli PS, Parida AK, Parani M (2021) A reference-grade genome identifies salt-tolerance genes from the salt-secreting mangrove species Avicennia marina. Communications Biology 4(1):851. https://doi.org/10.1038/s42003-021-02384-8

Narayan S, Beck MW, Wilson P, Thomas CJ, Guerrero A, Shepard CC, Reguero BG, Franco G, Ingram JC, Trespalacios D (2017) The value of coastal wetlands for flood damage reduction in the Northeastern USA. Sci Rep 7(1):9463. https://doi.org/10.1038/s41598-017-09269-z

Neumann B, Vafeidis AT, Zimmermann J, Nicholls RJ (2015) Future coastal population growth and exposure to sea-level rise and coastal flooding - a global assessment. PLoS ONE 10(3):e0118571. https://doi.org/10.1371/journal.pone.0118571

Ngo D, Nguyen H, Nguyen K, Dang C, Nguyen H, Dang N, Pham T (2023) Application of multi-spectral UAV to estimate mangrove biomass in Vietnam: a case study in Dong Rui commune. Quang Ninh Province. One Ecosys 8:e103760. https://doi.org/10.3897/oneeco.8.e103760

Ngo DT, Quoc KN, Dang NT, Dang CH, Tran LL, Nguyen HD (2024) Analysis of the mangrove structure in the dong rui commune based on multispectral unmanned aerial vehicle image data. Geog Environ Sustain 16(4):14–25. https://doi.org/10.24057/2071-9388-2023-2641

Nguyen KL, Nguyen HA, Richter O, Pham MT, Nguyen VP (2016) Ecophysiological responses of young mangrove species *Rhizophora apiculata* (Blume) to different chromium contaminated environments. Sci Total Environ 574:369–380. https://doi.org/10.1016/j.scitotenv.2016.09.063

Niwa H, Ise H, Kamada M (2023) Suitable LiDAR platform for measuring the 3D structure of Mangrove Forests. Remote Sens 15(4):1033. https://doi.org/10.3390/rs15041033

Nizam A, Meera SP, Kumar A (2022) Genetic and molecular mechanisms underlying mangrove adaptations to intertidal environments. Iscience 25(1):103547. https://doi.org/10.1016/j.isci.2021.103547

Okarda B, Purnomo H, Andrianto A, Kusumadewi SD, Nadhira S, Shuhada MI (2025) Bridging technology and local participation: a collaborative approach for peatland and Mangrove restoration. IOP conference series: earth and environmental science 1506:012010. https://doi.org/10.1088/1755-1315/1506/1/012010

Pacheco-Pascagaza AM, Gou Y, Louis V, Roberts JF, Rodríguez-Veiga P, da Conceição BP, Espírito-Santo FDB, Robb C, Upton C, Galindo G et al (2022) Near real-time change detection system using sentinel-2 and machine learning: a test for mexican and colombian forests. Remote Sens 14(3):707. https://doi.org/10.3390/rs14030707

Pádua L, Hruška J, Bessa J, Adão T, Martins LM, Gonçalves JA, Peres E, Sousa AM, Castro JP, Sousa JJ (2017) Multi-temporal analysis of forestry and coastal environments using UASs. Remote Sens 10(1):24. https://doi.org/10.3390/rs10010024

Paliyavuth C, Patanaponpaiboon P, Ninomiya I (2009) Responses of gas exchange characteristics and morphological features of two mangrove species, Avicennia alba and Bruguiera gymnorrhiza, to different salinity and light environments. Tropics 18(1):23–33. https://doi.org/10.3759/tropics.18.23

Parida AK, Jha B (2010) Salt tolerance mechanisms in mangroves: a review. Trees 24:199–217. https://doi.org/10.1007/s00468-010-0417-x

Parri L, Parrino S, Peruzzi G, Pozzebon A (2019) Low Power Wide Area Networks (LPWAN) at sea: performance analysis of offshore data transmission by means of LoRaWAN connectivity for marine monitoring applications. Sensors 19(14):3239. https://doi.org/10.3390/s19143239

Parton WJ (1996) The CENTURY model. In: Powlson DS, Smith P, Smith JU (eds) Evaluation of soil organic matter models. NATO ASI Series, vol 38. Springer, Berlin, Heidelberg. https://doi.org/10.1007/978-3-642-61094-3_23

Paul N, Sunil G, Horvath D, Sun X (2025) Deep learning for plant stress detection: a comprehensive review of technologies, challenges, and future directions. Comput Electron Agric 229:109734. https://doi.org/10.1016/j.compag.2024.109734

Paul SS, Mahmood TH (2025) Long-term Landsat time series reveals mangrove vegetation fluctuations and drivers in the Sundarbans. Ecol Ind 181:114401. https://doi.org/10.1016/j.ecolind.2025.114401

Payus CM, Herman F, Sentian J (2022) Combined climate impacts and vulnerability index on coastal ecosystems in prediction of future scenarios: extended sustainable indicator tool for adaptive strategy. SN Appl Sci 4(8):222. https://doi.org/10.1007/s42452-022-05112-x

Peng C, Song Y, Huang Q, Niu Z, Zeng G, Kang H, Ying M (2025) Estimating coastal mangrove aboveground biomass of Zhejiang province based on sentinel-2 images and multiple machine learning algorithms. Front Forests Global Change 8:1662546. https://doi.org/10.3389/ffgc.2025.1662546

Perera N, Lokupitiya E, Halwatura D, Udagedara S (2022) Quantification of blue carbon in tropical salt marshes and their role in climate change mitigation. Sci Total Environ 820:153313. https://doi.org/10.1016/j.scitotenv.2022.153313

Pham TD, Le NN, Ha NT, Nguyen LV, Xia J, Yokoya N, To TT, Trinh HX, Kieu LQ, Takeuchi W (2020) Estimating Mangrove above-ground biomass using extreme gradient boosting decision trees algorithm with fused sentinel-2 and ALOS-2 PALSAR-2 data in can gio biosphere reserve Vietnam. Remote Sens 12(5):777. https://doi.org/10.3390/rs12050777

Pham TD, Xia J, Ha NT, Bui DT, Le NN, Tekeuchi W (2019a) A review of remote sensing approaches for monitoring blue carbon ecosystems: mangroves, seagrasses and salt marshes during 2010–2018. Sensors 19(8):1933. https://doi.org/10.3390/s19081933

Pham TD, Yokoya N, Bui DT, Yoshino K, Friess DA (2019b) Remote sensing approaches for monitoring Mangrove species, structure, and biomass: opportunities and challenges. Remote Sens 11(3):230. https://doi.org/10.3390/rs11030230

Phan VH, Pham DP, Pham TV, Qureshi KN (2023) An IoT System and MODIS images enable smart environmental management for Mekong Delta. Future Internet 15(7): 245. https://doi.org/10.3390/fi15070245

Pillai UPA, Pinardi N, Alessandri J, Federico I, Causio S, Unguendoli S, Valentini A, Staneva J (2022) A digital twin modelling framework for the assessment of seagrass nature based solutions against storm surges. Sci Total Environ 847:157603. https://doi.org/10.1016/j.scitotenv.2022.157603

Pinkeaw S, Boonrat P, Koedsin W, Huete A (2024) Semi-automated mangrove mapping at National-Scale using Sentinel-2, Sentinel-1, and SRTM data with Google Earth Engine: a case study in Thailand. Egyptian J Remote Sens Space Sci 27(3):555–564. https://doi.org/10.1016/j.ejrs.2024.07.001

Pirasteh S, Zenner EK, Mafi-Gholami D, Jaafari A, Nouri Kamari A, Liu G, Zhu Q, Li J (2021) Modeling mangrove responses to multi-decadal climate change and anthropogenic impacts using a long-term time series of satellite imagery. Int J Appl Earth Obs Geoinf 102:102390. https://doi.org/10.1016/j.jag.2021.102390

Pourzangbar A, Jalali M, Brocchini M (2023) Machine learning application in modelling marine and coastal phenomena: a critical review. Front Environ Eng 2:1235557. https://doi.org/10.3389/fenve.2023.1235557

Pradisty NA, Schlund M, Horstman EM, Willemen L (2025) Estimating canopy height and above-ground biomass in tropical mangrove restoration areas through multisource remote sensing. Ecol Informat 103522. https://doi.org/10.1016/j.ecoinf.2025.103522

Prakash A, McGlade K, Roxy MK, Roy J, Some S, Rao N (2022) Climate adaptation interventions in coastal areas: a rapid review of social and gender dimensions. Front Climate 4:785212. https://doi.org/10.3389/fclim.2022.785212

Purwanto AD, Wikantika K, Deliar A, Darmawan S (2023) Decision tree and random forest classification algorithms for Mangrove forest mapping in Sembilang National Park. Indonesia. Remote Sens 15(1):16. https://doi.org/10.3390/rs15010016

Ramadhan AW, Adytia D, Saepudin D, Husrin S, Adiwijaya A (2021) Forecasting of sea level time series using RNN and ISTM case study in sunda strait. Lontar Komputer: J Ilmiah Teknologi Informasi 12(3):130–140. https://doi.org/10.24843/LKJITI.2021.v12.i03.p01

Reef R, Feller IC, Lovelock CE (2010) Nutrition of mangroves. Tree Physiol 30(9):1148–1160. https://doi.org/10.1093/treephys/tpq048

Reinert M, Lepper R, Kösters F (2025) Coastal digital twin of the world's largest tidal flat system, EGU general assembly 2025, Vienna, Austria, 27 Apr–2 May 2025, EGU25–19359, https://doi.org/10.5194/egusphere-egu25-19359

Reusch TB, Ehlers A, Hämmerli A, Worm B (2005) Ecosystem recovery after climatic extremes enhanced by genotypic diversity. Proc Natl Acad Sci 102(8):2826–2831. https://doi.org/10.1073/pnas.0500008102

Richards DR, Friess DA (2016) Rates and drivers of mangrove deforestation in Southeast Asia, 2000–2012. Proc Natl Acad Sci 113(2):344–349. https://doi.org/10.1073/pnas.1510272113

Rosen PA, Bawden GW, Barela P, Chapman B, Fattahi H, Jones CE, Shrikant CV (2025) The NASA-ISRO SAR mission: a summary. IEEE Geosci Remote Sens Mag 13:8–34. https://doi.org/10.1109/MGRS.2025.3578258

Ruwaimana M, Satyanarayana B, Otero V, M. Muslim A, Syafiq A. M, Ibrahim S, et al. (2018) The advantages of using drones over space-borne imagery in the mapping of mangrove forests. PLoS ONE 13(7): e0200288. https://doi.org/10.1371/journal.pone.0200288

Saccotelli L, Verri G, De Lorenzis A, Cherubini C, Caccioppoli R, Coppini G, Maglietta R (2024) Enhancing estuary salinity prediction: a machine learning and deep learning based approach. Appl Comput Geosci 23:100173. https://doi.org/10.1016/j.acags.2024.100173

Sagar A, Rai S, Ilyas N, Sayyed RZ, I, A, El Enshasy HA, Simarmata T (2022) Halotolerant rhizobacteria for salinity-stress mitigation: diversity, mechanisms and molecular approaches. Sustainability 14(1): 490. https://doi.org/10.3390/su14010490

San DC, Duong NB, Phong NC, Tu LX, Guillou S, Nguyen KD (2022) Numerical study on measures for protecting the go-cong coastlines (Vietnam) from Erosion. Water 14(23):3850. https://doi.org/10.3390/w14233850

Scala P, Manno G, Cozar LC, Ciraolo G (2025) COAST-PROSIM: A model for predicting shoreline evolution and assessing the impacts of coastal defence structures. Water 17(2):269. https://doi.org/10.3390/w17020269

Seddon N, Smith A, Smith P, Key I, Chausson A, Girardin C, House J, Srivastava S, Turner B (2021) Getting the message right on nature-based solutions to climate change. Glob Change Biol 27(8):1518–1546. https://doi.org/10.1111/gcb.15513

Seydi ST, Ahmadi SA, Ghorbanian A, Amani M (2024) Land cover mapping in a Mangrove ecosystem using hybrid selective kernel-based convolutional neural networks and multi-temporal sentinel-2 imagery. Remote Sens 16(15):2849. https://doi.org/10.3390/rs16152849

Singh A, Mahajan S (2022) A review of SAR image processing for Mangrove above ground biomass estimation. J Plant Sci Res 38(2). https://doi.org/10.32381/JPSR.2022.38.02.31

Skákala J, Menon PP, Wang K, Lessin G (2023) Future digital twins: Emulating a highly complex marine biogeochemical model with machine learning to predict hypoxia. Front Mar Sci 10:1058837. https://doi.org/10.3389/fmars.2023.1058837

Slinger J, Stive M, Luijendijk A (2021) Nature-based solutions for coastal engineering and management. Water 13(7):976. https://doi.org/10.3390/w13070976

Smolders S, João Teles M, Leroy A, Maximova T, Meire P, Temmerman S (2020) Modeling storm surge attenuation by an integrated nature-based and engineered flood defense system in the Scheldt Estuary (Belgium). J Marine Sci Eng 8(1):27. https://doi.org/10.3390/jmse8010027

Stachew E, Houette T, Gruber P (2021) Root systems research for bioinspired resilient design: a concept framework for foundation and coastal engineering. Front Robotics AI 8:548444. https://doi.org/10.3389/frobt.2021.548444

Stella LR, Alejandra BM, Andrea LX, Elkin A, Neyla B (2025) Plant growth-promoting traits of Rhizophora mangle-associated bacteria and their effects on tomato (*Solanum lycopersicum*). Biotechnol Reports 48:e00923. https://doi.org/10.1016/j.btre.2025.e00923

Subbiah A, Maniam VT, Katim NIA (2024) Project BlueForest: IoT sensor applications in mangrove ecosystem management–an In-depth pilot study analysis. In: 2024 IEEE 15th Control and System Graduate Research Colloquium (ICSGRC), pp 216–221. IEEE. Published in: 2024 IEEE 15th Control and System Graduate Research Colloquium (ICSGRC) https://doi.org/10.1109/ICSGRC62081.2024.10691055

Swan A, Williams SA, Brown K, Chambers A, Creque J, Wick J, Paustian K (2015) COMET-Planner. C and greenhouse gas evaluation for NRCS conservation practice planning. A companion report to www.comet-planner.com

Syvitski JP, Kettner AJ, Overeem I, Hutton EW, Hannon MT, Brakenridge GR, Day J, Vörösmarty C, Saito Y, Giosan L, Nicholls RJ (2009) Sinking deltas due to human activities. Nat Geosci 2(10):681–686. https://doi.org/10.1038/ngeo629

Taddia Y, Pellegrinelli A, Corbau C, Franchi G, Staver LW, Stevenson JC, Nardin W (2021) High-resolution monitoring of tidal systems using UAV: a case study on poplar island, MD (USA). Remote Sens 13(7):1364. https://doi.org/10.3390/rs13071364

Tengö M, Brondizio ES, Elmqvist T, Malmer P (2014) Spierenburg M (2014) Connecting diverse knowledge systems for enhanced ecosystem governance: the multiple evidence base approach. Ambio 43:579–591. https://doi.org/10.1007/s13280-014-0501-3

Thai TH, Tri DQ, Anh NX, Hoa VV, Nguyen HV, Nhat NV, Tuyet QT, Pham HT, Chung PH, Thang VV, Thuc TD (2022) Numerical simulation of the flood and inundation caused by typhoon noru downstream from the Vu Gia-Thu Bon River Basin. Sustainability 15(10):8203. https://doi.org/10.3390/su15108203

Tian Y, Huang H, Zhou G, Zhang Q, Xie X, Ou J, Zhang Y, Tao J, Lin J (2023) Mangrove biodiversity assessment using UAV lidar and hyperspectral data in China's Pinglu Canal Estuary. Remote Sensing 15(10):2622. https://doi.org/10.3390/rs15102622

Torres R, Snoeij P, Geudtner D, Bibby D, Davidson M, Attema E, Potin P, Rommen B, Floury N, Brown M, Traver IN, Deghaye P, Duesmann B, Rosich B, Miranda N, Bruno C, L'Abbate M, Croci R, Pietropaolo A, Rostan F (2012) GMES Sentinel-1 mission. Remote Sens Environ 120:9–24. https://doi.org/10.1016/j.rse.2011.05.028

Triest L, Phan TT, Luong QD, Do BT, Sierens T, Koedam N (2025) Migration history of Avicennia marina populations: A legacy of mangrove expansion on the Sunda Shelf. Front Mar Sci 12:1565908. https://doi.org/10.3389/fmars.2025.1565908

Tsiakos C, Chalkias C (2023) Use of machine learning and remote sensing techniques for shoreline monitoring: a review of recent literature. Appl Sci 13(5):3268. https://doi.org/10.3390/app13053268

Tysiąc P, Moskalewicz D, Janowski Ł (2025) Development of UAV-based LiDAR and multispectral measurement techniques for monitoring sediment intrusion in coastal wetlands. Measurement 253:117459. https://doi.org/10.1016/j.measurement.2025.11745

Van Wesenbeeck B, Balke T, Van Eijk P, Tonneijck F, Siry H, Rudianto M, Winterwerp J (2015) Aquaculture induced erosion of tropical coastlines throws coastal communities back into poverty. Ocean Coast Manag 116:466–469. https://doi.org/10.1016/j.ocecoaman.2015.09.004

Van Wesenbeeck BK, Van Zelst VT, Antolinez JA, De Boer WP (2025) Quantifying uncertainty in wave attenuation by mangroves to inform coastal green belt policies. Commun Earth Environ 6(1):258. https://doi.org/10.1038/s43247-025-02178-4

Van Zelst VT, Dijkstra JT, Van Wesenbeeck BK, Eilander D, Morris EP, Winsemius HC, Ward PJ, De Vries MB (2021) Cutting the costs of coastal protection by integrating vegetation in flood defences. Nat Commun 12(1):6533. https://doi.org/10.1038/s41467-021-26887-4

Veettil BK (2022) Current status of mangrove vegetation in Batticaloa Lagoon, Sri Lanka, using high-resolution satellite imagery. J Coast Conserv 26(6):74. https://doi.org/10.1007/s11852-022-00923-x

Veettil BK, Costi J, Marques WC, Tran X, Quang NX, Van DD, Hoai PN (2020a) Coastal environmental changes in Southeast Asia: a study from Quang Nam Province, Central Vietnam. Regional Stud Marine Sci 39:101420. https://doi.org/10.1016/j.rsma.2020.101420

Veettil BK, Tran XL, Lopes MS (2024) Mangrove vegetation changes and shoreline erosion along the central coastline of Vietnam: a study from Quang Ngai province. Environ Develop 52:101065. https://doi.org/10.1016/j.envdev.2024.101065

Veettil BK, Ward RD, Dung NTK, Van DD, Quang NX, Hoai PN, Hoang N (2021) The use of bioshields for coastal protection in Vietnam: current status and potential. Regional Stud Marine Sci 47:101945. https://doi.org/10.1016/j.rsma.2021.101945

Veettil BK, Ward RD, Lima MDAC, Stankovic M, Hoai PN, Quang NX (2020b) Opportunities for seagrass research derived from remote sensing: a review of current methods. Ecol Ind 117:106560. https://doi.org/10.1016/j.ecolind.2020.106560

Veettil BK, Ward RD, Quang NX, Trang NTT, Giang TH (2019) Mangroves of Vietnam: Historical development, current state of research and future threats. Estuar Coast Shelf Sci 218:212–236. https://doi.org/10.1016/j.ecss.2018.12.021

Verrelst J, Camps-Valls G, Muñoz-Marí J, Rivera JP, Veroustraete F, Clevers JG, Moreno J (2015) Optical remote sensing and the retrieval of terrestrial vegetation bio-geophysical properties—a review. ISPRS J Photogramm Remote Sens 108:273–290. https://doi.org/10.1016/j.isprsjprs.2015.05.005

Villanueva R, Paul M, Schlurmann T (2022) Anchor forces on coir-based artificial seagrass mats: dependence on wave dynamics and their potential use in seagrass restoration. Front Mar Sci 9:802343. https://doi.org/10.3389/fmars.2022.802343

Vogel A, Seeger K, Brill D, Brückner H, Kraas F (2022) Identifying land-use related potential disaster risk drivers in the Ayeyarwady Delta (Myanmar) during the last 50 Years (1974–2021)

using a hybrid ensemble learning model. Remote Sens 14(15):3568. https://doi.org/10.3390/rs14153568
Vos K, Splinter KD, Harley MD, Simmons JA, Turner IL (2019) CoastSat: A Google Earth Engine-enabled Python toolkit to extract shorelines from publicly available satellite imagery. Environ Model Softw 122:104528. https://doi.org/10.1016/j.envsoft.2019.104528
Wang D, Wan B, Qiu P, Tan X, Zhang Q (2022) Mapping mangrove species using combined UAV-LiDAR and Sentinel-2 data: feature selection and point density effects. Adv Space Res 69(3):1494–1512. https://doi.org/10.1016/j.asr.2021.11.020
Wang Q, Pacheco-Crosetti G, Calderon CV, Cohen J, Smyth E (2025) An integrative framework for AI-supported coastal hydrodynamics monitoring and analysis. Sci Rep 15:12403. https://doi.org/10.1038/s41598-025-94791-8
Wang X, Gao X, Zhang Y, Fei X, Chen Z, Wang J, Zhang Y, Lu X, Zhao H (2019) Land-cover classification of coastal wetlands using the RF algorithm for worldview-2 and Landsat 8 Images. Remote Sens 11(16):1927. https://doi.org/10.3390/rs11161927
Wdowinski S (2016) InSAR-Based mapping of tidal inundation extent and amplitude in Louisiana coastal wetlands. Remote Sens 8(5):393. https://doi.org/10.3390/rs8050393
Wee AKS, Noreen EAM, Ono J, Takayama K, Kumar PP, Tan WHT, Saleh MN, Kajita T, Webb EL (2020) Genetic structures across a biogeographical barrier reflect dispersal potential of four Southeast Asian Mangrove plant species. J Biogeogr 47(6):1258–1271. https://doi.org/10.1111/jbi.13813
Wei Y, Cheng Y, Yin X, Xu Q, Ke J, Li X (2023) Deep learning-based classification of high-resolution satellite images for Mangrove mapping. Appl Sci 13(14):8526. https://doi.org/10.3390/app13148526
Weinstein BG, Marconi S, Aubry-Kientz M, Vincent G, Senyondo H, White EP (2020) DeepForest: A Python package for RGB deep learning tree crown delineation. Methods Ecol Evol 11(12):1743–1751. https://doi.org/10.1111/2041-210X.13472
Williams AT, Giardino A, Pranzini E (2016) Canons of coastal engineering in the United Kingdom: seawalls/groynes, a century of change? J Coastal Res 32(5):1196–1211. https://doi.org/10.2112/JCOASTRES-D-15-00213.1
Windle AE, Staver LW, Elmore AJ, Scherer S, Keller S, Malmgren B, Silsbe GM (2023) Multi-temporal high-resolution marsh vegetation mapping using unoccupied aircraft system remote sensing and machine learning. Front Remote Sens 4:1140999. https://doi.org/10.3389/frsen.2023.1140999
Wu W, Gao Y, Chen C (2024) CLSER: a new indicator for the social-ecological resilience of coastal systems and sustainable management. J Clean Prod 435:140564. https://doi.org/10.1016/j.jclepro.2024.140564
Wyborn C (2015) Co-productive governance: a relational framework for adaptive governance. Glob Environ Chang 30:56–67. https://doi.org/10.1016/j.gloenvcha.2014.10.009
Xie D, Schwarz C, Kleinhans MG, Zhou Z (2022) Implications of coastal conditions and sea-level rise on Mangrove vulnerability: a bio-morphodynamic modeling study. J Geophys Res Earth Surf 127(3): e2021JF006301. https://doi.org/10.1029/2021JF006301
Xie J, Liang S, Sun Z, Chang J, Sun J (2016) Design and implementation of a robust decision support system for marine space resource utilization. ISPRS Int J Geo Inf 5(8):140. https://doi.org/10.3390/ijgi5080140
Xu C, Wang J, Sang Y, Li K, Liu J, Yang G (2022) An effective deep learning model for monitoring Mangroves: a case study of the indus delta. Remote Sens 15(9):2220. https://doi.org/10.3390/rs15092220
Xu G, Shen W, Wang X (2014) Applications of wireless sensor networks in marine environment monitoring: a survey. Sensors 14(9):16932–16954. https://doi.org/10.3390/s140916932
Yan Y, Duke NC, Sun M (2016) Comparative analysis of the pattern of population genetic diversity in three Indo-West Pacific Rhizophora Mangrove species. Front Plant Sci 7:213405. https://doi.org/10.3389/fpls.2016.01434

Zaka MM, Samat A (2023) Advances in remote sensing and machine learning methods for invasive plants study: a comprehensive review. Remote Sens 16(20):3781. https://doi.org/10.3390/rs16203781

Zamboni NS, Prudêncio MDC, Amaro VE, Matos MDFAD, Verutes GM, Carvalho AR (2022) The protective role of mangroves in safeguarding coastal populations through hazard risk reduction: a case study in northeast Brazil. Ocean Coast Manag 229:106353. https://doi.org/10.1016/j.ocecoaman.2022.106353

Zhang Z, Shen X, Yan C, Li R, Li B (2025) Unveiling seaward expansion pattern in mangrove forests using UAV remote sensing and deep learning. Ecol Ind 178:114054. https://doi.org/10.1016/j.ecolind.2025.114054

Zhong W, Chu T, Tissot P, Wu Z, Chen J, Zhang H (2022) Integrated coastal subsidence analysis using InSAR, LiDAR, and land cover data. Remote Sens Environ 282:113297. https://doi.org/10.1016/j.rse.2022.113297

Zhou X, Wang H, Chen C, Nagy G, Jancso T, Huang H (2023) Detection of growth change of young forest based on UAV RGB images at single-tree level. Forests 14(1):141. https://doi.org/10.3390/f14010141

Zhu X, Chen J, Gao F, Chen X, Masek JG (2010) An enhanced spatial and temporal adaptive reflectance fusion model for complex heterogeneous regions. Remote Sens Environ 114(11):2610–2623. https://doi.org/10.1016/j.rse.2010.05.032

Zhu XX, Tuia D, Mou L, Xia GS, Zhang L, Xu F, Fraundorfer F (2017) Deep learning in remote sensing: a comprehensive review and list of resources. IEEE Geosci Remote Sens Mag 5(4):8–36. https://doi.org/10.1109/MGRS.2017.2762307

Zhu Y, Myint SW, Liu K, Liu L, Cao J (2024) Integration of UAV LiDAR and WorldView-2 images for modeling mangrove aboveground biomass with GA-ANN wrapper. Ecol Proc 13(1):85. https://doi.org/10.1186/s13717-024-00566-w

Chapter 16
Towards Sustainable Coastal Futures

Abstract The sustainability of coastal protection strategies ultimately depends on the deliberate integration of ecological integrity, social inclusiveness, and adaptive governance. This chapter situates the future of bioshields within the wider agenda of sustainable coastal development and climate resilience. It synthesizes key insights from the preceding chapters, underscoring the imperative to move beyond discrete, project-based interventions toward systemic and long-term coastal resilience planning. The discussion outlines pathways for embedding bioshields within national adaptation policies, blue economy strategies, and global sustainability frameworks, including the United Nations Sustainable Development Goals (SDGs) and the Paris Agreement. Emphasis is directed toward the integration of scientific advances, technological innovation, and traditional knowledge systems to co-produce adaptive solutions that enhance both human well-being and ecosystem health. The chapter foregrounds emerging concepts—such as hybrid coastal infrastructure, ecosystem accounting, and resilience-based management—that link ecological restoration with broader socio-economic transformation. It further examines financial and policy instruments, including green bonds, carbon markets, and ecosystem service valuation approaches, that hold potential for sustaining bioshield initiatives over the long term. Ultimately, the chapter argues that the realization of sustainable coastal futures hinges on collaborative governance, inclusive participation, and iterative learning processes supported by robust scientific evidence and digital innovation. By positioning bioshields as catalysts for ecological restoration, climate adaptation, and social resilience, the chapter concludes that advancing sustainable coastal futures requires harmonizing human development with nature within an adaptive, equitable, and science-informed governance framework.

Keywords Blue economy · Climate resilience · Ecosystem-based management · Hybrid coastal infrastructure · Resilience-based planning · Sustainable coastal development

B. Kozhikkodan Veettil, *The Potential of Bioshields for Coastal Protection*, Advances in Natural and Technological Hazards Research 55, https://doi.org/10.1007/978-3-032-22888-8_16

16.1 Introduction

The twenty-first century has marked a period of unprecedented environmental and socio-economic transformation, with coastal regions emerging as critical frontlines where the impacts of global change are most acutely manifested (Griggs and Reguero 2021). Nowhere is this more evident than in Asia, where dense populations, accelerating urbanization, and climate change converge to create a complex mosaic of risks and opportunities. Coastal zones across the continent are facing intensifying pressures, varying from rising sea levels and stronger storm surges to saltwater intrusion, subsidence, and accelerating shoreline erosion (Veettil et al. 2021). These changes not only endanger fragile coastal ecosystems but also threaten the economic vitality, food security, and cultural heritage of millions who depend on these dynamic environments for their livelihoods (Cooley et al. 2023).

In response to these escalating threats, the idea of sustainable coastal futures has emerged as a central paradigm for research, policy, and practice. This concept envisions coastal systems that are ecologically resilient, socially equitable, and economically viable over the long term (Baumann et al. 2023). It advances beyond short-term, engineering-focused interventions toward integrated, nature-aligned approaches that harmonize human development with ecological processes (Slinger et al. 2021). Achieving such futures requires careful integration of ecological restoration, technological innovation, and community empowerment, acknowledging that the long-term sustainability of coastal regions constitutes not only an environmental concern but also a fundamentally social and institutional challenge.

Within this evolving framework, nature-based solutions (NbS) have gained increasing attention as effective, adaptive, and cost-efficient strategies for coastal management (Slinger et al. 2021). Among these, bioshields, which are protective belts of coastal vegetation composed of mangroves, casuarinas, salt marshes, and other salt-tolerant species, represent a particularly promising approach (Veettil et al. 2021). Unlike hard-engineering structures such as seawalls or groynes, bioshields harness the protective capacity of nature itself (Falkenrich et al. 2021). They buffer coastal communities against storm surges and erosion, stabilize shorelines, enhance sediment retention, and serve as nurseries for diverse aquatic species (Falkenrich et al. 2021; Veettil et al. 2021). Moreover, by sequestering carbon and supporting local livelihoods through fisheries, forestry, and ecotourism, bioshields contribute simultaneously to climate mitigation, biodiversity conservation, and socio-economic resilience (Donato et al. 2011; Zu Ermgassen et al. 2025).

The adoption of bioshields thus reflects a broader paradigm shift from controlling nature to collaborating with it. However, realizing their full potential requires careful consideration of ecological design, land-use planning, governance, and community participation (Key et al. 2022). Bioshields are not merely ecological interventions but socio-ecological systems whose effectiveness depends on integrating scientific knowledge with local experience, traditional practices, and inclusive, community-centered decision-making (Chee et al. 2021).

This chapter positions the concept of bioshields within the broader discourse on sustainable coastal development, with particular attention to the Asian context, where both vulnerabilities and opportunities are especially pronounced. It examines the scientific foundations, institutional arrangements, and socio-economic conditions required to integrate bioshields into coastal adaptation and management frameworks. Drawing on regional experiences from countries such as Vietnam, India, Indonesia, Bangladesh, Thailand, and the Philippines, the chapter highlights the varied ways in which bioshields have been conceptualized, implemented, and governed. Through these case studies, it demonstrates how bioshields can function as multifunctional assets that safeguard coastlines, enhance livelihoods, and support ecological recovery, ultimately contributing to a vision of sustainable coastal futures that harmonize human well-being with the integrity of the planet's most dynamic and vulnerable margins.

16.2 Understanding Sustainable Coastal Futures

16.2.1 Conceptual Framework

The concept of sustainable coastal futures reflects a forward-looking, systems-based perspective for managing the world's coastal zones—sites where land, sea, and human activities intersect in complex and dynamic socio-ecological relationships (Spencer et al. 2023). Rather than treating coastal management as a sectoral or short-term challenge, this approach envisions the coast as a living socio-ecological system that must sustain ecological integrity, support economic productivity, and ensure social equity across generations (Spencer et al. 2023). Achieving such a balance demands not only technical and scientific innovation but also adaptive governance, participatory decision-making, and a deep respect for local and traditional knowledge (Jordan and Fröhle 2022).

At its core, the framework for sustainable coastal futures is anchored in global sustainability principles and informed by international commitments. It aligns closely with the United Nations Sustainable Development Goals (SDGs), particularly SDG 13 (Climate Action), SDG 14 (Life Below Water), and SDG 15 (Life on Land), which collectively emphasize resilience to climate impacts, conservation of marine and terrestrial ecosystems, and the restoration of degraded environments. Furthermore, it resonates with the Sendai Framework for Disaster Risk Reduction (2015–2030) and the Paris Agreement on Climate Change (2015), both of which advocate integrated approaches to adaptation, mitigation, and sustainable development (UNISDR 2015; UNFCCC 2015). These global agendas provide a policy scaffold for reimagining coastal governance that is anticipatory, inclusive, and resilient to change.

In practical terms, achieving sustainable coastal futures involves the convergence of several interrelated dimensions:

a. ***Ecosystem-Based Management (EBM)***: This principle recognizes the coast as a mosaic of interacting ecosystems (mangroves, seagrasses, coral reefs, estuaries, and beaches) that collectively provide vital ecological services (Long et al. 2015; Malinconico 2025). EBM calls for planning and decision-making processes that respect ecological boundaries rather than administrative ones, integrating natural dynamics into management strategies (Aburto et al. 2012). By maintaining ecological functions such as sediment transport, nutrient cycling, and habitat connectivity, EBM enhances both environmental and societal resilience (Foley et al. 2010).
b. ***Community Resilience***: Coastal sustainability cannot be achieved without strong, empowered communities. Building resilience at the local level means enhancing the capacity of coastal populations to anticipate, absorb, and recover from shocks—whether they arise from cyclones, flooding, or economic disruptions (Masselink and Lazarus 2019; Choudhury et al. 2021). This involves not only physical preparedness but also strengthening social capital, livelihood diversification, and equitable access to resources. Locally led adaptation ensures that interventions are socially acceptable and grounded in the lived realities of those most affected (McNamara and Buggy 2017).
c. ***Climate Adaptation and Mitigation Synergy***: In coastal contexts, the boundaries between adaptation and mitigation are increasingly blurred (Zagare 2022). Nature-based interventions, such as mangrove reforestation and wetland restoration, fulfil dual, interconnected functions. They enhance coastal protection by reducing storm-surge impacts and mitigating shoreline erosion, while simultaneously contributing to climate-change mitigation through substantial carbon sequestration (Veettil et al. 2021). This integrative capacity highlights the strategic value of nature-based solutions as essential instruments for strengthening climate resilience and advancing ecological restoration.
d. ***Multi-stakeholder Collaboration***: Coastal sustainability requires governance systems capable of bridging the gap between science, policy, and community action. Governments, research institutions, the private sector, and civil society organizations must work together to co-create solutions (Rölfer et al. 2024). Such collaboration enhances policy coherence, mobilizes diverse forms of knowledge, and ensures that local priorities are reflected in broader development agendas.

Within this holistic framework, bioshields emerge as more than just vegetative barriers against coastal hazards, as they are key components of an integrated socio-ecological system. When properly designed and managed, bioshields exemplify core principles of sustainability: they safeguard communities and infrastructure, rehabilitate degraded ecosystems, and support local livelihoods (Zamboni et al. 2022; Gijón Mancheño et al. 2024; Khan et al. 2025). Their long-term effectiveness, however, hinges on integrating them into comprehensive coastal planning frameworks that recognize ecological connectivity, promote meaningful community participation, and enable adaptive, evidence-based governance.

16.2.2 The Role of Nature-Based Solutions in Sustainability

Over the past decade, nature-based solutions (NbS) have gained international recognition as transformative approaches to managing environmental challenges in a cost-effective and ecologically sensitive manner (Cohen-Shacham et al. 2016, 2019). Grounded in the recognition that healthy ecosystems serve as vital allies in addressing climate and disaster risks, nature-based solutions aim to harness and enhance natural processes rather than override them. In coastal settings, NbS encompass a range of interventions, varying from mangrove and seagrass restoration to coral reef rehabilitation, dune stabilization, and the creation of living shorelines (Jordan and Fröhle 2022). Each of these strategies contributes to resilience by enhancing ecosystem functions that buffer human communities against natural hazards while simultaneously delivering environmental, economic, and social benefits.

Coastal NbS operate through multiple mechanisms. Vegetation and reef systems, for example, attenuate wave energy, thereby reducing the destructive force of storm surges and minimizing erosion (Daro Justine and Seenath 2025). Wetlands and mangroves trap sediments, improve water quality, and provide critical breeding and nursery grounds for fish and crustaceans, sustaining local fisheries and food security (Paxton et al. 2024). Beyond these direct benefits, NbS contribute to global climate goals by storing substantial amounts of blue carbon in soils and biomass (Hilmi et al. 2021). They also enrich human well-being by providing spaces for recreation, tourism, and cultural connection—values that are often ignored in conventional engineering approaches (Ruslan et al. 2022).

Within this broad NbS framework, bioshields represent a promising model for sustainable coastal protection. Composed of carefully selected coastal vegetation (e.g., mangroves, casuarinas, saltmarsh grasses, and other salt-tolerant species), bioshields are designed to combine the protective functions of natural ecosystems with the planning precision of engineered structures. Unlike rigid coastal defenses like seawalls or revetments, bioshields evolve, adapting naturally to changing sea levels and sediment dynamics (Jayson-Quashigah et al. 2025). Their ecological flexibility reduces long-term maintenance costs and enhances sustainability.

However, the performance and resilience of bioshields depend on several key factors. Ecological integrity is crucial: species composition, planting density, and hydrological conditions must be compatible with local environmental parameters (Su et al. 2021; Juanico 2022; Jayson-Quashigah et al. 2025). Landscape connectivity determines the capacity of bioshields to interact with adjacent ecosystems, such as wetlands, dunes, or coral reefs, creating a continuum of defense and biodiversity (Guannel et al. 2016; Mishra et al. 2025). Equally important is long-term management, including community stewardship, monitoring, and adaptive maintenance to ensure the bioshield's continued vitality under evolving climate conditions (Primavera and Esteban 2008).

16.3 The Asian Coastal Context

16.3.1 *Coastal Vulnerabilities in Asia*

Asia's coastal regions stand at the convergence of extraordinary opportunity and profound risk (Fig. 16.1). As the most populous and economically dynamic continent, Asia hosts approximately 60% of the global population residing within 100 km of its coastal zones (Neumann et al. 2015). These densely inhabited coastal areas are essential to the region's economies, supporting agriculture, aquaculture, industry, trade, and tourism. Major deltas and estuaries, such as the Mekong Delta in Vietnam, the Ganges–Brahmaputra Delta spanning Bangladesh and India, the Chao Phraya Delta in Thailand, and the Irrawaddy Delta in Myanmar, represent some of the most fertile and productive landscapes on Earth (Becker et al. 2024). They are vital for regional food security and global supply chains, yet they are simultaneously among the most vulnerable zones to climate change and human-induced stress.

The geographic characteristics that make these areas so productive—flat terrain, rich sediments, and proximity to waterways—also render them acutely sensitive to environmental pressures (Becker et al. 2024). Rising sea levels, driven by global warming, threaten to inundate millions of hectares of farmland and displace millions of people. The Intergovernmental Panel on Climate Change (IPCC) projects that by 2100, sea levels in parts of Asia could rise by up to one meter, placing low-lying coastal plains at high risk of permanent submersion (IPCC 2007). Combined with land subsidence, which is often accelerated by groundwater extraction and infrastructural loading, the effective relative sea-level rise in some Asian cities and deltas is even more severe.

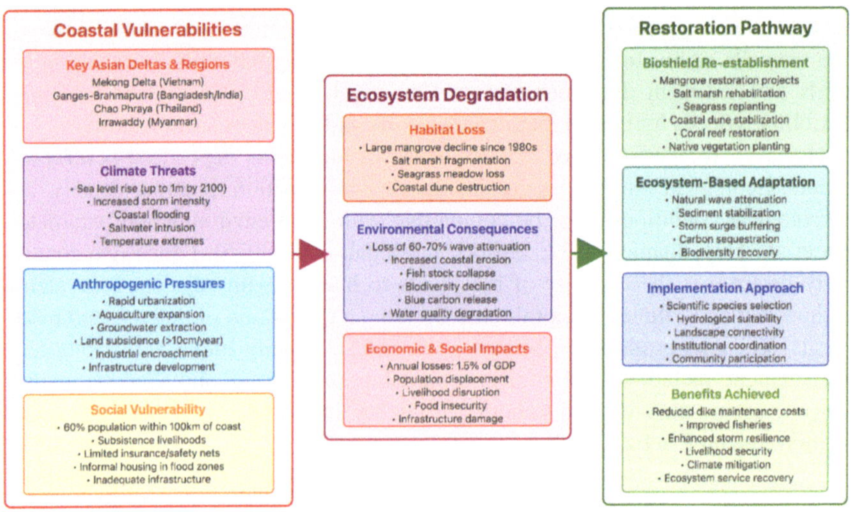

Fig. 16.1 Coastal vulnerabilities, ecosystem degradation and restoration pathways in Asian context

These physical threats translate directly into economic and social impacts. In Vietnam, for example, coastal erosion and flooding cause annual losses equivalent to 1.5% of the national GDP (Zhou and Botzen 2021). The Sundarbans in Bangladesh and eastern India, which is the world's largest mangrove forest and a UNESCO World Heritage Site, faced recurrent devastation from cyclones such as Aila (2009), Amphan (2020), and Remal (2024). These events have displaced thousands of families, destroyed agricultural lands through saltwater intrusion, and disrupted livelihoods dependent on fishing and forest resources (Subhani and Ahmad 2019). Indonesia's northern Java coast presents another striking example: rapid urbanization, industrial encroachment, and aquifer depletion have led to land subsidence rates exceeding 10 cm per year in some areas, resulting in severe flooding and the loss of entire coastal settlements (Susilo et al. 2023; Widodo et al. 2024).

Beyond the physical hazards, socio-economic vulnerability amplifies the risks. Many coastal populations in Asia are engaged in small-scale or subsistence livelihoods with limited access to insurance, credit, or formal safety nets. Informal housing in flood-prone zones, coupled with inadequate infrastructure for drainage and waste management, increases exposure to disasters (UN ESCAP 2019). Urban coastal centers such as Manila, Jakarta, Bangkok, and Ho Chi Minh City face the dual pressures of population growth and environmental degradation, straining both governance systems and public resources (Rosenzweig et al. 2018; Monioudi et al. 2025).

The convergence of these factors, including biophysical exposure, ecological degradation, and social vulnerability, creates a complex and multi-dimensional challenge. Addressing it requires integrated strategies that blend scientific understanding, participatory governance, and ecosystem-based adaptation. The concept of sustainable coastal futures, as discussed earlier, provides an essential framework for navigating this complexity, emphasizing resilience-building approaches that work with, rather than against, natural systems.

16.3.2 Degradation of Coastal Ecosystems

Over the past half-century, rapid coastal development, industrial expansion, and land conversion for agriculture and aquaculture have drastically reduced natural coastal habitats (Gilman et al. 2008). Mangrove forests, salt marshes, seagrass meadows, and coastal dunes, once abundant natural defenses, have been cleared or fragmented to make way for shrimp ponds, paddy fields, ports, and urban settlements (FAO 2018a).

Southeast Asia alone had a large decline in its mangrove cover since the 1980s (Richards and Friess 2016; Bunting et al. 2022), with certain regions such as Indonesia and Myanmar experiencing even higher rates of decline. This loss resulted in diminished coastal protection, reduced biodiversity, and weakened ecosystem services that underpin local economies. Mangroves, for example, act as natural wave attenuators, dissipating up to 60–70% of wave energy over short distances (Menéndez et al. 2020). Their removal exposes shorelines to direct erosion and storm impacts.

Similarly, coral reef degradation due to bleaching, overfishing, and pollution compromises another critical line of defense against coastal flooding (e.g., Zhao et al. 2019; Veettil et al. 2022).

The consequences of ecosystem degradation extend beyond physical protection. The collapse of fish stocks, loss of nursery habitats, and decline in water quality directly affect food security and livelihoods for millions of coastal residents (FAO 2018b). Moreover, the destruction of blue carbon ecosystems undermines global climate mitigation efforts. These habitats sequester carbon at rates up to four times greater than terrestrial forests; their loss not only releases stored carbon but also reduces future sequestration potential (Hussain et al. 2025).

Nonetheless, despite these challenges, an increasing body of evidence indicates that ecosystem restoration and the re-establishment of coastal bioshields can mitigate environmental degradation and enhance ecological resilience (Su et al. 2021). Across Asia, governments, NGOs, and local communities have initiated large-scale planting and rehabilitation programs that combine ecological restoration with disaster risk reduction. In Vietnam's Mekong and Red River Deltas, mangrove restoration projects initiated in the 1990s by organizations such as the Red Cross demonstrated the dual benefits of ecosystem recovery and community protection (Hai et al. 2020). These programs reduced dike maintenance costs, improved local fisheries, and enhanced storm resilience, especially during major typhoon events (Andreasson and Bechauf 2025).

The re-establishment of bioshields represents not merely a technical restoration exercise but a transformative approach to coastal governance. It redefines the relationship between people and nature, emphasizing that long-term safety and prosperity depend on ecological health. For rapidly developing Asian nations, the challenge is to scale these efforts while ensuring ecological integrity and social inclusion (Gerona-Daga and Salmo III 2022). Effective bioshield implementation demands sound scientific guidance on species selection, hydrological suitability, and landscape connectivity, coupled with institutional coordination and community participation (Kusumadewi et al. 2024). In essence, the Asian coastal context underscores both the urgency and the promise of pursuing sustainable coastal futures.

16.4 Bioshields as Cornerstones of Sustainable Coastal Futures

16.4.1 Ecological Functions and Climate Regulation

Bioshields represent one of the most effective and multifunctional nature-based solutions for building coastal resilience. Comprising vegetation belts of mangroves, casuarinas, palms, salt marshes, and other salt-tolerant species, bioshields function as living infrastructure that mediates a wide range of ecological and physical processes

(Veettil et al. 2021). Their ecological functions extend beyond mere shoreline protection—they help sustain the dynamic equilibrium of coastal systems by stabilizing sediments, regulating hydrological flows, and maintaining biodiversity.

At the most fundamental level, bioshields influence sediment dynamics through their extensive root systems, which trap suspended particles and enhance sediment accretion (Hongwiset et al. 2022). This process gradually elevates the coastal profile, counteracting erosion and relative sea-level rise. The dense above-ground structures, such as trunks, stems, and canopies, further act as frictional barriers that dissipate wave energy and attenuate storm surges (Daro Justine and Seenath 2025). Empirical studies in Vietnam and India have demonstrated that dense mangrove belts can reduce wave heights by 50–80% over distances of 100–300 m (e.g., Mazda et al. 1997; Bao 2011). These findings underscore that bioshields offer a cost-effective and self-sustaining form of coastal protection, in contrast to conventional hard-engineering structures such as seawalls and embankments, which typically entail high maintenance demands and may disrupt natural sediment transport processes (Veettil et al. 2021).

The protective role of bioshields extends to mitigating inland flooding and saltwater intrusion—phenomena increasingly aggravated by sea-level rise and extreme weather events. By absorbing wave energy and stabilizing coastal topography, vegetated buffers lower inundation depths during storms and reduce the salinization of agricultural lands (Alongi 2008; Narayan et al. 2017). This ecological service is particularly valuable in deltaic and low-lying regions such as the Mekong, Ganges–Brahmaputra, and Irrawaddy Deltas, where millions of people depend on coastal agriculture and aquaculture.

Beyond physical protection, bioshields make a profound contribution to climate regulation through their role in carbon sequestration (Hussain et al. 2025). The high productivity of mangroves, seagrasses, and salt marshes, combined with their capacity to trap organic matter in anaerobic sediments, results in significant long-term carbon storage (Alongi 2020). This dual function (i.e., coastal protection and climate mitigation) positions bioshields as climate-smart infrastructure, addressing both adaptation to environmental hazards and mitigation of greenhouse gas emissions.

Moreover, bioshields enhance local microclimates by moderating temperature extremes and acting as windbreaks (Enescu et al. 2025). They improve air and water quality by filtering pollutants, support groundwater recharge, and provide habitats for a range of flora and fauna (Sharma and Naik 2024). Biodiversity supported by mangroves and associated ecosystems, including fish, crustaceans, mollusks, and birds, plays a critical role in maintaining ecological balance and ensuring the resilience of coastal food webs (Veettil et al. 2018, 2019). The capacity of bioshields to regenerate, adapt, and self-sustain under changing environmental conditions makes them indispensable to the vision of sustainable coastal futures.

16.4.2 Socioeconomic and Cultural Benefits

While the ecological functions of bioshields are well recognized, their social and economic benefits are equally critical in ensuring the long-term sustainability of coastal management. Sustainable coastal futures cannot be achieved through ecological measures alone; they depend fundamentally on the well-being, inclusion, and active participation of local communities (Jennings et al. 2023; Marzo et al. 2023). Bioshields, when designed and managed in partnership with coastal residents, can generate multiple co-benefits that extend beyond environmental protection (Jennings et al. 2023; Marzo et al. 2023; Dulyakasem et al. 2026).

One of the most tangible advantages of coastal bioshields lies in livelihood enhancement (Dulyakasem et al. 2026). Mangroves and other coastal vegetation serve as breeding and nursery grounds for a wide variety of fish and shellfish species, supporting artisanal and small-scale fisheries that are essential sources of income and food security (Veettil et al. 2019). The detritus from mangrove leaf litter fuels coastal food chains, while the structural complexity of roots and canopies provides shelter for juvenile organisms (Hossain and Hoque 2008). In Vietnam, mangrove restoration initiatives led by the Vietnam Red Cross have yielded notable socio-economic benefits, including reduced dike maintenance costs and substantial increases in local fish and shrimp production resulting from enhanced water quality and habitat conditions (IFRC 2012).

Bioshields also provide opportunities for sustainable resource use through the production of honey, fuelwood, tannins, and timber (Veettil et al. 2023). When managed under community forestry schemes, these resources can supplement household incomes while fostering a sense of ownership and responsibility for ecosystem stewardship (Dulyakasem et al. 2026). In addition, bioshields have proven valuable for ecotourism and environmental education (Moussa et al. 2024).

Beyond economics, bioshields carry profound cultural and spiritual significance. Many coastal societies in Asia have long traditions of maintaining sacred groves, community forests, or vegetated coastal buffers as part of local belief systems and customary governance (e.g., Dudley et al. 2009; Veettil et al. 2018). These natural features often function as both physical and symbolic shields, protecting villages from storms while serving as spaces for communal gathering, rituals, and cultural continuity. Recognizing and integrating these cultural dimensions within contemporary restoration initiatives enhances legitimacy and local engagement. It reconnects communities to their heritage and reinforces the moral imperative to safeguard nature as a shared legacy.

From a governance perspective, community-based bioshield management offers a practical model for inclusive and participatory coastal governance (Dulyakasem et al. 2026). Empowering local communities to plan, plant, and monitor bioshields encourages collective stewardship, reduces conflict over resource use, and strengthens local capacity to adapt to changing environmental conditions (Lhosupasirirat et al. 2023). Such participatory models demonstrate that sustainability is not merely an environmental goal but a social process rooted in equity, trust, and collaboration.

16.4.3 Integration with Hybrid Engineering Approaches

Recent innovations in coastal protection have led to the emergence of hybrid engineering approaches, which combine the strengths of ecological and structural solutions (Narayan et al. 2016; Marino et al. 2025). These systems integrate bioshields with engineered features such as permeable breakwaters, low-crested barriers, and eco-revetments to achieve greater stability and resilience. While natural systems provide adaptability and regeneration, engineered components offer immediate protection and design predictability (Anderson et al. 2022).

Notable examples come from Indonesia (e.g., Demak District), where hybrid mangrove–dyke systems have been implemented to restore eroded mudflats and rebuild lost coastal land (Winterwerp et al. 2014; Sidik et al. 2021; van Bijsterveldt 2023). In Demak, permeable brushwood structures were installed parallel to the coast to reduce wave energy and encourage sediment deposition, creating favorable conditions for natural mangrove colonization (van Bijsterveldt 2023). Once vegetation was established, the mangroves took over the role of sediment stabilization and wave attenuation, significantly reducing erosion and flood risks.

For many Asian countries facing budgetary constraints and competing development priorities, hybrid bioshield approaches offer a pragmatic pathway toward sustainable coastal protection (Temmerman et al. 2013; Veettil et al. 2021). By blending engineering reliability with the regenerative power of ecosystems, they enable a transition toward infrastructure that not only safeguards against climate risks but also restores ecological integrity and enhances human well-being.

16.5 Governance and Policy Dimensions

16.5.1 Policy Frameworks for Nature-Based Coastal Management

The realization of sustainable coastal futures depends not only on ecological and technological innovation but also on robust and coherent governance frameworks that enable long-term planning, coordination, and investment (Elrick-Barr et al. 2024). In the context of nature-based solutions (NbS) such as bioshields, effective policy support is critical for mainstreaming these approaches into national adaptation and development agendas (Elliott et al. 2023). This requires policies that bridge environmental conservation, climate resilience, and socio-economic development—domains that are often managed in isolation within traditional bureaucratic systems.

Across Asia, there has been growing recognition of the role that coastal ecosystems play in climate adaptation, disaster risk reduction, and sustainable livelihoods (Pomeroy et al. 2006; Munang et al. 2013; Spalding et al. 2014). Several countries have taken important steps toward incorporating bioshields and ecosystem-based management into their national policy landscapes. For instance, Vietnam's

National Target Program on Climate Change (2008) identifies mangrove restoration and protection as a core strategy for adaptation to sea-level rise and storm impacts (PM 2008). The program promotes community participation in mangrove management and integrates coastal vegetation into broader land-use and disaster-preparedness plans. Complementing this, the Vietnam Forest Development Strategy (2006–2020) emphasizes mangrove afforestation and rehabilitation to enhance both ecological and socio-economic resilience in vulnerable coastal provinces (PM 2007).

In India, the Coastal Regulation Zone (CRZ) Notification, first issued in 1991 and periodically revised, provides a legal framework for the protection and sustainable use of coastal zones (Joseph and Balchand 2000; Panigrahi and Mohanty 2012). The CRZ recognizes the ecological importance of coastal vegetation, particularly mangroves, dunes, and coral reefs, and places restrictions on development activities in sensitive areas (Panigrahi and Mohanty 2012). It also encourages the conservation of greenbelts and shelterbelts as natural buffers against erosion and storms. Additionally, India's National Action Plan on Climate Change (NAPCC) and its state-level climate adaptation plans have begun to highlight the potential of NbS, including bioshields, as cost-effective adaptation tools that align with both environmental and developmental priorities (Dhyani et al. 2024).

Indonesia offers another compelling example through its National Mangrove Strategy, which promotes integrated coastal management and community-based stewardship of mangrove ecosystems (Damastuti and De Groot 2017; Miller and Tonoto 2023; Gunawan et al. 2025). The strategy links conservation with local livelihood enhancement, recognizing that sustainable management is only possible when communities derive tangible benefits from ecosystem protection (Miller and Tonoto 2023). The Indonesian government has also established a National Mangrove Rehabilitation Program, aiming to restore hundreds of thousands of hectares of degraded mangroves by 2045, in line with the country's net-zero emissions commitments (Nugroho et al. 2023).

Beyond national efforts, several regional and intergovernmental frameworks reinforce the policy momentum for ecosystem-based coastal management. The ASEAN Working Group on Coastal and Marine Environment, for instance, facilitates regional cooperation in mangrove restoration, pollution control, and marine spatial planning (Do et al. 2023). Similarly, the Partnerships in Environmental Management for the Seas of East Asia (PEMSEA) has been instrumental in promoting integrated coastal management (ICM) principles across more than a dozen Asian countries (Chua 2013). These collaborative initiatives emphasize knowledge exchange, capacity building, and cross-border and policy harmonization, which are crucial components in managing shared coastal ecosystems.

However, despite these advances, institutional fragmentation and policy incoherence remain significant challenges. In many countries, responsibilities for coastal management are divided among multiple agencies, typically environment, forestry, fisheries, and disaster management departments, leading to overlapping mandates and inefficient resource allocation (Botero et al. 2025). For example, in Vietnam, mangrove areas may simultaneously fall under the jurisdiction of forestry departments (for conservation), fisheries ministry (for aquaculture regulation), and disaster management authorities (for protective functions), resulting in conflicting priorities (Veettil et al. 2019).

To overcome these barriers, there is a pressing need for cross-sectoral coordination mechanisms that align objectives across ministries and administrative levels. Integrated coastal zone management (ICZM) and ecosystem-based management (EBM) frameworks offer practical pathways to harmonize these efforts (Delacámara et al. 2020; Malinconico 2025). Effective governance also requires strengthening local institutions, enabling community participation in decision-making, and ensuring that national policies translate into actionable local plans (De Oliveira et al. 2024). Only through such coherence can bioshields be systematically integrated into coastal adaptation, development, and conservation strategies.

Moreover, policy frameworks must explicitly recognize long-term maintenance and monitoring as essential components of bioshield success. Many coastal vegetation restoration projects fail after initial implementation due to the absence of sustained institutional and financial support (Bayraktarov et al. 2016; Lovelock et al. 2022). Embedding bioshield management into legal and regulatory systems, through zoning laws, environmental impact assessments, and disaster risk reduction strategies, can help ensure continuity and accountability (Veettil et al. 2021).

16.5.2 Financing Mechanisms and Economic Valuation

The successful implementation and scaling of bioshields depend on the establishment of sustainable financing mechanisms and the capacity to demonstrate their economic value relative to conventional engineering approaches (Narayan et al. 2016). Decision-makers often demand clear cost–benefit evidence to justify public investment in ecosystem-based interventions. Fortunately, a growing body of research indicates that bioshields not only deliver high returns on investment but also provide multiple co-benefits that are often unaccounted for in traditional infrastructure appraisals (Getzner and Islam 2020; Van Zelst et al. 2021; Huynh et al. 2024; Lee et al. 2025).

Economic valuation studies have shown that the benefit–cost ratios of bioshields can vary from 6:1 to 15:1 when the full suite of ecosystem services (e.g., coastal protection, fisheries enhancement, carbon sequestration, and water purification) is considered (Su et al. 2021; Zhong et al. 2022). In contrast, hard-engineering solutions like seawalls or levees typically provide a single service (protection) and often degrade over time (Hosseinzadeh et al. 2022). Bioshields, on the other hand, grow stronger as they mature, offering increasing ecological and social dividends (Veettil et al. 2021). These findings underscore that investing in nature is not only an environmental imperative but also an economically prudent strategy.

To operationalize this potential, innovative financing instruments are increasingly being explored. Payment for ecosystem services (PES) schemes reward communities for maintaining or restoring coastal vegetation that delivers protective or regulatory functions (Rakotomahazo et al. 2023). For instance, coastal villagers who manage mangrove forests could receive compensation from downstream beneficiaries such as aquaculture operators or tourism enterprises that rely on the ecological health of these ecosystems (Rahman et al. 2018; Malik et al. 2025).

Similarly, blue carbon finance represents a promising frontier. Because mangroves and other coastal ecosystems store large quantities of carbon, they are eligible for carbon credit schemes under voluntary and compliance markets (Karpowicz et al. 2024; Mitra et al. 2024). Countries like Indonesia, the Philippines, and Vietnam have begun developing blue carbon projects that link mangrove restoration to carbon offset programs, creating financial incentives for conservation (Gevaña et al. 2018). By quantifying and monetizing the carbon sequestration potential of bioshields, such initiatives can attract investment from both public and private sectors.

Public–private partnerships (PPPs) also offer viable pathways for financing large-scale restoration of coastal vegetation (Bennett et al. 2024). Governments can provide regulatory support and seed funding, while private enterprises contribute capital and technical expertise in exchange for long-term ecosystem service benefits. The tourism industry, for example, has a vested interest in preserving attractive and resilient coastlines, while insurance companies increasingly recognize the risk-reduction value of healthy ecosystems (Reguero et al. 2020). These synergies can be harnessed to develop innovative co-financing models.

Regional cooperation further enhances financing prospects. Organizations such as PEMSEA and the Asian Development Bank (ADB) facilitate funding and capacity-building initiatives that enable countries to share best practices and mobilize international resources (Whisnant and Vandeweerd 2019). Green and climate finance mechanisms, such as the Green Climate Fund (GCF) and the Global Environment Facility (GEF), have also begun supporting nature-based coastal adaptation projects, providing both grants and concessional loans for bioshield development (Herr 2015; Terton et al. 2022).

Ultimately, ensuring the financial sustainability of bioshields requires their integration into broader economic planning frameworks. National adaptation plans, disaster risk management budgets, and coastal zone development programs should incorporate dedicated allocations for ecosystem restoration and long-term maintenance (Temmerman et al. 2013). The establishment of trust funds or revolving financing mechanisms can further promote continuity and stability across political circles, thereby safeguarding the longevity of bioshield initiatives (Narayan et al. 2016).

16.6 Science, Technology, and Innovation for the Future

16.6.1 Remote Sensing and Modelling Tools

Advances in science and technology are fundamentally reshaping how coastal ecosystems are understood, monitored, and managed. Among these, remote sensing and modelling technologies have emerged as powerful enablers for the design, evaluation, and adaptive management of coastal bioshields (Veettil et al. 2021). The capacity to observe coastal dynamics continuously and at multiple spatial scales offers unprecedented opportunities for evidence-based decision-making, allowing managers to anticipate risks and optimize interventions (Narayan et al. 2016).

Modern remote sensing technologies, employing optical, radar, and LiDAR sensors, have become indispensable tools in mapping and monitoring coastal bioshield performance (Veettil et al. 2021). Optical imagery, such as Landsat, Sentinel-2, and MODIS data, enables the assessment of vegetation health, canopy density, and species distribution through indices like the Normalized Difference Vegetation Index (NDVI) (Veettil et al. 2020, 2024). Meanwhile, synthetic aperture radar (SAR) systems such as Sentinel-1 can penetrate cloud cover and operate day or night, making them particularly effective in humid tropical regions where optical visibility is often limited (Thomas et al. 2015). LiDAR technology further enhances this capacity by providing high-resolution topographic and canopy elevation data, essential for modeling storm-surge attenuation and sediment dynamics (Niwa et al. 2023; Wang et al. 2025).

In Vietnam, for example, Vu et al. (2022) used Landsat time-series imagery to monitor mangrove extent over time, offering valuable insights into the success of restoration initiatives and the effects of climate-related stressors. Such applications illustrate how spatially explicit datasets can be integrated into national adaptation strategies, enabling continuous monitoring and early detection of ecosystem degradation.

The power of remote sensing is amplified when coupled with hydrodynamic and ecological modelling tools. Models such as SWAN (Simulating WAves Nearshore), MIKE 21, and XBeach simulate the interaction among waves, tides, and coastal vegetation, helping scientists and policymakers predict how bioshields influence coastal hydrodynamics under various sea-level rise and storm scenarios (e.g., Cuc et al. 2015; Taylor-Burns et al. 2023; Marino et al. 2025). These tools are increasingly used to evaluate design configurations, such as vegetation density, species mix, and belt width, that optimize protection benefits while maintaining ecological integrity (Maiolo et al. 2020; Marino et al. 2025).

Beyond physical dynamics, ecological models are being developed to predict vegetation responses to changing salinity, inundation frequency, and sediment supply (Fagherazzi et al. 2012; LaFond-Hudson and Sulman 2023). When integrated with socio-economic data, these models can assess trade-offs between conservation and development goals, supporting multi-objective planning (Herr 2015). For instance, combined modelling frameworks can evaluate how bioshield expansion might simultaneously reduce disaster risk, enhance fisheries productivity, and contribute to carbon sequestration targets (Temmerman et al. 2013).

The future of bioshield science lies in integrated Earth observation systems that merge satellite, drone, and ground-based data with predictive models. As sensor resolution improves and data costs decline, near-real-time monitoring of coastal ecosystems will become increasingly feasible. These innovations will be vital for adaptive management, allowing restoration practitioners and decision-makers to respond dynamically to environmental feedback and changing climate conditions.

16.6.2 Genetic and Ecophysiological Research

While technological innovation enhances monitoring and modelling, understanding the biological and physiological foundations of bioshields remains equally critical. The resilience of coastal vegetation to stressors such as salinity, inundation, and temperature extremes is determined largely by species-specific adaptations and genetic diversity (Nizam et al. 2022). As climate change accelerates, identifying and cultivating resilient species and genotypes will be essential to ensure that bioshields can thrive under future conditions.

Research in India and Thailand has provided valuable insights into the adaptive traits of key mangrove species such as *Avicennia marina* and *Rhizophora apiculata*—two of the most salt- and flood-tolerant species found in tropical coastal systems (Kathiresan and Bingham 2001; Ruang-areerate et al. 2022). These species exhibit specialized physiological mechanisms, including salt-excreting glands, pneumatophores for oxygen uptake in waterlogged soils, and viviparous propagules that enhance seedling survival (Natarajan et al. 2021; Nizam et al. 2022). Such traits make them ideal candidates for bioshield establishment in degraded or high-stress environments where other vegetation might fail.

Advancements in genetic and biotechnological tools are opening new frontiers in bioshield science. Techniques such as DNA barcoding, microsatellite analysis, and genome sequencing are increasingly used to assess genetic diversity within and between mangrove populations (Maguire et al. 2000; Mao et al. 2021; Ma et al. 2022). Maintaining genetic diversity is vital for long-term adaptability, as it enables populations to respond to environmental shifts and disease pressures (Chung et al. 2023). This research also supports the identification of locally adapted ecotypes (genetically distinct populations suited to specific microhabitats), which can inform site-specific restoration strategies.

Furthermore, ecophysiological studies that examine plant responses to temperature and salinity gradients help refine species selection and planting design (Méndez-Alonzo et al. 2016). Controlled experiments and field trials reveal tolerance thresholds that guide restoration under different hydrological regimes (Hu et al. 2025). For example, understanding the limits of seedling establishment under increased tidal flooding or elevated soil salinity allows for adaptive planting schedules and improved survival rates (Kimera et al. 2024).

Emerging biotechnological innovations, such as microbial inoculants that enhance nutrient uptake, bioengineering of salt-tolerant strains, and the use of tissue culture for mass propagation, promise to improve the scalability and success of bioshield projects (Minh et al. 2020). However, such applications must be guided by ethical and ecological considerations, ensuring that genetic interventions complement rather than replace natural resilience processes (Sandler 2020).

Integrating genetic and physiological knowledge into bioshield planning bridges the gap between ecological theory and practical implementation. It ensures that restoration efforts are not only ecologically sound but also future-proof, able to withstand the environmental uncertainties of the twenty-first century.

16.6.3 Digital Platforms and Decision Support Systems

The growing complexity of coastal management calls for digital innovations that can integrate diverse data sources, facilitate collaboration, and support transparent decision-making. Digital platforms and decision support systems (DSS) are increasingly being used to synthesize information from remote sensing, field observations, and participatory mapping, thereby transforming raw data into actionable knowledge for policymakers and communities alike (Westmacott 2001; Koutalakis and Zaimes 2024).

One of the most significant developments has been the creation of open-access geospatial platforms dedicated to mangrove and coastal ecosystem monitoring. The World Mangrove Atlas (Spalding et al. 2010), for example, provides a comprehensive repository of spatial data on mangrove distribution, species composition, and ecosystem health globally. Similarly, the Global Mangrove Watch (GMW) integrates satellite data from multiple missions to produce high-resolution, regularly updated maps of mangrove extent and change (Bunting et al. 2018). These resources have become invaluable for tracking deforestation, restoration progress, and climate-related impacts, thereby supporting both local management and international reporting under frameworks such as the UNFCCC and SDGs.

At the national and sub-national levels, customized decision-support tools are being developed to aid coastal planning. These systems often integrate Geographic Information Systems (GIS) with ecological and socio-economic data, allowing users to visualize hazard exposure, land-use conflicts, and potential restoration sites. InVEST (Integrated Valuation of Ecosystem Services and Tradeoffs) is an example of open-source software models used to map and value the goods and services from nature (Galdino et al. 2023). Participatory components, such as mobile applications for community-based data collection, ensure that local knowledge is incorporated into formal decision-making processes.

Future innovation will depend on advancing interoperability and accessibility, ensuring that data systems can communicate across institutions and that communities can benefit from user-friendly tools. Cloud-based platforms and artificial intelligence (AI) are expected to play a major role in automating data analysis, forecasting coastal changes, and optimizing restoration designs. For example, machine-learning algorithms can already classify vegetation types from satellite imagery, detect early signs of degradation, and simulate the effectiveness of bioshield configurations under projected storm intensities (e.g., Kalinaki et al. 2023).

However, the success of digital tools depends not only on technological capacity but also on institutional collaboration and knowledge sharing. Building regional networks that connect scientists, planners, and local communities can ensure that technological advances translate into tangible on-the-ground outcomes. In the long term, advances in science, technology, and innovation, ranging from genetic research to satellite monitoring and digital decision-support systems, will shape the next generation of bioshield planning. Leveraging these tools will enable adaptive, data-driven, and participatory coastal management, ensuring that bioshields remain resilient components of sustainable coastal futures amid accelerating environmental change.

16.7 Community and Stakeholder Engagement

16.7.1 Co-management and Participatory Governance

The long-term sustainability of bioshields depends as much on the strength of community involvement and local governance as on ecological and technical factors. Coastal protection is inherently a socio-ecological challenge—one that requires a shared responsibility among government institutions, local communities, and other stakeholders (UNEP 2021). The concept of co-management, or shared stewardship, has therefore become central to modern approaches for managing coastal ecosystems (Ho 2023). It recognizes that local people are not merely beneficiaries of environmental projects but active agents with invaluable knowledge, experience, and vested interests in sustaining their environments (Berkes 2004).

Community-Based Forest Management (CBFM) models in countries such as Vietnam (Tran et al. 2019) and the Philippines (Pulhin et al. 2017) provide powerful examples of how participatory approaches can enhance both ecological outcomes and social equity. In these models, communities are formally recognized as co-managers of coastal forests, with defined rights and responsibilities over the use, protection, and rehabilitation of mangrove and coastal vegetation (Bown et al. 2013). This institutional arrangement encourages local stewardship while maintaining oversight and technical support from government agencies. In Vietnam, for instance, CBFM programs have led to the restoration of thousands of hectares of mangroves, reduced illegal cutting, and improved income diversification through sustainable aquaculture and eco-tourism ventures to some extent (Veettil et al. 2024; Ebeler et al. 2025).

Co-management arrangements succeed because they build on mutual accountability and shared incentives. When communities participate directly in decision-making through consultations, local committees, or participatory planning exercises, they are more likely to commit to long-term conservation (Campbell and Vainio-Mattila 2003; Ananda 2007; Chirenje et al. 2013). Conversely, when decisions are made without their input, restoration projects often face resistance, neglect, or failure once external funding ends (Hai et al. 2020). Participatory governance ensures that the benefits and responsibilities of bioshield management are equitably distributed, strengthening social cohesion and local ownership.

Participatory mapping is one of the most effective tools that allows local communities to delineate areas for protection, restoration, and livelihood use based on their own spatial knowledge and priorities (Yen et al. 2018; Kyem 2021). This process not only enhances transparency but also reduces conflict over resource boundaries. When combined with geospatial technologies such as GPS and GIS, participatory mapping can bridge the gap between traditional knowledge and scientific data, enabling planners to design bioshields that align with both ecological and social realities (Loc et al. 2021). This method has been successfully applied in Vietnam (Yen et al. 2018) and Indonesia (Damastuti and De Groot 2019).

Another key component of participatory governance is the establishment of benefit-sharing mechanisms (Santarlacci et al. 2024). For communities to remain invested in bioshield maintenance, they must derive tangible and equitable benefits from these ecosystems. Benefit-sharing may include sustainable harvesting rights for timber and non-timber products, access to eco-tourism revenues, or payments for environmental services (PES) (Pham et al. 2013; Wunder 2015). In the Philippines, for example, community associations involved in mangrove reforestation are often granted legal tenure or usufruct rights, allowing them to profit from sustainable resource use while maintaining conservation standards (Ramos and Osorio 2013; Theriault 2025). Such arrangements not only incentivize local engagement but also provide a safeguard against elite capture and resource overexploitation.

Inclusive governance structures are essential for ensuring that all community members, especially women, youth, and marginalized groups, have a voice in coastal management. Women play vital roles in resource management, aquaculture, and household-level adaptation strategies, yet they are frequently underrepresented in decision-making processes (Pravalprukskul and Resurreccion 2018). Encouraging gender-balanced participation in bioshield governance not only improves equity but also brings diverse perspectives that enhance adaptive capacity (De la Torre-Castro et al. 2017; UNEP-GWA 2022).

In sum, co-management represents a practical and ethical approach to coastal resilience. It decentralizes authority, builds trust between communities and governments, and integrates local wisdom into formal management systems. Ultimately, the sustainability of bioshields and coastal ecosystems depends on people seeing themselves as guardians of their environment rather than passive recipients of external aid.

16.7.2 Education, Awareness, and Capacity Building

While participatory governance provides the institutional foundation for sustainable bioshield management, education and capacity building provide the human foundation. Empowering communities with knowledge, skills, and awareness is essential for ensuring that bioshield initiatives are effectively implemented and maintained over time (Thupalli and Deen 2018; Hajar 2025). Local people must understand not only how to plant and protect vegetation but also why such actions matter for their long-term safety, livelihoods, and climate resilience (Hai et al. 2020).

Educational programs can take many forms, ranging from formal curricula and community workshops to informal awareness campaigns and hands-on field training. Environmental education that integrates coastal ecology, climate change, and sustainable resource management helps nurture a generation of informed citizens capable of making responsible decisions (Miao and Nduneseokwu 2025). In schools and universities, experiential learning, such as mangrove planting days, ecological monitoring projects, and student-led restoration initiatives, can encourage a sense of environmental stewardship from an early age.

In practice, capacity building often involves technical training in key areas such as mangrove nursery management, site selection, planting techniques, and biodiversity monitoring (Wickramasinghe 2017; Sigit et al. 2019; Hai et al. 2020). For example, in the Mekong Delta, collaborative efforts between universities, non-governmental organizations (NGOs), and local authorities have developed training programs tailored to the needs of farmers, fishers, and community leaders (Triyanti et al. 2024). These programs not only transfer scientific knowledge but also promote local innovations in mixed-use systems that integrate mangrove conservation with aquaculture and agriculture (McSherry et al. 2023). By linking education with livelihoods, such initiatives make restoration both ecologically and economically sustainable.

Another important dimension is extension services, which bridge scientific expertise and community practice (Agbayani and Corre 2000). Extension officers and NGO facilitators play a crucial role in translating complex ecological concepts into accessible and actionable guidance. Through regular site visits, participatory workshops, and peer-to-peer exchanges, they help build a network of local champions capable of leading bioshield initiatives independently. This approach fosters a culture of continuous learning and adaptation, which is vital in the face of evolving environmental challenges.

Public awareness campaigns also contribute to building widespread societal support for bioshields (O'Leary et al. 2023). Media outreach, community events, and storytelling that highlight local success stories can inspire broader participation and policy backing. Visual tools, such as maps, videos, and interactive models, make abstract concepts like sea-level rise or carbon sequestration more tangible, helping people connect their daily experiences to global environmental processes (Sheppard 2012).

A particularly promising approach involves engaging the youth and schools as catalysts for change. Young people are often the most enthusiastic participants in restoration projects and can act as bridges between scientific institutions and communities (Chawla 2024). Programs that combine environmental education with volunteerism and citizen science, such as water-quality monitoring or drone-based mangrove mapping, empower youth to contribute meaningfully to conservation while acquiring valuable skills (Ballard et al. 2017).

Capacity building should also extend to local institutions and governance structures. Training municipal officials, planners, and technical staff in ecosystem-based management, climate adaptation, and participatory planning ensures that bioshield initiatives are supported by informed policy implementation (CARE 2015). Building institutional capacity helps move beyond project-based interventions toward sustained, programmatic approaches integrated into national and local development agendas (CARE 2015).

Ultimately, education and capacity development are the engines that drive long-term change. They cultivate a sense of ownership, enhance technical competence, and foster a shared vision of sustainable coastal stewardship. As seen in regions like the Mekong Delta (Triyanti et al. 2024), where collaborative learning between universities, NGOs, and local communities has revitalized degraded landscapes, knowledge and participation are inseparable from ecological restoration.

16.8 Challenges and Future Directions

Despite their proven ecological and socio-economic value, the widespread adoption of bioshields as a cornerstone of sustainable coastal management continues to face numerous challenges. These obstacles span ecological, institutional, financial, and socio-political dimensions, reflecting the complexity of coastal systems and the competing demands placed upon them. Recognizing and addressing these challenges is essential for scaling up bioshield implementation and ensuring effectiveness in a rapidly changing climate.

16.8.1 Persistent Challenges

One of the most pressing issues faced by the bioshield establishment is land-use conflict in coastal areas. In many Asian coastal regions, mangrove and other bioshield areas have been cleared to make way for aquaculture ponds, agricultural expansion, tourism facilities, and infrastructure development (Veettil et al. 2020, 2021). These competing land uses often reflect short-term economic priorities that undervalue the long-term protective and ecological functions of coastal vegetation. Reversing this trend requires policies and incentives that integrate conservation goals into broader land-use planning, as well as mechanisms to compensate or incentivize landholders who maintain protective greenbelts (Veettil et al. 2019).

Funding shortfalls are another major barrier to sustainable coastal bioshield management (Abelson et al. 2020). While many restoration projects receive initial donor or government support, long-term maintenance, monitoring, and adaptive management are frequently underfunded. Without sustained investment, planted areas may degrade or fail to reach full maturity, thereby undermining protective capacity (Hai et al. 2020). Innovative financing mechanisms, such as blue carbon credits, payments for ecosystem services (PES), or public–private partnerships (PPP), are needed to provide continuous funding streams that align economic interests with conservation outcomes.

At the ecological level, the establishment of bioshields faces several biophysical limitations. In highly degraded or eroded coastal zones, seedling survival rates can be low due to unstable substrates, high salinity, or inadequate freshwater inflows (Fillyaw et al. 2021; Canning and Duke 2025). Monoculture planting, often used for expediency, may also reduce ecosystem resilience by limiting species diversity (Hutchison et al. 2018; Li et al. 2025). To overcome these constraints, restoration strategies must be based on sound ecological principles, emphasizing native species selection, hydrological restoration, and site-specific adaptive management. Incorporating scientific monitoring into project design can help identify early warning signs of stress and guide timely interventions.

Governance fragmentation further complicates bioshield implementation. Coastal management responsibilities are often dispersed across multiple ministries, such as environment, forestry, fisheries, and disaster management, resulting in overlapping mandates and weak coordination (Botero et al. 2025). This institutional fragmentation can lead to inefficiencies, conflicting regulations, and inconsistent enforcement. Strengthening interagency coordination and promoting integrated coastal zone management (ICZM) frameworks are crucial to overcoming these structural barriers (Botero et al. 2025).

16.8.2 Pathways for the Future

Looking ahead, several strategic directions can enhance the effectiveness and sustainability of bioshields. First, there is a pressing need for integrated coastal landscape planning that reconciles conservation, development, and livelihood objectives (Estrada-Carmona et al. 2014). By adopting a landscape-scale approach, policymakers can ensure that bioshields form part of a connected network of ecosystems that deliver multiple benefits, ranging from shoreline protection and carbon sequestration to biodiversity conservation and community resilience (Yang et al. 2024). Spatial planning tools that incorporate ecological corridors, zoning, and stakeholder consultation can help balance competing land uses in a transparent and participatory manner (Ros-Tonen et al. 2021).

Second, advances in hybrid engineering offer exciting opportunities to enhance the protective efficiency of bioshields. Combining natural vegetation with eco-friendly infrastructure, such as permeable dykes or sediment traps, can accelerate land accretion and strengthen long-term stability (Van Zelst et al. 2021; Geng et al. 2025). Such hybrid systems integrate the adaptability of nature-based solutions with the reliability of engineered defenses, providing a more resilient and cost-effective alternative to traditional hard infrastructure (Van Zelst et al. 2021; Geng et al. 2025).

Third, the future of bioshield management depends on regional collaboration and transboundary governance. Many coastal ecosystems, such as the mangrove belts along the Bay of Bengal or the South China Sea (East Sea), span multiple national boundaries. Therefore, effective management requires coordinated regional frameworks for knowledge exchange, monitoring, and policy alignment. Platforms such as the Partnerships in Environmental Management for the Seas of East Asia (PEMSEA) (Chua 2013) can play a vital role in fostering such cooperation, promoting shared learning, and mobilizing resources for cross-border restoration initiatives.

Finally, incorporating climate projections into bioshield design is essential to ensure long-term resilience. Restoration projects must anticipate future sea-level rise, temperature increases, and salinity shifts rather than relying solely on historical conditions (Griggs and Reguero 2021). Scenario-based modelling can guide decisions about species selection, planting elevation, and buffer width to optimize performance under future climate scenarios (Grenfell et al. 2019; Zhang et al. 2023). By aligning bioshield planning with climate adaptation strategies and nationally determined contributions (NDCs) under the Paris Agreement, countries can simultaneously advance their coastal protection and climate mitigation goals (Taillardat et al. 2018).

16.9 Vision for Sustainable Coastal Futures in Asia

A sustainable coastal future for Asia envisions shorelines that are resilient, inclusive, and ecologically balanced—landscapes in which human development is strategically planned and managed in accordance with natural processes, thereby fostering long-term socio-ecological coexistence (Mimura 2008). In this vision, coastal areas are not merely zones of economic activity or settlement but dynamic ecosystems that sustain biodiversity, support livelihoods, and protect from the growing threats of climate change (Harvey 2007). The foundation of such a future lies in recognizing that natural systems, when properly managed and restored, are not obstacles to development but essential allies in securing long-term prosperity and safety for coastal populations (Giffin et al. 2021).

Within this framework, coastal vegetation bioshields are envisioned as the backbone of green infrastructure for coastal resilience (Kamil et al. 2021; Thoi et al. 2025). They represent a living, self-sustaining form of protection that complements engineered defenses such as dikes, seawalls, and flood barriers (Morris et al. 2023; Jayson-Quashigah et al. 2025). By buffering wave energy, stabilizing sediments, and storing carbon, bioshields not only safeguard human settlements but also regenerate the ecological integrity of coastal zones (Jayson-Quashigah et al. 2025). Moreover, they create opportunities for sustainable livelihoods through fisheries enhancement, ecotourism, and small-scale forest product enterprises, ensuring that conservation directly contributes to human well-being (Luom et al. 2021; Basyuni et al. 2022). To realize this vision, a set of strategic directions must guide the transformation of policy, practice, and regional cooperation as discussed below.

a. Scaling Up Restoration Through National and Regional Partnerships

The priority is to expand the scale and scope of bioshield restoration efforts. While many successful pilot projects exist across Asia, varying from the mangrove reforestation programs in Vietnam's Mekong Delta (Tinh et al. 2022) to the community-led initiatives in the Philippines (Valenzuela et al. 2020) and Indonesia (Damastuti et al. 2023), these need to be scaled up and coordinated through national and regional partnerships. Governments should integrate bioshield restoration into their broader climate adaptation and biodiversity strategies, ensuring alignment with the UN Sustainable Development Goals (SDGs), the Paris Agreement, and the Kunming–Montreal Global Biodiversity Framework (UNEP 2021).

Regional platforms, such as the ASEAN Working Group on Coastal and Marine Environment (Sabatira 2020) and PEMSEA (Partnerships in Environmental Management for the Seas of East Asia) (Chua 2013), can play pivotal roles in facilitating knowledge exchange, standardizing monitoring frameworks, and mobilizing joint financing mechanisms. Shared data platforms and capacity-building programs will help countries learn from each other's successes and challenges, creating a collective movement towards sustainable coastal resilience (Van Dyke and Broder 2016).

b. Integrating Bioshields into Climate Adaptation Finance Mechanisms

Securing long-term financial sustainability is critical for the success of bioshield programs. Integrating bioshields into climate adaptation and mitigation finance mechanisms can unlock new sources of funding while strengthening their policy relevance (Li and Liu 2025). Instruments such as blue carbon credits, PES, and green bonds can channel both public and private investment into coastal restoration.

For example, recognizing mangroves and other coastal vegetation as blue carbon assets under global carbon markets allows governments and communities to monetize the carbon sequestration benefits of bioshields (Bertram et al. 2021; Montenegro et al. 2025). Similarly, adaptation funds, such as the Green Climate Fund (GCF), can support large-scale restoration efforts that align with national climate commitments (Owens et al. 2025). Embedding bioshields within these financing frameworks not only ensures continuity but also repositions them as integral to economic and environmental resilience.

c. Enhancing Cross-disciplinary Collaboration

Building sustainable coastal futures requires collaboration across disciplines that have traditionally worked in isolation. Ecologists, engineers, economists, and social scientists must work together to design and implement bioshields that are both technically effective and socially inclusive (Temmerman and Kirwan 2015). Ecologists provide insight into species selection and habitat restoration; engineers contribute to hybrid design innovations; social scientists facilitate participatory governance and community engagement; and economists assess cost-effectiveness and long-term returns (Narayan et al. 2016; Rohr et al. 2018).

Cross-disciplinary research and education programs, such as joint university initiatives, regional research networks, and multi-sector task forces, should be promoted to generate integrated solutions (Shusheng et al. 2023; Mittal and Bansal 2024). By fostering such collaboration, Asia can lead in developing coastal protection models that merge science, culture, and technology in pursuit of shared sustainability goals.

d. Embedding Bioshields in National Development and Disaster Risk Reduction Frameworks

Coastal bioshields must be embedded in the core frameworks of national development planning and disaster risk reduction (DRR) (Thupalli and Deen 2018). Too often, coastal greenbelts are treated as peripheral environmental projects rather than as critical infrastructure. By recognizing bioshields as part of national adaptation infrastructure, governments can secure policy continuity, attract investment, and ensure integration with land-use planning, agriculture, fisheries, and tourism sectors (Rudianto et al. 2022; Singha et al. 2024).

Aligning coastal bioshields with DRR strategies also enhances their role in safeguarding lives and assets (Rudianto et al. 2022). For instance, incorporating vegetated buffer zones into hazard zoning regulations can reduce vulnerability to storm surges and flooding, while also providing co-benefits such as biodiversity conservation and recreation (Lan and Hsu 2021).

In essence, the vision for sustainable coastal futures in Asia rests on a transformative paradigm—one that values nature not as an expendable resource but as a partner in resilience and development. Through strategic collaboration, innovative financing, scientific integration, and policy mainstreaming, bioshields can evolve from isolated restoration projects into cornerstone elements of a holistic, climate-resilient coastal landscape. In this envisioned future, Asia's coasts stand not only as frontlines of vulnerability but as symbols of regeneration and sustainability, where people and nature thrive together.

16.10 Conclusions

Achieving sustainable coastal futures requires a paradigmatic shift from resisting natural processes to aligning management practices with them. Bioshields exemplify this transition, functioning as dynamic, living interfaces between land and sea that offer protection, enhance productivity, and support ecological regeneration. Beyond serving as physical buffers against storms and erosion, bioshields reflect a broader ethos of coexistence: they contribute to coastal resilience, sustain local livelihoods, and promote the restoration of ecological integrity.

In Asia, where densely populated and economically critical coastlines face escalating pressures from sea-level rise, land subsidence, and increasingly frequent extreme weather events, bioshields present a multifunctional and future-oriented intervention. Their benefits extend beyond disaster risk reduction to include climate mitigation through carbon sequestration, biodiversity conservation, and the enhancement of community well-being. By integrating ecological and social priorities, bioshields illustrate how nature-based solutions can support equitable, resilient, and adaptive pathways for coastal development.

Realizing this potential, however, requires sustained commitment and coordinated action. Governments, scientists, engineers, and local communities must collaborate to embed bioshields within broader national adaptation and development frameworks. Ongoing research and technological innovation are essential for improving design, implementation, and monitoring, while inclusive governance structures and long-term financing mechanisms are necessary to ensure that restoration initiatives persist beyond individual project cycles. Moreover, empowering communities through education and co-management will help anchor bioshields within the social and cultural fabric of coastal life, thereby enhancing their long-term effectiveness and legitimacy.

Ultimately, the future of Asia's coasts hinges on a collective acknowledgment that nature is not an impediment to development but a critical partner in fostering resilience and renewal. Investing in bioshields represents not only an environmentally responsible course of action but also an ethical and economic necessity—one that offers the prospect of safer, more sustainable, and more prosperous coastlines for future generations. By advancing this partnership between human society and natural systems, Asia is well-positioned to lead global efforts toward truly sustainable coastal futures.

References

Abelson A, Reed DC, Edgar GJ, Smith CS, Kendrick GA, Orth RJ, Airoldi L, Silliman B, Beck MW, Krause G, Shashar N, Stambler N, Nelson P (2020) Challenges for restoration of coastal marine ecosystems in the Anthropocene. Front Mar Sci 7:544105. https://doi.org/10.3389/fmars.2020.544105

Aburto MO, de los Angeles Carvajal M, Barr B, Barbier EB, Boesch DF, Boyd J, Worcester T et al (2012) Ecosystem-based management for the oceans. Island Press, p 368

Agbayani RF, Corre KG (2000) SEAFDEC/AQD experience in mangrove-friendly aquaculture training and extension. In: Primavera JH, Garcia LMB, Castaños MT, Surtida MB (eds) Mangrove-friendly aquaculture. Proceedings of the Workshop on mangrove-friendly aquaculture organized by the SEAFDEC Aquaculture Department, January 11–15, 1999, Iloilo City, Philippines. Southeast Asian Fisheries Development Center, Aquaculture Department, Tigbauan, Iloilo, Philippines, pp 185–190

Alongi DM (2008) Mangrove forests: resilience, protection from tsunamis, and responses to global climate change. Estuar Coast Shelf Sci 76(1):1–13. https://doi.org/10.1016/j.ecss.2007.08.024

Alongi DM (2020) Carbon balance in salt marsh and mangrove ecosystems: a global synthesis. J Mar Sci Eng 8(10):767. https://doi.org/10.3390/jmse8100767

Ananda J (2007) Implementing participatory decision making in forest planning. Environ Manag 39:534–544. https://doi.org/10.1007/s00267-006-0031-2

Anderson CC, Renaud FG, Hanscomb S, Gonzalez-Ollauri A (2022) Green, hybrid, or grey disaster risk reduction measures: what shapes public preferences for nature-based solutions? J Environ Manag 310:114727. https://doi.org/10.1016/j.jenvman.2022.114727

Andreasson E, Bechauf R (2025) Sustainable asset valuation (Savi) of mangroves in Vietnam. https://www.iisd.org/publications/report/sustainable-asset-valuation-mangroves-vietnam

Ballard HL, Dixon CG, Harris EM (2017) Youth-focused citizen science: examining the role of environmental science learning and agency for conservation. Biol Cons 208:65–75. https://doi.org/10.1016/j.biocon.2016.05.024

Bao QT (2011) Effect of mangrove forest structures on wave attenuation in coastal Vietnam. Oceanologia 53(3):807–818. https://doi.org/10.5697/oc.53-3.807

Basyuni M, Sasmito SD, Analuddin K, Ulqodry TZ, Saragi-Sasmito MF, Eddy S, Milantara N (2022) Mangrove biodiversity, conservation and roles for livelihoods in Indonesia. In: Mangroves: biodiversity, livelihoods and conservation. Springer Nature Singapore, Singapore, pp 397–445. https://doi.org/10.1007/978-981-19-0519-3_16

Baumann L, Riechers M, Celliers L, Ferse SC (2023) Anticipating and transforming futures: a literature review on transdisciplinary coastal research in the Global South. Ecosyst People 19(1):2288957. https://doi.org/10.1080/26395916.2023.2288957

Bayraktarov E, Saunders MI, Abdullah S, Mills M, Beher J, Possingham HP, Mumby PJ, Lovelock CE (2016) The cost and feasibility of marine coastal restoration. Ecol Appl 26(4):1055–1074. https://doi.org/10.1890/15-1077

Becker M, Seeger K, Paszkowski A, Marcos M, Papa F, Almar R, Bates P, France-Lanord C, Hossain MS, Khan UMJ, Karegar MA, Karpytchev M, Long N, Minderhoud JPS, Neal J, Nicholls RJ, Syvitski J (2024) Coastal flooding in Asian megadeltas: recent advances, persistent challenges, and call for actions amidst local and global changes. Rev Geophys 62(4):e2024RG000846. https://doi.org/10.1029/2024RG000846

Bennett M, March A, Failler P (2024) Blue economy financing solutions for the fisheries and aquaculture sectors of Caribbean Island states. Fishes 9(8):305. https://doi.org/10.3390/fishes9080305

Berkes F (2004) Rethinking community-based conservation. Conserv Biol 18(3):621–630. https://doi.org/10.1111/j.1523-1739.2004.00077.x

Bertram C, Quaas M, Reusch TB, Vafeidis AT, Wolff C, Rickels W (2021) The blue carbon wealth of nations. Nat Clim Chang 11(8):704–709. https://doi.org/10.1038/s41558-021-01089-4

Botero CM, Suman DO, Milanes CB (2025) The multiple challenges faced by coastal and marine governance. Water 17(15):2322. https://doi.org/10.3390/w17152322

Bown N, Gray TS, Stead SM (2013) Contested forms of governance in marine protected areas: a study of co-management and adaptive co-management. Routledge

Bunting P, Rosenqvist A, Lucas RM, Rebelo L, Hilarides L, Thomas N, Hardy A, Itoh T, Shimada M, Finlayson CM (2018) The global mangrove watch—a new 2010 global baseline of mangrove extent. Remote Sens 10(10):1669. https://doi.org/10.3390/rs10101669

Bunting P, Rosenqvist A, Hilarides L, Lucas RM, Thomas N, Tadono T, Worthington TA, Spalding M, Murray NJ, Rebelo L (2022) Global mangrove extent change 1996–2020: global mangrove watch version 3.0. Remote Sens 14(15):3657. https://doi.org/10.3390/rs14153657

Campbell LM, Vainio-Mattila A (2003) Participatory development and community-based conservation: opportunities missed for lessons learned? Human Ecol 31:417–437. https://doi.org/10.1023/A:1025071822388

Canning AD, Duke NC (2025) Assessing threats and rehabilitation opportunities for mangrove-saltmarsh blue carbon ecosystems. Ocean Coast Manag 271:107974. https://doi.org/10.1016/j.ocecoaman.2025.107974

CARE (2015) Building coastal resilience in Vietnam: an integrated, community-based approach to mangrove management, disaster risk reduction, and climate change adaptation. CARE International in Vietnam, 80 pp

Chawla L (2024) Young people's experiences of participation in ecological restoration. Environ Educ Res 30(6):881–899. https://doi.org/10.1080/13504622.2023.2203426

Chee SY, Firth LB, Then AY, Yee JC, Mujahid A, Affendi YA, Amir AA, Lau CM, Ooi JL, Quek YA, Tan CE, Yap TK, Yeap CA (2021) Enhancing uptake of nature-based solutions for informing coastal sustainable development policy and planning: a Malaysia case study. Front Ecol Evol 9:708507. https://doi.org/10.3389/fevo.2021.708507

Chirenje LI, Giliba RA, Musamba EB (2013) Local communities' participation in decision-making processes through planning and budgeting in African countries. Chin J Popul Resour Environ 11(1):10–16. https://doi.org/10.1080/10042857.2013.777198

Choudhury M, Haque CE, Hostetler G (2021) Transformative learning and community resilience to cyclones and storm surges: the case of coastal communities in Bangladesh. Int J Disaster Risk Reduct 55:102063. https://doi.org/10.1016/j.ijdrr.2021.102063

Chua TE (2013) Coastal and ocean governance in the seas of East Asia: PEMSEA's experience. Coast Manag 41(2):99–119. https://doi.org/10.1080/08920753.2013.768517

Chung MY, Merilä J, Li J, Mao K, Tsumura Y, Chung MG (2023) Neutral and adaptive genetic diversity in plants: an overview. Front Ecol Evol 11:1116814. https://doi.org/10.3389/fevo.2023.1116814

Cohen-Shacham E, Walters G, Janzen C, Maginnis S (eds) (2016) Nature-based solutions to address global societal challenges. IUCN, Gland, xiii + 97 pp

Cohen-Shacham E, Andrade A, Dalton J, Dudley N, Jones M, Kumar C, Maginnis S, Maynard S, Nelson CR, Renaud FG, Welling R, Walters G (2019) Core principles for successfully implementing and upscaling nature-based solutions. Environ Sci Policy 98:20–29. https://doi.org/10.1016/j.envsci.2019.04.014

Cooley S, Schoeman D, Bopp L, Boyd P, Donner S, Kiessling W, Martinetto P, Ojea E, Racault MF, Rost B, Skern-Mauritzen M (2023) Oceans and coastal ecosystems and their services. Supplementary material. See IPCC 2022, pp. 3SM-1–68. https://www.ipcc.ch/report/ar6/wg2/downloads

Cuc NTK, Suzuki T, van Steveninck EDDR, Hai H (2015) Modelling the impacts of mangrove vegetation structure on wave dissipation in Ben Tre Province, Vietnam, under different climate change scenarios. J Coast Res 31(2):340–347. https://doi.org/10.2112/JCOASTRES-D-12-00271.1

Damastuti E, De Groot R (2017) Effectiveness of community-based mangrove management for sustainable resource use and livelihood support: a case study of four villages in Central Java, Indonesia. J Environ Manag 203:510–521. https://doi.org/10.1016/j.jenvman.2017.07.025

Damastuti E, De Groot R (2019) Participatory ecosystem service mapping to enhance community-based mangrove rehabilitation and management in Demak, Indonesia. Reg Environ Chang 19:65–78. https://doi.org/10.1007/s10113-018-1378-7

Damastuti E, Van Wesenbeeck BK, Leemans R, De Groot RS, Silvius MJ (2023) Effectiveness of community-based mangrove management for coastal protection: A case study from Central Java, Indonesia. Ocean Coast Manag 238:106498. https://doi.org/10.1016/j.ocecoaman.2023.106498

Daro Justine YE, Seenath A (2025) Vegetative nature-based solutions for coastal flood risk management: benefits, challenges, and uncertainties. Ocean Coast Manag 261:107520. https://doi.org/10.1016/j.ocecoaman.2024.107520

De la Torre-Castro M, Fröcklin S, Börjesson S, Okupnik J, Jiddawi NS (2017) Gender analysis for better coastal management—increasing our understanding of social-ecological seascapes. Mar Policy 83:62–74. https://doi.org/10.1016/j.marpol.2017.05.015

De Oliveira M, Morrison T, O'Brien KR, Lovelock CE (2024) Governance of coastal wetlands: beyond the community conservation paradigm. Ocean Coast Manag 255:107253. https://doi.org/10.1016/j.ocecoaman.2024.107253

Delacámara G, O'Higgins TG, Lago M, Langhans S (2020) Ecosystem-based management: moving from concept to practice. In: Ecosystem-based management, ecosystem services and aquatic biodiversity: theory, tools and applications. Springer International Publishing, Cham, pp 39–60. https://doi.org/10.1007/978-3-030-45843-0_3

Dhyani S, Thummarukudy M, Gupta AK (2024) Nature based solutions for disaster risk reduction: concepts and overview. In: Disaster risk and management under climate change. Springer Nature Singapore, Singapore, pp. 557–579. https://doi.org/10.1007/978-981-99-4105-6_27

Do HA, Pham TV, Van Le TT, Vu TM, Truong TL, Nguyen TH, Le TS et al (2023) ASEAN's soft law on marine environmental pollution. VMOST J Soci Sci Human 65(3):99–108. https://doi.org/10.31276/VMOSTJOSSH.65(3).99-108

Donato DC, Kauffman JB, Murdiyarso D, Kurnianto S, Stidham M, Kanninen M (2011) Mangroves among the most carbon-rich forests in the tropics. Nat Geosci 4(5):293–297. https://doi.org/10.1038/ngeo1123

Dudley N, Higgins-Zogib L, Mansourian S (2009) The links between protected areas, faiths, and sacred natural sites. Conserv Biol 23(3):568–577. https://doi.org/10.1111/j.1523-1739.2009.01201.x

Dulyakasem R, Brooks C, Lehnert SL, Newton P (2026) Community-based natural resource management in coastal communities: the contribution of mangroves to household livelihoods in southern Thailand. Environ Dev 57:101340. https://doi.org/10.1016/j.envdev.2025.101340

Ebeler L, Balke T, Laurie E, Le H, Nguyen T, Renaud FG (2025) Local-scale impacts of mangrove restoration and conservation on coastal communities in two Vietnamese deltas: socio-economic and institutional dynamics. Ocean Coast Manag 269:107812. https://doi.org/10.1016/j.ocecoaman.2025.107812

Elliott M, Borja Á, Cormier R (2023) Managing marine resources sustainably—ecological, societal and governance connectivity, coherence and equivalence in complex marine transboundary regions. Ocean Coast Manag 245:106875. https://doi.org/10.1016/j.ocecoaman.2023.106875

Elrick-Barr CE, Thomsen DC, Smith TF (2024) Governance innovations in the coastal zone: towards social-ecological resilience. Environ Sci Policy 153:103687. https://doi.org/10.1016/j.envsci.2024.103687

Enescu CM, Mihalache M, Ilie L, Dinca L, Constandache C, Murariu G (2025) Agricultural benefits of shelterbelts and windbreaks: a bibliometric analysis. Agriculture 15(11):1204. https://doi.org/10.3390/agriculture15111204

Estrada-Carmona N, Hart AK, DeClerck FA, Harvey CA, Milder JC (2014) Integrated landscape management for agriculture, rural livelihoods, and ecosystem conservation: an assessment of experience from Latin America and the Caribbean. Landsc Urban Plan 129:1–11. https://doi.org/10.1016/j.landurbplan.2014.05.001

Fagherazzi S, Kirwan ML, Mudd SM, Guntenspergen GR, Temmerman S, Rybczyk JM, Reyes E, Craft C, Clough J (2012) Numerical models of salt marsh evolution: ecological, geomorphic, and climatic factors. Rev Geophys 50(1). https://doi.org/10.1029/2011RG000359

Falkenrich P, Wilson J, Nistor I, Goseberg N, Cornett A, Mohammadian A (2021) Nature-based coastal protection by large woody debris as compared to seawalls: a physical model study of beach morphology and wave reflection. Water 13(15):2020. https://doi.org/10.3390/w13152020

FAO (2018a) The state of the world's forests 2018—forest pathways to sustainable development. Rome. Licence: CC BY-NC-SA 3.0 IGO

FAO (2018b) The state of world fisheries and aquaculture. Food and Agriculture Organization of the United Nations

Fillyaw RM, Donnelly MJ, Litwak JW, Rifenberg JL, Walters LJ (2021) Strategies for successful mangrove living shoreline stabilizations in shallow water subtropical estuaries. Sustainability 13(21):11704. https://doi.org/10.3390/su132111704

Foley MM, Halpern BS, Micheli F, Armsby MH, Caldwell MR, Crain CM, Prahler E, Rohr N, Sivas D, Beck MW, Carr MH, Crowder LB, Emmett Duffy J, Hacker SD, McLeod KL, Palumbi SR, Peterson CH, Regan HM, Ruckelshaus MH, Steneck RS et al (2010) Guiding ecological

principles for marine spatial planning. Mar Policy 34(5):955–966. https://doi.org/10.1016/j.marpol.2010.02.001
Galdino S, Silva R, Zolin C, Tosto S, Quartaroli C, Pereira L, Gomes M, Renan Ricardo Silva ET, Kevin Hyslop ET (2023) InVEST-integrated valuation of ecosystem services and tradeoffs. https://www.sidalc.net/search/Record/dig-infoteca-e-doc-1157748/Description
Geng L, Dou Z, Gu Y, Cao H, Yao P, Zhou Z, Shao J (2025) The synergistic effect between engineering measures and ecological measures in coastal erosion prevention. Front Mar Sci 12:1671752. https://doi.org/10.3389/fmars.2025.1671752
Gerona-Daga MEB, Salmo SG III (2022) A systematic review of mangrove restoration studies in Southeast Asia: challenges and opportunities for the United Nation's decade on ecosystem restoration. Front Mar Sci 9:987737. https://doi.org/10.3389/fmars.2022.987737
Getzner M, Islam MS (2020) Ecosystem services of mangrove forests: results of a meta-analysis of economic values. Int J Environ Res Public Health 17(16):5830. https://doi.org/10.3390/ijerph17165830
Gevaña DT, Camacho LD, Pulhin JM (2018) Conserving mangroves for their blue carbon: insights and prospects for community-based mangrove management in Southeast Asia. In: Threats to mangrove forests: hazards, vulnerability, and management. Springer International Publishing, Cham, pp 579–588. https://doi.org/10.1007/978-3-319-73016-5_26
Giffin AL, Brown CJ, Nalau J, Mackey BG, Connolly RM (2021) Marine and coastal ecosystem-based adaptation in Asia and Oceania: review of approaches and integration with marine spatial planning. Pac Conserv Biol 27:104–117. https://doi.org/10.1071/PC20025
Gijón Mancheño A, Jafino BA, Hofland B, Van Wesenbeeck BK, Kazi S, Urrutia I (2024) Nature meets infrastructure: the role of mangroves in strengthening Bangladesh's coastal flood defenses. Sustainability 17(4):1567. https://doi.org/10.3390/su17041567
Gilman EL, Ellison J, Duke NC, Field C (2008) Threats to mangroves from climate change and adaptation options: a review. Aquat Bot 89(2):237–250. https://doi.org/10.1016/j.aquabot.2007.12.009
Grenfell SE, Fortune F, Mamphoka MF, Sanderson N (2019) Coastal wetland resilience to climate change: modelling ecosystem response to rising sea level and salinity in a variable climate. Anthropocene Coasts 2(1):1–20. https://doi.org/10.1139/anc-2018-0004
Griggs G, Reguero BG (2021) Coastal adaptation to climate change and sea-level rise. Water 13(16):2151. https://doi.org/10.3390/w13162151
Guannel G, Arkema K, Ruggiero P, Verutes G (2016) The power of three: coral reefs, seagrasses and mangroves protect coastal regions and increase their resilience. PLoS ONE 11(7):e0158094. https://doi.org/10.1371/journal.pone.0158094
Gunawan H, Basyuni M, Subarudi Suharti S, Kustanti A, Wahyuni T, Effendi R et al (2025) Empowering conservation: the transformative role of mangrove education in Indonesia's climate strategies. For Sci Technol 21(4):374–396. https://doi.org/10.1080/21580103.2025.2519475
Hai N, Dell B, Phuong V, Harper RJ (2020) Towards a more robust approach for the restoration of mangroves in Vietnam. Ann For Sci 77:18. https://doi.org/10.1007/s13595-020-0921-0
Hajar S (2025) Community empowerment in strengthening coastal ecosystems based on ocean guardian village in realizing an eco-friendly village in Paluh Manan village. Int J Health Econ Soc Sci (IJHESS) 7(2):972–978. https://doi.org/10.56338/ijhess.v7i2.7881
Harvey N (ed) (2007) Global change and integrated coastal management: the Asia-Pacific region, vol 10. Springer Science & Business Media, p 352
Herr D (coord.) (2015) Coastal "blue" carbon: a revised guide to supporting coastal wetland programs and projects using climate finance and other financial mechanisms. International Union for Conservation of Nature and Natural Resources, Conservation International, Wetlands International, The Nature Conservancy, Forest Trends and the Nicholas Institute for Environmental Policy Solutions, Gland. https://bvearmb.do/handle/123456789/2841
Hilmi N, Chami R, Sutherland MD, Hall-Spencer JM, Lebleu L, Benitez MB, Levin LA (2021) The role of blue carbon in climate change mitigation and carbon stock conservation. Front Clim 3:710546. https://doi.org/10.3389/fclim.2021.710546

Ho N (2023) A review of coastal resources co-management in Southeast Asia. In: Baldwin C, van Bommel S (eds) Rural development for sustainable social-ecological systems. Palgrave studies in environmental sustainability. Palgrave Macmillan, Cham. https://doi.org/10.1007/978-3-031-34225-7_14

Hongwiset S, Rodtassana C, Poungparn S, Umnouysin S, Suchewaboripont V (2022) Synergetic roles of mangrove vegetation on sediment accretion in coastal mangrove plantations in Central Thailand. Forests 13(10):1739. https://doi.org/10.3390/f13101739

Hossain M, Hoque AF (2008) Litter production and decomposition in mangroves—a review. Indian J For 31(2):227–238. https://doi.org/10.54207/bsmps1000-2008-8TS8TD

Hosseinzadeh N, Ghiasian M, Andiroglu E, Lamere J, Rhode-Barbarigos L, Sobczak J, Sealey KS, Suraneni P (2022) Concrete seawalls: a review of load considerations, ecological performance, durability, and recent innovations. Ecol Eng 178:106573. https://doi.org/10.1016/j.ecoleng.2022.106573

Hu K, Wang W, Qian W, Sheng N, Cheng J, Xiong Y (2025) Responses of biomass and allometric growth equations of Juvenile mangrove plants to salinity, flooding, and aboveground competition. Horticulturae 11(7):712. https://doi.org/10.3390/horticulturae11070712

Hussain FS, Bilal H, Al-Ansari T (2025) Quantifying the role of mangroves in carbon sequestration in Qatar. Eco Inform 90:103332. https://doi.org/10.1016/j.ecoinf.2025.103332

Hutchison C, Gravel D, Guichard F, Potvin C (2018) Effect of diversity on growth, mortality, and loss of resilience to extreme climate events in a tropical planted forest experiment. Sci Rep 8(1):15443. https://doi.org/10.1038/s41598-018-33670-x

Huynh LT, Su J, Wang Q, Stringer LC, Switzer AD, Gasparatos A (2024) Meta-analysis indicates better climate adaptation and mitigation performance of hybrid engineering-natural coastal defence measures. Nat Commun 15(1):2870. https://doi.org/10.1038/s41467-024-46970-w

IFRC (2012) Mangrove plantation in Viet Nam: measuring impact and cost benefit. International Federation of Red Cross and Red Crescent Societies, Hanoi

IPCC (2007) Impacts, adaptation and vulnerability. Contribution of working group II to assessment report four of the intergovernmental panes of climate change (IPCC). https://archive.ipcc.ch/publications_and_data/ar4/wg2/en/ch10s10-4-3.html

Jayson-Quashigah PN, Staneva J, Chen W, Djath B, Mahu E, Appeaning Addo K (2025) Evaluating mangroves as nature-based solutions for coastal protection under current and future sea level rise scenarios. Front Mar Sci 12:1526082. https://doi.org/10.3389/fmars.2025.1526082

Jennings V, San Antonio KM, Brown MJ, Choice L, Simpson Q, Ford I, Cho HJ, Solis P, Lacey A, Robinson D (2023) Place-based conservation in coastal and marine ecosystems: the importance of engagement with underrepresented communities. Sustainability 16(22):9965. https://doi.org/10.3390/su16229965

Jordan P, Fröhle P (2022) Bridging the gap between coastal engineering and nature conservation? J Coast Conserv 26:4. https://doi.org/10.1007/s11852-021-00848-x

Joseph K, Balchand A (2000) The application of coastal regulation zones in coastal management—appraisal of Indian experience. Ocean Coast Manag 43(6):515–526. https://doi.org/10.1016/S0964-5691(00)00041-7

Juanico DE (2022) Does mangrove restoration imply coastal protection? A prospective simulation study. Front Mar Sci 9:968420. https://doi.org/10.3389/fmars.2022.968420

Kalinaki K, Malik OA, Lai DTC, Sukri RS, Wahab RBHA (2023) Spatial-temporal mapping of forest vegetation cover changes along highways in Brunei using deep learning techniques and Sentinel-2 images. Eco Inform 77:102193. https://doi.org/10.1016/j.ecoinf.2023.102193

Kamil EA, Takaijudin H, Hashim AM (2021) Mangroves as coastal bio-shield: a review of mangroves performance in wave attenuation. Civ Eng J 7(11):1964–1981. https://doi.org/10.28991/cej-2021-03091772

Karpowicz DA, Mohan M, Watt MS, Montenegro JF, King SA, Selvam PP, Nithyanandan M, Robyn B, Ali T, Abdullah MM, Doaemo W, Ewane EB (2024) Mangrove-based carbon market projects: 15 considerations for engaging and supporting local communities. Diversity 16(9):574. https://doi.org/10.3390/d16090574

Kathiresan K, Bingham BL (2001) Biology of mangroves and mangrove ecosystems. Adv Mar Biol 40:81–251. https://doi.org/10.1016/S0065-2881(01)40003-4

Key IB, Smith AC, Turner B, Chausson A, Girardin CA, Macgillivray M, Seddon N (2022) Biodiversity outcomes of nature-based solutions for climate change adaptation: characterising the evidence base. Front Environ Sci 10:905767. https://doi.org/10.3389/fenvs.2022.905767

Khan WR, Mohamad AL, Ibrahim FH (2025) Biodiversity, health, and ecosystem services of mangroves. Forests 16(8):1278. https://doi.org/10.3390/f16081278

Kimera F, Sobhi B, Omara M, Sewilam H (2024) Impact of salinity gradients on seed germination, establishment, and growth of two dominant mangrove species along the Red Sea coastline. Plants 13(24):3471. https://doi.org/10.3390/plants13243471

Koutalakis P, Zaimes G (2024) The development of an online decision support system to select optimal nature-based solutions to protect streams and the sea. Platforms 2(3):118–137. https://doi.org/10.3390/platforms2030008

Kusumadewi SD, Purnomo H, Nadhira S, Putrizulfan R (2024) Systematic review on the implementation of mangrove community-based restoration in Indonesia and beyond. IOP Conf Ser Earth Environ Sci 1315:012052. https://doi.org/10.1088/1755-1315/1315/1/012052

Kyem PAK (2021) Managing natural resource conflicts with participatory mapping and PGIS applications. Springer, Cham, p 260

LaFond-Hudson S, Sulman B (2023) Modeling strategies and data needs for representing coastal wetland vegetation in land surface models. New Phytol 238(3):938–951. https://doi.org/10.1111/nph.18760

Lan Y, Hsu T (2021) Planning and management of coastal buffer zones in Taiwan. Water 13(20):2925. https://doi.org/10.3390/w13202925

Lee H, Kim H, Park E, Lee B (2025) Beyond carbon: a systematic review of multiple ecosystem services of mangroves. J Coast Conserv 29:58. https://doi.org/10.1007/s11852-025-01148-4

Lhosupasirirat P, Dahdouh-Guebas F, Hugé J, Wodehouse D, Enright J (2023) Stakeholder perceptions on community-based ecological mangrove restoration (CBEMR): a case study in Thailand. Restor Ecol 31(5):e13894. https://doi.org/10.1111/rec.13894

Li P, Liu D (2025) How blue carbon financing can sustain blue carbon ecosystems protection and restoration: a proposed conceptual framework for the blue carbon financing mechanism. Ocean Coast Manag 265:107644. https://doi.org/10.1016/j.ocecoaman.2025.107644

Li Z, Li Y, Fan B (2025) Diverse restored mangrove stands enhance carbon storage compared to monospecific plantation: a meta-analysis. For Ecol Manag 578:122446. https://doi.org/10.1016/j.foreco.2024.122446

Loc HH, Park E, Thu TN, Diep NTH, Can NT (2021) An enhanced analytical framework of participatory GIS for ecosystem services assessment applied to a Ramsar wetland site in the Vietnam Mekong Delta. Ecosyst Serv 48:101245. https://doi.org/10.1016/j.ecoser.2021.101245

Long RD, Charles A, Stephenson RL (2015) Key principles of marine ecosystem-based management. Mar Policy 57:53–60. https://doi.org/10.1016/j.marpol.2015.01.013

Lovelock CE, Barbier E, Duarte CM (2022) Tackling the mangrove restoration challenge. PLoS Biol 20(10):e3001836. https://doi.org/10.1371/journal.pbio.3001836

Luom TT, Phong NT, Smithers S, Van Tai T (2021) Protected mangrove forests and aquaculture development for livelihoods. Ocean Coast Manag 205:105553. https://doi.org/10.1016/j.ocecoaman.2021.105553

Ma D, Ding Q, Guo Z, Xu C, Liang P, Zhao Z, Zheng HL et al (2022) The genome of a mangrove plant, Avicennia marina, provides insights into adaptation to coastal intertidal habitats. Planta 256(1):6. https://doi.org/10.1007/s00425-022-03916-0

Maguire TL, Saenger P, Baverstock P, Henry R (2000) Microsatellite analysis of genetic structure in the mangrove species *Avicennia marina* (Forsk.) Vierh.(Avicenniaceae). Mol Ecol 9(11):1853–1862. https://doi.org/10.1046/j.1365-294x.2000.01089.x

Maiolo M, Mel RA, Sinopoli S (2020) A simplified method for an evaluation of the effect of submerged breakwaters on wave damping: the case study of Calabaia Beach. J Mar Sci Eng 8(7):510. https://doi.org/10.3390/jmse8070510

Malik A, Ichsan Ali M, Rasyid Jalil A, Mannan A, Musyawarah R (2025) Promoting sustainable mangrove tourism through payments for ecosystem services: insights from Tongke-Tongke Village, South Sulawesi, Indonesia. Reg Sustain 6(2):100213. https://doi.org/10.1016/j.regsus.2025.100213

Malinconico N (2025) Towards beach spatial planning: ecosystem-based management as a background for holistic frameworks. PhD thesis, Universidade de São Paulo, Brazil

Mao X, Xie W, Li X, Shi S, Guo Z (2021) Establishing community-wide DNA barcode references for conserving mangrove forests in China. BMC Plant Biol 21(1):571. https://doi.org/10.1186/s12870-021-03349-z

Marino M, Nasca S, Alkharoubi AI, Cavallaro L, Foti E, Musumeci RE (2025) Efficacy of nature-based solutions for coastal protection under a changing climate: a modelling approach. Coast Eng 198:104700. https://doi.org/10.1016/j.coastaleng.2025.104700

Marzo RR, Chen HW, Anuar H, Abdul Wahab MK, Ibrahim MH, Ariffin IA, Ahmad AI, Kawuki J, Aljuaid M (2023) Effect of community participation on sustainable development: an assessment of sustainability domains in Malaysia. Front Environ Sci 11:1268036. https://doi.org/10.3389/fenvs.2023.1268036

Masselink G, Lazarus ED (2019) Defining coastal resilience. Water 11(12):2587. https://doi.org/10.3390/w11122587

Mazda Y, Magi M, Kogo M, Nguyen PH (1997) Mangroves as a coastal protection from waves in the Tong King delta, Vietnam. Mangroves Salt Marshes 1:127–135. https://doi.org/10.1023/A:1009928003700

McNamara KE, Buggy L (2017) Community-based climate change adaptation: a review of academic literature. Local Environ 22(4):443–460. https://doi.org/10.1080/13549839.2016.1216954

McSherry M, Davis RP, A D, Ahmadia GN, Van Kempen M, Wingard Brian S (2023) Integrated mangrove aquaculture: the sustainable choice for mangroves and aquaculture? Front For Glob Change 6:1094306. https://doi.org/10.3389/ffgc.2023.1094306

Méndez-Alonzo R, López-Portillo J, Moctezuma C, Bartlett MK, Sack L (2016) Osmotic and hydraulic adjustment of mangrove saplings to extreme salinity. Tree Physiol 36:1562–1572. https://doi.org/10.1093/treephys/tpw073

Menéndez P, Losada IJ, Narayan S, Beck MW (2020) The global flood protection benefits of mangroves. Sci Rep 10(1):4404. https://doi.org/10.1038/s41598-020-61136-6

Miao Q, Nduneseokwu C (2025) Advancing environmental leadership: education, development, and training. In: Environmental leadership in a VUCA era: an interdisciplinary handbook, pp 467–538. https://doi.org/10.1007/978-981-96-0324-4_10

Miller MA, Tonoto P (2023) Leveraging plural valuations of mangroves for climate interventions in Indonesia. Sustain Sci 18:1533–1547. https://doi.org/10.1007/s11625-023-01297-1

Mimura N (ed) (2008) Asia-Pacific coasts and their management: states of environment, vol 11. Springer Science & Business Media, p 383

Minh D, Nguyen TT, Hung S, Huang E, Huang C (2020) Roles of plant growth-promoting rhizobacteria (PGPR) in stimulating salinity stress defense in plants: a review. Int J Mol Sci 22(6):3154. https://doi.org/10.3390/ijms22063154

Mishra AK, Dey A, Mishra A, Mohakud SK, Farooq SH (2025) Seascape connectivity with mangroves positively influences tropical saltmarsh blue carbon stocks. Sci Total Environ 969:178929. https://doi.org/10.1016/j.scitotenv.2025.178929

Mitra A, Zaman S, Pramanick P, Ahmed S (2024) Carbon credit: concept and structure in context to mangroves of lower Gangetic Delta. In: Mangrove carbon trading in the lower Gangetic Delta: drawbacks and the potential. Springer Nature Switzerland, Cham, pp 149–208. https://doi.org/10.1007/978-3-031-76316-8_4

Mittal P, Bansal R (2024) Role of cross-disciplinary collaborations for holistic approach to sustainability. In: Community engagement for sustainable practices in higher education. Palgrave Macmillan, Cham. https://doi.org/10.1007/978-3-031-63981-4_9

Monioudi IN, Vousdoukas MI, Giardino A, Stocchino A, Mentaschi L, Feyen L (2025) Impacts of sea level rise and adaptation across Asia and the Pacific. Sci Rep 15(1):35742. https://doi.org/10.1038/s41598-025-11517-6

Montenegro JF, Mohan M, Ewane EB, Friess DA, Selvam PP, Dutta Roy A, Althauser A, Tidwell J, Gamboa-Cutz J, Shoot C, Watt MS, Macreadie PI, Karpowicz DA, Perera N, Atwood T, Chatting M, Burt JA, Udagedara S, Hendy I, Alongi DM et al (2025) Mangrove-based carbon market projects: current trends and future perspectives. For Policy Econ 181:103658. https://doi.org/10.1016/j.forpol.2025.103658

Morris RL, Fest B, Stokes D, Jenkins C, Swearer SE (2023) The coastal protection and blue carbon benefits of hybrid mangrove living shorelines. J Environ Manag 331:117310. https://doi.org/10.1016/j.jenvman.2023.117310

Moussa LG, Mohan M, Burmeister N, King SA, Burt JA, Rog SM, Watt MS, Udagedara S, Sujud L, Montenegro JF, Heng JE, Carvalho SA, Ali T, Veettil BK, Arachchige PS, Albanai JA, Sidik F, Shaban A, Peñaranda ML, Ewane EB et al (2024) Mangrove ecotourism along the coasts of the Gulf Cooperation Council countries: a systematic review. Land 13(9):1351. https://doi.org/10.3390/land13091351

Munang R, Thiaw I, Alverson K, Liu J, Han Z (2013) The role of ecosystem services in climate change adaptation and disaster risk reduction. Curr Opin Environ Sustain 5(1):47–52. https://doi.org/10.1016/j.cosust.2013.02.002

Narayan S, Beck MW, Reguero BG, Losada IJ, van Wesenbeeck B, Pontee N et al (2016) The effectiveness, costs and coastal protection benefits of natural and nature-based defences. PLoS ONE 11(5):e0154735. https://doi.org/10.1371/journal.pone.0154735

Narayan S, Beck MW, Wilson P, Thomas CJ, Guerrero A, Shepard CC, Reguero BG, Franco G, Ingram JC, Trespalacios D (2017) The value of coastal wetlands for flood damage reduction in the Northeastern USA. Sci Rep 7(1):9463. https://doi.org/10.1038/s41598-017-09269-z

Natarajan P, Murugesan AK, Govindan G, Gopalakrishnan A, Kumar R, Duraisamy P, Balaji R, Shyamli PS, Parida AK, Parani M (2021) A reference-grade genome identifies salt-tolerance genes from the salt-secreting mangrove species Avicennia marina. Commun Biol 4(1):851. https://doi.org/10.1038/s42003-021-02384-8

Neumann B, Vafeidis AT, Zimmermann J, Nicholls RJ (2015) Future coastal population growth and exposure to sea-level rise and coastal flooding—a global assessment. PLoS ONE 10(3):e0118571. https://doi.org/10.1371/journal.pone.0118571

Niwa H, Ise H, Kamada M (2023) Suitable LiDAR platform for measuring the 3D structure of mangrove forests. Remote Sens 15(4):1033. https://doi.org/10.3390/rs15041033

Nizam A, Meera SP, Kumar A (2022) Genetic and molecular mechanisms underlying mangrove adaptations to intertidal environments. Iscience 25(1):103547. https://doi.org/10.1016/j.isci.2021.103547

Nugroho HY, Indrajaya Y, Astana S, Suharti S, Basuki TM, Yuwati TW, Putra PB, Narendra BH, Abdulah L, Setyawati T, Krisnawati H, Saputra MH, Lisnawati Y, Garsetiasih R, Sawitri R, Putri IA, Setiawan O, Octavia D, Tata HL, Rahmila YI et al (2023) A chronicle of Indonesia's forest management: a long step towards environmental sustainability and community welfare. Land 12(6):1238. https://doi.org/10.3390/land12061238

O'Leary BC, Fonseca C, Cornet CC, De Vries MB, Degia A, Failler P, Furlan E, Garrabou J, Gil A, Hawkins JP, Krause-Jensen D, Le Roux X, Peck MA, Pérez G, Queirós AM, Różyński G, Sanchez-Arcilla A, Simide R, Sousa Pinto I, Roberts CM et al (2023) Embracing nature-based solutions to promote resilient marine and coastal ecosystems. Nat Based Solut 3:100044. https://doi.org/10.1016/j.nbsj.2022.100044

Owens K, Carter G, Park S, Viney G (2025) Bridging the adaptation-finance gap: pathways for the green climate fund in the Pacific. Earth Syst Gov 24:100247. https://doi.org/10.1016/j.esg.2025.100247

Panigrahi JK, Mohanty PK (2012) Effectiveness of the Indian coastal regulation zones provisions for coastal zone management and its evaluation using SWOT analysis. Ocean Coast Manag 65:34–50. https://doi.org/10.1016/j.ocecoaman.2012.04.023

Paxton AB, Riley TN, Steenrod CL, Puckett BJ, Alemu IJB, Paliotti ST, Silliman BR et al (2024) Evidence on the performance of nature-based solutions interventions for coastal protection in biogenic, shallow ecosystems: a systematic map. Environ Evid 13(1):28. https://doi.org/10.1186/s13750-024-00350-5

Pham TT, Bennett K, Vu TP, Brunner J, Le ND, Nguyen DT (2013) Payments for forest environmental services in Vietnam: from policy to practice. Occasional Paper 93. CIFOR, Bogor. https://doi.org/10.17528/cifor/004247. https://www.cifor-icraf.org/knowledge/publication/4247/

Pomeroy RS, Ratner BD, Hall SJ, Pimoljinda J, Vivekanandan V (2006) Coping with disaster: rehabilitating coastal livelihoods and communities. Mar Policy 30(6):786–793. https://doi.org/10.1016/j.marpol.2006.02.003

Pravalprukskul P, Resurreccion B (2018) Gender in coastal and marine resource management: a literature review. Stockholm Environment Institute, Stockholm

Primavera JH, Esteban JMA (2008) A review of mangrove rehabilitation in the Philippines: successes, failures and future prospects. Wetl Ecol Manag 16(5):345–358. https://doi.org/10.1007/s11273-008-9101-y

Prime Minister of Vietnam (PM) (2007) Approving Vietnam's forestry development strategy in the 2006–2020 period. Decision No. 18/2007/QD-TTg. Hanoi

Prime Minister of Vietnam (PM) (2008) Decision No. 158/2008/QD-TTg dated December 2, 2008, on Approving the National Target Program. https://chinhphu.vn/?pageid=27160&docid=81233&tagid=4&type=1 (in Vietnamese)

Pulhin JM, Gevaña DT, Pulhin FB (2017) Community-based mangrove management in the Philippines: experience and challenges in the context of changing climate. In: DasGupta R, Shaw R (eds) Participatory mangrove management in a changing climate. Disaster risk reduction. Springer, Tokyo. https://doi.org/10.1007/978-4-431-56481-2_16

Rahman MM, Jiang Y, Irvine K (2018) Assessing wetland services for improved development decision-making: a case study of mangroves in coastal Bangladesh. Wetl Ecol Manag 26:563–580. https://doi.org/10.1007/s11273-018-9592-0

Rakotomahazo C, Ranivoarivelo NL, Razanoelisoa J, Todinanahary GGB, Ranaivoson E, Remanevy ME, Ravaoarinorotsihoarana LA, Lavitra T (2023) Exploring the policy and institutional context of a payment for ecosystem services (PES) scheme for mangroves in southwestern Madagascar. Mar Policy 148:105450. https://doi.org/10.1016/j.marpol.2022.105450

Ramos GE, Osorio RLE (2013) REDD+ in the Philippines: legal status and conservation of mangrove forests in the Philippines. Int J Rural Law Policy 1:1–12. https://doi.org/10.5130/ijrlp.i1.2013.3359

Reguero BG, Beck MW, Schmid D, Stadtmüller D, Raepple J, Schüssele S, Pfliegner K (2020) Financing coastal resilience by combining nature-based risk reduction with insurance. Ecol Econ 169:106487. https://doi.org/10.1016/j.ecolecon.2019.106487

Richards DR, Friess DA (2016) Rates and drivers of mangrove deforestation in Southeast Asia, 2000–2012. Proc Natl Acad Sci 113(2):344–349. https://doi.org/10.1073/pnas.1510272113

Rohr JR, Bernhardt ES, Cadotte MW, Clements WH (2018) The ecology and economics of restoration. Ecol Soc 23(2):15. https://doi.org/10.5751/ES-09876-230215

Rölfer L, Celliers L, Fernandes M, Rivers N, Snow B, Abson DJ (2024) Assessing collaboration, knowledge exchange, and stakeholder agency in coastal governance to enhance climate resilience. Reg Environ Change 24(1):6. https://doi.org/10.1007/s10113-023-02163-7

Rosenzweig C, Solecki WD, Romero-Lankao P, Mehrotra S, Dhakal S, Ibrahim SA (eds) (2018) Climate change and cities: second assessment report of the urban climate change research network. Cambridge University Press, p 350

Ros-Tonen MAF, Willemen L, McCall MK (2021) Spatial tools for integrated and inclusive landscape governance: toward a new research agenda. Environ Manag 68:611–618. https://doi.org/10.1007/s00267-021-01547-x

Ruang-areerate P, Naktang C, Kongkachana W, Sangsrakru D, Narong N, Maknual C, Pravinvongvuthi T, Promchoo W, Yamprasai S, Tangphatsornruang S et al (2022) Assessment of the genetic diversity and population structure of Rhizophora apiculata Blume (Rhizophoraceae) in Thailand. Biology 11(10):1449. https://doi.org/10.3390/biology11101449

Rudianto R, Darmawan V, Isdianto A, Bintoro G (2022) Restoration of coastal ecosystems as an approach to the integrated mangrove ecosystem management and mitigation and adaptation to climate changes in north coast of East Java. J Coast Conserv 26(4):37. https://doi.org/10.1007/s11852-022-00865-4

Ruslan NFN, Goh HC, Hattam C, Edwards-Jones A, Moh HH (2022) Mangrove ecosystem services: contribution to the well-being of the coastal communities in Klang Islands. Mar Policy 144:105222. https://doi.org/10.1016/j.marpol.2022.105222

Sabatira F (2020) Southeast Asia regional cooperation on tackling marine plastic litter. Lampung J Int Law 2(2):69–84. https://doi.org/10.25041/lajil.v2i2.2033

Sandler R (2020) The ethics of genetic engineering and gene drives in conservation. Conserv Biol 34(2):378–385. https://doi.org/10.1111/cobi.13407

Santarlacci AD, Angelo H, Souza ÁN, Lima MD, Joaquim MS, Miguel EP, Carneiro JD (2024) Benefit sharing governance framework: pathways for financial benefit sharing in traditional communities. Sustainability 16(7):2650. https://doi.org/10.3390/su16072650

Sharma LK, Naik R (2024) Wetland ecosystems. In: Conservation of saline wetland ecosystems. Springer, Singapore. https://doi.org/10.1007/978-981-97-5069-6_1

Sheppard SR (2012) Visualizing climate change: a guide to visual communication of climate change and developing local solutions. Routledge

Shusheng X, Jingqian X, Mianrun C (2023) Integrated management improves emerging coastal industries and ecological restoration with the participation of social capital. Front Mar Sci 9:1015262. https://doi.org/10.3389/fmars.2022.1015262

Sidik F, Kusuma DW, Priyono B, Proisy C, Lovelock CE (2021) Managing sediment dynamics through reintroduction of tidal flow for mangrove restoration in abandoned aquaculture ponds. In: Dynamic sedimentary environments of mangrove coasts. Elsevier, pp 563–582. https://doi.org/10.1016/B978-0-12-816437-2.00004-5

Sigit DV, Miarsyah M, Komala R, Suryanda A, Fadrikal R, Ichsan IZ (2019) Improvement of knowledge and attitude in conservation of mangrove and coral reefs through environmental education community network model. J Phys Conf Ser 1317:12201. https://doi.org/10.1088/1742-6596/1317/1/012201

Singha P, Dasgupta R, Hashimoto S, Mitra BK (2024) An integrated coastal exposure modelling approach to assist mangrove ecosystem-based disaster risk reduction (Eco-DRR) in Tamil Nadu, India. Prog Disaster Sci 24:100388. https://doi.org/10.1016/j.pdisas.2024.100388

Slinger J, Stive M, Luijendijk A (2021) Nature-based solutions for coastal engineering and management. Water 13(7):976. https://doi.org/10.3390/w13070976

Spalding M, Kainuma M, Collins L (2010) World atlas of mangroves. Routledge

Spalding MD, Ruffo S, Lacambra C, Meliane I, Hale LZ, Shepard CC, Beck MW (2014) The role of ecosystems in coastal protection: adapting to climate change and coastal hazards. Ocean Coast Manag 90:50–57. https://doi.org/10.1016/j.ocecoaman.2013.09.007

Spencer T, Adams J, Le Tissier M, Murray AB, Splinter K (2023) Coastal futures: new framings, many questions, some ways forward. Camb Prisms Coast Futures 1:e32. https://doi.org/10.1017/cft.2023.22

Su J, Friess DA, Gasparatos A (2021) A meta-analysis of the ecological and economic outcomes of mangrove restoration. Nat Commun 12(1):5050. https://doi.org/10.1038/s41467-021-25349-1

Subhani R, Ahmad MM (2019) Socio-economic impacts of cyclone Aila on Bangladesh. Geosciences 9(11):482. https://doi.org/10.3390/geosciences9110482

Susilo S, Salman R, Hermawan W, Widyaningrum R, Wibowo ST, Ardhitasari Y, Meilano I, Yun S (2023) GNSS land subsidence observations along the northern coastline of Java, Indonesia. Sci Data 10(1):421. https://doi.org/10.1038/s41597-023-02274-0

Taillardat P, Friess DA, Lupascu M (2018) Mangrove blue carbon strategies for climate change mitigation are most effective at the national scale. Biol Lett 14(10):20180251. https://doi.org/10.1098/rsbl.2018.0251

Taylor-Burns R, Nederhoff K, Lacy JR, Barnard PL (2023) The influence of vegetated marshes on wave transformation in sheltered estuaries. Coast Eng 184:104346. https://doi.org/10.1016/j.coastaleng.2023.104346

Temmerman S, Kirwan ML (2015) Building land with a rising sea. Science 349(6248):588–589. https://doi.org/10.1126/science.aac8312

Temmerman S, Meire P, Bouma TJ, Herman PM, Ysebaert T, De Vriend HJ (2013) Ecosystem-based coastal defence in the face of global change. Nature 504(7478):79–83. https://doi.org/10.1038/nature12859

Terton A, Tsioumani E, Förster J, Morchain D (2022) Synergies between biodiversity- and climate-relevant policy frameworks and their implementation. Deutsche Gesellschaft für Internationale Zusammenarbeit (GIZ), International Institute for Sustainable Development (IISD), Helmholtz Centre for Environmental Research (UFZ)

Theriault N (2025) A forest of dreams: capitalism, conservation, and indigenous rights in the Philippines. University of Hawaii Press, p 232

Thoi HV, Quang LT, Kiet BNT, Hoan NM, Vy DTP (2025) Research on techniques to improve biomass and quality of mangroves in response to climate change in Can Gio coastal area, Ho Chi Minh City. J For Sci. https://doi.org/10.70169/VJFS.1035

Thomas N, Lucas R, Itoh T, Simard M, Fatoyinbo L, Bunting P, Rosenqvist A (2015) An approach to monitoring mangrove extents through time-series comparison of JERS-1 SAR and ALOS PALSAR data. Wetl Ecol Manag 23(1):3–17. https://doi.org/10.1007/s11273-014-9370-6

Thupalli R, Deen TA (2018) An investment strategy for reducing disaster risks and coastal pollution using nature based solutions. In: Nagabhatla N, Metcalfe C (eds) Multifunctional wetlands. Environmental contamination remediation and management. Springer, Cham. https://doi.org/10.1007/978-3-319-67416-2_5

Tinh PH, MacKenzie RA, Hung TD, Vinh TV, Ha HT, Lam MH, Hanh NT, Tung NX, Hai PM, Huyen BT (2022) Mangrove restoration in Vietnamese Mekong Delta during 2015–2020: achievements and challenges. Front Mar Sci 9:1043943. https://doi.org/10.3389/fmars.2022.1043943

Tran TA, Pittock J, Tuan LA (2019) Adaptive co-management in the Vietnamese Mekong Delta: examining the interface between flood management and adaptation. Int J Water Resour Dev 35(2):326–342. https://doi.org/10.1080/07900627.2018.1437713

Triyanti A, Dieperink C, Hegger D, Vu TT, Luu TT, Nguyen DC, Nguyen HQ (2024) Enhancing the role of International NGOs in promoting the implementation of ecosystem-based adaptation policies: insights from an International Union for Conservation and Foundation of Netherlands Volunteers led project in the Vietnamese Mekong Delta. Ecol Soc 29(1). https://doi.org/10.5751/ES-14727-290131

UNFCCC (2015) Paris Agreement. United Nations framework convention on climate change. https://unfccc.int/process-and-meetings/the-paris-agreement

UNISDR (2015) Sendai framework for disaster risk reduction 2015–2030. In: Proceedings of the 3rd United Nations world conference on DRR, Sendai, Japan, vol 1

United Nations Environment Programme (UNEP) (2021) Guidelines for integrating ecosystem-based adaptation into national adaptation plans: supplement to the UNFCCC NAP technical guidelines. Nairobi

United Nations Environment Programme and Gender and Water Alliance (UNEP-GWA) (2022) Gender mainstreaming in coastal and marine ecosystems management: principles, case studies and lessons learned. Nairobi

UN ESCAP (2019) The disaster riskscape across Asia-Pacific: pathways for resilience, inclusion and empowerment. https://repository.unescap.org/handle/20.500.12870/1578

Valenzuela BR, Yeo-Chang Y, Park MS, Chun J-N (2020) Local people's participation in mangrove restoration projects and impacts on social capital and livelihood: a case study in the Philippines. Forests 11(5):580. https://doi.org/10.3390/f11050580

van Bijsterveldt CEJ (2023) Mangrove restoration for coastal protection. PhD thesis, Utrecht University

Van Dyke JM, Broder SP (2016) Regional maritime cooperation in the South China Sea: COBSEA and PEMSEA. In: Major law and policy issues in the South China Sea. Routledge, pp 17–23
Van Zelst VT, Dijkstra JT, Van Wesenbeeck BK, Eilander D, Morris EP, Winsemius HC, Ward PJ, De Vries MB (2021) Cutting the costs of coastal protection by integrating vegetation in flood defences. Nat Commun 12(1):6533. https://doi.org/10.1038/s41467-021-26887-4
Veettil BK, Pereira SFR, Quang NX (2018) Rapidly diminishing mangrove forests in Myanmar (Burma): a review. Hydrobiologia 822:19–35. https://doi.org/10.1007/s10750-018-3673-1
Veettil BK, Ward RD, Quang NX, Trang NTT, Giang TH (2019) Mangroves of Vietnam: historical development, current state of research and future threats. Estuar Coast Shelf Sci 218:212–236. https://doi.org/10.1016/j.ecss.2018.12.021
Veettil BK, Costi J, Marques WC, Tran X, Quang NX, Van DD, Hoai PN (2020) Coastal environmental changes in Southeast Asia: a study from Quang Nam Province, Central Vietnam. Reg Stud Mar Sci 39:101420. https://doi.org/10.1016/j.rsma.2020.101420
Veettil BK, Ward RD, Dung NTK, Van DD, Quang NX, Hoai PN, Hoang N (2021) The use of bioshields for coastal protection in Vietnam: current status and potential. Reg Stud Mar Sci 47:101945. https://doi.org/10.1016/j.rsma.2021.101945
Veettil BK, Ward RD, Van DD, Quang NX, Hoai PN (2022) Seagrass ecosystems along the Vietnamese coastline: current state of research and future perspectives. Estuar Coast Shelf Sci 277:108085. https://doi.org/10.1016/j.ecss.2022.108085
Veettil BK, Wickramasinghe D, Amarakoon V (2023) Mangrove forests in Sri Lanka: an updated review on distribution, diversity, current state of research and future perspectives. Reg Stud Mar Sci 62:102932. https://doi.org/10.1016/j.rsma.2023.102932
Veettil BK, Tran X, Lopes MS (2024) Mangrove vegetation changes and shoreline erosion along the central coastline of Vietnam: a study from Quang Ngai province. Environ Dev 52:101065. https://doi.org/10.1016/j.envdev.2024.101065
Vu TT, Pham TD, Saintilan N, Skidmore A, Luu HV, Vu QH, Le NN, Nguyen HQ, Matsushita B (2022) Mapping multi-decadal mangrove extent in the Northern Coast of Vietnam using Landsat time-series data on Google Earth Engine platform. Remote Sens 14(18):4664. https://doi.org/10.3390/rs14184664
Wang D, Wu M, Chen Z, Cheng C, Huang L, Gao C, Wan B, Qiu P (2025) Mapping mangrove forest degradation using LiDAR and Landsat data. Int J Appl Earth Obs Geoinf 145:104984. https://doi.org/10.1016/j.jag.2025.104984
Westmacott S (2001) Developing decision support systems for integrated coastal management in the tropics: is the ICM decision-making environment too complex for the development of a useable and useful DSS? J Environ Manag 62(1):55–74. https://doi.org/10.1006/jema.2001.0420
Whisnant R, Vandeweerd V (2019) Investing in the new Blue Economy: the changing role of international development organizations in catalyzing private sector investment in support of regional strategic action programmes for the sustainable development of coasts and oceans. J Ocean Coast Econ 6(1):8. https://doi.org/10.15351/2373-8456.1116
Wickramasinghe D (2017) Regreening the coast: community-based mangrove conservation and restoration in Sri Lanka. In: DasGupta R, Shaw R (eds) Participatory mangrove management in a changing climate. Disaster risk reduction. Springer, Tokyo. https://doi.org/10.1007/978-4-431-56481-2_10
Widodo J, Trihatmoko E, Khomarudin MR, Ardha M, Nugroho UC, Setyaningrum N, Dinanta GP, Arief R, Setiyoko A, Novresiandi DA, Handika R, Priyatna M, Sobue S, Sarah D, Hermawan W (2024) Dynamic geo-visualization of urban land subsidence and land cover data using PS-InSAR and Google Earth Engine (GEE) for spatial planning assessment. Urban Sci 8(4):234. https://doi.org/10.3390/urbansci8040234
Winterwerp H, Van Wesenbeeck B, Van Dalfsen J, Tonneijck F, Astra A, Verschure S, Van Eijk P (2014) A sustainable solution for massive coastal erosion in Central Java: towards regional scale application of hybrid engineering. Discussion paper. Deltares, Delft, The Netherlands.
Wunder S (2015) Revisiting the concept of payments for environmental services. Ecol Econ 117:234–243. https://doi.org/10.1016/j.ecolecon.2014.08.016

Yang L, Lu W, Liu J, Liu Z, Borja A, Tao Y, Wang X, Zeng R, Zuo G, Wang T (2024) Multi-objective spatial suitability evaluations for marine spatial planning optimization in Dalian Coast, China. Sustainability 17(21):9851. https://doi.org/10.3390/su17219851

Yen BT, Son NH, Tung LT, Amjath-Babu T, Sebastian L (2018) Development of a participatory approach for mapping climate risks and adaptive interventions (CS-MAP) in Vietnam's Mekong River Delta. Clim Risk Manag 24:59–70. https://doi.org/10.1016/j.crm.2019.04.004

Zagare V (2022) Nature-based solutions for climate adaptation and mitigation in deltas and coastal areas. J Delta Urban 3:26–45. https://doi.org/10.59490/jdu.3.2022.7298

Zamboni NS, Prudêncio MDC, Amaro VE, Matos MDFAD, Verutes GM, Carvalho AR (2022) The protective role of mangroves in safeguarding coastal populations through hazard risk reduction: a case study in northeast Brazil. Ocean Coast Manag 229:106353. https://doi.org/10.1016/j.ocecoaman.2022.106353

Zhang Y, Cui L, Xie D, Jiang J (2023) Simulation and prediction of sea level rise impact on the distribution of mangrove and Spartina alterniflora in coastal China. Forests 14(4):831. https://doi.org/10.3390/f14040831

Zhao M, Zhang H, Zhong Y, Jiang D, Liu G, Yan H, Zhang H, Guo P, Li C, Yang H, Chen T, Wang R (2019) The status of coral reefs and its importance for coastal protection: a case study of Northeastern Hainan Island, South China Sea. Sustainability 11(16):4354. https://doi.org/10.3390/su11164354

Zhong W, Chu T, Tissot P, Wu Z, Chen J, Zhang H (2022) Integrated coastal subsidence analysis using InSAR, LiDAR, and land cover data. Remote Sens Environ 282:113297. https://doi.org/10.1016/j.rse.2022.113297

Zhou F, Botzen W (2021) Firm level evidence of disaster impacts on growth in Vietnam. Environ Resour Econ 79:277–322. https://doi.org/10.1007/s10640-021-00562-0

Zu Ermgassen PS, Worthington TA, Gair JR, Garnett EE, Mukherjee N, Nagelkerken I, Abrantes K, Acosta A, Araujo AR, Baker R, Barnett A, Beitl CM, Benzeev R, Brookes J, Ching Chong V, Connolly RM, Diele K, Dwyer PG, Spalding MD et al (2025) Mangroves support an estimated annual abundance of over 700 billion juvenile fish and invertebrates. Commun Earth Environ 6(1):299. https://doi.org/10.1038/s43247-025-02229-w

Chapter 17
Bioshields for Coastal Protection: Conclusions and Recommendations

Abstract This concluding chapter synthesizes the key findings, thematic insights, and forward-looking perspectives on the potential of bioshields for coastal protection. It reflects on the interdisciplinary linkages among ecology, engineering, governance, and socioeconomics that have been explored throughout the book, emphasizing how coastal bioshields represent the principles of nature-based solutions and ecosystem-based adaptation. The chapter recaps that while bioshields offer a cost-effective and ecologically sound approach to coastal resilience, their long-term success depends on robust policy support, community engagement, and continuous scientific innovation. Emerging trends, such as digital monitoring, blue carbon financing, and hybrid infrastructure, are highlighted as critical components of next-generation bioshield strategies. The chapter calls for strengthened regional and global cooperation in research, capacity building, and policy harmonization to mainstream bioshields into coastal zone management and climate adaptation planning. It also identifies key research gaps, including the need for standardized performance metrics, socio-ecological modeling, and inclusive benefit-sharing mechanisms. By integrating ecological wisdom with technological and institutional innovation, the future of coastal protection lies in advancing bioshields as a cornerstone element of sustainable coastal futures worldwide.

Keywords Blue carbon · Coastal resilience · Ecosystem-based adaptation · Integrated policy frameworks · Nature-based solutions · Sustainable coastal management

17.1 Overview

Coastal zones are among the most dynamic and productive environments on Earth, yet they are also some of the most vulnerable to climate change, sea-level rise, and anthropogenic pressures (Veettil et al. 2020). Over the past decades, traditional hard-engineering solutions, such as seawalls, groynes, and breakwaters, have provided localized protection but often at high ecological and economic costs (Morris et al.

B. Kozhikkodan Veettil, *The Potential of Bioshields for Coastal Protection*, Advances in Natural and Technological Hazards Research 55,
https://doi.org/10.1007/978-3-032-22888-8_17

2018; Perricone et al. 2023). This book has highlighted the transformative potential of bioshields, which harness the natural capacity of coastal vegetation to mitigate wave energy, stabilize shorelines, and enhance ecosystem resilience.

Through interdisciplinary analysis encompassing ecology, engineering, socioeconomics, and governance, the preceding chapters have demonstrated that bioshields can serve as cost-effective, multifunctional, and sustainable alternatives or complements to conventional coastal protection measures. However, realizing their full potential requires integrating scientific understanding with community participation, supportive policies, and adaptive management approaches.

17.2 Key Conclusions

17.2.1 Coastal Bioshields Are Multifunctional, Nature-Based Solutions

Bioshields provide not only physical protection from waves, winds, and storm surges but also deliver a wide range of ecosystem services, including carbon sequestration, biodiversity enhancement, fisheries support, and aesthetic and cultural values (Veettil et al. 2021). This multifunctionality makes them integral components of integrated coastal zone management and climate adaptation strategies.

17.2.2 Ecological Integrity Determines Protective Effectiveness

The ability of bioshields to attenuate wave energy and stabilize sediments is directly linked to their species composition, density, structure, and health. Monoculture plantations or poorly adapted species reduce resilience and may even cause ecological imbalance (Chen et al. 2012). Therefore, a site-specific, ecosystem-based approach, considering local hydrodynamics, soil characteristics, and salinity regimes, is essential for the long-term success of coastal bioshields.

17.2.3 Hybrid Approaches Enhance Coastal Resilience

Combining bioshields (e.g., mangroves, marshes, dunes) with engineered structures (e.g., permeable barriers, eco-revetments) can provide enhanced and immediate protection while maintaining ecological functions (Veettil et al. 2021). These hybrid designs, which combine the grey and green coastal infrastructures (Waryszak et al. 2021), represent a promising path toward balancing safety, sustainability, and cost-effectiveness in coastal defense planning.

17.2.4 Social and Institutional Factors Are Critical to Success

The sustainability of bioshield projects depends heavily on community engagement, land tenure clarity, and inter-sectoral coordination. Indigenous knowledge on nature-based solutions in coastal protection has been practiced by several communities during the pre-colonial era (Redvers et al. 2023). Participatory planning, local stewardship, and capacity building empower communities to maintain and monitor bioshields, ensuring long-term viability. Policy coherence across environmental, forestry, and coastal disaster management sectors remains a crucial enabling factor.

17.2.5 Monitoring and Adaptive Management Are Indispensable

Natural systems evolve, and bioshields must be monitored and adaptively managed to respond to environmental changes, species mortality, or shifting sediment dynamics. The use of remote sensing, GIS, and AI-based monitoring tools can support evidence-based decision-making and improve project accountability. In addition, real-time coastal monitoring is essential in gathering information for policy-making in adaptive coastal management (Jacobson et al. 2014).

17.2.6 Climate Change Poses Both Challenges and Opportunities

While sea-level rise, salinity intrusion, and extreme weather conditions threaten coastal vegetation, bioshields themselves can serve as frontline defenses against climate impacts. Blue carbon initiatives offer climate mitigation benefits, linking coastal restoration with national commitments under the Paris Agreement. Moreover, blue carbon ecosystems, such as mangrove forests, can adapt to climate change (Schaeffer-Novelli et al. 2016).

17.3 Strategic Recommendations

As an outcome of this book, a set of strategic recommendations has been developed, addressing research and innovation, policy and governance, and implementation and community engagement.

17.3.1 For Research and Innovation

Expand long-term empirical studies to quantify the biophysical processes governing wave attenuation, sediment transport, and vegetation resilience across different coastal ecosystems.

Develop standardized methodologies for bioshield design, performance assessment, and cost–benefit evaluation.

Integrate modelling tools, remote sensing technologies, and genetic studies to identify climate-resilient plant species and optimize bioshield configurations.

Encourage interdisciplinary research connecting ecology, coastal engineering, and socioeconomics for a comprehensive understanding and application.

17.3.2 For Policy and Governance

Embed bioshields within national coastal protection frameworks and climate adaptation strategies, ensuring recognition as legitimate infrastructure investments.

Strengthen legal instruments and land-use planning to secure areas for bioshield development and prevent encroachment or conversion into aquaculture or agricultural areas.

Promote cross-sectoral coordination among ministries of environment, forestry, fisheries, and disaster management.

Introduce economic incentives (e.g., payments for ecosystem services, carbon credits) to encourage community and private sector participation.

17.3.3 For Implementation and Community Engagement

Ensure that bioshield projects are community-driven, respecting local knowledge and customary practices.

Prioritize capacity building and participatory monitoring, empowering local stakeholders to manage, maintain, and benefit from bioshields.

Combine education, awareness programs, and livelihood diversification to enhance social acceptance and reduce dependency on destructive coastal activities.

Adopt adaptive management frameworks that incorporate periodic review, feedback, and learning mechanisms to refine project strategies.

17.4 The Path Forward

The concept of bioshields represents a paradigm shift from resisting natural forces to working with nature. In an era marked by intensifying coastal hazards and environmental degradation, investing in bioshields is both an ecological necessity and a socio-economic opportunity. They offer a triple-win solution: protecting lives and property, sustaining livelihoods, and restoring the natural environment.

Future coastal protection strategies must embrace a systems approach, integrating bioshields within broader coastal landscape planning, supported by sound science, community participation, and strong governance. By doing so, societies can transition toward resilient and sustainable coastal futures, where human prosperity coexists with thriving ecosystems.

17.5 Final Remarks

The potential of bioshields extends beyond their physical role as natural barriers. They embody the principles of resilience, adaptation, and coexistence that are vital for navigating the uncertain challenges of the Anthropocene. Moving forward, success will depend not merely on planting trees or restoring mangroves but on fostering enduring harmony between people and the coasts that sustain them.

References

Chen L, Zeng X, Tam NF, Lu W, Luo Z, Du X, Wang J (2012) Comparing carbon sequestration and stand structure of monoculture and mixed mangrove plantations of *Sonneratia caseolaris* and *S. apetala* in Southern China. For Ecol Manag 284:222–229. https://doi.org/10.1016/j.foreco.2012.06.058

Jacobson C, Carter R, Thomsen D, Smith T (2014) Monitoring and evaluation for adaptive coastal management. Ocean Coast Manag 89:51–57. https://doi.org/10.1016/j.ocecoaman.2013.12.008

Morris RL, Konlechner TM, Ghisalberti M, Swearer SE (2018) From grey to green: efficacy of eco-engineering solutions for nature-based coastal defence. Glob Change Biol 24(5):1827–1842. https://doi.org/10.1111/gcb.14063

Perricone V, Mutalipassi M, Mele A, Buono M, Vicinanza D, Contestabile P (2023) Nature-based and bioinspired solutions for coastal protection: an overview among key ecosystems and a promising pathway for new functional and sustainable designs. ICES J Mar Sci 80(5):1218–1239. https://doi.org/10.1093/icesjms/fsad080

Redvers N, Aubrey P, Celidwen Y, Hill K (2023) Indigenous peoples: traditional knowledges, climate change, and health. PLOS Glob Public Health 3(10):e0002474. https://doi.org/10.1371/journal.pgph.0002474

Schaeffer-Novelli Y, Soriano-Sierra EJ, Vale CCD, Bernini E, Rovai AS, Pinheiro MAA, Cintrón-Molero G et al (2016) Climate changes in mangrove forests and salt marshes. Braz J Oceanogr 64(spe2):37–52. https://doi.org/10.1590/S1679-875920160919064sp2

Veettil BK, Costi J, Marques WC, Tran XL, Quang NX, Van DD, Hoai PN (2020) Coastal environmental changes in Southeast Asia: a study from Quang Nam province, central Vietnam. Reg Stud Mar Sci 39:101420. https://doi.org/10.1016/j.rsma.2020.101420

Veettil BK, Ward RD, Dung NTK, Van DD, Quang NX, Hoai PN, Hoang N (2021) The use of bioshields for coastal protection in Vietnam: current status and potential. Reg Stud Mar Sci 47:101945. https://doi.org/10.1016/j.rsma.2021.101945

Waryszak P, Gavoille A, Whitt AA, Kelvin J, Macreadie PI (2021) Combining gray and green infrastructure to improve coastal resilience: lessons learnt from hybrid flood defenses. Coast Eng J 63(3):335–350. https://doi.org/10.1080/21664250.2021.1920278X

MIX
Papier aus verantwortungsvollen Quellen
Paper from responsible sources
FSC® C105338

If you have any concerns about our products,
you can contact us on
ProductSafety@springernature.com

In case Publisher is established outside the EU,
the EU authorized representative is:
Springer Nature Customer Service Center GmbH
Europaplatz 3, 69115 Heidelberg, Germany

Printed by Libri Plureos GmbH
in Hamburg, Germany